Managing and Transforming Water Conflicts

All living things need water. . . . Where water crosses boundaries – be they economic, legal, political, or cultural – the stage is set for disputes among different users trying to safeguard access to a vital resource, while protecting the natural environment. Without strategies to anticipate, address, and mediate among competing users, intractable water conflicts are likely to become more frequent, more intense, and more disruptive around the world. In this book, Jerome Delli Priscoli and Aaron T. Wolf investigate the dynamics of water conflict and conflict resolution, from the local to the international. They explore the inexorable links among three facets of conflict management and transformation: alternative dispute resolution (ADR), public participation, and institutional capacity. This practical guide will be invaluable to water management professionals, as well as to researchers and students in engineering, economics, geography, geology, and political science who are involved in any aspect of water management.

JEROME DELLI PRISCOLI is a senior advisor at the U.S. Army Corps of Engineers, Institute for Water Resources. For the past 30 years he has designed and run social assessment, public participation, and conflict resolution research and training programs. Delli Priscoli has been a water policy advisor to the World Bank and the United Nations (UN) water-related agencies, and he works closely with international government water ministers. He is author of many articles and books and is the editor-in-chief of the peer-reviewed journal *Water Policy*. He was an original member of the U.S. delegation to the multi-lateral Middle East peace talks on water, and he has played pivotal roles in each of the five World Water Forums and most of the critical water resources policy meetings over the past 15 years. He serves on the Bureau and Board of Governors of the World Water Council. The American Water Resources Association awarded him the Icko Iben Award for achievement in cross-disciplinary communications in water in 2005.

AARON T. WOLF is a professor of geography in the Geosciences Department at Oregon State University. His research and teaching focus is on the interaction between water science and water policy, particularly as related to conflict prevention and resolution. He has acted as a consultant to the World Bank and several international governments and governmental agencies on various aspects of transboundary water resources and dispute resolution. Wolf is a trained mediator/facilitator and directs the Program in Water Conflict Management and Transformation, through which he has offered workshops, facilitations, and mediation in basins throughout the world. He coordinates the Transboundary Freshwater Dispute Database and is a codirector of the Universities Partnership on Transboundary Waters. He has been an author or editor of seven books, as well as almost fifty journal articles, book chapters, and professional reports on various aspects of transboundary waters.

i

INTERNATIONAL HYDROLOGY SERIES

The **International Hydrological Programme** (IHP) was established by the United Nations Educational, Scientific, and Cultural Organization (UNESCO) in 1975 as the successor to the International Hydrological Decade. The long-term goal of the IHP is to advance our understanding of processes occurring in the water cycle and to integrate this knowledge into water resources management. The IHP is the only UN science and educational program in the field of water resources. One of its outputs has been a steady stream of technical and information documents aimed at water specialists and decision makers.

The **International Hydrology Series** has been developed by the IHP in collaboration with Cambridge University Press as a major collection of research monographs, synthesis volumes, and graduate texts on the subject of water. Authoritative and international in scope, the books within the series all contribute to the aims of the IHP in improving scientific and technical knowledge of freshwater processes, in providing research know-how, and in stimulating the responsible management of water resources.

Managing and Transforming Water Conflicts

Jerome Delli Priscoli

Institute for Water Resources, U.S. Army Corps of Engineers

Aaron T. Wolf

Department of Geosciences, Oregon State University

CAMBRIDGE UNIVERSITY PRESS
Cambridge, New York, Melbourne, Madrid, Cape Town, Singapore, São Paulo, Delhi

Cambridge University Press
32 Avenue of the Americas, New York, NY 10013-2473, USA

www.cambridge.org
Information on this title: www.cambridge.org/9780521632164

First published 2009

Printed in the United States of America

A catalog record for this publication is available from the British Library.

Library of Congress Cataloging in Publication Data

Delli Priscoli, Jerome.
Managing and transforming water conflicts / by Jerome Delli Priscoli and
Aaron T. Wolf ; with contributions from Kristin M. Anderson, Joshua
T. Newton, Lisa J. Gaines, Kyoko Matsumoto, and Meredith A. Giordano.
   p.   cm. – (International Hydrology Series)
Includes bibliographical references and index.
ISBN 978-0-521-63216-4 (hardback)
1. Water supply.   2. Water resources development.
3. Water rights.   I. Wolf, Aaron T.   II. Title.
HD1691.D44   2009
333.91′17–dc22   2007037927

ISBN   978-0-521-63216-4 hardback

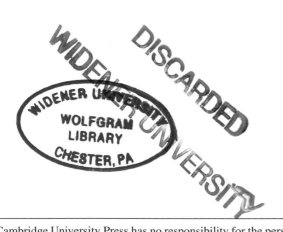

Jerome Delli Priscoli dedicates this book to the memory of Pierre Teilhard de Chardin, S. J., whose writings and spirit have been so helpful in trying to integrate thirty-five years of multidisciplinary research and practical work in water resources around the world.

Aaron T. Wolf dedicates this book to Professor John Ross, advisor and friend, who taught a generation of students how to see watersheds without boundaries.

Contents

Figures

Tables

Sidebars

Foreword

With the dramatically increasing number of users and the potential impact of climatic change, human systems and the hydrological system might be on an unsustainable path. Indeed, population rises, the climate seems to change, and water demands grow to quench the thirst of cities, suburbs, industry, and agriculture, often leaving ecosystem needs on the wayside. All of these factors might lead to potential conflicts among uses and users. However, water scarcity is not the only concern, as we are reminded by recent flooding events that have spelled disaster for the human communities living in affected areas. In the past decade, water-related pressures have resulted in media headlines foreboding a future wrought with "water wars." With 263 rivers and countless aquifers transversing national boundaries, the cultivation of such a somber image is not surprising. Indeed, water cuts across the boundaries of countries, cultures, and economic sectors, meaning that water planning and decision making in one jurisdiction has the potential to spill over to into others. The risk of disruption, conflict, and violence exists around transboundary waters. However, earlier work by the authors points to a lengthy history of cooperative interactions, rather than conflicts, over this precious resource.

Sparked by the concern over water security, in particular, the challenges of sharing water resources across political boundaries and of responding to the needs of many Member States, UNESCO initiated the project From Potential Conflict to Co-operation Potential (PCCP) in 2000. PCCP endeavors to increase the capacity of stakeholders to find conciliatory ways to reach mutually accepted solutions for the management of their shared water resources. Since its inception, PCCP has produced many relevant publications, training courses, and educational materials, increasing the knowledge base of issues pertaining to transboundary waters, conflict, and cooperation. In response to requests of Member States, the project has also focused on providing technical assistance. In spite of these advancements, further work is needed to broaden and deepen our understanding of how to manage shared water resources in an equitable manner that enables social and economic development, contributes to poverty eradication, and protects environmental systems.

In light of the growing complexity of water resources management, UNESCO is grateful for the outstanding contributions of Dr. Wolf and Dr. Delli Priscoli to the field of transboundary waters. In this comprehensive book, *Managing and Transforming Water Conflicts*, Wolf and Delli Priscoli, each a preeminent expert in the field, bring together their incisive and visionary thinking, providing a resource for achieving constructive interactions around water resources. The authors pose and dissect questions of stakeholder involvement and interest-based negotiations and provide tools to affect each phase of a conflict resolution process. They also build on earlier research, which underscores the importance of institutional capacity in preventing and resolving water-related conflicts.

Eloquently written, Wolf and Delli Priscoli's book is instructive and hopeful, taking us closer to realizing a more optimistic vision of water as a tool for promoting cooperation and peace. UNESCO, through PCCP, has cautiously embraced a positive view of managing shared water resources. Reading this book, one remains hopeful that even in these times of unprecedented ecological, social, and political change, the possibility exists to transform potential water conflicts into avenues for cooperation.

Water connects us and forces us to build bridges.

András Szöllösi-Nagy
Deputy Assistant Director General of UNESCO
Secretary of the International Hydrological Programme

Acknowledgments

This work would not have been possible without the generous contributions of time and advice from a number of people, and it is with pleasure that we are finally able to thank them publicly. When the two authors began initial discussions of this project in the mid-1990s, Jerome Delli Priscoli was in an exchange program with the World Bank, and Aaron T. Wolf was at the University of Alabama and working with the U.S. Agency for International Development's (USAID's) Irrigation Support Program for Asia and the Near East (ISPAN), and both were associated with the United Nations Educational, Scientific and Cultural Organization's (UNESCO's) International Hydrology Programme. We are grateful for initial support from all these institutions, especially to Dr. András Szöllösi-Nagy at UNESCO and Peter Reiss at ISPAN. Others who were tremendously helpful early on include Janos Bogardi and Léna Salamé at UNESCO, and Tracy Atwood, Herb Blank, and Marjorie Shovlin from USAID. We are grateful to the many professionals in the dispute resolution fields with whom we have worked and generously traded ideas over the years, especially to Dr. Chris Moore of CDR, Dr. James Creighton, and many others. In addition, the inputs received from the many professionals, from the U.S. government and other countries, who have attended training programs and used the materials presented herein have been most helpful.

Subsequently, we have drawn on a host of expertise from around the world to present the ideas in this book. The "Four Stages in Water Conflict Transformation" presented in Chapter 6 was drawn from a skills-building course developed over years within the World Bank on which Wolf had the honor to work. We are grateful to our co-instructors, Undala Alam, Inger Andersen, Terry Barnett, David Grey, Bo Kjellen, Stephen McCaffrey, Claudia Sadoff, Salman Salman, and Dale Whittington, for the discussions and experiences that led to this thinking, and to Len Abrams, who crafted the world (and maps) of the Sandus Basin.[1]

Many people have added substantively to the text or approach of the book. The case studies in Appendix C were substantially updated and expanded by Joshua T. Newton, currently at Tufts University, who also wrote the Guaraní, Kura–Araks, Senegal, and Lake Titicaca case studies. Appendix E, on groundwater, and Appendix F, on quality-related treaties, come from the excellent theses by Kyoko Matsumoto and Meredith Giordano, respectively. Appendix D, on international water pricing, was researched and authored by Kristin M. Anderson and Lisa J. Gaines of Oregon State University. Jakub Landovsky and Olga Zarubova-Pfeffermannova contributed to the research on the Kosi and Ganduk for Chapter 4. Some of the text draws from articles coauthored with Ariel Dinar, at the World Bank; Meredith Giordano, now at International Water Management Institute (IWMI); Sandra Postel, of the Global Water Policy Project; and Shira Yoffe, currently with the U.S. Department of State. Todd Jarvis and Marloes Bakker, both of Oregon State University, read the entire document critically and carried out a thorough update of relevant literature. The bibliography and some text were augmented and finalized thanks to Nathan Eidem, Eva Lieberherr, Michele Lizon, Patrick MacQuarrie, Olivia Odom, and Kate Zahnle-Hostetler, of Oregon State University, and Jamie Frey-Frankenfield of the University of Oregon. The maps and figures owe their appeal to the cartographic expertise of Sara Ashley Watterson, currently of Earthjustice, and Gretchen Bracher and Nathan Eidem, of Oregon State University. The transboundary waters team at Oregon State University has learned long ago never to allow any document to leave the building without the thorough crafting and enhancing of Caryn M. Davis of Cascadia Editing. Caryn's energy, attention to detail, grasp of the document as a whole, and passion for the work provide so much more than one has a right to expect from a "technical edit." If this book has any coherence and appeal, it is due in no small part to Caryn and her magic.

Wolf benefited profoundly from his relationship with his colleagues at the Universities Partnership for Transboundary

---

[1] A course workbook was published as Wolf (2008), which an interested instructor might use in a classroom to supplement this text.

Waters: Marcia Macomber (now at IWMI), Lisa J. Gaines, and Michael Campana at Oregon State University; Marilyn O'Leary of the University of New Mexico; Anthony Turton, Peter Ashton, and Anton Earle at the University of Pretoria, South Africa; Emmanuel Manzungu and Pieter van der Zaag (now at UNESCO-IHE Delft) at the University of Zimbabwe; Olli Varis at the Helsinki University of Technology; Jan Lundqvist at Linköping University, Sweden; Patricia Wouters, Sergei Vinogradov, and Alistair Rieu-Clarke at the University of Dundee; Daming He at Yunnan University; Ashim Das Gupta at the Asian Institute of Technology; Mikiyasu Nakayama at the University of Tokyo; Jennifer McKay at the University of South Australia; Alexander López Ramirez at the Universdad Nacional, Costa Rica; and Ofelia Clara Tujchneider at the Universidad Nacional de El Litoral, Argentina. Likewise, much of the thinking on conflict, cooperation, and international institutions draws from particularly productive collaborations with Alexander Carius, Geoff Dabelko, Mark Giordano, Meredith Giordano, Kerstin Stahl, and Shira Yoffe.

Many people agreed to be interviewed, offered to read one or another section of the work, or otherwise contributed valuable comments along the way. Thanks are therefore due, but not limited, to: Undala Alam, Jeremy Berkoff, Asit Biswas, Ariel Dinar, Itay Fischhendler, David Grey, Munther Haddadin, John Hayward, Fred Hof, John Kolars, Charles Lankester, Charles Lawson, Jonathan Margolis, Alan More, Masahiro Murakami, Bill Phelps, Keith Pitman, George Renkewitz, Claudia Sadoff, Uri Shamir, Yona Shamir, Muhammed Shatanawi, Miguel Solanes, John Waterbury, and Moshe Yisraeli. The Danube case study simply would not have been possible without the generous assistance of Kathy Alison of ISPAN. Wolf was generously sponsored by Professor Eran Feitelson and the Department of Geography at Hebrew University, and by Professor Suwit Laohasiriwong at Khon Kaen University's Institute for Dispute Resolution, during a particularly productive and enjoyable sabbatical year, for which he is tremendously grateful. Jesse Hamner and Shannon Wall were extremely valuable research assistants at the University of Alabama, as were Brian Blankespoor, Nathan Eidem, Greg Fiske, and Sam Littlefield at Oregon State University. The spring 2004 GEO 4/524: International Hydropolitics class at Oregon State University critically read and commented on an early draft, for which we are grateful.

At Cambridge University Press, we are more grateful than we can say to Matt Lloyd, Helen R. Morris, and others for tremendous patience and graciousness during deadlines regularly pushed back by career moves, sabbaticals, hurricanes, and the normal wear-and-tear of the lives of busy professionals. We are hopeful that the experience gained during these extensions offers a richer text. Special thanks go to Mary Paden and her colleagues at Aptara Inc. for their perseverance and dedication in shepherding the book through the final publishing and production processes, as well as to our U.S. Cambridge production controller, Shelby Peak.

Finally, Jerome Delli Priscoli acknowledges with gratitude and love the patience of his wife, Suzanne, and his sons, Stephen and Matthew, who have patiently listened to many of the arguments presented in this book. Aaron Wolf would like to acknowledge with gratitude and love the patience and perseverance of his family, Ariella, Yardena, and Eitan, who supported him despite perhaps one too many trips to Washington, DC, and abroad, and definitely one too many nights at the office, while we saw this thing through.

Special Contributors

Appendix C. Case studies of transboundary dispute resolution

Joshua T. Newton   South Burlington, Vermont

Appendix D. International water pricing: An overview and historic and modern case studies

Kristin M. Anderson   Researcher, Department of Geosciences, Oregon State University, Corvallis, Oregon
Dr. Lisa J. Gaines   Associate Director, Institute for Natural Resources, Oregon State University, Corvallis, Oregon

Appendix E. Treaties with groundwater provisions

Kyoko Matsumoto   Researcher, Freshwater Project, Institute for Global Environmental Strategies, Kanagawa, Japan

Appendix F. Treaties with water quality provisions

Dr. Meredith A. Giordano   Director, Research Impact, International Water Management Institute, Colombo, Sri Lanka

Introduction

Till taught by pain, men really know not what good water's worth.
— Lord Byron, "Don Juan"

In 1978 the Dead Sea turned over for the first time in centuries. For millennia, this terminal lake at the lowest point on the Earth's surface had been receiving the sweet waters of the Jordan River, losing only pure water to relentless evaporation and collecting the salts left behind. The result had been an inhospitably briny lake eight times saltier than the sea, topped by a thin layer of the Jordan's relatively less-dense fresh water. The two salinity levels of the river and the lake kept the Dead Sea in a perpetually layered state even while the lake level remained fairly constant – evaporation from the lake surface occurs at roughly the rate of the natural flow of the Jordan and other tributaries and springs.

These delicate balances were disrupted as modern nations – with all of their human and economic needs tied inexorably to the local supply of fresh water – built up along the shores of the Jordan. In the past century, as both Jewish and Arab nationalism focused on this historic strip of land, the two peoples locked in a demographic race for numerical superiority. As more and more of the Jordan was diverted for the needs of these new nations, the lake began to drop, most recently by about one-half meter per year. As it dropped, greater shoreline was exposed, the lake was cut in half by the Lisan Straits, the shallow southern half all but dried up, and the potash works and health spas built to take advantage of the lake's unique waters found themselves ever farther from the shore.

Along with the drop in lake level came a relative rise in the pycnocline – the dividing line between the less-saline surface water and its hypersaline fossil base. The division between the two layers was finally eradicated briefly in the winter of 1978–1979, and the Dead Sea turned over, effectively rolling in its grave – a hydrologic protest against the loss of the Jordan. The turnover brought water to the surface that had not seen light of day for three hundred years. Although it sterilized the lake, this turnover was not counted as an ecological disaster –

except for bacteria, fungi, and one type of algae, the Dead Sea is appropriately named – but the event was a symptom of a wider crisis of history-influencing proportions.

The fact is that the populated world is running out of "easy" water. Although the total quantity of water in the world is immense, the vast majority is either saltwater (97.5 percent) or locked in ice caps (1.75 percent). The amount economically available for human use is only 0.007 percent of the total, or about 13,500 km^3 (about 2,300 m^3 per a person – a 37 percent drop since 1970; United Nations, 2005). This number continues to fall as populations grow and as existing supplies become more polluted (Figure I.1).

Most of the world's population lives in areas where the majority of the rainfall comes in only a few short months of the year. The same poor who are continually flooded also suffer recurrent drought. Our variability in climate is indicating changing patterns and timing of runoffs. Hydrologists suggest that intensity of hydrological events could increase. No one can predict the regional variability of these projected changes. Thus the sources of potential water stress are varied and uncertain. Human responses will require use of mixed infrastructures and changes in our behavior.

In conjunction with these hydrological stresses, though, come the political stresses, which result as the people who have built their lives and livelihoods on a reliable source of fresh water are seeing the shortage of this vital resource impinge on all aspects of the tenuous relations that have developed over the years – among nations, among economic sectors, and among individuals and their environment. This book speaks to how people have, and have not, dealt with hydropolitics and its impact on people and the environment. It seeks to help us to better understand and deal with these stresses and their uncertainties (Sidebar I.1).

United Nations data of a decade ago showed that 25 percent of the world's human population lived in areas at high risk of drought and floods. The number of victims affected by droughts and floods is growing rapidly. Average annual losses

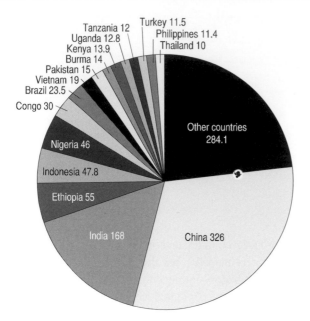

Figure I.1 Seventy-five percent of people without access to safe water reside in sixteen countries (United Nations, State of the World Population, 2004).

are now over US$40 billion, with economic losses that are ten times greater than in the 1950s (Munich Re Group, 2000). More recent data indicates that this trend is continuing (United Nations International Strategy for Disaster Reduction, 2004). Cheap and accessible electricity has traditionally been the key to economic development. However, two billion people lack access to electricity, and the demand is growing. Throughout the world, developed countries (occasionally defined by membership in the Organisation for Economic Co-operation and Development) have used about 70 percent of their hydropower

Sidebar I.1 Some of the Gloomy Arithmetic of Water

- 1.4 billion people lack safe water
- 80 percent of diseases are carried by water: One child dies every 8 seconds and 5 to 7 million people die annually; $125 billion in workday losses per year
- 50 percent of people lack adequate sanitation
- 20 percent of freshwater species are near extinction
- 76 percent of people live in water-stressed areas (less than 1,000 centimeters of rainfall per year), most in politically unstable regions
- We will lose irrigated land by 30 percent in 2025 and 50 percent by 2050
- 50 percent of people will depend on world markets for food
- Asia: More than two-thirds of the population live in areas where 80 percent of rainfall occurs in 20 percent of the year

potential, but in Africa, with its huge potential for hydropower, only 6 percent has been used. In Asia, 20 percent has been used, and in Latin America, about 35 percent (Sadoff and Grey, 2005). The poor pay a far higher percentage of their income or available wealth for water: an average of about US$1.00 to US$2.50 per cubic meter. This is in contrast to the United States, where citizens pay about US$0.30 to US$0.80 per cubic meter. The connected poor pay about US$1.00 per cubic meter, and the unconnected poor pay US$5.50 to US$16.50 per cubic meter on average. These are numbers no rich country would tolerate.

There has been a spectacular increase in groundwater development for irrigation in most arid and semiarid countries. This is a "silent revolution." Probably, about half of the value of irrigated agriculture is obtained with groundwater, but the volume of groundwater used is only a small fraction of the corresponding volume of surface water used for irrigation (Delli Priscoli, 2005a). A huge and growing literature speaks to the human and ecological disasters attendant on the global water crisis – essentially an ongoing deployment of a hydrological weapon of mass destruction (see especially the works of Peter Gleick [e.g., his biennial World's Water Series, 2003], Sandra Postel (1992, 1999), United Nations Environment Programme (UNEP) [UNEP and OSU, 2002; UNEP and the Woodrow Wilson Center, 2004; Wolf, 2006; Carius, Feil, and Taenzler, 2003], United Nations Educational, Scientific and Cultural Organization (UNESCO) [which has produced dozens of papers under the auspices of its PCCP Programme], and others).

We investigate the dynamics of water conflict stemming from such stresses, as well as the processes that are integral to capacity and institution building for managing waters that cross jurisdictional boundaries. We explore the intertwined nature of three facets of capacity and institution building processes: conflict management processes, public participation processes, and institutional capacity-building processes, as they are embedded in disputes over water resources around the world.

For many reasons documented in these pages and in many other reports, water is becoming the critical resource of the twenty-first century. The more we realize its importance, the more we also realize a fundamental reality of water management: water, which is created by nature, often crosses jurisdictional boundaries, which are created by humans, and this generates conflicts. Water behaves in an integrated way, whereas the institutions through which we manage it are fragmented. Fragmented institutional management seems to be the rule for water that crosses national boundaries and boundaries of states and provinces within national boundaries.

Water management, by nature, becomes a process of anticipating disputes and managing conflict so as to build new ways and means to deal with stresses. The industrialized world has prospered and grown because of its ability to do so, and societies around the world have developed in part through their successes in managing water resources. Indeed, we can argue that creating capacity and institutions to manage water across federal jurisdictional boundaries has been a major element in creating the United States as one nation on the North American continent. It is no accident that the European Union, in looking toward more integration, has put management for river basins that cross State boundaries in the center of its directives. In the past, countries such as Holland have integrated and evolved politically in ways heavily influenced by the experiences of local water boards (Reuss, 2002). Indeed, needs arising from cross-boundary water uses, such as navigation or pollution control, have consistently led the way to increased joint planning, diagnosing, and managing of Continental Europe's waters, often leading to cooperation in other areas.

The terms *transboundary* and *international waters* can be the sources of endless debate. Clearly, sovereign States must play out debates over water within relevant diplomatic and international rules. However, even when water crosses jurisdictional boundaries within nations, conflicts arise between entities, some of whom have legal status and sovereignty concerns close to those of international States, including, for example, Canadian or German provinces or Indian tribes within the United States. The institutional experience of such management can often provide models or at least concepts that are relevant to those in international waters. For our purposes, we use the term *international waters* to refer to those water resources that cross the boundaries of two or more countries, and *transboundary waters* as a more inclusive term to refer to water that crosses *any* jurisdictional or sectoral boundaries, including those within a nation. It is important that the experiences in managing transboundary disputes within nation-states and those that occur internationally across States better inform each other (Sidebar I.2).

Boundaries change. Sovereign States have emerged from regions, thus creating international waters, which were once better thought of as regional waters. So, too, international waters can become regional if incorporated into other states. Jurisdictional boundaries are more fluid than nature's watercourses, especially if we look over a time period of 50 to 100 years. The fact that the waters continually conflict with jurisdictional boundaries, however, does not change.

We couch the lessons of this book within this broad perspective. We do this to help expand the possibilities for the capacity of individuals, organizations and institutions, and society to manage water. Doing so opens up new possibilities for all of us who are struggling with the number-one need recognized by the World Water Forums and many other regional forums – the need for institution and capacity building for water management.

Sidebar I.2 Some Useful Definitions

Basin: We use *basin* synonymously with what is referred to in the United States as a *watershed* and in the United Kingdom as a *catchment*, or all waters, whether surface water or groundwater, that flow into a common terminus. The 1997 UN Convention on the Law of Non-Navigational Uses of International Watercourses (Appendix A; United Nations, 2005) similarly defines a *watercourse* as "a system of surface and underground waters constituting by virtue of their physical relationship a unitary whole and flowing into a common terminus." By definition, basins can include lakes, wetlands, and aquifer systems in addition to rivers. Colloquially, some use watersheds as smaller units, whereby many watersheds make up a river basin.

Conflict: "[T]wo or more entities, one or more of which perceives a goal as being blocked by another entity, and power [of some sort] being exerted to overcome the perceived blockage" (Frey, 1993). *Acute conflict* is defined as conflict that results in military or violent actions among competing parties.

Dispute: Conflicts that result in nonviolent tensions among parties, including political, legal, or economic actions.

Hydropolitical Resilience: The complex human-environmental system's ability to adapt to permutations and change within these systems; *hydropolitical vulnerability* is defined by the risk of political dispute over shared water systems.

International Waters: Following the 1997 Convention, we use "international waters" as a watercourse, parts of which are situated in different States (nations).

Transboundary Waters: Water that crosses *any* boundaries – be they economic sectors, legal or political jurisdictions, cultural divides, or international borders. This also includes water crossing boundaries of sovereign entities, whether these boundaries are within a federalist nation-state or among nation-states.

Institutions/Organizations: Keohane defines *institutions* as "persistent and connected sets of rules (formal and informal) that prescribe behavioral rules, constrain activity, and shape expectations" (Keohane, 1989, p. 3), whereas Lasswell (1971) uses "routinized patterns of behavior creating stable expectations over time." *Organizations* are generally the formal bodies that implement institutional arrangements. Although the authors recognize the distinctions between *institutions* and *organizations*, these terms are used interchangeably in many places in this book.

War: Including both formal and informal declarations of war; extensive acts of violence between two nations, or among more nations, causing deaths, dislocation, and high strategic costs (Azar, 1980).

We also examine issues that tend to recur in water conflicts: questions of equity in water allocations and characteristics of water that tend both to encourage cooperative management and to rend apart negotiations. The question we seek to answer is, in short: what aspects about water disputes are unique in the realm of conflict resolution, and how can these attributes be harnessed to help encourage cooperation?

Following this introduction, Chapter 1 describes background and trends that lead to and prescribe the need for more explicit managing of water conflicts. Chapter 2 revisits the thesis of "water wars" and reframes the issue of water conflict and war by looking at water realities in history and cooperation around water.

Chapter 3 reviews some theories of conflict management and what has been called "alternative dispute resolution" (ADR), as they might be applied to water conflicts. The principles of conflict management are described and applied to water issues, in both the unassisted and assisted settings. Some background is then provided on diagnosing the causes of conflict, and on generating creative value- and interest-based alternatives in water disputes. Chapter 4 looks at the means for conflict management and includes some discussion on participation processes. Although the two often use similar tools, they do not always hold the same purposes and ends. Both conflict management and participation do strive to enhance the institutional capacity for managing water conflicts. Possibilities for public participation in water resources are then explored, followed by a description of capacity for water institutions to resolve disputes.

Chapter 5 looks at some ends of negotiating over water conflicts to build transboundary arrangements or organizations to manage water conflicts. It moves into the realm of how conflict management is actually practiced in the world of water resources by using these theories as a context. The chapter begins with a general description of water conflict resolution in the interjurisdictional setting.

Chapter 5 also discusses ends or purposes of negotiating around water issues. Frequently we think of each negotiation as an event unto itself rather than a longer-term process. Transboundary and international water arrangements evolve and grow over long periods of time, however, with some of the best known – such as on the Delaware River in the United States – taking almost 50 years to create. Negotiations in areas with difficult problems are therefore usually a start. This chapter presents the idea that these ends are really institution and capacity building. It presents successful conflict management of water issues around transboundary waters as a longer-term process rather than simply a series of short-term events.

Chapter 6 discusses some means for negotiating transboundary water disputes. This chapter uses the theories presented in earlier chapters to frame discussions of how actual processes can evolve. It also emphasizes new views of interest-based negotiations and the use of outside or third parties.

Chapter 6 also includes one construct for understanding the process of water conflict transformation. Although there are no blueprints for resolving water disputes, some patterns do tend to emerge, particularly an organic evolution from thinking of water in terms of rights, to needs, to benefits, to equity.

Chapter 7 describes the lessons learned through an examination of patterns and issues in both national and international case studies and asks: what issues of the future may look nothing at all like the past?

Much of the popular debate around water issues is couched in apocalyptic and doomsday view of crisis. But what is crisis? *Crisis*, from its Greek roots *krisis*, refers to decision and not necessarily disaster – to a time of decisive action, to a turning point that may make things worse or better. Crisis also signifies opportunity as much as, or more than, disaster. Crisis is like a wake-up call for decision and action. Today, water crises are carrying wake-up calls but also a hope for creativity and opportunities for community building.

The debate about water and conflict is frequently heavy on problems and light on solutions. For example, we hear much about conservation, population control, and vaguely defined "better ways." More frequently we find critiques of what has been done or what exists, with little discussion of what would have happened without various projects or programs. In other words, the retrospective balance of benefits and costs is rarely clear. It is unclear where such retrospective balance would lead us, but what is clear is that this is a new level of discussion in which we must engage.

One way or another, humans are going to change societal and individual behavior around water – even under the best assumptions of population growth, conservation, and better pricing. We hope this book will help decision makers concerned with water, water professionals, nongovernmental organizations, and other users adapt to these changes. We can be reactive or choose to be proactive. To do nothing is likely to be an invitation for bad socioeconomic and environmental projects. To proactively codesign and coengineer our ecology with God and/or nature carries awesome responsibilities – and can be frightening. Pessimism and fear will not get us there. We need to tap our rich history of water resources experience. Such is the spirit guiding this book.

A Note on the OSU Transboundary Freshwater Dispute Database

To facilitate the comprehensive study of issues related to conflict and cooperation over shared water resources, researchers at the Oregon State University Department of Geosciences have collaborated with the Northwest Alliance for Computational Science and Engineering over the years to develop what has become known as the Transboundary Freshwater Dispute Database (see Wolf, 1999b; Yoffe et al., 2004).

The database currently includes digital mapping of the world's 263 international watersheds, along with geographic information system (GIS) mapping of many spatial parameters; a searchable compilation of all 400 water-related treaties and 39 U.S. interstate compacts, along with the full text of each; an annotated bibliography of the state-of-the-art of water conflict resolution, including approximately 1,000 entries; negotiating notes (primary or secondary) from the detailed case studies of water conflict resolution (see Appendix C); a "Water Event Dataset," which includes comprehensive news files of all reported cases of international water-related disputes and dispute resolution (1950–2000), along with similar datasets for Oregon and for the U.S. West; and descriptions of indigenous/traditional methods of water dispute resolution.

Work on the database has resulted in dozens of published articles (most available online at the database's Web site), and several master's theses and Ph.D. dissertations (including those by Kyoko Matsumoto and Meredith Giordano, from which we derive Appendices E and F). In this book, when we describe "our collection," "the treaties surveyed," or "our survey of events," we are referring either to the database itself (available at: http://www.transboundarywaters.orst.edu) or to resulting research.

1   Background, trends, and concepts

The sage's transformation of the World arises from solving the problem of water. If water is united, the human heart will be corrected. If water is pure and clean, the heart of the people will readily be unified and desirous of cleanliness. Even when the citizenry's heart is changed, their conduct will not be depraved. So the sage's government. . . . consists of talking to people and persuading them, family by family. The pivot (of work) is water.

                        – Lao Tze, ca. sixth century BCE

1.1 CONFLICT MANAGEMENT, PUBLIC PARTICIPATION, AND WATER MANAGEMENT

Water is likely to be the most pressing environmental concern of this century. As global populations continue to grow exponentially, and as environmental change shifts the location of the flow, timing, quality, and quantity of water, the ability of nations and states to peacefully manage and resolve conflicts over distributed water resources will increasingly be at the heart of both stable and secure international relations and of political stability within many countries. There are 263 watersheds and untold aquifers that cross or underlie the political boundaries of two or more countries (Figure 1.1). These international surface basins cover 45.3 percent of the land surface of the Earth, affect about 40 percent of the world's population, and account for approximately 60 percent of global river flow (Wolf et al., 1999). Water has been a cause of political tensions and occasional exchanges of fire between Arabs and Israelis, Indians and Pakistanis, and Americans and Mexicans and among all ten riparian states of the Nile River. Water is one of the few scarce resources for which there is no substitute, over which there is poorly developed international law, and the need for which is overwhelming, constant, and immediate (Bingham, Wolf, and Wohlgenant, 1994).

Within nations, too, there are many examples of internal water conflicts, ranging from inter-State violence and death along the Cauvery River in India (Baviskar, 1995; Anand,

2004) to the United States, where California farmers blew up a pipeline meant for Los Angeles (Reisner, 1986) to inter-tribal bloodshed between Maasai herdsmen and Kikuyu farmers in Kenya (*News24.com*, 2005; *BBC*, 2005). The inland, desert U.S. state of Arizona even commissioned a navy (made up of one ferryboat) and sent its state militia to stop a dam and diversion on the Colorado River in 1934 (Miller, 2001). Recent research on internal disputes suggests that as geographical scale drops, the likelihood, and the intensity, of violence rises (see, e.g., Giordano, Giordano, and Wolf, 2002).

These resource conflicts will gain in frequency and intensity as water resources become relatively scarcer and their use within jurisdictions can no longer be insulated from having an impact on neighboring jurisdictions. A clear understanding of the details of how water conflicts have been resolved historically will be vital to those responsible for bringing together the parties to resolve or to prevent these future conflicts.

Humans have been managing and resolving disputes for thousands of years (Biswas, 1970). Recently, the formal fields of dispute resolution and conflict management have emerged from attempts to find alternatives to expensive litigation, an adversarial and highly expensive means for resolving disputes or, bluntly, to avoid violence or war. Conflict management has been driven by traditional fields of labor management negotiations, contract settlements, community mediation, and, most recently, environmental and resource conflicts.

At the same time there is growing concern for public participation. This concern was highly visible in the western United States during debates on water and natural resources management in the 1970s and early 1980s. This was due primarily to a spate of legislation on resources, starting with the 1969 National Environmental Policy Act (NEPA) and the 1972 Clean Water Act. The salience of participation to technical water managers seemed to lapse during the 1980s and early 1990s, to the detriment of many U.S. water agencies. Recently, however, public participation is once again being seen as a useful tool in the industrial countries and throughout

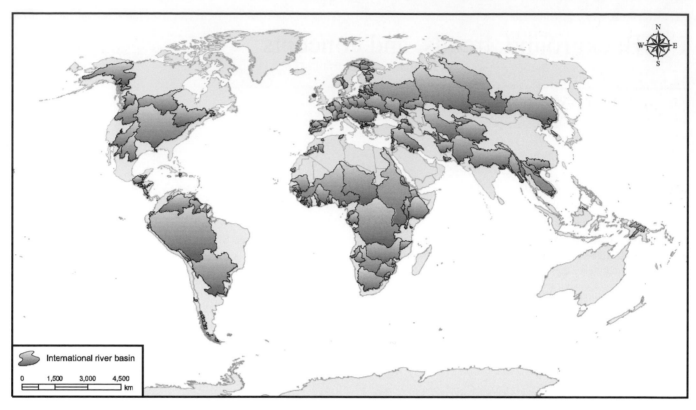

Figure 1.1 Map of the international river basins of the world (Transboundary Freshwater Dispute Database, 2006).

poorer countries. Traditionally, public participation emerged from concerns about open access to government, empowerment of people, and building democratic culture (Bruch et al., 2005). Both participation and conflict management advocate similar process procedures and thus can be included under the general rubric of integrative bargaining or collaboration. But although conflict management and alternative dispute resolution (ADR) speak of dealing with, anticipating, and avoiding conflict, they have much less to say about long-term institution building and structural change based on fundamental value change driving the behavior of water resources managers. Much can be gained by mixing the lessons from these fields, and all are central to efficient governance of water.

Social scientists tell us that institutions are routinized patterns of behavior creating stable expectations over time (Lasswell, 1971). These patterns are driven by values that, over time, are often latent and unexamined. Water resources institutions are being transformed by profound changes in the values of those societies they support. For example, the wealthy West has come to see pollution as critical; however, those who are poor, although they understand the problems of pollution, are more concerned with water's utilitarian value as an engine for growth. The institutions designed to deal with water in turn reflect these different values in their priorities. Bringing

new values and their attendant claims to bear on water institutions means a long-term shift in water resources managers' patterns of behavior. However, by focusing on the nation-state, the rich experience of building water institutions is often missed because much of it has fallen within and not among nation-states. What were once regional intrastate issues can become international. We have only to look at Central Asia and the Aral Sea for such an example. We choose to use the word *interjurisdictional* to cast a broad net to capture such water resources institution-building experiences.

The 1997 UN Convention on Non-Navigational Uses of International Watercourses (see Appendix A; United Nations, 2005) builds on the 1966 Helsinki rules (International Law Association, 1966), various UN deliberations, and the 20-year process of the International Law Association's deliberations. They have produced some sound principles for nonnavigational uses of international waters. In summary, they call for

- equitable and reasonable use
- obligations not to cause significant harm
- general obligation to cooperate
- regular exchange of data and information
- examination of relations between users

These are good principles. They could be useful in all transboundary water management, whether within a federal state with competing jurisdictions or among sovereign nations. However, they present operational questions: Which principle prevails when equitable use conflicts with the obligation not to cause appreciable harm? What is appreciable harm? What are the standards of responsibility for a breach of principles? What should we do when there is no internationally recognized legislation and no compulsory enforcement jurisdiction?

Water resources management requires collaboration across jurisdictions and sectors, whether within or among states. Indeed, much of the history of water institutions is about the conflict between geographic dictates of the resource versus the realities of political jurisdictions. Water resource institutions go to the heart of our changing notions of subsidiarity. *Subsidiarity* generally means the principle that none of the polity's tasks should be assigned to a body larger than the smallest that can satisfactorily perform it. For example, water resources management and administration in the United States must be seen as a result of the Federalist system of government, where the states within the United States have first sovereignty over the water.

Building water resources institutions for collaboration depends on how we see the principle of subsidiarity at work in water resources management. Building water resources institutions is also directly related to capacity building and governance. The most important factors in building cross-jurisdictional and sectoral institutions are creating the will and incentives to cooperate and the processes to do so.

Before we examine trends pushing for cooperation and reasons for process techniques and procedures, we should acknowledge two important points. First, that contrary to common thinking, dispute management is neither modern nor Western. Traditions of dispute resolution date back millennia. "*Acequia*," for example, is the term used in the United States' desert southwest and other Spanish-speaking parts of the world to denote both an irrigation ditch and the informal institution that manages it. *Acequias* have their roots in Spain. According to tradition, the Tribunal de las Aguas (Water Court) has been meeting to resolve the disputes over the *acequias* around Valencia in the same church-front square since medieval times, if not before (Glick, 1970). But the root of *acequia* is *al-saqia*, Arabic for a gear-driven waterwheel, the technology that made early irrigation possible along many of the rivers of the ancient Middle East (Oleson, 1984). From the Middle East to Spain to the New World – the roots of collaborative approaches run deep. This system teaches us much about the subsidiarity principle of dealing with conflicts and of cooperating for planning and operations at the lowest possible levels.

Second, the water resources field is rich with experiences and illustrations of collaboration approaches. Indeed, the water resources field is at the nexus of one of the oldest and most contemporary of public policy questions: How should specialized knowledge relate to power in a society? We can learn much from our water resources experience that can inform our current search for answers to this question in the water resources and other related areas.

1.2 SOME TRENDS PUSHING TOWARD COOPERATION

Like all trends, whether they are positive or negative is in the eyes of the beholder. We choose to be optimistic about what follows:

1. Water compels us to think regionally. Because it ignores legal delineations, and because technical information has and continues to play a crucial role in water resources decision making, the need for regional management and data-gathering bonds water professionals across jurisdictions.
2. There is growing realization that the price for having some control over agreements is sharing ownership and cooperating in both the process and outcome of those agreements.
3. As constraints on the resource grow, especially in an era of fiscal austerity, the opportunity costs for not cooperating become clearer. Indeed, negotiations can be seen as a social-learning process, and the need for cooperation is one of its lessons.
4. The movement for environmental justice will bring new environmental value claims directly to social claims and link them to per capita measurements.
5. Influential new actors are emerging that represent new claims on water resources that cross jurisdictional boundaries.
6. The politics of water is moving from that of distributing benefits of an expanding pie to the perception of redistributing a decreasing pie, now and in the future.
7. The transaction costs in time, dollars, resources, lost revenues, and even violence are escalating beyond traditional management methods and/or capacity to keep up. This is forcing the adoption of alternative approaches.
8. Available money relative to identified needs is contracting. Therefore, more must be done with less. A qualitative multiplier is needed for our management procedures. Cooperation built on a new ethic of informed consent, rather than an old ethic of paternalism, can provide such a multiplier, especially in terms of increased program effectiveness and enhanced implementation.
9. There is a growing moral imperative for more accountability, responsiveness, and intergenerational equity in water resources decisions.

10. There is a shift from a deterministic prediction of the future to the notion of jointly creating the future.

11. Traditional legal systems everywhere are seen as unable to cope with change. The reliance on precedent is insufficient if the problem is that current legal obligations are locked into allocation formulas that diverge dangerously from current demographic realities.

12. International lenders and donors are beginning to perceive their role as that of a facilitator to agreements rather than an expert dictator of agreements. These actors have resources that can be incentives for cooperation, even in a world with weak legal systems and sanctions.

13. New treaties and agreements that are multipurpose are growing. Old, single-purpose treaties and agreements are under pressure to expand.

14. There is a renewed interest in functional diplomacy and what is now called "second-track" diplomacy.

15. Technologies that are accessible to ordinary people and technologies that help rather than hinder dialogue, alternative generation, and sensitivity testing are rapidly emerging.

16. There is a growing and changing public awareness of water resources.

17. There is evidence from divergent fields of science that cooperation is and has been the key to growth and evolution. Such evidence can be found in computer science and game theory, evolutionary biology, social psychology, and hard and soft technology. Lewis Thomas (1992) notes, "The driving force in nature, on this planet and biosphere is cooperation . . . and that our bacterial ancestors learned, early on, to live in communities." Speaking about trench warfare in World War I, anthropologist Ashworth (1968, 1980) notes "how a kill or be killed strategy turned into something like live and let live." Computer scientist Axelrod (1984) finds the "Roots of Cooperation" in playing millions of prisoners' dilemma games. The result is that a tit-for-tat strategy – a strategy that starts with cooperation and repeats whatever moves the other player makes – works best.

1.2.1 Trends of cooperation at the institutional level

CHANGING CONTEXT OF GOVERNANCE
The world is changing. A renewed democratic spirit and a new ecological awareness are two of the principal forces driving change. This democratic spirit is calling us to new notions of individual freedom, transparency, and accountability in public decisions. The new ecological awareness reminds us of a collective responsibility and leads us to notions of holistic and comprehensive systems. With its long-term focus and its calls

to include stakeholders in decision making, sustainability has become a venue for this dialogue (United Nations, 2002a). Building the physical water infrastructure in a collaborative and participatory way is now an important means of "governance environment." Water resource management, with its current debates over markets, pricing, planning, participation, and environmental assessment, is a meeting ground for these forces.

Indeed there is growing recognition that the experience of solving and managing water conflicts can greatly influence the political structures of nations. Contrary to the old Wittfogel (1956) thesis that development begets large water infrastructure, which begets large bureaucracy, which begets control, which begets authoritarianism, the opposite is also true. For example, many have recognized how the experience of the Dutch water boards, over several hundred years, greatly influenced the current structure of Dutch government. Those boards, through the experience of being elected and making decisions on essential matters, helped to build a culture of democracy. The experience of managing water, which is close to and vital to people's lives, is an enormous opportunity for social learning on how to live together, as well as how to manage water.

EXPANDING OF ISSUES AND OF STAKEHOLDER NUMBER AND ASYMMETRY
Water management must now integrate new ecological values and criteria of sustainability. Both require more information, which, in turn, highlights additional new risk and uncertainty. Both require professionals to compare among incommensurable values and other values that are difficult to quantify. More explicit understanding of risk requires an active choosing of, rather than passive reacting to, risk by beneficiaries. All of this will push water resource professionals beyond traditional methodologies and into new process considerations.

More voices with competing views of the future must be involved in water development. Although the distribution of power among these parties is asymmetrical, the power to stop or delay is diffusing faster than incentives to create and cooperate. New ways to prioritize investments and manage conflicts among competing interests will be needed. Inertia toward negative and reactive attitudes must be countered with incentives for positive and creative development and with new ways to foster ownership in both the plans and the process of generating those plans among interested and affected parties. Impact assessments are crucial for both informed technical and good moral decision making because, to the best of our ability, we must know the consequences of our actions. However, we must move beyond being paralyzed by our understanding of consequences by simply looking at costs. Process techniques and procedures offer a route out of paralysis toward action.

GROWING GAP BETWEEN DEVELOPMENT NEEDS AND AVAILABLE CAPITAL

While the industrialized world debates reallocation and reapportionment within existing water systems, many in the world have little or nothing to reapportion and need new systems. At the same time capital is short. So doing more with less means, in part, being more efficient. But being more efficient confronts us with issues of distributive equity and fairness. In recent years, water managers have moved beyond the traditional structural interventions into natural systems to management of social systems and now biological systems as means for water management. Thus, cost recovery and project performance will become even more important to decision makers. For several years World Bank evaluations at the project level show how participatory processes can be effective in meeting these challenges (Nagle and Ghose, 1990).

Creative alternatives and new public/private partnerships must be found to develop and allocate water use. Without the strategic management of allocation, the transaction costs of managing water can escalate to unacceptable levels. Indeed, resource scarcity, when seen by some parties as a relative deprivation, whether perceived or real, can lead to violence and political authoritarianism (Gurr, 1985) and corruption (Rinaudo, 2002). Without operating agreements between and within nations and among users, the opportunity costs in lost economic benefits, poverty reduction, and public health could escalate to the point of social stagnation. We must begin to reinterpret our awareness of water interdependence as an opportunity to create cooperation rather than as inevitable zero-sum competition.

A key to such reinterpretation is in the way – or the processes – by which we anticipate and manage the competing and conflicting demands for the resource. Water resource development is becoming more dependent on integrative bargaining, agreement building, participation, collaborating, and using fair processes for managing conflict. To this extent, the international agencies have a stake in integrative bargaining, especially in the international system, where incentives for proactive collaboration are often weak.

GROWING WATER INTERDEPENDENCE WITH WEAK COMPLIANCE AND INCENTIVE SYSTEMS

Water policy reviews in the international development agencies have been documenting how water use and its allocation and reallocation are likely to drive development strategies. Water is central to poverty alleviation through food production and infrastructure development (Sullivan et al., 2003). A report on international environmental conflict resolution (IECR) notes that "Most current IEC's are related to international rivers" (Trolldalen, 1992). Stern and Druckman (2000) identify numerous conceptual, methodological, and inferential challenges of using a scientific approach to evaluate the effects of past conflict-resolution interventions.

As population and urbanization accentuate conflicting demands for the same resource – water – our interdependence becomes more evident. Everywhere the call for better water pricing and readjustment of agricultural subsidies is heard. But the question is "how?" The reality is that agreements on agricultural prices, as shown in the World Trade Organization (WTO) process and in the European Community (EC), are difficult if not impossible to reach. Thus, it is hard to see how food security interests, to say nothing of national ideological interest, will be met.

Most of the world's largest rivers are international, and more are becoming so because of political changes, such as the breakup of the Soviet Union and the Balkan states, as well as access to today's better mapping sources and technology, which can better trace a watershed. There were 214 international basins listed in 1978 (United Nations, 1978), the last time any official body attempted to delineate them, and there are 263 today.

Even more striking than the total number of basins is the percentage of each nation's land surface that falls within these watersheds. A total of 145 nations include territory within international basins. Twenty-one nations lie in their entirety within international basins; including these, a total of thirty-three countries have more than 95 percent of their territory within these basins. These nations are not limited to smaller countries, such as Lichtenstein and Andorra, but include such sizable countries as Hungary, Bangladesh, Belarus, and Zambia (Wolf et al., 1999; UNEP and OSU, 2002).

A final way to visualize the dilemmas posed by international water resources is to look at the number of countries that share each international basin: nineteen basins are shared by five or more riparian countries; the Danube has seventeen riparian nations; the Congo, Niger, Nile, Rhine, and Zambezi are each shared by between nine and eleven countries; and thirteen basins – the Amazon, Ganges-Brahmaputra-Meghna, Lake Chad, Tarim, Aral Sea, Jordan, Kura-Araks, Mekong, Tigris-Euphrates, Volga, La Plata, Neman, and Vistula (Wisla) – have between five and eight riparian countries. Likewise, the large countries of the world, such as the United States, Canada, India, Mexico, China, Nigeria, Russia, and Brazil, have states or provinces within them across whose boundaries water flows.

In the Middle East, two-thirds of Arabic-speaking people depend on transboundary waters that flow from non-Arabic

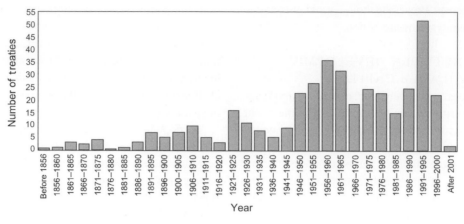

Figure 1.2 Number of international water treaties by year (Transboundary Freshwater Dispute Database, 2006).

areas (Kolars, 1992). Because the structure of international compliance to water quality, the environment, and other supply issues is weak, interdependence will have to be served through incentives. As the Oslo report states, international financial institutions with financial leverage will become critical to encouraging and leading new incentives (Trolldalen, 1992).

Cai and Rosegrant (2002) and Rosegrant and Cai (2002) modeled global water demand and supply projections and determined that water demand will grow rapidly for domestic and industrial uses with slowing growth for agricultural uses. Current water use in several of the shared basins is generating demands that either already exceed available supply or soon will. Some of the most pronounced deficits are likely to occur in regions already ripe for high-intensity conflict and with rivers of high flow variations, such as in the Middle East (Amery and Wolf, 2000; Allan, 2001). Other projected deficits are likely to occur in areas already prone to famine. Projected deficits in arid regions of the United States, despite their comparative wealth, are already causing significant political realignments (Miller, 2001).

GROWING DISCONTINUITY BETWEEN TRADITIONAL INSTITUTIONS AND NEW REALITIES

In recognition of growing interdependence, 286 international treaties concerning water were signed by 1970. By 1986 there were 324, and there are currently more than 400 (Figure 1.2 shows the continued growth in the rate of treaty development). Although the rate of agreements increases, only two-thirds of these treaties relate to river basins in Europe and North America (Nagy, 1987). Few exist in the developing world, where the need is rapidly growing (Nagy, 1987). For example, Europe, with 48 river basins, has 175 water-related treaties, whereas Africa, with 34 river basins, has 34 treaties (Linnerooth-Bayer, 1986). More important, almost 85 percent are bilateral rather

than multilateral, even on multilateral basins, and single purpose rather than multipurpose (Hamner and Wolf, 1998). For example, of eighteen agreements on the Danube since 1948, all but one has been bilateral (Linnerooth-Bayer, 1986).

The most frequent purposes of earlier treaties were navigation and hydropower production, which gave way to water allocation. Multipurpose use, water quality, and environmental aspects have now become more prominent. Flood control management is a major objective in about 10 percent of existing treaties. Most treaties relate to planning or preliminary surveys, whereas those relating to construction and joint operation are far fewer. Few relate to groundwater or water quality – about 25 percent mention water quality, but only four treaties are explicit in their requirements (Giordano, 2002). Also, few treaties use a basinwide approach, and most relate to specific sections of the rivers (Nagy, 1987). However, current agreements are beginning to reflect an interest in a comprehensive view of uses: basinwide management, multisectoral development, and water-quality control (Vlachos, 1991).

Changing demographics are demanding new priorities and flexibility in water use and are straining the capacity of traditional water institutions. Almost 15 percent of the World Bank's portfolio is water related, leading to overlapping sectoral jurisdictions within the Bank and overlapping geographic jurisdictions outside. Beyond the Bank, institutional means to achieve environmental health and development seem inadequate. In the end, no matter what organization is created, discontinuities will require more managed flexibility and planning.

DEVELOPMENT OF WATER RESOURCES MANAGEMENT AS A MEANS TO BROAD AGREEMENTS

Because many of the world's rivers are regional, not global, because their related social interdependencies are so tangible and so clearly shared, and because they have such a rich history of interdependence, management of these rivers offers opportunities for cooperation built on technical needs, which could produce further positive political, social, and economic cooperation (Conca and Dabelko, 2002). Although it is perhaps open to criticism as either geographical determinism or naive neofunctionalism, water resources management has helped and continues to help integrate social and political groups. The earliest U.S. Supreme Court decisions establishing federal power concerned water navigation. European rivers, such as Rhine, Rhône, and Danube, have been steadily moving from functional agreements around water to more administrative integration. In the midst of land grabs and war, some southern African nations, through mediation, discovered shared interest in irrigation and hydroelectric power. They signed a joint nonaggression pact and teamed up to gain international financing for a water development project (Hickey, 1992). Although commentators like to focus on water potential to ignite Middle East conflict, it is currently one of the few areas serving as a means for parties to talk. Senior technical/administrative water officials share a technical language that can be a powerful base for communication. In addition, at some level almost all cultures recognize the sanctity of water. Water as cleanser and healer is one of the paramount metaphors of human experience. Water has a deep, almost primordial significance and immense potential symbolic power to move people.

CHANGING ETHICAL BASIS OF PROFESSIONALISM

The ethical basis of professionalism is moving from a traditional paternalism to a newer notion of informed consent. Throughout societies, the very meaning of professionalism is changing. Some patients no longer say "cure me"; they participate with doctors in their own diagnosis and treatment. Clergy may no longer maintain strict distinctions between the "lay" and "religious" and may no longer consider themselves the sole salvation mediators between heaven and earth. Lawyers may no longer neglect alternatives to litigation or avoid linking their individual actions to the overall state of social justice. Water professionals should not be surprised when affected groups and beneficiaries of their works demand rights in influencing project design and locations.

Professionalism includes not only the final goods and services provided but also the means employed to deliver those goods and services. The means by which the goods and services are delivered establish a relationship with client and partners. Process procedures are means to help professional engineers cope with these changing demands emanating from a new understanding of professionalism throughout society.

GROWING NEED FOR BALANCE AMONG DEVELOPMENT, GROWTH, DEMAND MANAGEMENT, AND STRUCTURAL INVESTMENTS

After a period of unfettered development, followed by a period of sometimes indulgent introspection and assessment, the world is entering a new period of balancing management and structures. Once a certain level of wealth was attained in the West, the environmental and other costs became more evident. During periods of early growth or during a depression, the focus was on generating income, wealth, and social well-being. Once these issues were settled, the costs, often hidden, became evident. This understanding eventually led to new policies on growth and various forms of impact assessments. These requirements have spilled over to lending and granting to the Third World from the external support agencies (ESAs). These requirements for impact assessment have engendered great debate and have often looked like cultural imperialism from the developed world. Today, as witnessed by the new sector strategies of the World Bank, the call is for both management and development. This call essentially means negotiated approaches, which are more open and inclusive of both the people benefiting as well as those impacted and also of the distribution of risk sharing. In short, there is growing recognition that integrated water management or poverty reduction cannot be attained if either structures or management are taken off the table. The question is how to attain the appropriate mix – and this requires more process sensitivity and skills on the part of the water managers.

Nowhere is this clearer than in today's dialogue between the rich and the poor about water. Figure 1.3 illustrates the dilemma. As countries and societies develop, they must first invest in water infrastructure. At this point, investment in management is usually lower. Over time, as a country prospers, the investment in management increases as that in infrastructure decreases. Well-intended prescriptions from those in

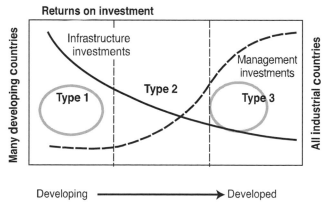

Figure 1.3 Need for new rich–poor dialogue (World Bank, 2006).

the rich countries to increase funding for water management schemes may not be appropriate for developing countries that have little or no water infrastructure. For example, Sidebar 1.1 shows some approximate investment data for water in recent U.S. history. No one really knows the extent of that investment, but it is huge when one adds the state and local investment and water-services investment. The Water Environment Federation estimates that the United States must spend about US$23 billion a year to meet its environmental standards. There are more than 100 countries in the world with little infrastructure and with GDPs less than US$23 billion. Water prescriptions for poor areas of the world based on the assumptions gained from the U.S. experience can lead to fundamental misunderstandings. In fact, it can often appear that rich nations are admonishing the poor not to use the resources in the same way the rich did during development. Such prescriptions can come across as a new form of imperialism. Instead, rich nations should reexamine how they used their water resources during development and at what costs – then, based on those costs, help currently developing countries design mitigative measures to help them avoid similar costs as they grow. But this dialogue is not prevalent.

1.2.2 The world of the professional water manger is changing

The world has changed for water resources managers, planners, and decision makers. Today, especially in the context of new demands for integrated water resources management (IWRM), water managers and planners often work in teams involving multiple disciplines rather than just engineering and associated technical fields (e.g., see Diplas, 2002). Increasingly they also work in multiagency teams that include a variety of public, nongovernmental organization (NGO), and private sponsors. Today's water managers and decision makers must consult with a broader range of stakeholders, publics, and NGOs locally, regionally, and often internationally. And, they must do all this while operating in a world of increasing demands on water.

Technical expertise remains necessary for creating sustainable water management decisions – perhaps even more necessary than ever – however, that alone is not enough. People all over the world need technical engineering competence, but the

---

Sidebar 1.1 History: U.S. Investment in Water Supply

- New Deal: Works Progress Administration (WPA) 2,600 water projects = US$312 million (in 1930s dollars).
- Federal Power Commission (FPC) [later the Federal Energy Regulatory Commission (FERC)]; Civil Works Administration (CWA); WPA US$112 million for municipal water (in 1930s dollars).
- 1972–1990 more than US$650 billion in federal grants for sewage treatment and US$20+ billion from states.
- World Economic Forum (WEF) estimates that US$23 billion per year is needed for 20 years to meet the U.S. Environmental Protection Agency (EPA) standards.
- *More than 100 countries that lack adequate sanitation have an annual budget less than US$23 billion.*

---

ability to put that competence in service of those who need it depends, in many cases, on changing the relationship between the experts and those whom they are serving. This book aims at helping to build, modify, or create such new functional relationships.

This new water resources decision-making environment requires at least two sets of skills. First, it requires broad, interdisciplinary technical skills, which reach across disciplines to allow consideration of alternatives that in the past were often not evaluated. Many water decisions rest on a scientific basis that is itself incomplete. This means that water decision makers may first need to get agreement on what studies should be conducted and what data collected to ensure that decisions are based on science, not rhetoric. As a result, water planners and managers need a breadth of technical knowledge that goes beyond traditional engineering.

Second, water planners and managers need another set of skills: the skills to design and conduct processes that draw together partners, stakeholders, and publics, resulting in decisions that enjoy broad cross-sectoral, and often transboundary, public support. The era where water planners and managers employ the "decide-announce-defend" approach is rapidly disappearing. In this new era, water management is "done with" (as opposed to being done "for" or "to") potentially affected agencies, public and private organizations, and individuals.

2 Water wars, water reality: Reframing the debate on transboundary water disputes, hydropolitics, and preventive hydrodiplomacy[1]

Fierce competition for fresh water may well become a source of conflict and wars in the future.

– Kofi Annan, March 2001

But the water problems of our world need not be only a cause of tension; they can also be a catalyst for cooperation.... If we work together, a secure and sustainable water future can be ours.

– Kofi Annan, January 2002

Before delineating appropriate measures for water conflict prevention and management, we first need to address the larger issues between people and their environment – that is, who affects whom? It is quite clear that people affect their environment, but to what extent is the opposite true: just how deep is the causal relationship between environmental stresses and the structure of human politics? This relationship is at the heart of understanding the processes of environmental conflict prevention and resolution. If, as the large and growing "water wars" literature would have it (see, for example, Cooley, 1984; Starr, 1991; Bulloch and Darwish, 1993; Remans, 1995; Amery, 2002), the greatest threat for water conflicts is that water scarcity can and will lead directly to warfare between nations. This lends itself to diversion of a potentially huge amount of resources, in attempts to arrest these processes at the highest levels. If the processes are actually both more subtle and more local in nature (as suggested by, among others, Elhance, 1999; Marty, 2001; Chatterji, Arlosoroff, and Guha, 2002; Wolf, Yoffe, and Giordano, 2003b; Carius, Dabelko, and Wolf, 2004) then so too are the potential solutions.

Throughout this book, we will note that shared water does lead to tensions, threats, and even to some localized violence – and we will offer strategies for preventing and mitigating these tensions – but not to war. Moreover, these tense "flash points" generally induce the parties to enter negotiations, often resulting in dialogue and, occasionally, to especially creative and resilient working arrangements. We will note also that shared water provides compelling inducements to dialogue and cooperation, even while hosilities rage over other issues.

But let's look at the evolution of the "water leads to war" thesis. Although the extreme "water wars" literature mostly began to fade in the late 1990s, a number of articles dating back decades argue quite persuasively for some degree of causality between environmental stress – reaching up against relative resource limits – and political decision making. One cannot discuss water institutions, for example, without invoking Wittfogel (1956) and his classic argument that the drive to manage water in semiarid environments led both to the dawn of institutional civilization – described by Delli Priscoli (1998a) as the "training ground for civilization" – and to particularly autocratic, despotic forms of government. This latter argument, and the generally enthusiastic reception he received, needs to be understood in the Cold War setting from which it sprang and was quite effectively challenged by Toynbee (1958), among others. Toynbee's vehemence (in his review he calls Wittfogel's book a "menace") is particularly interesting because many of Wittfogel's theories can be seen as extensions of a sort of Toynbee's (1946) "challenge-response" thesis in which he argues that the impetus toward civilization becomes stronger with greater environmental stress. Toynbee's objections are primarily with Wittfogel's "tribalistic" lens to history, aimed, as Toynbee charges, at demonizing the Soviet Union. Wittfogel (1956) in turn, distinguishing himself from Toynbee, writes of his own position, "causality yes, determinism no" (p. 504). However, the premise that there is a critical link between how society manages water and its social structure/political culture remains as an important and valid insight.

This thread of causality between the environment and politics has been taken up regularly over the years. When Sprout and Sprout (1957) describe the environmental factors inherent in international politics, it becomes the direct intellectual precursor to today's blossoming "environmental security" literature, as spearheaded by Homer-Dixon (1991). Homer-Dixon,

---

[1] This chapter draws from *Conflict Prevention and Resolution in Water Systems*, edited by A. Wolf. Cheltenham, UK: Elgar, 2002b.

9

like Wittfogel, was initially greeted enthusiastically by the defense establishment, this time in the setting of the post—Cold War redefinition of relevance and, again like Wittfogel, has been taken to task for the degree of causality in his arguments. (A summary of Homer-Dixon's findings, along with a debate on the topic is presented in Wolf, 2002b.) In his defense, Homer-Dixon's arguments, along with those of much of the "water wars" crowd, have become more muted over the past few years — in 1994, he wrote: "the renewable resource most likely to stimulate interstate resource war is river water," which he repeats in his 1996 article. He modifies the claim, elaborated in his 1999 book: "In reality, wars over river water between upstream and downstream neighbors are likely only in a narrow set of circumstances ... [and] ... there are, in fact very few river basins around the world where all these conditions hold now or might hold in the future."

In water systems, the dichotomy of causality is manifested as whether the stress on water resources lends itself more readily to conflict or cooperation. Both arguments are powerful and have been supported by a rich, if mostly anecdotal, history. Postel (1999) describes the roots of the problem at the subnational level. Water, unlike other scarce, consumable resources, is used to fuel *all* facets of society, from biologies to economies to aesthetics and religious practice. As such, there is no such thing as managing water for a single purpose – *all* water management is multiobjective and is therefore, by definition, based on conflicting interests. Within a nation, these interests include domestic use, agriculture, hydropower generation, recreation, and environment – any two of which are regularly at odds – and the chances of finding mutually acceptable solutions drop precipitously as more actors are involved.

Conceptually, and as described in case studies by Trolldalen (1992), these conflicting interests within a nation represent both a microcosm of the international setting and a direct influence on it. Trolldalen's work is particularly useful in that he sidesteps the common trap of treating nations as homogeneous, rational entities, and explicitly links internal with external interests. Bangladesh is not just the national government of Bangladesh when it negotiates a treaty with India over Ganges flow: it is its coastal population, inundated with saltwater intrusion; its farmers, dealing with decreasing quantities of water and increasing fluctuations; and its fishermen, competing for dwindling stocks.

This link between the internal and external is critical when we look at violent international conflicts (Conca, 2006). Gleick (1993) is widely cited as providing what appears to be a history replete with violence over water resources. But a close read of his article reveals greater subtlety and depth to the argument. Wolf (1998) points out that what Gleick and others have actually provided is a history rife with tensions,

exacerbated relations, and conflicting interests over water but *not* State-level violence, at least not between nations or over water as a scarce resource. It is worth noting Gleick's careful categorization because the violence he describes actually turns out to be water as a tool, target, or victim of warfare – *not* the cause of the violence. Wolf (1998) contrasts the results of a systematic search for interstate violence – one true water war in history, 4,500 years ago – with the much richer record of explicit, legal cooperation – 3,600 water-related treaties. In fact, a scan of the most vociferous enmities around the world reveals that almost all the sets of nations with the greatest degree of animosity between them, whether Arabs and Israelis, Indians and Pakistanis, or Azeris and Armenians, either have a water-related agreement in place or are in the process of negotiating one.

2.1 WHY IS THE WATER WAR ARGUMENT SO COMPELLING?[2]

If water is at the heart of most human activity, if it is shared between often hostile users, and if it is becoming relatively scarcer year by year, it is difficult to think of *alternatives* to inevitable warfare. Recent articles in the academic literature (Cooley, 1984; Starr, 1991; Remans, 1995; Amery, 2002) and popular press (Bulloch and Darwish, 1993; *World Press Review*, 1995) point to water not only as a cause of historic armed conflict but also as *the* resource that will bring combatants to the battlefield in the twenty-first century. Invariably, these writings on "water wars" point to the arid and hostile Middle East as an example of a worst-case scenario, where armies have in fact been mobilized and shots fired over this scarce and precious resource. Elaborate, if misnamed, "hydraulic imperative" theories have been developed for the region, particularly between Arabs and Israelis, citing water as the prime motivator for military strategy and territorial conquest.

Westing (1986), for example, suggests that, "competition for limited ... freshwater ... leads to severe political tensions and even to war"; Gleick (1993) describes water resources as military and political goals, using the Jordan and Nile as examples; Remans (1995) uses case studies from the Middle East, South Asia, and South America as "well-known examples" of water as a cause of armed conflict; Samson and Charrier (1997) write that "a number of conflicts linked to freshwater are already apparent" and suggest that "growing conflict looms ahead"; and Butts (1997) suggests that "history is replete with

---

[2] This section draws from Wolf (1997). International water conflict resolution: Lessons from comparative analysis. *International Journal of Water Resources Development*, **13**(3, September), 333–356.

examples of violent conflict over water" and names four Middle Eastern water sources particularly at risk. The basic argument for "water wars" is as follows: Water is a resource vital to all aspects of a nation's survival. The scarcity of water in an arid and semiarid environment leads to intense political pressures, often referred to as "water stress," a term coined by Falkenmark (1989), or "water poverty" as suggested by Feitelson and Chenoweth (2002). Furthermore, water not only ignores our political boundaries, it evades institutional classification and eludes legal generalizations. Interdisciplinary by nature, water's natural management unit – the watershed, where quantity, quality, surface water, and groundwater all interconnect – can strain both institutional and legal capabilities past capacity. When international water institutions step in, the result is often a lack of quality considerations in quantity decisions, a lack of specificity in rights allocations, disproportionate political power by special interest, and a general neglect for environmental concerns in water resources decision making.

The problems of water management are compounded in the international realm by the fact that international water law is poorly developed, viewed as contradictory, and often unenforceable. The 1997 Convention on the Non-Navigational Uses of International Watercourses Commission (see Appendix A; United Nations, 2005), which took 27 years to develop, reflects the difficulty of marrying legal and hydrologic intricacies: although the Convention provides many important principles for cooperation, including responsibility for cooperation and joint management, it also institutionalizes the inherent upstream–downstream conflict by calling for both "equitable use" and an "obligation not to cause appreciable harm." These two principles are in implicit conflict in the setting of an international waterway: upstream riparians have advocated that the emphasis between the two principles be on "equitable use," because that principle gives the needs of the present the same weight as those of the past. In contrast, downstream riparians have pushed for emphasis on "no significant harm," which effectively protects the preexisting uses generally found in the lower reaches of most major streams. The Convention also provides little practical guidelines for allocations – the heart of most water conflict. Allocations are to be based on seven relevant factors, which are to be dealt with as a whole. These factors are (1) geographic, hydrographic, hydrological, climatic, ecological, and other natural factors; (2) social and economic needs of each riparian state; (3) population dependent on the watercourse; (4) effects of use in one state on the uses of other states; (5) existing and potential uses; (6) conservation, protection, development and economy of use, and the costs of measures taken to that effect; (7) and the availability of alternatives, of corresponding value, to a particular planned or existing use.

Furthermore, international law only concerns itself with the rights and responsibilities of *nations*. Some political entities who might claim water rights, therefore, would not be represented, such as the Palestinians along the Jordan or the Kurds along the Euphrates. Cases are heard by the International Court of Justice (ICJ) only with the consent of the parties involved, and no practical enforcement mechanism exists to back up the Court's findings, except in the most extreme cases. A State with pressing national interests can therefore disclaim entirely the Court's jurisdiction or findings (Rosenne, 1995). Given all the intricacies and limitations involved, it is hardly surprising that water treaties are rarely explicitly informed by general legal principles. The International Court of Justice, for example, has decided only a single case regarding international water law. The ICJ came into being in 1946, with the dissolution of its predecessor, the Permanent Court of International Justice. That body did rule on four international water disputes during its existence from 1922 to 1946. The one case decided by ICJ was a 1997 ruling on the Gabçíkovo Dam on the Danube. Even in that case the court told each party that they had some fault and directed the parties to negotiate outside the court for a resolution.

Taken together – international water as a critical, nonsubstitutable resource that flows and fluctuates across time and space, for which legal principles are vague and contradictory, and is becoming relatively more scarce with every increase in population or standard of living – and one finds a compelling argument that, in the words of World Bank Vice-President Ismail Serageldin (1995), "the wars of the next century will be about water."

A close examination of the case studies cited as historic interstate water conflict, though, suggest that one problem is some looseness in classification. Samson and Charrier (1997), for example, list eighteen cases of water disputes, only one of which is described as "armed conflict," and that particular case (on the Cenepa River) turns out not to be about water at all, but rather about the location of a shared boundary, which happens to coincide with the watershed. Armed conflict did not take place in any of Remans's (1995) "well-known" cases (save the one between Israel and Syria, that we describe in this chapter), nor in any of the other lists of water-related tensions presented.

The examples most widely cited are wars between Israel and her neighbors, and Trottier (2003b) traces water wars as a "hegemonic concept" to writings on the region. Westing (1986) lists the Jordan River as a cause of the 1967 war and, in the same volume, Falkenmark (1986), mostly citing Cooley (1984), describes water as a causal factor in both the 1967 war and the 1982 Israeli invasion of Lebanon. Myers (1993), citing Middle East water as his first example of "ultimate security"; writes that "Israel started the 1967 war in part because the

Arabs were planning to divert the waters of the Jordan River system." In fact, in the years since Israel's invasion of Lebanon in 1982, a "hydraulic imperative" theory, which describes the quest for water resources as *the* motivator for Israeli military conquests, both in Lebanon in 1979 and 1982 and earlier, on the Golan Heights and West Bank in 1967, was developed in the academic literature and the popular press (see, for example, Davis, Maks, and Richardson, 1980; Stauffer, 1982; Schmida, 1983; Stork, 1983; Cooley, 1984; Dillman, 1989; and Beaumont, 1991).

The only problem with these theories is a paucity of evidence. Although shots were fired over water between Israel and Syria from 1951 to 1953 and 1964 to 1966, the final exchange, including both tanks and aircraft on July 14, 1966, stopped Syrian construction of the diversion project in dispute, effectively ending water-related tensions between the two States – the 1967 war broke out almost a year later. The 1982 invasion provides even less evidence of any relation between hydrologic and military decision making. In extensive papers investigating precisely such a linkage between hydrostrategic and geostrategic considerations, Libiszewski (1995) and Wolf (1995b) conclude that water was neither a cause nor a goal of any Arab–Israeli warfare.

To be fair, we should note that this analysis describes only the relationship between inter-State armed conflict and water resources *as a scarce resource*. We exclude here both internal disputes, such as those between interests or States (although we will pick up this issue later), as well as those where water was a means, method, or victim of warfare. We also exclude disputes where water is incidental to the dispute, such as those about fishing rights, access to ports, transportation, or river boundaries. Many of the authors that we cite, notably Gleick (1993), Libiszewski (1995), and Remans (1995), are very careful about these distinctions. The bulk of the articles cited above, then, turn out to be about political tensions or stability rather than about warfare, or about water as a tool, target, or victim of armed conflict – all important issues, just not the same as "water wars" (see Stucki (2005) for a lucid summary of the debate).

2.1.1 Hydromyth I: "Water wars" are prevalent and inevitable

To aid in the assessment of the process of water conflict resolution, researchers have over the past nine years developed the Transboundary Freshwater Dispute Database (TFDD), a project of the Oregon State University Department of Geosciences, in collaboration with the Northwest Alliance for Computational Science and Engineering. The database currently includes digital mapping of the world's 263 international watersheds; a searchable compilation of 400 water-related treaties, along with the full text of each; an annotated bibliography of the state-of-the-art of water conflict resolution, including approximately 1,000 entries; negotiating notes (primary or secondary) from detailed case studies of water conflict resolution (see Appendix C); a comprehensive news file of all reported cases of international water-related disputes and dispute resolution (1950–2000); and descriptions of indigenous/traditional methods of water dispute resolution (TFDD, 2006, p. 55).

Within the context of the TFDD project, in order to cut through the prevailing anecdotal approach to the history of water conflicts, researchers at Oregon State University undertook a 3-year research project that attempted to compile a dataset of *every* reported interaction between two or more nations, whether conflictive or cooperative, which involved water as a scarce and/or consumable resource or as a quantity to be managed – that is, where water is the *driver* of the events – over the past 50 years (see Wolf, Yoffe, and Giordano, 2003b; Yoffe, Wolf, and Giordano, 2003; and Yoffe et al., 2004). Excluded are events where water is incidental to the dispute, such as those concerning fishing rights, access to ports, transportation, or river boundaries. Also excluded are events where water is not the driver, such as those where water is a tool, target, or victim of armed conflict. The study documents a total of 1,831 interactions, both conflictive and cooperative, between two or more nations over water during the past 50 years, and found the following conclusions.

First, despite the potential for dispute in international basins, the record of acute conflict over international water resources is historically overwhelmed by the record of cooperation. Since the early 1950s only thirty-seven acute disputes have occurred (those involving violence) – of those, thirty are between Israel and one or another of its neighbors, violence that ended in 1970. Non-Mideast cases account for only five acute events, while, during the same period, 157 treaties were negotiated and signed. In fact, the only "water war" between nations on record occurred more than 4,500 years ago, between the city-states of Lagash and Umma in the Tigris–Euphrates basin (Wolf, 1988). The total number of water-related events among nations of any magnitude is likewise weighted toward cooperation: 507 conflict-related events versus 1,228 cooperative, implying that violence over water is not strategically rational, hydrographically effective, or economically viable (Table 2.1; Figure 2.1).

Second, despite the fiery rhetoric of politicians, often aimed at their own constituencies rather than at an enemy, most actions taken over water are mild. Seven hundred eighty-four events, or 42.8 percent of all events, fall between mild verbal support (+1) and mild verbal hostility (−1). If we add the next level on either side – official verbal support (+2) and official verbal hostility (−2) – we account for 1,138 events, or 62 percent of the total. Another way to look at this is that almost two-thirds of all events are only verbal and, of these verbal

Table 2.1 *Basins at risk (BAR) event intensity scale*

BAR scale	COPDAB scale	BAR event description
−7	15	Formal declaration of war; extensive war acts causing deaths, dislocation, or high strategic costs
−6	14	Extensive military acts
−5	13	Small-scale military acts
−4	12	Political−military hostile actions
−3	11	Diplomatic−economic hostile actions
−2	10	Strong verbal expressions displaying hostility in interaction
−1	9	Mild verbal expressions displaying discord in interaction
0	8	Neutral or nonsignificant acts for the international situation
1	7	Minor official exchanges, talks or policy expressions – mild verbal support
2	6	Official verbal support of goals, values, or regime
3	5	Cultural or scientific agreement or support (nonstrategic)
4	4	Nonmilitary economic, technological or industrial agreement
5	3	Military economic or strategic support
6	2	International freshwater treaty; Major strategic alliance (regional or international)
7	1	Voluntary unification into one nation

Source: Modified from E. Azar's COPDAB International Conflict and Cooperation Scale (Azar, 1980).

events, more than two-thirds are reported as having no official sanction at all.

Third, nations find many more issues of cooperation than of conflict. The distribution of cooperative events indicates a broad spectrum of issue types, including quantity, quality, economic development, hydropower, and joint management (Figure 2.2). In contrast, almost 90 percent of the conflictive events relate to quantity and infrastructure. Furthermore, if we look specifically at extensive military acts, the most extreme cases of conflict, almost 100 percent of events fall within these two categories.

Fourth, at the subacute level, water acts as both an irritant and as a unifier. As an irritant, water can make good relations bad and bad relations worse, but equally, international waters, despite their complexities, can also act as a unifier in basins where relatively strong institutions are in place. The historical

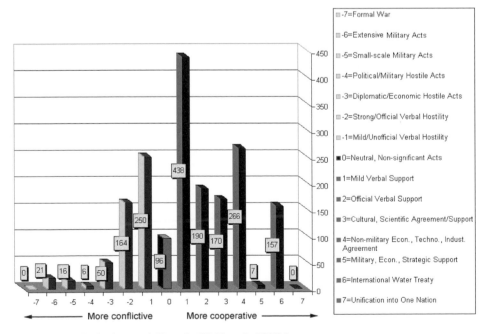

Figure 2.1 Number of events by BAR (basins-at-risk) scale (Wolf et al., 2003b).

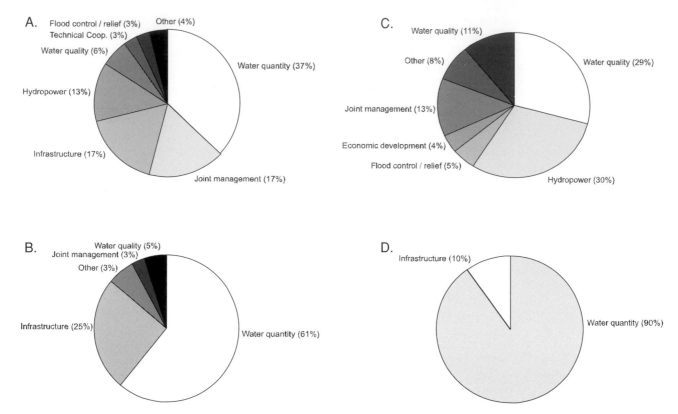

Figure 2.2 Number of events by issue area. Charts A and B show the distribution of all event issues, both cooperative (A) and conflictive (B). Charts C and D isolate the issue areas of extreme events, +6 and −6 on the BAR scale, respectively. Note the wide distribution of issue areas for cooperative events, while conflict is predominantly over water quantity and infrastructure (Wolf, Yoffe, and Giordano, 2003b).

record shows that international water disputes *do* get resolved, even among bitter enemies, and even as conflicts erupt over other issues. Some of the most vociferous enemies around the world have negotiated water agreements or are in the process of doing so, and the institutions they have created frequently prove to be resilient over time and during periods of otherwise strained relations. The Mekong Committee, for example, has functioned since 1957, exchanging data throughout the Vietnam War. Secret "picnic table" talks have been held between Israel and Jordan since the unsuccessful Johnston negotiations of 1953−1955, even as these riparians until only recently were in a legal state of war. The Indus River Commission survived through two wars between India and Pakistan. An agreement between China and Hong Kong survived strains between those two countries (Wolf, 1997). And all ten Nile riparians are currently involved in negotiations over cooperative development of the basin.

2.1.2 Hydromyth II: Everything is OK

So if there is little violence between nations over their shared waters, what's the problem? Is water actually a security concern at all? In fact, there are a number of issues where water causes or exacerbates tensions, and it is worth understanding

these processes to know both how complications arise and how they are eventually resolved. Noncooperation costs primarily in inefficient water management, leading to decreasing water quantity, quality, and environmental health. But political tensions can also be affected, leading to years or even decades of efficient, cooperative futures foregone.

TENSIONS AND TIME LAGS: CAUSES FOR CONCERN

The first complicating factor is the time lag between when nations first start to impinge on each other's water planning and when agreements are finally, arduously, reached. A general pattern has emerged for international basins over time. Riparians of an international basin implement water development projects unilaterally – first on water within their own territory – in an attempt to avoid the political intricacies of the shared resource. At some point, one of the riparians, generally the regional power,[3] will implement a project that affect at least one of its neighbors. This might be to continue to meet

---

[3] "Power" in regional hydropolitics can include riparian position, with an upstream riparian having more relative strength *vis-à-vis* the water resources than its downstream riparian, in addition to the more-conventional measures of military, political, and economic strength. Nevertheless, when a project is implemented that impacts one's neighbors, it is generally undertaken by

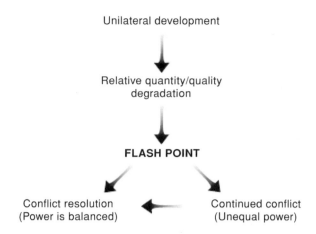

Unilateral development

↓

Relative quantity/quality
degradation

↓

FLASH POINT

Conflict resolution   Continued conflict
(Power is balanced)   (Unequal power)

Figure 2.3 Chronology of international water disputes. (*Source:* Authors).

existing uses in the face of decreasing relative water availability, as, for example, Egypt's plans for a high dam on the Nile, Indian diversions of the Ganges to protect the port of Calcutta, or to meet new needs reflecting new agricultural policy, such as Turkey's GAP project on the Euphrates. In the absence of relations or institutions conducive to conflict resolution, the project can become a flash point (Figure 2.3), heightening tensions and regional instability and requiring years or decades to resolve – the Indus treaty took 10 years of negotiations, the Ganges 30 years, and the Jordan 40 years – and, all the while, water quality and quantity degrades to where the health of dependent populations and ecosystems are damaged or destroyed.

A reread through the history of international waters suggests that the simple fact that humans suffer and die in the absence of agreement apparently offers little in the way of incentive to cooperate, even less so the health of aquatic ecosystems. This problem gets worse as the dispute gains in intensity: one rarely hears talk about the ecosystems of the lower Nile, the lower Jordan, or the tributaries of the Aral Sea – they have effectively been written off to the vagaries of human intractability. During such periods of low-level tensions, threats and disputes rage across boundaries with relations as diverse as those between Indians and Pakistanis and between Americans and Canadians. Water was the last and most contentious issue resolved in negotiations over a 1994 peace treaty between Israel and Jordan and was relegated to "final status" negotiations – along with other of the most difficult issues such as Jerusalem and refugees – between Israel and the Palestinians (Isaac and Selby, 1996).

The timing of water flow is also important; thus, the operation of dams is also contested. For example, upstream users might release water from reservoirs in the winter for

the regional power, as defined by traditional terms, *regardless* of its riparian position.

hydropower production, whereas downstream users might need it for irrigation in the summer. In addition, water quantity and water flow patterns are crucial to maintaining freshwater ecosystems that depend on seasonal flooding. Freshwater ecosystems perform a variety of ecological and economical functions and often play an important role in sustaining livelihoods, especially in developing countries. As awareness of environmental issues and the economic value of ecosystems increases, claims for the environment's water requirements are growing. For example, in the Okavango Basin, Botswana's claims for water to sustain the Okavango Delta and its lucrative ecotourism industry have contributed to a dispute with upstream Namibia, which wants to use some of the water passing through the Caprivi Strip on its way to the delta for irrigation.

Water-quality problems include excessive levels of salt, nutrients, or suspended solids. Salt intrusion can be caused by groundwater overuse or insufficient freshwater flows into estuaries. For example, dams in the South African part of the Incomati River basin reduced freshwater flows into the Incomati estuary in Mozambique and led to increased salt levels (Table 2.2). This altered the estuary's ecosystem and led to the disappearance of salt-intolerant flora and fauna important for people's livelihoods (the links between loss of livelihoods and the threat of conflict are described here).

Excessive amounts of nutrients or suspended solids can result from unsustainable agricultural practices, eventually leading to erosion. Nutrients and suspended solids pose a threat to freshwater ecosystems and their use by downstream riparians because they can cause eutrophication and siltation, respectively, which, in turn, can lead to loss of fishing grounds or arable land. Suspended solids can also cause the siltation of reservoirs and harbors: for example, Rotterdam's harbor had to be dredged frequently to remove contaminated sludge deposited by the Rhine River. The cost was enormous, and consequently led to conflict over compensation and responsibility among the river's users. Although negotiations led to a peaceful solution in that case, without such a framework for dispute resolution, siltation problems can lead to upstream–downstream disputes, such as those in the Lempa River basin in Central America (Lopez, 2004).

One of the main reasons for decreasing water quality is pollution, for example, through industrial and domestic wastewater or agricultural pesticides. In Tajikistan, for example, where environmental stress has been linked to civil war (1992–1997), high levels of water pollution have been identified as one of the key environmental issues threatening human development and security (Carius, Feil, and Taenzler, 2003). Water pollution from the tanning industry in the Palar Basin of the Indian state of Tamil Nadu makes the water within the basin unfit for irrigation and consumption. The pollution contributed

Table 2.2 *Selected examples of water-related disputes*

	Main issue
Location	Observation
Quantity	
Cauvery River, South Asia	The dispute on India's Cauvery River sprang from the allocation of water between the downstream state of Tamil Nadu, which had been using the river's water for irrigation, and upstream Karnataka, which wanted to increase irrigated agriculture. The parties did not accept a tribunal's adjudication of the water dispute, leading to violence and death along the river.
Mekong Basin, Southeast Asia	Following construction of Thailand's Pak Mun Dam, more than 25,000 people were affected by drastic reductions in upstream fisheries and other livelihood problems. Affected communities have struggled for reparations since the dam was completed in 1994.
Okavango Basin, Southern Africa	In the Okavango River basin, Botswana's claims for water to sustain the delta and its lucrative ecotourism industry contribute to a dispute with upstream Namibia, which wants to pipe water passing through the Caprivi Strip to supply its capital city with drinking water.
Quality	
Rhine River, Western Europe	Rotterdam's harbor had to be dredged frequently to remove contaminated sludge deposited by the Rhine River. The cost was enormous and consequently led to controversy over compensation and responsibility among Rhine users. Although in this case negotiations led to a peaceful solution, in areas that lack the Rhine's dispute resolution framework, siltation problems could lead to upstream−downstream disputes.
Quantity and quality	
Incomati River, Southern Africa	Dams in the South African part of the Incomati River basin reduced freshwater flows and increased salt levels in Mozambique's Incomati estuary. This altered the estuary's ecosystem and led to the disappearance of salt-intolerant plants and animals that are important for people's livelihoods.
Timing	
Syr Dar'ya, Central Asia	Relations between Kazakhstan, Kyrgyzstan, and Uzbekistan – all riparians of the Syr Dar'ya, a major tributary of the disappearing Aral Sea – exemplify the problems caused by water flow timing. Under the Soviet Union's central management, spring and summer irrigation in downstream Uzbekistan and Kazakhstan balanced upstream Kyrgyzstan's use of hydropower to generate heat in the winter. But the parties are barely adhering to recent agreements that exchange upstream flows of alternate heating sources (natural gas, coal, and fuel oil) for downstream irrigation, sporadically breaching the agreements.

Source: Wolf, Kramer, Carius, and Dabelko, 2005.

to an acute drinking water crisis, which led to protests by the local community and activist organizations, as well as to disputes and court cases between tanners and farmers.

REGIONAL INSTABILITY: POLITICAL DYNAMICS OF LOSS OF IRRIGATION WATER

The second set of security issues occurs at the subnational level. As noted in Chapter 1, if there is a history of water-related violence, and there is, it is a history of incidents at the subnational level, generally between tribes, water-use sectors, or states/provinces. Giordano and colleagues (2002) found that as the scale drops, the likelihood and intensity of violence goes up. As water quality degrades – or quantity diminishes – over time, the effect on the stability of a region can be unsettling (see also Ravnborg, 2004; and Conca, 2006). For example, for

30 years the Gaza Strip was under Israeli occupation. Water quality deteriorated steadily, saltwater intrusion degraded local wells, and water-related diseases took a rising toll on the people living there. In 1987, the intifada, or Palestinian uprising, broke out in the Gaza Strip and quickly spread throughout the West Bank. Was water quality the cause? It would be simplistic to claim direct causality. Was it an irritant exacerbating an already tenuous situation? Undoubtedly.

An examination of relations between India and Bangladesh demonstrate these internal instabilities can be both caused and exacerbated by international water disputes. In the 1960s, India built a barrage at Farakka, diverting a portion of the Ganges flow away from its course into Bangladesh, in an effort to flush silt away from Calcutta's seaport, some 100 miles to the south. In Bangladesh, the reduced upstream flow resulted in a number of adverse effects: degraded surface and ground-water, impeded navigation, increased salinity, degraded fisheries, and endangered water supplies and public health. Migration from affected areas further compounded the problem. Ironically, many of those displaced in Bangladesh have found refuge in India.

Two-thirds of the world's water use is for agriculture. When access to irrigation water is threatened, one result can be movement of huge populations of out-of-work, disgruntled men from the countryside to the cities – an invariable recipe for political instability. In pioneering, but unpublished, work, Sandra Postel identified those countries that rely heavily on irrigation and whose agricultural water supplies are threatened either by a decline in quality or quantity. The list coincides precisely with the world community's current security hot spots: India, China, Pakistan, Iran, Uzbekistan, Bangladesh, Iraq, and Egypt (summarized in Postel and Wolf, 2001).

Water management in many countries is also characterized by overlapping and competing responsibilities among government bodies. Disaggregated decision making often produces divergent management approaches that serve contradictory objectives and lead to competing claims from different sectors. And such claims are even more likely to contribute to disputes in countries where there is no formal system of water-use permits or where enforcement and monitoring are inadequate. Controversy also often arises when management decisions are formulated without sufficient participation by local communities and water users, thus failing to take into account local rights and practices. Protests are especially likely when the public suspects that water allocations are diverting public resources for private gain or when water-use rights are assigned in a secretive and possibly corrupt manner, as demonstrated by the violent confrontations in 2000 following the privatization of Cochabamba, Bolivia's water utility.

Finally, there is the human security issue of water-related disease. It is estimated that between five and ten million people die each year from water-related diseases or inadequate sanitation. More than half the people in the world lack adequate sanitation. Eighty percent of disease in the developing world is related to water. This is a crisis of epidemic proportions, and the threats to human security are self-evident.

COSTS OF NONCOOPERATION
Noncooperation effectively prohibits effective integration of water uses and the resulting impacts include both economic and noneconomic costs.

WATER QUANTITY ISSUES
Often, simply extrapolating water supply and demand curves will give an indication of when a conflict may occur, as the two curves approach each other and noncooperation prohibits the search for effective solutions. The mid-1960s, a period of water conflict in the Jordan basin saw demand approaching supply in both Israel and Jordan. Major shifts in supply might also indicate likely conflict, due to greater upstream use or, in the longer range, to global change. Greater upstream use is currently the case both on the Mekong and on the Ganges. Likewise, shifts in demand, due to new agricultural policies or movements of refugees or immigrants, can indicate problems. Water systems with a high degree of natural fluctuation can cause greater problems than relatively predictable systems.

WATER-QUALITY ISSUES
Any new source of pollution, or any new extensive agricultural developing resulting in saline return flow to the system, cannot be effectively mitigated in a state of noncooperation. Arizona return flow into the Colorado was the issue over which Mexico sought to sue the United States in the 1960s through the International Court of Justice and is currently a point of contention on the lower Jordan among Israelis, Jordanians, and West Bank Palestinians (Shmueli and Shamir, 2001).

WATER-RELATED DISEASE
Noncooperation also limits the effectiveness of responding to threats of water-related disease. As noted in the Introduction, millions of people die each year from water-related diseases or inadequate sanitation. Although much of this devastation is internal to nations, noncooperation on shared waters is an effective barrier to addressing these issues at the border.

MANAGEMENT FOR MULTIPLE USES
Water is managed for a particular use or a combination of uses: for example, a dam might be managed for storage of irrigation water, power generation, recreation, or a combination of all three. When the needs of riparians conflict, and there is no cooperative mechanism to coordinate interests, disputes are likely. Many upstream riparians, for instance, would manage the river within their territory primarily for hydropower whereas the primary needs of their downstream neighbors might be timely irrigation flows. Chinese plans for hydropower generation and/or Thai plans for irrigation diversions would have an impact on Vietnamese needs for both irrigation and better drainage in the Mekong Delta.

POLITICAL DIVISIONS
Shifting political divisions that reflect new riparian relations are a common indicator of water conflict in a noncooperative setting. For example, conflicts on the Ganges, Indus, and Nile rivers took on international complications as the central authority of a hegemon, the British Empire, dissipated. Such has

recently been the case throughout Central Europe as national water bodies, such as the Amu Dar'ya and the Syr Dar'ya, become international.

Along with clues useful in anticipating whether water conflicts might occur, patterns based on past disputes in varying noncooperative settings may provide lessons for determining both the type and intensity of impending conflicts. These indicators might include geopolitical setting, level of national development, the hydropolitical issue at stake, institutional control of water resources, and national water ethos.

GEOPOLITICAL SETTING
As mentioned above, relative power relationships, including riparian position, determine how a conflict unfolds. A regional power that also has an upstream riparian position is in a greater situation to implement projects that may become flash points for regional conflict. Turkey and India have been in such positions on the Euphrates and the Ganges, respectively. In contrast, the development plans of an upstream riparian, such as Ethiopia, may be held in check by a downstream power, such as Egypt.

The perception of unresolved issues with one's neighbors, both water-related and otherwise, is also an exacerbating factor in water conflicts. For example, Israel, Syria, and Turkey have difficult political issues outstanding, which make discussions on the Jordan and Euphrates more intricate.

LEVEL OF NATIONAL DEVELOPMENT
Relative development can inform the nature of water disputes in a number of ways. For example, a more-developed region may have better options to alternative sources of water, or to different water management schemes, than less-developed regions, resulting in more options once negotiations begin. In the Middle East Multilateral Working Group on Water Resources, for instance, a variety of technical and management options, such as desalination, drip irrigation, and moving water from agriculture to industry, have all been presented, which in turn supplement discussions over allocations of international water resources (see Jordan River Case Study in Appendix C).

Different levels of development within a watershed, however, can exacerbate the hydropolitical setting. Relative deprivation is a major generator of conflict (Gurr, 1969). It can occur even in situations where all parties receive benefits none had received before, if there is the perception that some are receiving more than others. This is why water is so closely linked to development and why development is so important for transboundary stability. As a country develops, personal and industrial water demand tends to rise, as does demand for previously marginal agricultural areas. Although this demand can

be somewhat balanced by more access to water-saving technology, a developing country often will be the first to develop an international resource to meet its growing needs. Thailand has been making these needs clear with its relatively greater emphasis on Mekong development.

THE HYDROPOLITICAL ISSUE AT STAKE
In a survey of fourteen river basin conflicts, Mandel (1992) offers interesting insights by relating the issue at stake to the intensity of a water conflict. He suggests that issues that include a border dispute in conjunction with a water dispute, such as on the Shatt al Arab waterway between Iran and Iraq and the Rio Grande between the United States and Mexico, can induce more severe conflicts than issues of water quality, such as with the Colorado, Danube, and La Plata rivers. Likewise, conflicts triggered by human-initiated technological disruptions – dams and diversions, as on the Euphrates, Ganges, Indus, and Nile – are more severe than those triggered by natural flooding, such as on the Columbia and Senegal rivers.

Mandel (1992) found a lack of correlation between the number of disputants and intensity of conflict. He suggests that this challenges the common notion that the more limited, in terms of number of parties involved, river disputes are easier to resolve. Another surprising and somewhat counterintuitive finding is that climate seems not to be a major variable in water disputes. This finding may be because water has multiple uses, but these uses vary in critical importance, depending on climatic conditions. The hydropower or transportation offered by a river in a humid climate is no less important to its riparians than is the irrigation water provided by a river in an arid zone.

INSTITUTIONAL CONTROL OF WATER RESOURCES
An important aspect of international water conflicts is how water is controlled *within* each of the countries involved. Whether control of the resource is vested at the national level, as in the Middle East, the state level, as in India, or at the substate level, as in the United States, informs the complication of international dialogue. Also, *where* control is vested institutionally is important. In Israel, for example, the Water Commissioner for years was under the authority of the Ministry of Agriculture, whereas in Jordan, the Ministry of Water has the authority. These respective institutional settings can make internal political dynamics quite different for similar issues.

NATIONAL WATER ETHOS
This term incorporates several somewhat ambiguous parameters that together determine how a nation "feels" about its water resources. This can in turn help determine how much it "cares" about a water conflict. Some factors of a water ethos might include:

- "mythology" of water in national history, for example, has water been the "lifeblood of the region?" as is core to identity in the U.S. West? Was the country built up around the heroic *fellah*, as in most of the Middle East? Is "making the desert bloom" a national aspiration as is central to Israeli history? In most countries, in contrast, water plays little role in the national history.
- importance of water/food security in political rhetoric
- relative importance of agriculture versus industry in the national economy

2.1.3 Hydromyth III: Political tensions are caused by water scarcity

Most authors who write about hydropolitics, and especially those who explicitly address the issue of water conflicts, hold to the common assumption that it is scarcity of such a critical resource that drives people to conflict (see, for example, the work by Homer-Dixon (1991, 1994, and 1996), and Beaumont, 2000). It feels intuitive – the less there is of something, especially something as important as water, the more dear it is held and the more likely people are to fight over it. Yet one simply cannot evaluate scarcity in an absolute way, without taking into account either the uses to which water is put (i.e., comparing supply and demand) or spatial variability (nations are *not* hydrologic monoliths). Arid nations, and arid regions within nations, have developed a whole series of needs and institutions based on aridity, where humid nations and regions have entirely different demands. In the former, the needs may be primarily irrigation, where in the latter, the focus may be transportation, fisheries, and/or hydropower, but depending on the setting, a small fluctuation may or may not have human repercussions. It is pointless to examine water's impact on politics without also considering the relationship among demand, supply, variations in both, and the economic and institutional capacity of the region to respond to change. Moreover, "virtual" water is constantly traded across sometimes very long distances in the form of food and agricultural products, thus making the internal supply of water absolutely a useless gauge of anything (see Allan, 1998a, for a summary).

There have been attempts to address the many complexities surrounding international water issues and to discern indicators of conflict (TFDD, 2006). One study built a 100-layer, 50-year geographic information system (GIS) – a spatial database of all the parameters that might prove part of the conflict–cooperation story, including parameters that are physical (runoff, droughts), socioeconomic (GDP, rural/urban populations), and geopolitical (government type, votes on water-related UN resolutions). With this GIS in place, a statistical snapshot of each setting for each of the events of conflict

or cooperation since the early 1950s that had been collected was developed (Wolf, Yoffe, and Giordano, 2003b; Yoffe et al., 2004).

Their results are surprising and often counterintuitive. *None* of the physical parameters was statistically significant – arid climates were no more conflictive than humid climates, and international cooperation actually *increased* during droughts. In fact, almost no single variable proved causal – democracies were as conflictive as autocracies, rich countries as poor countries, densely populated countries as sparsely populated ones, and large countries the same as small countries.

The study's close examination of aridity indicated that institutional capacity was the key. Naturally arid countries were cooperative: if one lives in a water-scarce environment, one develops institutional strategies for adapting to that environment. The focus on institutions – whether defined by formal treaties, informal working groups, or generally warm relations – and their relationship to the physical environment gave a clear picture of the settings most conducive to political tensions in international waterways. The results showed that the likelihood of conflict increased significantly whenever two factors came into play. The first factor was the occurrence of some large or rapid change in the basin's physical setting – typically the construction of a dam, river diversion, or irrigation scheme – or in its political setting, especially the breakup of a nation, resulting in new international rivers. The second factor was the inability of existing institutions to absorb and effectively manage that change. This inability was typically the case when there was no treaty spelling out each nation's rights and responsibilities with regard to the shared river, nor any implicit agreements or cooperative arrangements. However, the study also found that even the existence of technical working groups can provide some capability to manage contentious issues, as they have in the Middle East. The working hypothesis, which was borne out by the study, was as follows:

"The likelihood of conflict rises as the rate of change within the basin exceeds the institutional capacity to absorb that change" (Wolf, Yoffe, and Giordano, 2003b).

The Oregon State University (OSU) study findings suggest that there are two sides to the dispute setting: the rate of change in the system and the institutional capacity. In general, most of the parameters regularly identified as indicators of water conflict are actually only weakly linked to dispute. Institutional capacity within a basin, however, whether defined as water-management bodies or treaties or generally positive international relations, is as important, if not more so, than the physical aspects of a system. It turns out, then, that very rapid changes, either on the institutional side or in the physical system that outpace the institutional capacity to absorb those changes are at the root of most water conflict. For example, the rapid institutional change in "internationalized" basins, that is, basins that include the management structures of newly independent states, has resulted in disputes in areas formerly under British administration (e.g., the Nile, Jordan, Tigris—Euphrates, Indus, and Ganges—Brahmaputra), as well as in the former Soviet Union (e.g., the Aral tributaries and the Kura—Araks). On the physical side, rapid change most outpaces institutional capacity in basins that include unilateral development projects *and* the absence of cooperative regimes, such as treaties, river basin organizations (RBOs), or technical working groups, or when relations are especially tenuous over other issues (Wolf, Yoffe, and Giordano, 2003b).

The general assumption, then, is that rapid change tends to indicate vulnerability, whereas institutional capacity tends to indicate resilience *and* that the two sides need to be assessed in conjunction with each other for a more accurate gauge of hydropolitical sustainability. Building on these relationships, the characteristics of a basin that would tend to enhance resilience to change include international (and intrastate, cross-jurisdictional) agreements and institutions, such as

- River basins organizations
- A history of collaborative projects
- Generally positive political relations
- Higher levels of economic development[4]

In contrast, facets that would tend toward vulnerability would include

- Rapid environmental change
- Rapid population growth or asymmetric economic growth
- Major unilateral development projects
- The absence of institutional and/or organizational capacity
- The potential for "internationalization" of a basin
- Generally hostile relations

In practice, the overarching lesson was that unilateral actions to construct a dam or river diversion *in the absence* of a treaty or institutional mechanism that safeguards the interests of other countries in the basin is highly destabilizing to a region, often spurring decades of hostility before cooperation is pursued. The red flag for water-related tension between countries and/or across jurisdictions is not water stress per se, as it is within countries, but the unilateral exercise of domination of an international river, usually by a regional power.

In the Jordan River basin, for example, violence broke out in the mid-1960s over an "all-Arab" plan to divert the river's headwaters (itself a preemptive move to thwart Israel's intention to siphon water from the Sea of Galilee). Israel and Syria sporadically exchanged fire between March 1965 and July 1966. Water-related tensions in the basin persisted for decades and only recently have begun to dissipate (Wolf, 1995b; Haddadin and Shamir, 2003; Jägerskog, 2003).

A similar sequence of events transpired in the Nile basin, which is shared by ten countries – of which Egypt is last in line. In the late 1950s, hostilities broke out between Egypt and Sudan over Egypt's planned construction of the High Dam at Aswân. The signing of a treaty between the two countries in 1959 defused tensions before the dam was built. But no water-sharing agreement exists between Egypt and Ethiopia, where some 85 percent of the Nile's flow originates, and a war of words has raged between these two nations for decades. As in the case of the Jordan, in recent years the Nile nations have begun to work cooperatively toward a solution thanks in part to unofficial dialogues among scientists and technical specialists that have been held since the early 1990s and more recently a ministerial-level "Nile Basin Initiative" facilitated by the United Nations and the World Bank (Waterbury, 2002; Nicol, 2003).

These conflicts share a common trajectory: unilateral construction of a big dam or other development project, leading to a protracted period of regional insecurity and hostility, typically followed eventually by a long and arduous process of dispute resolution, by which the issue is eventually resolved.

2.1.4 Basins at risk: Water on the horizon[5]

The TFDD study (Wolf, Yoffe, and Giordano, 2003b; Yoffe et al., 2004) identified seventeen river basins that were ripe for the onset of tensions or conflict over a period of 10 years, where dams or diversions were planned or under construction that could negatively affect other countries, and where there was no mechanism for resolving resulting disputes. The study also identified four basins in which serious unresolved water disputes existed or were being negotiated (Aral, Nile, Jordan, and Tigris—Euphrates). The basins at risk included fifty-one nations on five continents in just about every climatic zone. Eight of the basins were in Africa, primarily in the south, while six were in Asia, mostly in the southeast. Few of them were on the radar screens of water-and-security analysts. (It is critical to note that the TFDD study's "Basins at Risk" component developed a *process* for understanding conflict and identifying risk, which is more important over time than the specific

---

[4] Higher levels of economic development enhance resilience because these countries can afford alternatives as water becomes relatively more scarce or degraded. Contrast developing and developed countries, for example – whereas the former may struggle for a safe, stable supply of basic water resources, the latter might utilize greenhouses, expensive drip-irrigation systems, bioengineered crops, or desalination.

[5] This section draws from Postel and Wolf (2001).

basins named in the initial study. Many "basins-at-risk" named are, fortunately, no longer at risk, precisely because the world and local communities were able to invest in the institutional capacity building necessary to reduce the possibility of tensions.)

Consider, for example, the Salween River, which rises in southern China, then flows into Myanmar (Burma) and Thailand. Each nation plans to construct dams and development projects along the Salween – and no two sets of plans are compatible. China, moreover, has not lately been warm to notions of water sharing. It was one of just three countries that voted against a 1997 United Nations convention that established basic guidelines and principles for the use of international rivers (see Appendix A; United Nations, 2005). Add in other destabilizing factors in the Salween basin – including the status of Tibet, indigenous resistance movements, opium production, and a burgeoning urban population in Bangkok, and the familiar conflict trajectory emerges. Without a treaty in place, or even regular dialogue between the nations about their respective plans, there is little in the way of institutional capacity to buffer the inevitable shock as construction begins (Onta, Gupta, and Loof, 1996; Paoletto and Uitto, 1996).

Consider, too, the Okavango, the fourth largest river in southern Africa. Its watershed spans portions of Angola, Botswana, Namibia, and Zimbabwe, and its vast delta in northern Botswana offers world-renowned wildlife habitat – the "jewel of the Kalahari." In 1996, drought-prone Namibia revived colonial plans to divert Okavango water to its capital city of Windhoek. Angola and Botswana objected to the scheme because of its potential harm to the people and ecosystems that depend on the Okavango's flow for their existence. The main institution that can help manage the dispute is the fledgling Okavango Commission formed in 1994 to coordinate plans in the basin. The commission has recently received renewed support from the Southern Africa Development Community, the U.S. Bureau of Reclamation, and other agencies, but the water dispute continues to simmer (Nakayama, 2003; Turton, Ashton, and Cloete, 2003).

Several river basins are at risk of future disputes more because of rapid changes in their political settings than any specific dam or development scheme. Indeed, the development schemes can actually encourage creation of new benefits. The breakup of the Soviet Union resulted in several new international river basins almost overnight, and, not surprisingly, institutional capacity for managing water disputes in them is still weak. The Kura–Araks river system, for example, runs through the politically volatile Caucasus, which includes the now-independent countries of Armenia, Georgia, and Azerbaijan. The river system is the source of drinking water for large portions of these nations, but millions of tons of untreated sewage and industrial waste regularly push water quality to 10 to 100 times above international standards for levels of contaminants. On top of the pollution problems, some forecasts project severe water shortages within 10 years. These water strains exacerbate – and are exacerbated by – relations over other contentious issues in the region, notably those of Nagorno–Karabakh and the oil pipeline being built to transport Caspian crude oil across the region to Turkey (Allouche, 2005). In light of this region's strategic importance, the strengthening of its water institutions takes on new urgency.

WHY AREN'T WATER WARS MORE COMMON?[6]

Basing an argument about the future on history alone would be disingenuous. Part of the argument for future "water wars," after all, is that we are reaching unprecedented demand on relatively decreasing clean water supplies. Other arguments against the possibility of "water wars" follow, although, because we are discussing the future, each has less evidence in its favor than the historic argument.

STRATEGIC ARGUMENT

If we were to launch a war over water, what would be the goal? Presumably, the aggressor would have to be both downstream and the regional hegemon – an upstream riparian would have no cause to launch an attack and a weaker state would be foolhardy to do so. Foolhardiness apparently does not preclude such "asymmetric conflicts" over nonwater issues. Paul (1994) describes eight such case studies from 1904 to 1982 but points out that in none did the weaker power achieve its goals. An upstream riparian, then, would have to launch a project that decreases either quantity or quality, knowing that it will antagonize a stronger downstream neighbor.

The downstream power would then have to decide whether to launch an attack – if the project were a dam, destroying it would result in a wall of water rushing back on downstream territory; were it a quality-related project, either industrial or waste treatment, destroying it would probably result in even worse quality than before. Furthermore, the hegemon would have to weigh not only an invasion, but an occupation and depopulation of the entire watershed in order to forestall any retribution; otherwise, it would be extremely simple to pollute the water source of the invading power. Both countries could not be democracies because the political scientists tell us that democracies do not go to war against each other, and the international community would have to refuse to become involved (this, of course, is the least far-fetched aspect of the scenario).

---

[6] Draws from Wolf (1998). Conflict and cooperation along international waterways. *Water Policy*, **1**(2), 251–265.

All of this effort would be expended for a resource that costs about a U.S. dollar per cubic meter to create from seawater.

There are 263 international watersheds – there are only a handful on which the above scenario is even feasible (e.g., the Nile, La Plata, and Mekong), and many of those either have existing treaties or ongoing negotiations toward a treaty. Finding a site for a "water war" turns out to be as difficult as accepting the rationale for launching one.

SHARED INTEREST ARGUMENT

The treaties negotiated over international waterways provide insights into what it is about water that tends to induce cooperation, even among riparians who are hostile over other issues. The treaties often show exquisite sensitivity to the unique setting and needs of each basin, and many detail the shared interests a common waterway will bring. Along larger waterways, for instance, the better dam sites are usually upstream at the headwaters where valley walls are steeper and, incidentally, the environmental impact of dams is not as great, because they have an impact on and block off smaller areas. The prime agricultural land is generally downstream, where gradient drops off and alluvial deposits enrich the soil. A dam in the headwaters, then, can not only provide hydropower and other benefits for the upstream riparian but also can be managed to even out the flow for downstream agriculture or even to enhance water transportation for the benefit of both riparians and more.

Other examples of shared interests abound: no development of a river that acts as a boundary can take place without cooperation; farmers, environmentalists, and beachgoers all share an interest in seeing a healthy stream system; and all riparians share an interest in high water quality.

These shared interests are regularly exemplified in treaties. In conjunction with the 1957 Mekong Agreement, Thailand helped fund a hydroelectric project in Laos in exchange for a proportion of the power to be generated (Le-Huu and Nguyen-Duc, 2003; Swain, 2004). In the particularly elaborate 1986 Lesotho Highlands Treaty, South Africa agreed to help finance a hydroelectric – water diversion facility in Lesotho. South Africa acquired rights to drinking water for Johannesburg and Lesotho receives all of the power generated (Nakayama, 2003). Under the 1998 Agreement on the Use of Water and Energy Resources of the Syr Dar'ya Basin, Uzbekistan and Kazakhstan make in-kind compensation to the Kyrgyz Republic for the transfer of excess power generated during the growing season (Dukhovny and Sokolov, 2003). Similar arrangements have been suggested in China on the Mekong, Nepal on the Ganges, and between Syria and Jordan on the Yarmuk.

The unique interests in each basin, whether hydrological, political, or cultural, stand out in the creativity of many of the treaties. A 1969 accord on the Cunene River allows for "humanitarian" diversions solely for human and animal requirements in Southwest Africa as part of a larger project for hydropower (Figure 2.4; Turton and Earle, 2005). Water loans are made from Sudan to Egypt (1959) and from the United States to Mexico (1966); Jordan stores water in an Israeli lake (Haddadin and Shamir, 2003), whereas Israel leases Jordanian land and wells (1994); and India plants trees in Nepal to protect its own water supplies (1966) (Swain, 2004). In a 1964 agreement, Iraq "gives" water to Kuwait "in brotherhood" and without monetary compensation. A 1957 agreement between Iran and the Soviet Union even had a clause that allowed for cooperation in identifying corpses found in their shared rivers (Wolf, 1999a).

The changes in local needs over time are seen in the boundary waters between Canada and the United States. Even as the boundary waters agreements of 1910 were modified in 1941 to allow for greater hydropower generation along the Niagara to bolster the war effort, Canada and the United States nevertheless reaffirmed that protecting the "scenic beauty of this great heritage of the two countries" is their primary obligation. A 1950 revision continued to allow hydropower generation but added a provision to allow a greater minimum flow over the famous falls during summer daylight hours, when tourism is at its peak.

INSTITUTIONAL RESILIENCY ARGUMENT

In general, concepts of "resilience" and "vulnerability" as related to water resources are often assessed within the framework of "sustainability" (e.g., Blaikie et al., 1994) and relate to the ability of biophysical systems to adapt to change (e.g., Gunderson and Pritchard, 2002). As the sustainability discourse has broadened to include human systems in recent years, so too has work been increasingly geared toward identifying indicators of resilience and vulnerability within this broader context (e.g., Bolte et al., 2004; Lonergan, Gustavson, and Carter, 2000; Turner et al., 2003). In parallel, dialogue on "security" has migrated from traditional issues of war and peace to begin incorporating the human–environment relationship in the relatively new field of "environmental security" (see UNEP, 2004, and Vogel and O'Brien, 2004).[7]

The term *hydropolitics* (coined by Waterbury, 1979) came about as substantial new attention has been paid to the potential for conflict and violence to erupt over international waters and relates to the ability of geopolitical institutions to manage shared water resources in a politically sustainable manner, that is, without tensions or conflict between political entities.

---

[7] "Environmental security," the securitization or conflict potential of environmental issues, should not be confused with either "food security" or "water security," which are defined as self-sufficiency in food and water, respectively.

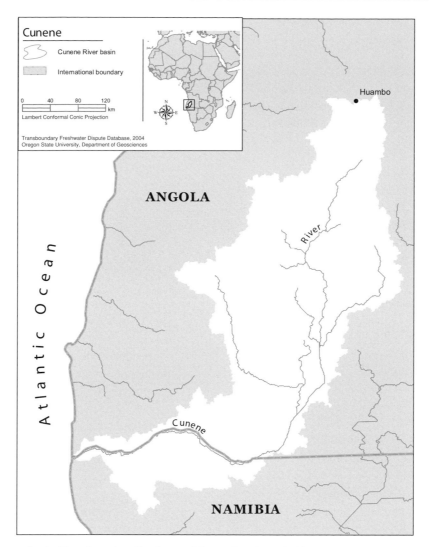

Figure 2.4 Map of the Cunene basin (Transboundary Freshwater Dispute Database, 2004).

Hydropolitical resilience, then, is defined as the complex human–environmental system's ability to adapt to permutations and change within these systems, and "hydropolitical vulnerability" is defined by the risk of political dispute over shared water systems.

Another factor adding to the stability of international watersheds, then, is that once cooperative water regimes are established through treaty, they turn out to be resilient over time, even between otherwise hostile riparians and even as conflict is waged over other issues.

ECONOMIC ARGUMENT?
It is tempting to add an economic argument against "water wars." Water is neither a particularly costly commodity nor, given the financial resources to treat, store and deliver it, is it particularly scarce. Full-scale warfare, however, is

tremendously expensive. A "water war" simply would not cost out.

This point was probably best made by the Israeli Defense Forces analyst responsible for long-term planning during the 1982 invasion of Lebanon. When asked whether water was a factor in decision making, he noted, "Why go to war over water? For the price of one week's fighting, you could build five desalination plants. No loss of life, no international pressure, and a reliable supply you don't have to defend in hostile territory" (cited in Wolf, 1995b).

To make such a case convincingly, though, one would have to show times when war *was* cost-effective and, if such a thing is possible, it is well beyond the scope of this book.

J. A. Allan has made an important and compelling argument that nations facing water scarcity find it more economical (and less harrowing) simply to increase food imports, thereby importing "virtual water," that is, the water embedded in growing the crop elsewhere (see, for example, Allan 1998a, 2002).

SUMMARY OF GLOBAL STUDIES

Many researchers have been compiling global datasets of various aspects of political conflict and adding substantive knowledge to trends in shared waterways. These include:

- Azar's Conflict and Peace Data Bank (COPDAB) (Azar, 1980), 1948–1978
- Davies' Global Event Data System (GEDS) Project, 1979–1994
- International Crisis Behavior (ICB) dataset, collected by Jonathan Wilkenfeld and Michael Brecher (1997)
- Pennsylvania State University's Correlates of War (Correlates of War, 2006)

Also dedicated to water resources and hydropolitics include:

- African Transboundary Water Law (http://www.africanwaterlaw.org)
- International Water Law Project, (http://www.internationalwaterlaw.org)
- International Water Law Research Institute (IWRLI), http://www.dundee.ac.uk/law/iwlri/index.html)
- Center for International Earth Science Information Network
- World Resources Institute
- University of Kassel, Germany
- Balazs Fekete, Complex Systems Research Center, University of New Hampshire
- Michael D. Ward, Department of Political Science, University of Washington (Seattle)
- Jerome E. Dobson, Oak Ridge National Laboratory
- Jeff Danielson and Kent Lethcoe, EROS Data Center, (The Center for Earth Resources Observation and Science, South Dakota)
- Oregon State University's Transboundary Freshwater Dispute Database (TFDD) (http://www.transboundarywaters.orst.edu/)
- National Geographic Society

A number of studies, in addition to those mentioned above, have been able to report on global trends in water conflict and cooperation. What follows is a summary of those findings, with an important caveat: global studies, by nature, are generalizations, based often on incomplete or inaccurate data. Moreover, studies based on general datasets (rather than those based in water resources), report *only* statistical significance and should not be used to allude to causality. All statistical findings should be used as only intended: to point out *possible* sets of relations and likely directions for more focused case study approaches.

As mentioned, the Transboundary Freshwater Dispute Database (TFDD, 2006) has been tapped extensively for lessons learned in hydropolitics, including:

1. History-based indicators of settings with a potential for conflicting interests include those regions where the rate of change in a basin exceeds the institutional capacity to absorb that change (Wolf, Yoffe, and Giordano, 2003b; Yoffe, Wolf, and Giordano, 2003). This is most likely where change is exceedingly rapid, either on the institutional side or within the physical system, notably in:
   a. "Internationalized" basins. The clearest examples of this "internationalizing" process is the breakup of empires, notably the British empire in the 1940s and the Soviet Union in the late 1980s. Conflicts on the world's most tense basins – the Jordan, Nile, Tigris–Euphrates, Indus, and Aral – were all precipitated by these breakups. Recent internationalization seems to be one of the most significant indicators of dispute.
   b. Basins where unilateral development in the absence of a cooperative transboundary institution is imminent.
   c. Basins where there is general animosity over other, non-water issues.
2. Other general relationships (or lack of relationships) noted, include (for details, see Wolf, Yoffe, and Giordano, 2003b; Yoffe, Wolf, and Giordano, 2003):
   a. Countries that cooperate in general cooperate about water; countries that dispute in general, dispute over water.
   b. Higher GDPs are not statistically correlated with greater cooperation. [Interestingly, Yoffe (2001) found a better relationship between the rate of population growth and general (nonwater) levels of conflict/cooperation.]
   c. Regardless of how water stress is measured, it is not a statistically significant indicator of water dispute (Figure 2.5).
   d. Neither government type nor average climate shows any patterns of impact on water disputes.[8]

Yoffe and colleagues (2004) reported on subsequent findings from TFDD, including:

- Work by Giordano, Giordano, and Wolf (2002), which quantitatively explored the linkages between internal and

---

[8] Wolf, Yoffe, and Giordano (2003b) note that prevailing wisdom seems to be challenged in both items; the first appears to suggest that democracies seem not to be more cooperative than other types of government (in fact, autocratic countries are only barely less cooperative than the strongest democracies) and the latter disputes the commonly held perception that disputes are more common in arid environments – there is little perceptible difference between most climate types (with the notable exception of humid mesothermal, apparently the most cooperative climate). They also do not seem to find disputes in "creeping problems," such as gradual degradation of water quality or climate-change-induced hydrologic variability.

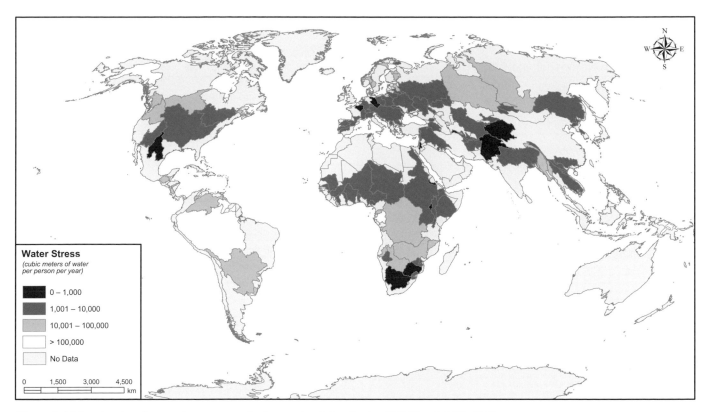

Figure 2.5 Map showing degree of water stress by international river basins (UNEP and OSU, 2002).

international water and nonwater events for three specific regions (the Middle East, Southern Africa, and Southeast Asia) and found generally synchronous chronologies (i.e., similar periods identified as conflictive and cooperative, for both internal and international relations) for Israel and its neighbors in the Middle East and for India and its neighbors in Southeast Asia, but not for Southern Africa (see Figures 2.6 and 2.7).

• Work by Stahl and Wolf (2003), which refined the question of climate to look specifically at the relationship between variability and conflict. The research found that, historically, extreme events of conflict were more frequent in marginal climates with highly variable hydrologic conditions, while the riparians of rivers with less extreme natural conditions

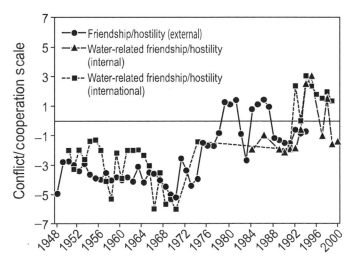

Figure 2.6 Israel: Water-related friendship/hostility (external) and (internal). Note general correlation between internal (Israel) and international relations.

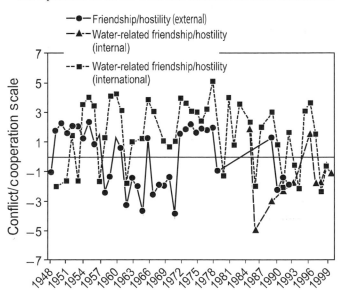

Figure 2.7 India: Water-related friendship/hostility (external) and (internal). Note general correlation between internal (India) and international relations.

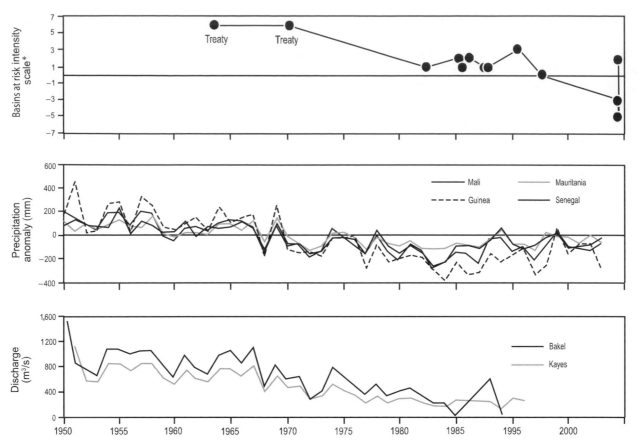

Figure 2.8 Time series of events of conflict and cooperation of Mali over the Senegal River, precipitation anomaly in the Senegal basin portion of the four riparian countries, and annual mean discharge at two gauging stations along the Senegal River. Note general decrease in hydropolitical relations with decreasing precipitation and discharges. (Modified from Wolf et al., 2003a). Note: *Negative numbers denote more conflict; positive numbers denote more cooperation.

have been more moderate in their conflict–cooperation relationship. The entire causal relationship between hydroclimatology and water-related political relations also depends on socioeconomic conditions and institutional capacity as well as the timing and occurrence of changes and extremes in a country and basin (Figure 2.8).

Nils Petter Gleditsch of the Center for the Study of Civil War, International Peace Research Institute, Oslo (PRIO) leads another group of researchers who have conducted "large-*n*" studies relating water to conflict. This group relates issues of hydropolitics to the Correlates of War dataset, and their findings to date include the following:

- Toset, Gleditsch, and Hegre (2001) show that two countries that share a river statistically, but moderately, have an increased probability of a militarized inter-State dispute over and above mere contiguity. They also find that the upstream–downstream relationship appears to be the form of shared river most frequently associated with conflict, in contrast to adjacent streams.

- Furlong, Gleditsch, and Hegre (2006) find that these relationships hold even when controlling for the length of the land boundary between countries, using a new dataset on boundary length developed by Furlong and Gleditsch (2003).

- Gleditsch and colleagues (2006) confirm that the relationship between shared rivers and militarized disputes holds for an improved database on shared rivers, derived from the database on river basins, developed by Wolf, Yoffe, and Giordano (2003b). They find little support for the idea that "fuzzy" river boundaries provide a source of conflict, limited support for the upstream–downstream scenario, and more support for the importance of the size of the basin.

- Gleditsch and Hamner (2001) found, on the basis of events data for the period 1948–1992, that shared rivers and water scarcity were associated with increased cooperation between countries, as well as conflict. A similar finding is recorded by Brochmann (2006) using data for trade and joint membership in international organizations as indicators in cooperation.

Of course, not all "large-*n*" empirical studies come from research groups; individual researchers have also been applying their analytical skills to these global datasets:

- Song and Whittington (2004) developed a typology of international rivers that relates pairs of coriparians by their "power" (as measured by per capita GDP), their "size" (population), and their "upstream–downstream" relationship. The authors then draw preliminary findings about the likelihoods of treaty development, based on their typology, finding that basins with countervailing riparians, one with large size and one with high GDP, were marginally more likely to enter into treaties than those in other settings.
- Shlomi Dinar (2004) did an extensive assessment connecting river geography, water scarcity, and treaty cooperation and suggests that, counter to the "water scarcity leads to conflict" claim, long-term water scarcity has a significant influence on levels of *cooperation*. Additional variables that are considered in explaining cooperation patterns include trade, level of governance among the basin countries, and the geography of the basin.

MULTISCALAR STUDIES AND INSTITUTIONAL CAPACITY

Multiscalar studies are regularly ignored in water resources management research (Trottier, 2003a). Much literature on transboundary waters treats political entities as homogeneous monoliths – "Canada feels…" or "Jordan wants…." Analysts are only recently highlighting the pitfalls of this approach, often by showing how different subsets of actors relate very different "meanings" to water (see, for example, Blatter and Ingram, 2001). Rather than being simply another environmental input, water is regularly treated as a security issue, a gift of nature, or a focal point for local society. Disputes, therefore, need to be understood as more than "simply" over a quantity of a resource but also over conflicting attitudes, meanings, and contexts. In the U.S. West, as elsewhere, local water issues revolve around core values that often date back generations. Irrigators, Native Americans, and environmentalists, for example, can see water as tied to their very ways of life – which are increasingly threatened by newer water uses, such as cities and hydropower (Smith, 2003; Rothfelder, 2003).

This shift means that water management must be understood in terms of the specific, local context. History matters, as do power flows – the "meaning" of water to its users is as critical to understanding disputes, and sometimes more so, than its quantity, quality, and timing. For this new world, new tools for analysis are being added to the traditional arsenal, including network analysis, discourse analysis, and historical and ethnographic analysis, each of which can be bolstered and made more robust through the judicious application of appropriate information technologies.

One highlight of these new approaches is that the results of conflict analysis are very different depending on the scale being investigated (Swallow, Garrity, and van Noordwijk, 2001). To clearly understand the dynamics of water management and conflict potential, then, thorough assessments would investigate dynamics at multiscales simultaneously. María Rosa García-Acevedo (2001), for example, puts nominally a "United States–Mexico" dispute over the Colorado into its specific historic context and tracks water's changing meanings to the local populations involved, primarily indigenous groups and U.S. and Mexican farm communities, throughout the twentieth century. The local setting strongly influences international dynamics and vice versa.

This relationship between internal and international scales are borne out both in qualitative and in quantitative analyses. In a wonderfully nuanced study, Trottier (2000) examines hydropolitics at multiple political scales in the West Bank. Similarly, Giordano and colleagues (2002) report synchronous conflictive and cooperative periods in both internal and international relationships, both between Israel and Jordan and between India and Pakistan.

What we notice in the global record of water negotiations is that, regardless of scale, some patterns do emerge: initiating a process of conflict management is regularly brought on by a crisis, and parties often base their initial positions in terms of rights – the sense that a riparian is entitled to a certain allocation based on hydrography or chronology of use (Sherk, 2003). Irrigators in the Klamath basin in Oregon, for example, invoke rights under the 1902 Reclamation Act, while environmentalists refer to the 1973 Endangered Species Act (Walden, 2004).

Upstream riparians often invoke some variation of the Harmon Doctrine, or "absolute sovereighty," claiming that water rights originate where the water falls. The Harmon Doctrine, named for the U.S. attorney general who suggested this stance in 1895 regarding a dispute with Mexico over the Rio Grande, argues that a state has absolute rights to water flowing through its territory (LeMarquand, 1993; McCaffrey, 1996a, 1996b). Downstream riparians often invoke absolute river integrity, claiming rights to an undisturbed system or, if on an exotic stream, historic rights based on their history of use.

The Columbia basin offers another case in point. It has become one of the world's leading hydropower rivers with huge impacts on fish, navigation, irrigation, recreation, and indigenous cultures. It includes parts of Oregon, Montana, Idaho, Washington, and Canada. The basin is the fourth largest in United States and equal to the size of France. The Columbia has 10 times the flow of the Colorado and 2.5 times the flow of the Nile and includes seventy-nine facilities and thirteen large dams – eleven in the United States and two in Canada. The basin has high variability and depends on snow mass and complex drainage paths in and out of Canada. Canada has 15 percent of the basin area but 30 percent of the flow. In 1944, as Canadian and U.S. planners recognized that cooperative development might well be superior to individual actions, both countries requested that the International Joint Commission (IJC) study the feasibility of cooperative development in the Columbia basin (Muckleston, 2003; Kenney, 2005). The treaty was driven by droughts and floods. From 1944 to 1959, the IJC studied cooperation options, and in 1964 the Columbia River Treaty and Protocol were ratified by the governments of Canada and the United States. The treaty set up a complex system of selling downstream power benefits in exchange for upstream storage benefits on a 30-year basis. The treaty provides new storage to optimize flows for power and flood control and the Canadians obtained "rights to power." The flow in the river is evened out, as peak flows from snowmelt are held back then released as needed. The treaty did not have instream flow provisions. A mechanism for the parties to coordinate "as if" they were one owner was put in place, with an annual operating plan and an obligation to cooperate in shortage. It regulates reimbursements for upstream releases that benefit downstream generators and "in lieu" energy payments when holding water for decreases in power generation.

The treaty is one of the most sophisticated in the world, particularly because it circumvents the zero-sum approach to allocating fixed quantities of water by instead allocating to each country an equal share of benefits derived from the multiple uses of water in the shared basin. Hydropower production, flood control, and other benefits are quantified and shared annually, and there is little dispute across international boundaries. It is an example of how joint diagnosis can lead to joint benefits versus fighting over the allocation of flows (Muckleston, 2003).

2.2 PREVENTIVE DIPLOMACY[9]

The causal argument, then, seems both more complex and more subtle in water systems than has been argued, affecting primarily issues of stability, rather than violence, and tied intractably to the surrounding political setting. The real lessons of history turn out to be that, although water can act as an irritant, making good relations bad, and bad relations worse, it rarely induces acute violence and often acts as a catalyst to cooperation, even between bitter enemies. Moreover, those institutions that are created turn out to be extremely resilient and flexible over time, even as conflicts rage over other issues (Wolf, 1998). What, then, does this knowledge suggest about the most productive path toward conflict prevention and resolution? Spec-

[9] This section draws from the Introduction to Wolf (2002b).

tor (2000) offers detailed lessons for what has been termed "preventive diplomacy," a concept based on the premise that it is easier and cheaper to prevent disputes before they begin. Although seemingly self-evident, preventive diplomacy has proven difficult in practice, primarily because of the barriers within the international community of mobilizing crisis-level interest and resources before a crisis actually occurs. As Spector describes it, though, the concept is gaining momentum, particularly within the Western defense establishment, and he offers cases for how it has been used effectively, as well as the processes of preventive negotiations for problem solving.

Coming full circle from the local to the international and back to the local, Painter (1995) and Clark, Bingham, and Orenstein (1991) and Delli Priscoli (1996) describe how the tools used by alternative dispute resolution (ADR) – mediation, facilitation, and arbitration – can be effective in resolving environmental disputes, an application of which is termed EDR (environmental dispute resolution). The rationale for ADR and EDR is similar to that of preventive diplomacy – that is, it is cheaper and the solutions are more robust when issues are resolved through dialogue rather than litigation (or combat) – and Clark, Bingham, and Orenstein (1991) offer settings and cases to back up the argument. Painter (1995), a healthily skeptical advocate (and practitioner), offers a brief history of EDR from its roots in labor negotiations, suggests some problems with the approach, and concludes with "poststructural alternatives."

Most often, international attention, and resultant financing, is focused on a basin only *after* a crisis or flash point or events such as drought and floods. The focus is on reconstruction and recovery, and very little is put into planning for avoiding future events. Although short-term humanitarian impacts are achieved, long-term capacity to deal with future events is not enhanced (Figure 2.9).

Such has also been the case on the Indus, Jordan, Nile, and Tigris–Euphrates basins, for example. It is worth noting, however, that in the exceptions to this pattern, such as the Mekong and La Plata commissions, an institutional framework for joint management and dispute resolution was established well in advance of any likely conflict. It is also worth noting the Mekong Committee's impressive record of continuing its work throughout intense political disputes between the riparian countries, as well as the fact that data conflicts, common and contentious in all of the other basins presented, have not been a factor in the Mekong. In fact, the experience of the commission such as those of the Amazon, La Plata, or Mekong may suggest that when international institutions and organizations are established well in advance of water stress, they can help preclude such dangerous flash points. As noted earlier, other basins have equally resilient institutions, which have survived even when relations on other issues were strained.

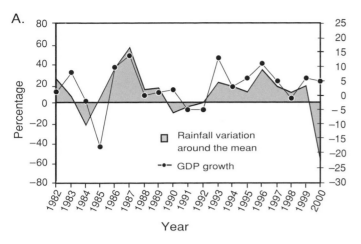
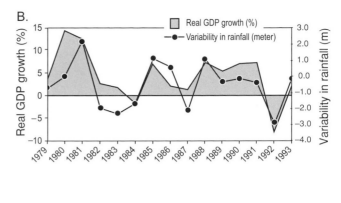

Figure 2.9 Economy-wide impacts. (A) Rainfall and GDP growth: Ethiopia; (B) Rainfall and GDP growth: Zimbabwe. Developing countries that have no infrastructure to store flood water remain hostage to fluctuation in rainfall. As much as 25 percent of the fluctuation in a country's GDP can be accounted for by lack of capacity to manage for floods and drought. This lack of capacity can easily spill into frustration and violence (World Bank, 2006).

Early intervention is also beneficial to the process of conflict management helping to shift the mode of dispute from costly, impasse-oriented dynamics to less costly, problem-solving dynamics. In the heat of some flash points, such as the Nile, the Indus, and the Jordan, as armed conflict seemed imminent, tremendous energy was spent just getting the parties to talk to each other. Hostilities were so pointed that negotiations inevitably began confrontationally, usually resulting in a hurried and inefficient solutions being the only ones viable.

In contrast, discussions in the Mekong Committee, the multilateral working group in the Middle East, the Environmental Programme for the Danube, and the Nile Basin Initiative, have all moved beyond the causes of immediate disputes on to actual, practical projects that may be implemented in an integrative framework. To be able to entice early cooperation, however, the incentives have to be made sufficiently clear to the riparians. In all of the cases mentioned above, not only was there strong third-party involvement in encouraging the parties to come together, but extensive funding was made available on the part of the international community to help finance projects that would come from the process.

Successful agreements are "organic" – that is once some agreement is made, it provides a framework for further action, and it can grow and often does. Thus agreements do not have to be a perfect treaties or organizations from the start. Rather, they must provide enough to allow parties some enhanced ability to talk with each other and provide a safe environment within which to communicate. This experience of communicating in such an environment will in the best sense create some new awareness and ability to negotiate. They will help foster the experience of working together, thus fostering social learning and growth.

2.3 REFRAMING THE DEBATE: WATER SHARING AND POLITICAL VULNERABILITY

Water is forcing us to rethink the notions of security, dependency, and interdependency. Increased interdependence through water-sharing plans and infrastructure networks is often viewed as increasing vulnerability and dependence and thereby reducing security. However, there is an alternative way to look at interdependence. It can be seen as building networks that will increase our flexibility and capacity to respond to exigencies of nature and reduce our vulnerability to events such as droughts and floods – and thereby increase security. Indeed, this perspective has been central to the evolution of civilization. It may strike deeper primordial fears and instincts than we might imagine. Interdependence also plays into fundamental beliefs found in most major religions – namely that in sharing our vulnerabilities we find strength.

This flexibility addresses the basic, almost primordial, fear and insecurity that has driven humans to become toolmakers and engineers. That is, reducing the uncertainty and building predictability and safety into what was often experienced as a harsh environment. Although often challenging the engineering mentality, this same fear – that we might kill life – inspires environmental concerns. Both relate to the fear of death we all carry. Both carry the instinct to life even though they produce conflicting views of what we should do. Somehow water forces us to go deeper than familiar adversarial positions and confront what we really share – this instinct to life.

Water carries a symbolic and subconscious power that is coupled to this fear and instinct. Water as a carrier of memory, as poets and scientists attest, may ultimately be telling us: stop,

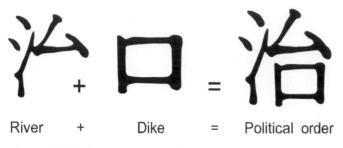

River   +   Dike   =   Political order

Figure 2.10 Chinese characters: River + dike = political order (Grey, 2008).

you don't just kill the other but all when you use water to make war. This stop sign, although unseen and rarely acknowledged, overcomes the instinct to fight and to destroy. It carries the symbol of a higher order, or superordinate value, which drives deeply into our identity as humans. Water, in effect constantly calls us, kicking and screaming, to higher notions of social integration and connection.

Following this psychological interpretation, conflict among water uses becomes highly functional to humans. Conflict becomes the opportunity to bring to consciousness unexamined fear: of change, of the future, or even of death. Indeed Jung viewed water in dreams as a symbol of the unconscious (Jung, 1968). In so doing water forces us to reflect on ourselves and our behavior and to internalize these reflections – this is the definition of growth and learning. And, thus, it forces us into more integration.

We might go further and say that the continued search for integrated and multipurpose river basin and watershed management, itself, is an outward social manifestation of the lifelong growth process of searching for integration. After all, we were all nurtured in water – the womb – and once we are outside of its safety, we face the constant challenge of overcoming a sense of being alone and being cut off. Our encounters with water are both a symbol of this and a powerful facilitator of our growth.

Many years ago Lao Tze wrote,

The sage's transformation of the world arises from solving the problem of water. If water is united, the human heart will be corrected. If water is pure and clean, the heart of the people will readily be unified and desirous of cleanliness. Even when the citizenry's heart is changed, their conduct will not be depraved. So the sage's government does not consist of talking to people and persuading them, family by family. The pivot (of work) is water. (quoted in Warshall, 1985, p. 5)

This is reflected in the ancient Chinese symbols for river, dike, and political order. Truly water management is linked to political order and civic culture (Figure 2.10).

To bring this into our practical world today, here is an example: in the 1990s, on the Hungarian and Slovakian border,

citizens on both sides of the Danube, Hungarians and Slovaks, on their own initiative, in a region fraught with ethnic violence, came together to meet and discuss how to clean up the pollution, to manage the water and to reduce terrible health risks to themselves and their children. Water, as a superordinate goal, facilitated a dialogue among dangerously conflicted ethnic peoples in an explosive area of the world. The dialogue resulted in agreement on cleanup and management, the first open border crossing in the current era, and a variety of joint projects still being carried out (Delli Priscoli and Montville, 1994). Similar dialogue takes place between Israelis and Palestinians (Friends of the Earth, 2005) and Azeris, Armenians, and Georgians (UNEP, UNDP, and OSCE, 2003). The sense of joint ownership and moral imperative from these actions, taken without and, indeed, contrary to the desires of national governments, forced the governments to follow.

The symbolic content of water as cleansing, healing, regenerating, and reconciling can provide a powerful tool for cooperation and symbolic acts of reconciliations so necessary to conflict resolution in other areas of society. In a sense, negotiations over water use, itself, could be seen as a secular and ecumenical ritual of reconciliation and creativity. This adds to the earlier, "rational" reasons that, although water can lead to political tensions and flash points, and even to occasional violence, it also leads to a preponderance of cooperation and dialogue. We believe in life, we want to survive and prosper, we know we are connected and water will not let us forget it. In fact water is calling us to learn its lessons so we can grow and prosper.

2.4 POLITICS AND HYDROCOOPERATION

Environmental deficiencies, not abundances, explain the development of irrigation technology – and irrigation permitted the emergence of urban civilization. One anthropologist states, "the remarkable fact about the origins of advanced agricultural economy and urban civilization in the ancient world was its location in regions of limited in water supply" (Bennett, 1974). Researchers have noted that the quantity of available water may be paramount in determining the sociopolitical structures. For example, the temperate and humid climate along European rivers did not force population nucleation and thus urban civilization appeared late (Clark, 1952; Waterbolk, 1962). Others suggest that the constant shifting of centers of power in Mesopotamian history were associated with the degradation of irrigation systems as well as military and economic situations (Adams, 1974). Wittfogel (1956) attributed the growth of centralized bureaucracy and autocratic rule to increasing connection of water through irrigation and navigation. The combination of hydraulic agriculture, a hydraulic government, and a single-centered society constitutes the institutional essence of hydraulic civilization. This permitted an accumulation of rural and urban populations that, though paralleled in a few non-hydraulic territories of small-scale irrigation, such as Japan, has not been matched by the higher agrarian civilizations based on rainfall farming. These hydraulic civilizations covered a vastly larger proportion of the surface of the globe than all other significant agrarian civilizations taken together.

Other researchers support these views. Some note that the centralized authority of Sasanid rule, in the Sistan region, which is in the southwestern corner of present Afghanistan, made the establishment of a complex irrigation network possible. This was the area where Zarathushtra found refuge (Gyuk, 1977). Others argue that the ability to manage water lies at the center of vigorous debate over the rise and fall of the Mayan civilizations. They theorize that intense agriculture, coupled with centralized water management, probably required a high degree of social organization. They use archaeological work at Tikal as evidence to further speculate that lack of sufficient water reserves in drought, rather than military or political conflict, may have caused abandonment of lowlands (Booth, 1991).

At the other end of the spectrum, researchers talk of how community irrigation engendered a democratic spirit and a sense of community (Glick, 1970). For example, sixteenth- and seventeenth-century Spanish irrigation was generally initiated, organized, financed, built, and maintained by local communities (Smith, 1975, p. 24). Some suggest that the change in political organization toward greater or lesser centralization may better be seen as social responses to environmental degradation. Although initial responses to increased environmental degradation might have been increased centralization, long-term degradation resulted in decentralization. Populations have moved from sedentary agriculture to nomadic pastoralism and back. The conclusion is that irrigation in and of itself does not necessitate political centralization. Also political centralization does not require the use of canal irrigation. In fact, the major civilizations seem to have experienced repeated expansions and collapses of political empires (Lees, 1973).

The Wittfogel thesis has been used to partially explain the development of the irrigated western United States. The western United States is seen as an example of the movement to large-scale bureaucracy, if not centralization of arid societies, based on large-scale irrigation (Worster, 1992). Another political scientist finds that regardless of which political framework is used, distributive systems, collective goods and so forth, the results are the same. Those with power will gain the access to the water whether through prices, participation, or, administrative procedures (Ingram, 1990). This is certainly borne out in American literature and films dealing with western water, such as *The Milagro Beanfield War* (Nichols, 1974) and *Chinatown* (1974).

One of the main reasons that no more entities like the Tennessee Valley Authority (TVA) were begun in the United States is because of the resistance from other large-scale water bureaucracies, which felt threatened (Leuchtenburg, 1952). By implication this notion of irrigation's tendency to big bureaucracy, or "impulse to empire," is sometimes extended to the history of foreign aid given by Western nations. Some of that aid, in part, helped create "clones" of large-scale irrigation bureaucracies, which today are now being asked to change.

But strong community and participatory traditions have also flourished amid the large-scale movement of bureaucratic irrigation discussed above. For example, in the United States, there is a rich history of farmers' associations. The Soil Conservation Service (SCS), which was the child of a large central bureaucracy, existed to foster community management of soil. The agricultural extension services are another such example. Likewise, small-scale water markets and trading have also flourished throughout the arid areas. Even the U.S. examples are best understood, like much of what we know historically, as a mixed system.

Building the physical water infrastructure in a collaborative and participatory way is now an important means for building the civic infrastructure and the civil society, or what many call the governance environment. Water resource management, with its current debates over markets, pricing, planning, participation, and environmental assessment, is a meeting ground for these forces. Such issues have historically been at the center of water resources administration and the rise and fall of civilizations. The fountains of ancient Rome, like standpipes in small villages today or medieval cities of Europe, played roles in building civic culture, as well as to quench thirst. They have become occasions for civic dialogue and meeting places central to creating sense of civic belonging and responsibility. Indeed the fountain was truly a civic work. It was the gathering place of the nations, believers and unbelievers (Schama, 1995, p. 288). We should not forget that civil society, civic culture, and civil engineering share commons roots. Whether it be irrigation associations, community water and sewage, and even large-scale multipurpose river operations, water management forces us to connect and balance rights and responsibilities. Although this process is imperfect, balancing is undertaken, and the exercise is often useful in and of itself. Most democratic theorists see the experience of such balancing as central to development of civic society (Barber, 1985).

Today there are many signs of how specific technologies are subtly transforming conflict resolution, negotiations, and decision dynamics in water conflicts. For example, software and visual displays facilitate the joint creation of models of water resources by political and technical stakeholders (USACE, 2004). They also raise the real potential for expanding options for political negotiators and decision makers. And as negotiation theory tells us, the ability to expand options is often the key to successful negotiations.

Remote sensing technology, although not replacing the need for "ground truthing," gives countries and jurisdictions the ability to build a fairly accurate picture of water flow in other jurisdictions, regardless of the level of data sharing. This technological capability transforms the relationships and negotiations among jurisdictions and will continue to do so. Trying to keep it all secret or giving misleading data just won't work like it used to; more people have more access to data. And all of this technology is disseminating, democratizing, faster than anyone predicted.

Virtually all of the world's viable river basin organizations evolved, usually over a period of several decades, in response to extreme hydrologic events. The achievement of shared data and trusted technical expertise has been central to their success. The interplay between the political and technical in achieving this state is complicated. But RBO viability, often demanded by the populations served, has ultimately depended in great part on such trusted technical agents.

Learning more about the wisdom and viability of traditional water management methods are important payoffs of surveying water and civilization. These range from old technologies, such as found in the Negev or other areas in North Africa, to various procedures for irrigation release management and hierarchy of rights revealed in court records in medieval Spain and other areas.

The history of social organization around river basins and watersheds is humanity's richest records of our dialogue with nature. It is among the most fertile areas for learning about how political and technical realms interact.

There is a large and growing literature warning of future "water wars" – these authors point to water not only as a cause of historic armed conflict but also as *the* resource that will bring combatants to the battlefield in the twenty-first century.

The historic reality has been quite different. In the modern times, only minor skirmishes have been waged over international waters – invariably other interrelated issues also factor in. Conversely, more than 3,600 treaties have been signed historically over different aspects of international waters (400 in this century on water *qua* water), many showing tremendous elegance and creativity for dealing with this critical resource. This is not to say that armed conflict has not taken place over water, only that such disputes generally are among tribe, water-use sector, or state. What we seem to be finding, in fact, is that geographic scale and intensity of conflict are *inversely* related.

War over water is not strategically rational, hydrographically effective, or economically viable. Shared interests along a waterway seem to overwhelm water's conflict-inducing characteristics and, once water management institutions are in place, they tend to be tremendously resilient. The patterns described in this book suggest that the more valuable lesson of international water is as a resource whose characteristics tend to induce cooperation and incite violence only in the exception. However, a new sense of ethics and new skills of management are needed to work in this reality.

3 Water conflict management: Theory and practice

Water is an eloquent advocate for reason.

– Admiral Lewis Strauss

3.1 WATER CONFLICT MANAGEMENT THEORY: ALTERNATIVE DISPUTE RESOLUTION AND THE FLOW OF BENEFITS

3.1.1 The principles

The field of conflict management and alternative dispute resolution (ADR) has brought new insights to negotiation and bargaining, adding much to the theory and practice of assisted negotiations, facilitation, and mediation. It has added practical tools to diagnose the causes of conflict and relate diagnosis to ADR techniques (see Delli Priscoli and Moore, 1988; Moore, 2003; and Shamir, 2003) The ADR field has codified a new language of interest-based bargaining. And much of these insights have arisen from environmental and natural resources cases.

Much of the ADR literature is found among works written by mediators or negotiators themselves about their own work, case studies by outside observers, and a growing body of theoretical work (see, for example, Fisher and Ury, 1981, Fisher and Ury, 1991; Susskind and Cruikshank, 1987; Lewicki et al., 1994; Mnookin, Peppet, and Tulumello, 2000; and Kaufmann, 2002, as representative works that combine the three approaches). One distinction important in ADR is that between distributive (also known as zero-sum or win–lose) bargaining – negotiating over one set amount, where one party's gain is the other's loss – and integrative (positive-sum or win–win) bargaining, where the solution is to everyone's gain. Reaching a collaborative arrangement is the goal of integrative bargaining. It depends on identifying values and interests that underlie positions; using these interests as building blocks for durable agreements; diagnosing the causes of conflict and designing processes appropriate to these causes; and focusing on procedural and psychological, as well as substantive satisfaction of parties. Interest-based bargaining or negotiations is the preferred way to accomplish this.

In traditional positional, or distributive, bargaining, parties open with high positions while keeping a low position in mind, and they negotiate to some space in between. Sometimes this is all that can be done. In contrast, interest-based or integrative bargaining involves parties in a collaborative effort to jointly meet each other's needs and satisfy their mutual interests. Rather than moving from positions to counter positions toward a compromise settlement, negotiators pursuing an interest-based bargaining approach attempt to identify the interests or needs of other parties *prior* to developing specific solutions. Often, outside help is needed to facilitate dialogue rather than to dictate solutions. It essentially is a process of social learning. Parties actually educate each other in their interests and thus become reeducated in their own interests in the process.

After the interests are identified, the negotiators jointly search for a variety of settlement options that might satisfy all interests rather than argue for any single position. This encourages creativity from the parties, especially in technical water management negotiations. Engineers may use their technical knowledge to liberate creativity rather than simply applying it to defending solutions. The process can actually generate solutions that no one person may have thought of before negotiations. The parties select a solution from these jointly generated options. This approach to negotiation is frequently called *integrative bargaining* because of its emphasis on cooperation, meeting mutual needs, and the efforts by the parties to expand the bargaining options so that a wiser decision, with more benefits to all, can be achieved.

Susskind and Cruikshank (1987) divide negotiations into three phases – prenegotiation, negotiation, and implementation – and offer concrete suggestions, such as "joint fact-finding" and "inventing options for mutual gain" in order to build consensus in an unassisted process. In assisted negotiations (facilitation, mediation, and arbitration), they observe

33

that whether the outcome is distributive or integrative depends primarily on the personal style of the negotiator. They also offer the interesting note that "negotiation researchers have established that cooperative negotiators are not necessarily more successful than competitive negotiators in reaching satisfactory agreement."

Lewicki and Litterer (1985) identify five styles of conflict management in a "dual-concern model" along a ratio of the degree of concern for one's own outcome, compared with the degree of concern of the other's outcome. The five styles possible are avoidance, compromise, and collaboration, as equal concern for both parties, and competition and accommodation as completely selfish and selfless, respectively.

In their classic, *Getting to Yes*, Fisher and Ury (1981) offer guidelines to reach this ideal, positive-sum solution. In language that is now common to much of the ADR literature, including Lewicki and Litterer (1985), whose terminology for similar concepts is presented in parentheses), Fisher and Ury suggest the following principles:

- Separate the people from the problem (identify the problem).
- Focus on interests, not positions (generate alternative solutions).
- Invent options for mutual gain (generate viable solutions).
- Insist on objective criteria (evaluate and select alternatives).

Although a collaborative arrangement is frequently seen as superior to any other, Lewicki and Litter (1985) offer a series of common pitfalls that preclude such an agreement. These factors that make integrative bargaining difficult include the failure to perceive a situation as having integrative potential, the history of the relationship between the parties, and polarized thinking. Ury (1991) offers specific advice on how to get past historically difficult and value-based conflicts – "getting past NO." And Donahue and Johnston (1998), Faure and Rubin (1993), and Blatter and Ingram (2001) describe cultural differences in approaches to water disputes.

Amy (1987) provides an altogether different approach to ADR, one of harsh criticism. He suggests that, because most studies of mediation are carried out by mediators, there is relatively little criticism of the fundamental claims made by the field. He begins by reviewing the advantages claimed by mediation over legislature, bureaucracy, and the courts to resolve environmental conflicts and concludes that mediation only tends to be justified when (1) there is a relative balance of power between the disputants and (2) an impasse has been reached in the conflict such that neither side can move unilaterally toward what they perceive as their best interest.

Restricting himself to intranational disputes, he also contests the common assertions that environmental mediation is cheaper, faster, and more satisfying than other approaches, particularly litigation. Amy (1987) approaches his critique from the perspective of power politics, and his most important observations are of power distributions throughout the process of mediation and of some resulting drawbacks. He argues that the same power relationships existing in the real world are brought into the negotiating process. In the classic environmental dispute of developer versus conservationist, for example, the former will usually have the power advantage. As such, the developer will only enter into negotiations if he or she somehow has that power blocked through, for example, a restraining order. The mediator, then, usually approaches a conflict looking for a compromise. The assumption is that the compromise will be found between the two initial positions. The problem may be rooted in fundamental differences in values or principles, though – for example, whether development should even take place – which may represent alternatives that are not even on the table.

Furthermore, if one party believes strongly one way or the other, any compromise may seem like capitulation. In other words, positions or interests can be compromised, but not principles. A mediator is usually not entrusted with finding the right solution, only the best compromise – and a mediator who becomes an advocate, either against disproportionate power or in favor of any specific worldview, will not likely find ready employment.

3.1.2 High politics and low politics

International relations theory has long grappled with the conflict between the unilateral sovereignty needs of states, and the requirement for cooperation for transboundary transactions. Because the flow of water does not respect political boundaries, it has been clear that regional management, at the watershed level at least, would be a much more efficient approach, at least from a management perspective. Nevertheless, water has regularly been "securitized," primarily due to internal politics, but has regularly had international repercussions. The question has historically been posed repeatedly, whether issues of regional water resources, considered a "low" political issue, can be addressed in advance of larger, "high," political issues of nationalism and diplomacy. Both sides have been argued in the past.

The "functionalist theory" of international politics, an alternative to the fairly self-explanatory "power politics," claims that states will willingly transfer sovereignty over matters of public concern to a common authority (Mitrany, 1975). Cooperation over resources, then, may induce cooperation over other, more contentious and emotional issues. In the Middle East, this thinking was the rationale for (1) the extensive

Johnston negotiations over a regional water-sharing plan for the riparians of the Jordan River from 1953 to 1955 (Wishart, 1990; Wolf, 1995b); (2) under President Johnson's worldwide program called "Water for Peace," for cooperative projects for immense agro-industrial complexes fueled by nuclear energy and desalination in the late 1960s; (3) multilateral negotiations over the Yarmuk River and the Unity Dam in the 1970s and 1980s (Bingham, Wolf, and Wohlgenant, 1994); and (4) an attempt at a Global Water Summit Initiative including Middle Eastern participation in 1991.

It has also been argued that one need only wait for the cessation of hostilities before developing regional water-sharing plans and projects but that cooperation over these projects may advance the pace of resolution of larger issues: "A regional water plan need not await the achievement of peace. To the contrary, its preparation, before a comprehensive peace settlement is attained, could help clarify objectives to be aimed for in achieving peace" (Ben-Shachar, 1989).

Elisha Kally, an architect of many regional water projects in the Middle East, has also contended that "the successful implementation of cooperative projects . . . will strengthen and stabilize peace" (Kally in Fishelson, 1989, p. 325).

In contrast to the functionalist argument, realist critics respond that states that are antagonists in the "high" politics of war and diplomacy tend not to be able to cooperate in the realm of "low" politics of economics and welfare. Until the Arab—Israeli peace negotiations began in 1991, attempts at Middle East conflict resolution had either endeavored to tackle political or resource problems, always separately. By separating the two realms of "high" and "low" politics, some have argued, each process was doomed to fail (Lowi, 1993; Waterbury, 1993). In water resource issues – the Johnston Negotiations attempts at "water-for-peace," negotiations over the Yarmuk River and the Unity Dam, and the Global Water Summit Initiative – all addressed water *qua* water, separate from the political differences between the parties (for more detail of these issues in the region's hydropolitical history, see Wolf, 1995b). All failed to one degree or another. In the most detailed argument in support of the realists regarding Middle East water resources, Lowi (1993) suggests that issues of regional water sharing simply could not be successfully broached in the Jordan basin until the larger political issues of territory and refugees are resolved.

The Arab—Israeli Peace Talks of the early 1990s, however, were the first time that both bilateral and multilateral tracks took place simultaneously. The design was explicitly to provide venues for issues of both high politics and low politics, with the premise that each might help catalyze the pace of the other. As Secretary of State James Baker, architect of the negotiating structure, described the relationship in his opening of the organizational meeting of the multilateral talks in Moscow:

Only the bilateral talks can address and one day resolve the basic issues of territory, security, and peace, which the parties have identified as the core elements of a lasting and comprehensive peace between Israel and its neighbors. But it is true that those bilateral negotiations do not take place in a vacuum, and that the condition of the region at large will affect them. In short, the multilateral talks are intended as a complement to the bilateral negotiations: each can and will buttress the other. (Baker, quoted in Peters, 1994)

Or, as Joel Peters describes it, "Whereas the bilaterals would deal with the problems inherited from the past, the multilaterals would focus on the future shape of the Middle East" (Peters, 1994).

The multilateral talks included five issues of regional importance. The only set that has survived the collapse of the peace negotiations and the renewed violence of the early 2000s, and continues to function to this day, is the Multilateral Working Group on Water Resources.

3.2 ADR AND WATER RESOURCES CONFLICTS

Alternative dispute resolution, with its subfield of environmental dispute resolution, uses example of water disputes quite widely as, for example, in Amy (1987) and Bingham and Orenstein (1989). Although international relations, in general, are treated extensively in the ADR literature by Kriesberg (1988), Stein (1988), and Ury (1987), application of ADR techniques to international resource conflicts are rare. Dryzek and Hunter (1987) describe mediation as a mechanism to resolve international environmental problems and Zartman (1992) discusses the challenges presented in international environmental negotiations. Excellent summaries of the potential of ADR in the context of international water resources conflicts can be found in Vlachos, Webb, and Murphy, 1986; Anderson, 1994; Bingham, Wolf, and Wohlgenant, 1994; Vlachos, 1994; Wolf, 2002b; and Shamir, 2003).

3.2.1 The unassisted setting

When the water demand of a population in a water basin begins to approach its available supply, the inhabitants have two choices (see Falkenmark, Lundquist, and Widstrand, 1989, and LeMarquand, 1977, for related work). These options are equally true for the inhabitants of a single basin that includes two or more political entities:

1. They can work unilaterally within the basin (or state) pursuing strategies of (a) making no changes in planning or

infrastructure and facing each cycle of drought with increasing hardship; (b) increasing supply – through wastewater reclamation, desalination, or increasing catchment or storage; and/or (c) decreasing demand – through conservation or greater efficiency in agricultural practices.

2. They can cooperate with the inhabitants within the basin or of other basins for a more efficient distribution of water resources.

Examples of unilateral development, unfortunately, abound. Some of the most vociferous examples include development in the Euphrates and the Jordan basins, along which troops have actually been mobilized by downstream riparians in response to upstream unilateral developments. Syria and Iraq came close to armed conflict over the issue of Euphrates flow in 1975. Along the Jordan, sporadic shooting over water developments occurred between Israel and Syria both in the mid-1950s and from 1965 to1967 (Wolf, 1994).

Examples of cooperation are somewhat more prevalent. The Rhine has treaties for water use dating back to 1814. Since that time, more than 400 treaties have been signed to legislate the water use of more than 200 international river basins (Caponera, 1985; UNEP and OSU, 2002).

For the game theorist, this dichotomy between two parties of whether to work unilaterally (defect) or to cooperate is recognizable as a familiar two-player, two-strategy game (Rogers, 1969; Bennett, Ragland, and Yolles, 1998; and Just and Netanyahu, 1998, discuss game theoretical aspects of water resources). The strategies chosen by each player often depend on the geopolitical relationship between them and the differing impetus toward conflict or cooperation. For two water basins within the same political entity, with clear water rights and a strong government interest, the game may resemble a "stag hunt," where mutual cooperation is the rational strategy (Axelrod, 1984). Interbasin transfers in Spain, where water rights are vested in a national authority and transfers occur with relative ease between basins, might be cited as examples (Gonzalez and Rubio, 1992). Between somewhat hostile players, either within a State or more often internationally, the game becomes a "Prisoner's Dilemma," where, in the absence of strong incentives to cooperate, each player's individual self-interest suggests defection as the rational approach (Axelrod, 1984). This would be the case in both the river basins of the Nile (Dinar and Wolf, 1994a) and the Nestos (Giannias and Lekakis, 1996). In cases of high levels of hostility a game of chicken can develop, with each player competing to divert or degrade the greatest amount of water before the opponent can do the same. The southern-most part of the Jordan River might be used as an example, with Syrian, Jordanian, and Israeli unilateral diversions all impeding basinwide cooperation (Wolf, 1995b).

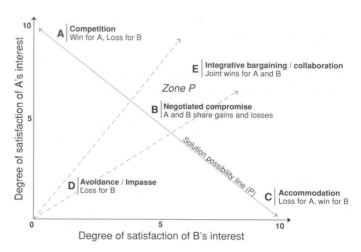

Figure 3.1 Strategies and outcomes of two-party disputes (Thomas, 1976).

As the amount of water surplus decreases over time, though, the impetus toward conflict or cooperation (payoffs) might change, depending on such political factors as relative power, level of hostility, legal arrangements, and form and stability of government.

3.2.2 The assisted setting

Figure 3.1 outlines a simple two-party dispute and shows Lewicki and Litterer's (1985) five styles from Party A's perspective. Frequently, we view negotiating as movement along Line P, which, using indifference-curve analogy, could be called the "solution possibility line." Point B is a caricature of the negotiated compromise where parties share equally in losses and gains – we split the pie. Point A represents a competitive win for Party A and point C an accommodating loss. If no agreement or conclusion is reached, then we often find ourselves inside P at a point D, due either to avoidance or to impasse.

Interestingly, point D is the situation in a variety of river basins needing water investments. For example, the World Bank examined the funding of Karnali Dam in Nepal, and an argument was made that no appreciable harm from the project would result to downstream Bangladesh (Delli Priscoli, 1996); however, if the Bank does not encourage involvement of Bangladesh in the project, what will this do to possibilities for broader systemwide negotiations on the whole river? Middle East competition has, until very recently, failed to deliver benefits to any of the parties and has alternatively resulted in avoidance and impasse. As an external actor with significant influence, it is tempting for an individual country to try to bring a third party into a point D situation to create a benefit for one party while avoiding direct bargaining. As LeMarquand (1990) shows in Senegal, the Bank avoided such a situation, whereas other donors did not. German and French donors, by supporting separate pieces of the project, fostered a point B compromise in which the sharing of economic costs might vastly outweigh gains.

Frequently, technical professionals prematurely define solution possibility curve P, albeit for noble reasons. Too early a use of deterministic analysis can have the effect of using our expertise to stifle rather than create options. Point E, in zone P, goes beyond the traditional possibility curve. Getting to point P usually requires some form of integrative bargaining and often the use of external assistance.

This zone is built on the assumptions that dispute management can be creative (Coser, 1959), that negotiations are a social learning process (Lincoln, 1986), and the rationale is a necessary but not sufficient condition – we seek a reasonable and acceptable outcome. At this point, solutions emerge that were not previously imagined by any one party. When this happens, it is clear that the process of dialogue adds significant value to the situation.

The water field can already point to a variety of point E outcomes and the number of examples is growing. The Indus and the Columbia river treaties are instructive early international water resources examples. Rogers observes how the Columbia treaty, by rejecting an originally proposed "Pareto optimal solution," has forgone significant benefits (Rogers, 1992a). Mehta (1986) and others note that the economic benefits from the Indus were suboptimal by some economic rationality. However, in both cases the economic and other costs of no agreement – or, in planning terms, the expected cost of the no agreement – clearly outweighed benefits forgone in a suboptimal economic solution. Evaluators of both cases see the experience of negotiating and living with the treaties as contributing to more positive relationships within which to carry out water development, clearly indicating social learning. In both cases original optimal-based solutions were rejected and creative new options emerged.

Fisher and Ury (1981) call this the "best alternative to negotiated agreement" (BATNA). In many ways it parallels the economist's notion of opportunity costs. For example, using the "Pareto improvement method," Rogers (1992a) calculates that more than US$2 billion of benefits have been forgiven in the Ganges–Brahmaputra region without an agreement. Creating incentives for parties to explicitly discuss their BATNA requires conscious design and frequently the help of a neutral party. Susskind and Cruikshank (1987) point out, however, that no one should be at a bargaining table to begin with if their BATNA away from the table is likely to be higher than what can be gained through negotiations. A clear understanding of one's own BATNA and, if possible, of the opponent's, gives a pretty clear idea of what the bargaining range is likely to be.

An interprovincial water allocation agreement in Pakistan is an example of how this notion can be applied within a country to break through traditional competitive behavior patterns that were resulting in poor allocation. This process asked provinces to look at their individual costs, with no allocation agreement. The provinces thus examined their crop loss due to uncertainty of incremental game playing and compared that against the assured production due to the certainty of a yearly allocation. The most technically rational or perfect solution is not always the one that the parties find most acceptable or feasible to implement. Obviously the reasonable solution should not require a compromise of ethical or legal standards, but the degree of purity of a solution should be weighed against the desirability of agreement and the long-term impacts of a stalemate.

There are multiple satisfactory and genuinely elegant solutions to most problems. Managing conflicts and resolving disputes is not always a zero-sum game or a question of slicing up and allocating a limited pie. Obviously, slicing the pie and zero-sum gaming are present in many disputes. However, this need not be the dominant approach, especially with water and its great potential for multiple uses that can increase the negotiating pie through creation of benefits. Integrative bargaining seeks to create a whole solution that is greater than the sum of its parts. It tries to create the environment in which synergy and creativity can prosper.

3.3 DIAGNOSING CAUSES OF CONFLICT

If we are to more consciously design dispute management and collaborative management systems, we need a means to diagnose or describe the causes of disputes in given situations. The "Circle of Conflict" (Figure 3.2) is one way of thinking about the sources of conflict, regardless of whether they are at the interpersonal, intraorganizational, communal, societal, or international level (see Moore, 1986, 2003).

Although the figure portrays causes as ideal types, any given dispute will contain pieces of each cause – but frequently one or two causes will be more dominant. The theory is that collaboration and management strategies should be based on the understanding of causes and that various intervention strategies are appropriate to different causes, as suggested outside each segment on the circle.

The circle identifies five central causes of conflict

1. Disagreements over data
2. Problems with the people's relationships

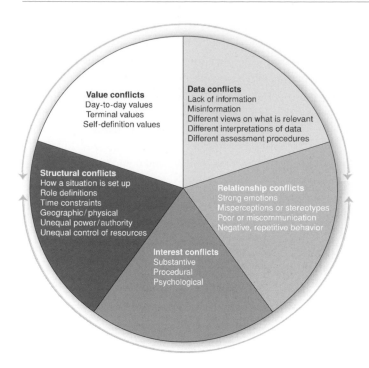

Figure 3.2 Circle of conflict/causes of disputes (Delli Priscoli and Moore, 1985).

3. Perceived or actual incompatible interests
4. Problems with structural forces
5. Perceived or actual competing values

Data conflicts occur when people lack the information necessary to make wise decisions, are misinformed, disagree over what data are relevant, interpret information differently, or have competing assessment procedures. Some data conflicts may be unnecessary, such as those caused by poor communication between the people in conflict. Other data conflicts may be genuine in that the information and procedures used to collect or assess data are not compatible.

Relationship conflicts occur because of the presence of strong negative emotions, misperceptions, stereotypes, poor communication, or repetitive negative behavior. These problems often result in what has been called unrealistic or unnecessary conflict, in that it may occur even when more objective conditions for a dispute, such as limited resources or mutually exclusive goals, are not present. Relationship problems often fuel disputes to an unnecessary escalatory spiral of destructive conflict (see Coser, 1959, and Moore, 1986). Interest conflicts are caused by competition over perceived or actual incompatible needs. Conflicts of interest result when one party believes that the needs of an opponent must be sacrificed to satisfy its own needs. Interest-based conflicts occur over substantive issues (money, physical resources, time), procedural issues (the

way the dispute is to be resolved), or psychological issues (perceptions of trust, fairness, desire for participation, respect). For an interest-based dispute to be resolved, all parties must have a significant number of their interests addressed or met in each of these three areas. Interests are based on and driven by values. However, the relative importance of many (not all) values are likely to change in given circumstances. The notion of interest thus captures the rank or salience of values in a given circumstance.

Structural conflicts are caused by patterns of human relationships. These patterns are often shaped by forces external to the people in dispute. Limited physical resources or authority, geographic constraints (distance or proximity), time (too little or too much), organizational structures, and so forth often promote structural conflict.

Value conflicts are caused by perceived or actual incompatible belief systems. Values are beliefs that give meaning to life. Values explain what is good or bad, right or wrong, just or unjust. Differing values need not cause conflict. People can live together with quite different value systems. Value disputes arise when people attempt to force one set of values on others, often without realizing it, or lay claims to exclusive value systems that do not allow for divergent beliefs.

3.4 GENERATING VALUE- AND INTEREST-BASED ALTERNATIVES IN WATER DISPUTES

The relationship between causes of disputes and intervention strategies is important to the water resource field. Frequently, technical agencies and engineers will consciously and unconsciously try to reduce most allocation conflicts to the level of data problems. Although data availability and data sharing are critical problems in both the industrial and developing world, disputes over data are often surrogates for interest, value, and relationship conflict. This is particularly true as the uncertainties surrounding data, such as with ecological impact or development projections, become more explicit.

For example, some time ago, the U.S. Army Corps of Engineers sought projections for electrical energy needs to the year 2000 in the U.S. Pacific Northwest. What they found is not surprising. One projection – by utility companies – showed steady growth in electrical energy needs to the year 2000. Another projection – by environmental groups – showed a steady downtrend to the year 2000. One or two projections were found somewhat nearer to the center; these were done by consultant groups. Each projection was done in a statistically "pedigreed" fashion. Each was logical and internally elegant, if not flawless (Delli Priscoli, 1989). The point is, once we know the stakeholders, we know the relative position of their projections: the group, organization, or institution embodies a set of values. These values are visions of the way the world ought to be. These visions become assumptions that, in turn, play out into different numerical results. Therefore, the trend lines go in different directions. Although we probably could not know the exact number, we could tell the relative position of these projections.

The problem water-resource professionals face is to foster negotiations around the assumptions and not simply the position. But how can we do this when playing the assumption game requires highly technical knowledge? What percentage of the population will follow the statistics necessary to understand value projection, and how much incentive is there for technocrats to make their arguments cogent to the population? We must not call our projections objective, value-free facts when they are really an elegant extension of our values or a vision of how we think the future ought to look. So to start engineering design, the water-resources planner or manager must find processes that mediate and somehow negotiate among the value-driven assumptions behind projections.

Figure 3.3 shows values from another perspective (Delli Priscoli and Creighton, 2004.) The bottom axis shows tradeoffs between environmental quality (EQ) and economic development (ED). The vertical axis shows tradeoffs between high government control and low government control. Both axes describe familiar perceptions of value tradeoffs found in water resource development. Where would we find organizations commonly involved in water conflicts, such as a ministry of agriculture, environmental NGOs, and irrigation districts?

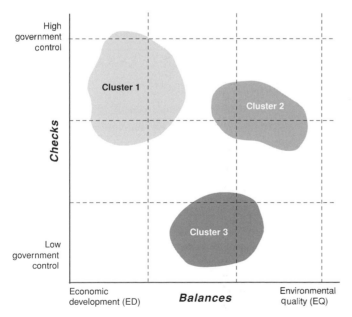

Figure 3.3 Developing value-based alternatives (Creighton, 1998a).

For the sake of argument, take a water problem such as urban flooding. Suppose twenty agencies, interest groups, and other organizations are stakeholders. How different will the planning be if we do the following? First, we identify where such groups fall in Figure 3.3 by placing a dot with their name on the chart. This has been done hypothetically without including names. Once we have distributed twenty dots around that chart, we would probably find various clusters. Circles are drawn around these hypothetical clusters on the figure. Having drawn these clusters, we then design specific alternatives, each of which can solve the flooding problem, for each of the three value clusters.

The technical professional is asked to understand the values, find how those values cluster, and design alternatives to serve those values. By using such a thought process, the professional can design based on existing values, as opposed to presenting solutions that themselves include an unexamined and frequently too narrow range of values. In this way, fewer alternatives may be developed while still representing a broader range of values. Designs that flow from such a thought process will greatly reduce the time spent on unacceptable alternatives. Technical professionals often need process procedures to understand competing values and to provide a road map for turning such competing values into the creative generation of alternatives and successful implementation of water plans.

3.5 INCENTIVES AND SHARED INTERESTS

In the end there must be incentives to cooperate. Contrary to popular belief, water provides many such incentives because it flows, can be used in so many vital ways, and can be reused. Indeed, several observers have noted that incentives for cooperation in river basins do exist. Rogers (1993) outlined the classic technical argument for incentives in multipurpose upstream and downstream cooperation. It links the effects of upstream and downstream activities (Table 3.1).

As Rogers (1993) notes, there is a pervasive unidirectional flow of effects, upstream or downstream, of water use in river basins. For example, upstream hydropower dams affect downstream flows. This can be positive by, for example, evening out flows so that water is available when it might not have been naturally or negative by trapping sediment or disrupting ecosystem requirements. Similarly, upstream storage of water can help protect downstream lands and activities from flooding. However, it is possible that some effects may flow in the opposite direction, such as the movement of migrating fish or barge transportation, and those stemming from water use that affect price levels or the availability of other resources. These

Table 3.1 *Possible downstream effects of upstream water use*

Water use	Downstream effects
Hydropower	Helps regulate river (+)
Base load	Creates additional peaks (−)
Peak load	Downstream flood protection (+)
Irrigation diversion	Removes waters from system (−)
Flood storage	Adds pollution to river (−)
Municipal and industrial diversions	Keeps water in river (+)
Wastewater treatments	
Navigation	Keeps water out of system (−)
Recreational storage	
Ecological maintenance	Keeps low flows in river (+)
Groundwater development	Reduces groundwater available (−)
Indirect use	
Agriculture	Sediment and air chemical (−)
Forestry	Sediment and chemical runoff (−)
Animal husbandry	Adds sediments and nutrients (−)
Filling wetlands	Reduces ecological carrying capacity increases floods (−)
Urban development	Induces flooding, adds pollutants (−)
Mineral deposits	Chemicals to surface- and groundwater (−)

Source: Rogers (1993).

effects can be positive or negative. He notes that because water is the universal solvent and the major geomorphological transport mechanism, the effects are caused not only by water use but other natural and anthropogenic activities occurring in the upstream reaches.

McCaffrey (2001b) notes that many incentives for cooperation are evident in the history of transboundary waters. Some involve reciprocal advantages, such as flooding an upstream state in return for sharing hydroelectric power or provision of water to one state in return for electricity from another. There have also been political and economic benefits flowing from water agreements, some of which may be indirect to the agreement. Reciprocal disadvantages have also provided incentives for cooperation. For example, a dam in a downstream state could cut off navigation or fish migration to an upstream state.

Pressure from the international community or outside parties can also be an incentive. For example, mediators with resources as discussed below have played roles in the Indus and other water disputes. In today's world, the need for private and public capital for construction and management can force riparians to look beyond pure allocation of water to the creation of benefits or revenue streams as a means to get needed resources. Indeed, this ability to transform pure allocation concerns into creating benefits is often a key to conflict management on transboundary rivers. It is the basis of the World Bank's Nile initiative of the late 1990s and early 2000s (Figure 3.4, Sidebar 3.1).

For years, water professionals have recognized that basin-wide development of international waters can produce the most optimal solutions to water needs. This can also be an incentive if the riparians are aware of it and if the technical and

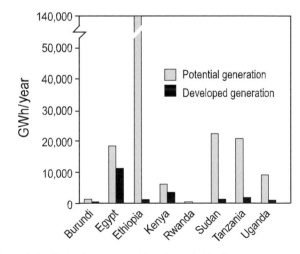

Figure 3.4 Nile basin opportunities. Potential hydropower generation is one set of possible benefits from cooperation.

Sidebar 3.1 Nile Basin Opportunities

Major potential win−win benefits from cooperative development

- power production/trade
  - about 90 percent hydropower potential undeveloped
  - about 85 percent population not served
- food production/trade
  - about 60 percent of irrigable land not irrigated
- multipurpose storage very low despite high rainfall variability
- environmental sustainability of watersheds, soils, wetlands, and lakes
- conflict prevention
- reduced tensions promote integration

political forces are able to work together. Some of these areas noted by scholars are sharing of data and information, transboundary environmental and impact studies, prior notification and consultation allocation of freshwater resources, pollution of freshwater resources, fishing, dispute settlement, and joint institutions.

Recent trends in water agreements have begun to characterize the waters as being "shared" resources of "common" interest. McCaffrey (2001b) notes that these ideas are prominent in the protocol on Shared Watercourse Systems in the Southern African Development Community (SADC) Region on August 28, 1995. They are prominent in an Agreement between Namibia and South Africa on the Establishment of a Permanent Water Commission on September 14, 1992.

3.6 FROM FLOWS TO BENEFITS: ECONOMIC CRITERIA[1]

Another emerging principle incorporated into water conflict resolution theory is the allocation of water resources according to economic value. Here we distinguish between "efficiency" – the allocation of water to its highest value use – and "equity" – the distribution of gains from an allocation (Howe, 1996). The idea of an efficient distribution is that different uses and users of the water along a given waterway may place differing values on the resource. Therefore, water sharing should take into consideration the possibility of increasing the overall efficiency of water utilization by reallocating the water according to these values. This principle alone may not be accepted as equitable or fair by the parties involved; however, the inclu-

[1] This section draws from Wolf and Dinar (1994).

sion of economic aspects in water resource allocation may enhance cooperation and collaboration in joint projects in the region of concern. Moreover, by recognizing the concept of "virtual water," a developed nation can often mitigate both the economic and political impacts of internal water shortages through trade (Allan, 1996).

3.6.1 Central planning versus market approaches

Allocation according to the economic value of water has usually been demonstrated using two approaches. The long-standing approach assumes a hypothetical central planning authority who knows what is "best" for society – a "social planner" in economic terms – who views the region as one planning unit. The social planner maximizes regional welfare subject to all available water resources in the region and given all possible water utilizing sectors. In some instances the social planner (government) also includes preferences (policy). The "water market" approach employs the market mechanism to achieve an efficient allocation of scarce water resources among competing users.

Examples of these approaches can be found in several studies that consider institutional and economic aspects of international cooperation for interbasin development. Goslin (1977) followed by Kenney (1995) examined the economic, legal, and technological aspects of the Colorado River basin allocation between the U.S. riparian states and Mexico. Krutilla (1969) analyzed the economics of the Columbia River Agreement between the United States and Canada. LeMarquand (1976, 1977) developed a framework to analyze economic and political aspects of a water basin development. Haynes and Whittington (1981) suggested a social planner solution for the entire Nile Basin. One team of researchers has been working to monetize the water dispute on the Jordan River, arguing that it will be easier to negotiate responsibility for a sum of money than over a scarce and emotionally charged natural resource (Fisher et al., 2002).

These studies generally argue that to cooperatively solve the problem of water allocations within a basin, the parties involved should realize some mutual benefit that can be achieved only through cooperation. Each party must participate voluntarily and accept the joint outcome from the cooperative project. Once a cooperative interest exists, the only remaining problem to solve is the allocation of the associated joint costs or benefits. The requirements for a cooperative solution to be accepted are that (a) the joint cost or benefit is partitioned such that each participant is better off than with a noncooperative outcome, (b) the partitioned cost or benefit to participants are preferred in the cooperative solution compared with subcoalitions that include part of the potential participants, and (c) all the cost or benefit is allocated.

It is important to note that the planning function need not be autocratic or deterministic. It can be, as it generally is today, open and collaborative. In addition, no significant examples exist of major water transfers among uses and sectors using markets. Small transfers usually within local areas do exist. In most cases, the markets are actually administratively run markets, which set prices and account for third-party transaction impacts and costs.

Recent studies have questioned the equity and justice associated with market allocations (see, for example, Margat, 1989; London and Miley, 1990; Tsur and Easter, 1994; and Frohlich and Oppenheimer, 1994), whereas others (e.g., Wolf, 1995b; Dellapenna, 2001) question whether related issues of property rights, externalities, transaction costs, and intangible values can be resolved to the point necessary for a functional water market. The conclusion from these studies is that economic considerations alone may not provide an acceptable solution to water allocation problems, especially between nations. Although the social planner and the market approaches may provide unique solutions to a problem of regional water allocation, they have drawbacks that may affect the efficiency and the acceptability of their proposed solutions. In its "pure" type, the social planner approach assumes that all social preferences are known and incorporated into the regional objective function. This, of course, is unlikely, especially when dealing with regional water allocations involving many countries with cultural differences and preferences.

In its "pure" type, the market approach assumes the existence of many parties in the region, each acting independently, so that the market price for water reflects its true value for each party. If, in that market, one party's decision does not affect the outcome of other individuals, then the self-interest of the parties lead to an efficient outcome for the whole region. In the case of water, one party's decision may affect another party's outcome, creating what is called an externality or third-party effect. If the externality effect (cost) is not included in the supply curve of water, the market mechanism collapses. This introduces inefficiency into the system and results in what economists call "market failure." In the case of water (in a water basin), the externality effects might be multidirectional. This is particularly true for water basins shared by more than one country, and for water used for more than one purpose. Also, water allocation problems are not exactly similar to market setups with which we are familiar (e.g., the market for cars), because they are characterized by a relatively small number of agents with different objectives and water-related perspectives.

Although "pure" systems are debated, what is most often implemented is usually some sort of hybrid.

3.6.2 Game theory

The game-theory approach allows the incorporation of economic and political aspects into a regional water-sharing analysis with a relatively small number of participants, each with different objectives and perspectives (for a more detailed discussion, see Shubik, 1984, 2002). The economic literature dealing with application of game theory solutions does not provide many examples of regional – international water-sharing problems. Rogers (1969) applied a game-theory approach to the disputed Ganges–Brahmaputra subbasin involving different uses of the water by India and Pakistan. He found a range of strategies for cooperation between the two riparian nations that will result in significant benefits to each. Rogers (1991) further discussed cooperative game theory approaches applied to water sharing in the Columbia basin between the United States and Canada; the Ganges—Brahmaputra basin between Nepal, India, and Bangladesh; and the Nile basin between Ethiopia, Sudan, and Egypt. In-depth analysis was conducted for the Ganges—Brahmaputra case to determine where each country's welfare is better off in a joint solution compared with any non-cooperative solution (Rogers, 1993).

Dinar and Wolf (1994a, 1994b), using a game-theory approach, evaluated the idea of trading hydrotechnology for interbasin water transfers among neighboring nations. They attempted to develop a broader, more realistic approach that addressed both the economic and political problems of the process. A conceptual framework for efficient allocation of water and hydrotechnology between two potential cooperators provided the basis for trade of water against water-saving technology. A game-theory model was then applied to a potential water trade in the western Middle East, involving Egypt, Israel, the West Bank, and the Gaza Strip. The model allocated potential benefits from trade between the cooperators. Dinar and Wolf's (1994a, 1994b) main findings were that economic merit exists for water transfer in the region, but political considerations may harm the process, if not block it. Part of the objection to regional water transfer might be due to unbalanced allocations of the regional gains and part to regional considerations other than those directly related to water transfer. In the real world, parties often do not have a clear sense of the interests of others. The benefits identified in game theory might not even be clear to parties themselves.

Perhaps because of the concerns we have mentioned, economic criteria have never been explicitly used to determine water allocations in an international treaty and, although states have compensated coriparians for water in some cases, no international water market has ever been established. Nevertheless, harnessing market forces for efficient and equitable allocations has become the focus of much debate within the water world, notably in the principles embedded in "full cost recovery" and the entire process of globalization, as will be explored later (see, for example, Dinar and Subramanian, 1997; Anderson and Snyder, 1997; Dinar, 2000; Finger and Allouche, 2002; and Appendix D, which includes principles and cases of international water transfers.)

3.7 WATER CONFLICT MANAGEMENT PRACTICE: PROCESS AND INSTITUTIONS

3.7.1 Beyond zero-sum: Overview of the continuum of procedures

Procedures for collaboration and dispute management can be placed on a continuum from more directed initiatives by the parties toward increased involvement and to interventions by third parties that provide various types of resolution assistance. In Figure 3.5, point A represents what some affectionately call the "hot tub" approach. That is, all parties jump into the hot tub and somehow agree (Delli Priscoli and Moore, 1985). Point B represents the opposite extreme, that is, parties go to war or use a highly adversarial approach. The left of the contin-

uum covers unassisted procedures, the middle covers assisted procedures, and the right covers third-party decision-making procedures. Most of the procedures have some elements of relationship building, procedural assistance, substantive assistance, or advice giving as a means of facilitating resolution, but they differ significantly in degree and emphasis.

Moving from point A to point B: the power and the authority to settle is gradually given to outside parties. A dividing line, point C, shows that point at which power to resolve disputes moves out of the hands of the disputants and into the hands of an outside party. This is a critical distinction. Fundamentally different relationships and communication patterns are established by procedures to the right as compared with those to the left of point C (Figure 3.5).

With third-party decision making, the primary communication pattern is between parties and the arbiter, judge, or panel. Each party presents a case to the arbiter, judge, or panel who makes the decision, which may or may not be binding. With assisted procedures, the facilitator or mediator seeks to encourage a primary and direct communication pattern between the parties. In this way, the parties can jointly diagnose problems, create alternatives, and own agreements (see, for example, Permanent Court of Arbitration, 1991).

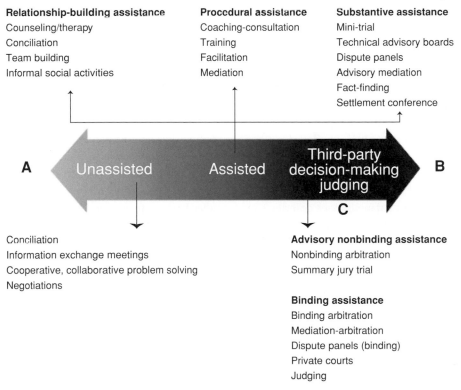

Figure 3.5 A continuum of alternative dispute resolution techniques. Point A represents what some affectionately call the "hot tub" approach (Delli Priscoli and Moore, 1985).

Though individuals can accomplish unassisted, integrative bargaining, as the number of stakeholders in water resources grow, the issues become more complex, and resources dwindle, and third- or neutral-party assistance is often needed. Few evaluations exist of interest-based negotiations used in water resources. They show how shared interests, which seem obvious after agreement, are hard for parties to discover during negotiations without assistance. For example, developers, oil companies, and environmentalists discovered that they shared interests of time and money in wetland use conflicts in the southern United States. Developers whose positions were to build unconstrained condominiums or to do offshore drilling saw that stabilizing building permits over 5-year periods could mean assured profit; so too with exploratory oil drilling in the Gulf Coast. Uncertainty of project stoppage was reduced. Environmentalists, whose position was that not another inch of wetland would be used or another estuary endangered, then saw that a stabilized permit situation would free their scarce resources, time, and money, which could be thrown into other priority fights. Though at first skeptical, parties used assisted, integrative bargaining to jointly understand their shared interests and reach agreements that allowed them to preserve their values and integrity (Delli Priscoli, 1988). The major premise of these procedures is that, by separating the process of dialogue and the content of dialogue, we can better manage the discussions and promote agreement. This separation of process and content is what leads to the use of third parties, sometimes called "interveners." These third-party facilitators or mediators become caretakers of the process of dialogue in the disputes.

Much of the dispute management literature encourages the use of procedures to the left of point C (Figure 3.5). These procedures, whether done as planning or regulating, emphasize the anticipation and prevention of high-conflict situations. In the United States, the growing experience of litigation, threat of litigation, and processes to the right of point C are becoming incentives to move to the techniques on the left. Reviews of hundreds of international mediations describe similar experiences. Bercovitch (1986) finds that mediations of high-intensity international conflicts are more effective when they follow, rather than precede, tests of strength and that the best time to enter is at points of stalemate and/or exhaustion. Indeed, the willingness to move to the left of point C is an indicator of social learning spawned by experience of conflict management.

Procedures to the left ultimately allow parties more control over the outcome. These procedures enhance the probability that parties will be able to break through positions and negotiate around interests. The price for these possibilities, direct dialogue, is often lower than the expected cost of highly adversarial battles.

Mediation developed from areas where the number of parties and issues are limited, such as in labor-management negotiations and some international disputes. Facilitation developed from multi-issue/multiparty situations such as resource controversies. However, with the growing practice of environmental mediation, the terms and practice overlap (Bingham, 1986). Facilitators are caretakers of the process. Although they don't have to be outsiders, they must remain impartial to the substance discussed. They suggest ways to structure dialogue, help stakeholders listen to each other, and encourage creative thinking (Moore and Delli Priscoli, 1989).

Mediators are generally outsiders to the stakeholders. Like a facilitator, a mediator primarily makes procedural suggestions, but occasionally, through caucuses or other means, may suggest substantive options. Some mediators are "orchestrators" and set the stage for bargaining. Others are "deal-makers" and are more involved in forging the details of a settlement (Delli Priscoli and Moore, 1988). Studies of mediations in violent international conflicts find that the mediators' active participation in substance and procedure is useful (Bercovitch, 1986). Mediation can be used in more polarized situations than can facilitation to break impasse and to initiate dialogue. Bercovitch (1986) shows that from 1816 to 1960, mediations were attempted, on average, every 4.5 months in highly polarized international situations. Reviews of hundreds of international mediations describe high frequency and high effectiveness of the procedure. Mediation has been more successful in security disputes than in primarily ideological and independence disputes (Bercovitch, 1986).

Once parties begin to prepare and posture as if they will go to point B, they begin an inertia that could create the reality – adversarial battle – they otherwise seek to avoid. Legal rules of evidence and disclosure separate, rather than integrate, information sharing. Substantive and technical experts, on all sides of the problem, move to the background and are further separated. Fortunes are spent on information gathering to get to a point – litigation – where lawyers spend their time keeping other lawyers from learning what they know!

Similar scenarios occur internationally. Analysts have documented a spiraling of conflict that occurs as parties posture and caricature. Often substantive experts are separated and move to the background behind the political and legal issues. In tracing the Del Plata Basin negotiations among Argentina, Brazil, Paraguay, Bolivia, and Uruguay, Cano (in Vlachos, Webb, and Murphy, 1986) describes how negotiation based too much on politics can drive the technical to the background and reduce the chance for success. In the end, most signed agreements were negotiated by the senior technical professionals. Reviews of managing international water resources echo the same point and often emphasize the collaboration of experts (Conca, 2006).

We could argue that failure of the Salmon Summit of the 1990s in the U.S. Pacific Northwest was due, in part, to being convened and driven too clearly by the political. Experts in environmental mediation were used for procedural assistance to bring together representation of a variety of interests. The operating agencies, especially the U.S. Army Corps of Engineers, became of the focus of controversy. Had the operating agencies (with political participation) convened the sessions and offered the commitment to operate according to a negotiated agreement, if one emerged, the results may have been different. Such an approach was used successfully to mediate operations of the Harry S. Truman Dam in Missouri (Moore, 1991), a dispute thought intractable for many years. Procedures to the left of point C have evolved in multiparty and multi-issue situations. Although procedures to the right work better for ripened and polarized disputes, they have limited capacity to deal with multiparty and multi-issue disputes and to encourage the generation of creative options. This is important to the water field, where the need is clearly for multiobjective and multiparty agreements. It is also important to the policy of international organizations, such as the World Bank which, through OD 7.50, tend to emphasize variations of procedures to the right of the continuum, such as expert boards. Expert panels or commissions have actually been common in the water-resources field. For example, there are technical committees on the Nile, the Euphrates, the Indus, and other rivers. Technical committees have been central to the workings of the International Joint Commission and the International Boundary Waters Commission and a variety of river basin commissions in the United States and Canada.

Staying on the left of the continuum, water banking as done in California and now in Texas can be seen as institutional mediation combined with market approaches. A mediating state institution buys water from agriculture at a set price and sells it to other users who put a higher value on the water. As a mediating institution, the World Bank can anticipate and manage third-party impacts and transaction costs, while still relying on the market.

New software technologies are creating interesting combinations of technical fact finding and facilitation. Software that allows technical and nontechnical personnel to jointly build models is now being used in the United States for drought contingency planning. These simulations are inexpensive and avoid the often unnecessary expenses of feeding huge models that only one or two people can manipulate and that often contribute only marginally to decision making. They create a sense of ownership in the algorithm used to generate and test sensitivity of alternatives.

Looking to the right of point C, the United States has experience beyond court and judicial decisions. For example, State

Water Masters and water engineers can exercise considerable power over allocating water in arid zones of the United States.

Since the 1970s, the UN and other international organizations have recognized this trend. The UN review (1975) of international institutions for managing international water resources called for use of conciliation, mediation, and procedures left of point C. However the same study's documentation reveals that many basin organizations and treaties have a variety of provisions for techniques to the right of point C, such as expert technical panels and forms of arbitration and little elaboration of those to the left. A Norwegian analysis of international environment conflict resolution found "most legal instruments relating to environment lack formal compulsory dispute resolution settlement mechanisms" (Trolldalen, 1992).

However, this may be changing. Chapter IV of the International Law Association's (ILA) "Berlin Rules," which deals with dispute resolution, encourages fact finding commissions composed of one member from each affected state, where appropriate, other "competent international organizations." This is similar to the successful model of disputes review boards used on construction projects throughout the United States. It also suggests a process of disputes management: start with fact finding, then move to conciliation, then mediation, and, finally, to arbitration and judicial settlement.

The search for cooperation over water in the Middle East has included approaches across the continuum. The on-again/off-again peace process includes traditional bilateral negotiations and multilateral negotiations on technical areas, of which water is one. The purpose of the multilaterals is to help professionals explore ideas and to support the bilaterals. The early Johnston negotiations along the Jordan basin in the early 1950s can be seen as a mediation effort by a third party with technical competence and resources. The current multilaterals have used a variety of relationship building and procedural-assistance measures. Study tours, joint information seminars, and other research by a variety of donors and lenders have dramatically enhanced the dialogue. Both these tracks have been surrounded by numerous other second track dialogues and academic-related fora. All of these are activities that fall to the left of point C on the continuum. They are providing an arena for expanded negotiation and even an outlet to keep the peace process moving.

But in the end, incentives become critical. In the Indus, the possibility of war (point B on the continuum) in the subcontinent was real enough to motivate use of mediation. Although some argue that the Middle East is another case, not all cases are so dramatic. However, the awareness of development benefits forgone and damages sustained (such as environmental damages) due to lack of agreements may become an incentive.

This is clearly reflected in growing attempts to create multi-purpose water agreements (Vlachos, 1991).

As the Oslo report notes, development banks and financial institutions will play increasingly important roles in prevention of conflict (Trolldalen, 1992). Access to capital will require review by international financial organizations, which will generate critical information about transboundary environmental and operational effects of projects. This is particularly true regarding rivers and water resources. The early participation of stakeholders, both intranational and international, will become a necessity for presenting workable plans. Thus, the leverage of financial institutions can become incentive for parties to use procedures on the left of the continuum.

The intersectoral dialogue and three-way agreement process in California was one of the more dramatic illustrations of seeking to participate, collaborate, and prevent further highly adversarial battles over water allocation. Ultimately the stakes are the reapportionment of water use among environmental, agriculture, and urban interests.

Even with a sophisticated system of water rights, laws, technical expertise, and articulate public interest groups, California water development had been at an impasse. Going to war, courts, and all-out positional bargaining had not worked. The drought of the early 1990s, coupled with the impasse, raised the stakes of no agreement (Peabody, 1991). The three-way dialogue was developed to look at alternative water futures and to develop a consensus-based framework for future development. It explicitly encourages interest-based negotiation leading to joint solutions.

Similar patterns developed on the Missouri River basin and even in humid areas of the United States, such as around Georgia, Florida, and Alabama. Formal mediation was used to reach agreement on the operations of Truman Dam in the Missouri basin. The Truman Dam had generated controversy since it was completed in 1981. Hydropower interests sought increased power generation and were being thwarted by environmental interests seeking fish and wildlife protection and by landowners seeking to reduce downstream effects of pool fluctuation. The U.S. Army Corps of Engineers, authorized to operate the project, was challenged no matter what approach it took. Therefore, it convened a mediation process that involved representations of all the stakeholders, including senior political officials. Once again, part of the incentive was impasse; part was the possibility of designing an agreement. The mediator designed an interest-based negotiation, which produced an agreement that no one party had thought of before the process. It included new hydropower units and preservations of instream values.

The World Bank and other donors and lenders have adopted procedures both to the right and left of point C (Figure 3.5) as the continuum of procedures. For example, the Bank formed its first expert Board under OD 7.50 to examine the international aspect of a dam project involving Somalia and Ethiopia. Neither country expressed much procedural or psychological satisfaction with the process, which is often the case with procedures to the right of the continuum. However, on the Komati River, between Swaziland and the Republic of South Africa, and on the Orange River, between Lesotho and the Republic of South Africa, the Bank adopted a more advisory role, similar to conciliation and team-building procedures on the left of the continuum (Rangeley and Kirmani, 1992). Using United Nations Development Programme (UNDP) financing, the Bank assisted Swaziland in preparation of its plans. The process resulted in two treaties. One would set up a technical advisory board, and the other, cost-sharing arrangements for two projects (cited in McDonald, 1988). On the Lesotho Highlands Water Treaty, an agreement was reached between the Republic of South Africa and Lesotho to create two national authorities and a permanent Joint Technical Commission to build and operate a multipurpose water project. Although they agreed on how to define benefits, the lack of hydrological data made it difficult to agree on annual yields of the project, so a contingent agreement was used. The parties agreed on the data that would be collected, who would collect the data, how to resolve disputes about the data, and how the benefit of the project would be calculated (cited in McDonald, 1988).

Substantive assistance and third-party judging techniques are probably closest to many donors' and lenders' traditional role and self-image – after all, as lenders they must evaluate projects according to some criteria. Institutions such as the World Bank are centers of expertise; however, as the Orange and Komati negotiations show, more than these techniques are likely to be needed. Water resources allocation is likely to demand the use of facilitation and mediation techniques, and the question will be how and by whom?

Do the substantive expert roles and images conflict with potential process roles for donors and lenders? The multiparty/multi-issue facilitating approach says that reaching agreement, to a point, becomes more important than the substantive terms of agreement. It is not necessary to abandon all notions of objectivity to play the role. However, in such roles, lenders and donors must become less deterministic. They must also be more willing to accept the possibility of agreements that they would not normally choose – as long as the agreement falls within some broadly defined professional bounds. The question is, what rationality will determine which bounds? Typically, professional engineers, lawyers, economists, and others begin with narrow notions of bounds, but given the inherent uncertainties of water management, and will ultimately admit that the bounds are usually far wider and less determined than originally presented. The water resources field has traditionally resisted placing bounds of probability on benefit–cost ratios and on the projected accruing of those benefits. The willingness to be flexible and accept agreements crafted by the parties can be enough to legitimize a procedural assistance role. It may even encourage subsequent substantive assistance in response to parties' needs.

Even if donors and lenders adopted this flexibility when situations called for it, do their development objectives or interests conflict with the capacity to catalyze or perform facilitation and mediation? Process theory is not built on the idea of value-free objectivity, but rather on the social/psychological notion of role clarification and the process and content distinction.

Process assistance can work because it liberates parties to engage in content negotiations without simultaneous procedural posturing. Process assistance does have a value bias – trying to help the parties reach agreements. If donors and lenders are advocates for a particular substantive agreement or alternative project configurations, the procedural assistance role would be meaningless. If they feel agreements are needed, but are open to a variety of alternative approaches, including a "no project" option, they can play an assistance role. In the Indus, once the World Bank moved away from its preferred option to facilitating joint options among the parties, its assistance role became more effective.

We usually think of moving from the left to the right of continuum, but the Indus experience can be seen as a movement from right to left. The first intervention for arbitration was rejected. The World Bank initially intervened and offered its preferred solution. This was both a procedural and a substantive role but also had strong elements of a third-party expert judging role – to the right of point C. After parties rejected this initial solution, the Bank adopted clearer procedural and substantive assistance roles – to the left of point C. India and Pakistan became more engaged in the creation of options. Once they produced and agreed to a solution, the Bank expanded its procedural assistance role and worked with other funding sources toward implementation of the agreement (Kirmani, 1990; Mehta, 1986).

The fact that the World Bank had financial resources and the capacity to generate more was crucial to the intervention. In studying violent international conflict, Zartman (1991) and others make the same point: effective mediation in international relations is greatly dependent on the ability to command resources. Other international water resources cases confirm this experience. The entire premise of the Global Environment Facility, which has a program specifically for international waterways, is that major resources need to be generated for transboundary environmental protection (Uitto and Duda,

2001). In other examples, United Nations Environment Programme (UNEP) funds were used as incentives to reluctant countries for participation in developing the Mediterranean Action Plan and to help establish a working group of experts to develop the Zambezi Action Plan (ZACPLAN). The Vatican used its resources of moral authority and confidentiality to promote agreement on the Beagle Channel. The Italians, through ITALCONSULT, brought resources to study dangers of unconditional national projects (or BATNAs) for riparians in the Niger Basin, which provided a common reference and substantive basis for subsequent agreements (these four examples are cited in McDonald, 1988). On the Nam Ngum wetlands restoration project in Laos, the United Nations and other donor financing provided a feasibility study and mobilized construction grants among adversarial riparians for mutually beneficial endeavors (Kirmani, 1990). And the ongoing Nile Basin Initiative, launched in 1999, has been facilitated with funding from the international community (Whittington and Sadoff, 2005).

3.7.2 Process techniques: BATNA, STN, and interest-based bargaining

A variety of techniques are emerging that can be used across most of the procedures in the continuum. Many are already used in the water-resources field. The best alternatives to negotiations (BATNA) were mentioned in previous sections. Single-text negotiation (STN) means developing a complete package, putting the package before parties, revising, and repeating the process. The technique helps parties to envision a whole and encourages them to work off the same page. Often, even negotiations within organizations require the assistance of a facilitator or mediator. The successful mediation of operating rules on the Truman Dam on the Osage River used STN to break impasse and generate agreement (Moore, 1991). This technique was crucial in developing the Camp David Accords (Raiffa, 1982; cited in McDonald, 1988). As we have seen in the Niger basin, Law of the Sea, and Antarctic Minerals negotiations, STN is effective for international natural resources issues.

Interest-based negotiations have become the preferred technique for integrative bargaining. This can be contrasted to what is traditionally called *positional bargaining* (Moore and Delli Priscoli, 1989). Positional bargaining is a negotiation strategy in which a series of positions or alternative solutions that meet particular interests or needs are selected by a negotiator, ordered sequentially according to preferred outcomes, and presented to another party in an effort to reach agreement. The first, or opening, position represents the maximum gain hoped for or expected in the negotiations. Each subsequent position demands less of the other party and results in fewer benefits for the person advocating it. Agreement is reached when the negotiators' positions converge and they reach an acceptable settlement range.

As we have discussed, interest-based bargaining involves parties in a collaborative effort to jointly meet each other's needs and satisfy mutual interests. After the interests are identified, the negotiators search jointly for a variety of settlement options that might satisfy all interests. The parties select a solution from these jointly generated options. This approach to negotiation is called integrative bargaining because of its emphasis on cooperation, meeting mutual needs, and the efforts by the parties to expand the bargaining options so that a wiser decision, with more benefits to all, can be achieved.

3.8 BUILDING INSTITUTIONAL CAPACITY FOR CONFLICT RESOLUTION

So what does our experience tell us? As we have previously noted, water resources institution building for collaboration and dispute management is forcing us to reexamine our notions of interdependencies, independence, and security.

We should be able to discern patterns in water resources negotiations and cooperation. For example, look at a short-term reactive pattern: there is a precipitating event (drought or flood), study, data gathering, general agreement on allocations or principles by treaty or court, specific agreements by jurisdictions, and implementations by subjurisdictional entities. Or look at gradual or long-term growth patterns: functional necessity creates limited-purpose organizations, such as for transportation, gradually being pressured to expand across sectors and to include new actors representing new interests.

We have learned that water institutions must include multiple purposes for water and participation of affected groups and users; improve realistic pricing of water; encourage integrative (win−win) rather than distributive (win−lose) bargaining, be flexible enough to react to short-term events but provide a stable mechanism for long-term visions, encourage meaningful allocation across sector interests but also efficient use at operating or retail levels, be driven also by nonmarket (instream) ecological values, and stay within reasonable bounds of distributive equity.

We have also learned that building institutions for cooperation over water resources takes time, that it frequently it starts with information exchange; that agreements continue to evolve after initial institutional frameworks are established, that the availability of credible technical assistance can be critical to facilitating cooperation; and that the more flexible and simple, the better chance for cooperation (see Wolf, Stahl, and Macomber, 2003a).

Interjurisdictional and cross-sectoral issues will become more critical to development generally, and to water investment specifically, especially on complex multipurpose projects (Blomquist, Heikkila, and Schlager, 2004). Experience indicates that the key to successful "multi objective projects" (MOP) will be the early generation of creative alternatives and the facilitation of a sense of ownership among stakeholders in both the alternatives and the process by which the alternatives are generated. Waiting to react to a few detailed and narrow alternatives or until a dispute ripens means acting too late because the alternatives become hardened positions. At this point, the process options – usually on the right of continuum – have limited ability to go beyond splitting differences and offer little hope for generating creative options. It will be in a donor's or lender's interest for early and meaningful collaboration and participation to occur in projects that they will be asked to finance. The probability of implementation will increase, transaction costs will go down, opportunities for future cooperation will go up, and the security of investments will be improved.

Building institutions for regional and interjurisdictional water resources cooperation takes time. It can start at any point on the continuum and need not end at the far right to achieve coordination that brings significant water-resources benefits. Frequently, the path to cooperation starts with information exchange (Chenoweth and Feitelson, 2001); however, agreements on allocation and sharing are not absolute – they continue to evolve after the establishment of the initial cooperative institutions. These institutions provide a secure context for negotiations.

Our knowledge of water resources is pushing toward a vision of developing methods for comprehensive analysis and operation so we can better integrate water uses. It is also moving us to integrate resources management across jurisdictions. As we begin to reach the limits of use, the ability of our organizations to respond to water-flow fluctuations becomes crucial. This flexibility is most needed to provide new forums for dealing with political tradeoffs that cross both time and space. Indeed, flexibility has been central to the recent successful negotiating of international environmental treaties (Conca, 2006).

Comprehensive planning can provide a "cloak of professionalism and objectivity and potential information useful in identifying the stakes of those not well represented and in the design of more equitable plans" (Allee and Abdalla, 1989). However, the essence of river basin management becomes the process and management of facilitated bargaining among stakeholders (Daniels and Walker, 2001). One major participant in the ebb and flow of water institutions in the United States offers a useful perspective (Allee and Abdalla, 1989).

He notes that, to a great degree, the river basin management concept has been driven by a rational analytical model as seen in the use of words such as "coordinated" and "comprehensive." Although this model might provide an ideal, it does not fit reality. The reality of river basin management goes beyond unified administration and rational analytic models to one of facilitated dialogue and negotiation among stakeholders in the basin. It leads to cooperation and integration, not just coordination. Rogers (1992a) notes, "Approaches based on game theory . . . ranging from pure conflict to pure cooperation do not directly yield norms for decisions regarding conflicts found in international river basins . . . consequently the field has relied increasingly on process oriented approaches."

After examining cases of international environmental negotiations, Oran Young (1992), a prominent theorist in international organizations, notes that building international regimes for natural resources management requires conscious design efforts beyond spontaneous intervention. He notes that "institutional design emerges as a process of steering complex bargaining toward coherent and socially desirable outcomes" (p. 230). Among the more important lessons for success are to seize windows of opportunity that are often exogenous to the bargaining process, to go beyond traditional distributive (positional) bargaining to integrative bargaining, to mobilize leadership, and to simplify implementation. This analysis and prescription of practical experience reflected above are the main messages of assisted negotiation and the ADR field.

We have learned that the role of technical information is critical to eventual legitimacy and acceptability. Technology is clearly a transforming agent. (Nandalal and Simonovic, 2003). As water issues become more prominent, the gray area between the technical and the political will expand. However, the fact that water professionals share a common technical language across jurisdictions will contribute to more than water negotiations; it will help more general relationship building. New interjurisdictional actors who represent new claims on water use are emerging to add to growing claims of traditional uses (Sidebar 3.2).

Sidebar 3.2 Some of the Important Lessons Being Learned from Using "Process" Tools

- Use process to build consensus.
- Create a commitment to implementation by participation in decisions.
- Accept the legitimacy of feelings and seek not only the rational but the reasonable.
- Start by defining the problem rather than proposing solutions or taking positions.
- Focus on interests.
- Identify numerous alternatives.
- Separate the generation of alternatives from their evaluation.
- Agree on principles or criteria to evaluate alternatives.
- Expect agreements to go through several refinements.
- Document agreement to reduce risk of later misunderstanding.
- Agree on the process by which agreement can be revised.
- Seek inclusiveness: jointly diagnose, jointly create options, jointly implement.
- Seek to link claims of rights with obligations for consequences of solutions.
- Move beyond adversarial science.
- Strive to keep decision close to those involved in the controversy and who have to live with outcome.
- Seek to "offload" – not replace – the formal legal systems.
- Anticipate and act to prevent disputes.
- Explicitly assess alternatives and costs of not having some form of river basin organization (RBO).
- Think of building an RBO and RBM as a creative process, as social learning, rather than as a contest.
- Negotiate and solve problems by satisfying interests rather than capitulating to positions.
- Design procedures to address causes of disputes not symptoms and build them into the arrangements and RBOs.
- Look to the "shadow of the future" – long term as well as short term.

4 Crafting institutions: Law, treaties, and shared benefits

It is not the strongest of the species that survive, nor the most intelligent, but the one most responsive to change.

– Commonly attributed to Charles Darwin

4.1 DISCIPLINES AND WORLDVIEWS[1]

Water is a powerfully unifying resource, so it is ironic that water education, management, and discourse are so fragmented (Goldfarb, 1997). To truly learn about water in its most holistic sense, one needs to understand the many aspects of the hydrologic cycle, from meteorology to surface hydrology to soil sciences to groundwater to limnology to aquatic ecosystems. One should also have an integral sense of the human dimensions, from economics to law, ethics, aesthetics, sociology, and anthropology (Freeman, 2000). Universities and management institutions are simply not organized along these lines; often they are fragmented to where even surface water and groundwater, quality and quantity, are separated out as if they were not inextricably interrelated. Yet each of these disciplines offers a particular perspective on conflict prevention, management, and resolution. Although each discipline is rooted in its own typologies and terminologies, there are again surprising similarities from discipline to discipline, particularly in that each strives to provide a more structured framework to the often chaotic processes of conflict resolution: *law* (see, for example, Wescoat, 1996; Bennett and Howe, 1998; McCaffrey, 1999, McCaffrey, 2001b; Wouters, 2001; and Paisley, 2003), through its clear delineation of the terms, boundaries, and solutions; *economics and game theory* (Howe, Schurmeier, and Shaw, 1986; Rogers, 1993), through the unifying concepts of rationality and efficiency; *engineering* (Bleed, 1990; Lancaster, 1990), by its depiction of present and future states, and how to get from one to the other; and *political economy* (Just and

Netanyahu, 1998; Allan, 1998a), through its position at the intersect between political and economic decision making.

Each discipline brings its distinctive set of tools to help the parties prevent disputes, resolve disputes, or visualize the problem in new ways to facilitate either prevention or resolution. Howe, Schurmeier, and Shaw (1986) and Anderson and Snyder (1997) offer ways in which market mechanisms can help with the problem of water allocations; Rogers (1993) and Dinar and Wolf (1994a, 1994b) describes through game theory how benefits might be equitably allocated across international boundaries; and, as previously noted, Allan (1998a) offers his useful and adroitly named concept of "virtual water," the water that moves between consumers and across nations embedded within the products it was used to produce, as an argument against the limiting concept of water security. Geography is best represented by Gilbert White (1974), who demonstrates geography's capabilities in interdisciplinary analysis, and by White's own prescience as he looks to the coming information age and its effects on systems analysis, risk assessment, and societal responses. Simonovic (1996) and Nandalal and Simonovic (2003) focus on the technology of the twenty-first century, describing how new modeling tools, visualization techniques, and information technologies can be packaged as decision support systems to aid parties in dispute in their decision making. Each worldview offers a lens to one perspective of water conflict management; collectively, these worldviews help inform the development and implementation of cooperative institutions.

4.2 INTERNATIONAL INSTITUTIONS AND DECLARATIONS

Just as the flow of water ignores political boundaries, so too does its management strain the capabilities of institutional boundaries. Although water managers generally understand and advocate the inherent powers of the watershed concept as a unit of management, where surface water and groundwater,

---

[1] This section draws from the Introduction to Wolf (2002b).

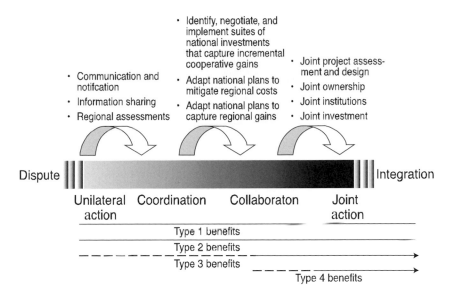

Figure 4.1 Types of cooperation: Cooperation continuum (Sadoff and Grey, 2002).

quantity and quality, are all inexorably connected, the institutions that have developed to manage the resource have rarely followed these tenets (Burchi and Spreij, 2003; Figure 4.1).

An agreement or institution may be thought of as a sociopolitical analogue to a vibrant ecosystem and thus vulnerable to the same categories of stresses that threaten ecosystem sustainability. Will the agreement and institutions that were crafted in the exercise sustain themselves through

- Biophysical stresses? Are there mechanisms for droughts and floods? Shifts in the climate or river course? Threats to ecosystem health?
- Geopolitical stresses? Will the agreement survive elections or dramatic changes in government? Political stresses, both internal and international?
- Socioeconomic stresses? Is there public support for the agreement? Does it have a stable funding mechanism? Will it survive changing societal values and norms?

The best management, which is similar to ecosystem management, is *adaptive* management, that is, the institution has mechanisms to adapt to changes and stresses and to mitigate their impact on its sustainability (see Lee, 1995, for the classic text on adaptive management). Crafting institutions requires a balance between the efficiency of integrated management with the sovereignty-protection of national interests. Along with greater integration of scope and authority may come greater efficiency but also greater potential for disagreements, greater infringement on sovereignty, and greater transaction costs (see Feitelson and Haddad, 1998, for more information). Some possible institutional models are offered in Figure 4.2. Nevertheless, for every set of political relations, there is some possible

institutional arrangement that will be acceptable (even if it is only to collect data separately, but in a unified format, in the hopes that they may one day be merged) and, if its management is iterative and adaptive, responsibility can be regularly "recrafted" to adapt or even lead political relations.

Hundreds of examples of transboundary conflict and cooperation exist throughout the world. Although several examples have already been referenced, some particularly salient cases are discussed below in greater detail, with the aim of further exploring themes already mentioned, while identifying new findings and lessons. Appendix C describes these cases in more detail.

Scholars of environmental institutionalism have developed an extensive literature over the years (see, for example, Keohane, 1989; Young, 1989; Ostrom, 1990; and Agrawal, 2002). Keohane defines "institutions" as a "persistent and connected sets of rules (formal and informal) that prescribe behavioral rules, constrain activity, and shape expectations" (Keohane, 1989, p. 3). In water resources, these institutional "rules of the game" can be as informal as an unwritten understanding, more formal as a working arrangement or river basin organization, or very formal as a legally binding treaty. Young (1989) suggests formality as being defined by whether an agreement is written or rather arises spontaneously. (Note, too, that a treaty can legally create a river basin organization.) Regardless of its formality, Zawahri (2006) reminds us "an institution's design or attributes has a direct impact on states' ability to facilitate and maintain cooperation."

Frederiksen (1992) describes principles and practice of water resources institutions from around the world. He argues that while, ideally, water institutions should provide for

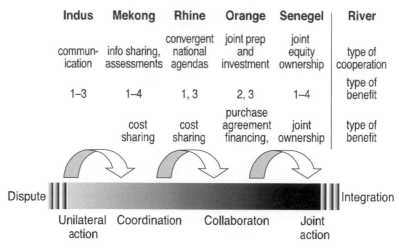

Indus	Mekong	Rhine	Orange	Senegel	River
commun-ication	info sharing, assessments	convergent national agendas	joint prep and investment	joint equity ownership	type of cooperation
1–3	1–4	1, 3	2, 3	1–4	type of benefit
	cost sharing	cost sharing	purchase agreement financing,	joint ownership	type of benefit

Figure 4.2 Types of cooperation: Some examples (Sadoff and Grey, 2002).

ongoing evaluation, comprehensive review, and consistency among actions, in practice this integrated foresight is rare. Rather, he finds rampant lack of consideration of quality considerations in quantity decisions, a lack of specificity in rights allocations, disproportionate political power held by power companies, and a general neglect for environmental concerns in water resources decision making. Buck, Gleason, and Sofuku (1993) describe an "institutional imperative" in their comparison of transboundary water conflicts in the United States and the former Soviet Union, whereas Gooch and Stålnacke (2006) focus on the Lake Peipsi region between Russia and Estonia to develop institutional lessons. Feitelson and Haddad (1995, 2000), Jarvis (2006), and Jarvis et al. (2006) take up the particular institutional challenges of transboundary groundwater. Gopalakrishnan, Tortajada, and Biswas (2005) offer institutional lessons from around the world, Conca (2006) sets institutional development within the context of global governance, and Saleth and Dinar (2005) describe the theory and practice of water institutional reform.

To address these deficiencies at the international level, some have argued that international agencies might take a greater institutional role. Lee and Dinar (1995) describe the importance of an integrated approach to river basin planning, development, and management. Young, Dooge, and Rodda (1994) provide guidelines for coordination among levels of management at the global, national, regional, and local levels. As far back as 1989, Delli Priscoli described the importance of public involvement in water conflict management, and Bruch and colleagues (2005) offer a current survey of the state of the art. In 1992, Delli Priscoli made a strong case for the potential of alternative dispute resolution (ADR) in the World Bank's handling of water resources issues. Trolldalen (1992) likewise chronicles environmental conflict resolution at the United Nations, including a chapter on international rivers.

4.3 DEVELOPMENTS IN INTERNATIONAL TRANSBOUNDARY WATER: CONTRIBUTIONS FROM THE INTERNATIONAL COMMUNITY[2]

Acknowledging the benefits of cooperative water management, the international community has long advocated institutional development in the world's international waterways and has focused considerable attention in the twentieth century on developing and refining principles of shared management. In 1911, the Institute of International Law published the Madrid Declaration on the International Regulation regarding the Use of International Watercourses for Purposes other than Navigation (Giordano and Wolf, 2003). The Madrid Declaration outlined certain basic principles of shared water management, recommending that coriparian states establish permanent joint commissions and discouraging unilateral basin alterations and harmful modifications of international rivers. Expanding on these guidelines, the International Law Association developed the Helsinki Rules of 1966 on the Uses of Waters of International Rivers. Since then, international freshwater law has matured through the work of these two organizations as well as the United Nations and other governmental and nongovernmental bodies (International Law Association, 1966).

The 1990s and early 2000s, however, have witnessed a perhaps unprecedented number of declarations, as well as organizational and legal developments to further the international community's objective of promoting cooperative river basin management. The decade began with the International Conference on Water and the Environment in the lead-up to the 1992 UN Conference on Environment and Development (UNCED) in Rio, referred to as the Rio Earth Summit. Subsequently, actions taken by the international community have included

---

[2] This section draws from Giordano and Wolf (2003).

the pronouncement of nonbinding conventions and declarations, the creation of global water institutions, and the codification of international water principles. Although more work is required, these initiatives have not only raised awareness of the myriad issues related to international water resource management but have also led to the creation of frameworks in which the issues can be addressed.

4.3.1 Conventions, declarations, and organizational developments

The 1992 UNCED served as a forum for world policy makers to discuss problems of the environment and development. Management of the world's water resources was only one of several topics addressed; however, water was the primary focus of the International Conference on Water and the Environment (ICWE), a preparatory conference held in advance of the Rio Earth Summit. The ICWE participants, representing governmental and nongovernmental organizations, developed a set of policy recommendations outlined in the Conference's Dublin Statement on Water and Sustainable Development, which the drafters entrusted to the world leaders gathering in Rio for translation into a plan of action (ICWE, 1992). Although it covers a range of water resource management issues, the Dublin Statement specifically highlights the growing importance of international transboundary water management and encourages greater attention to the creation and implementation of integrated water management institutions endorsed by all affected basin states. Moreover, the drafters outlined certain essential functions of international water institutions, including "reconciling and harmonizing the interests of riparian countries, monitoring water quantity and quality, development of concerted action programs, exchange of information, and enforcing agreements" (ICWE, 1992). The output has come to be called the Dublin Principles. These were arrived at through a highly interactive and facilitated process at the Dublin meeting. As a result the cross section of the world water community that was present has come to feel ownership and to embrace these principles. The Dublin Principles remain at the heart of the emerging world consensus on water resources management principles. They have held long after many of the numerous other conferences have come and gone.

At the Rio Conference, water resource management was specifically addressed in Chapter 18 of Agenda 21, a nonbinding action plan adopted by UNCED participants for improving the state of the globe's natural resources in the twenty-first century. The overall goal of Chapter 18 is to ensure that the supply and quality of water is sufficient to meet both human and ecological needs worldwide, and measures to implement this objective are detailed in the chapter's ambitious, seven-part action plan. Although transboundary water resource management is mentioned in Chapter 18, few specific and substantive references are made to water resource issues at the international scale. Indeed, the Dublin Principles as such were left out of the Rio Conference for technical diplomatic reasons. The Rio Conference did, however, generate a number of activities concerning freshwater management in general, with implications for international transboundary water management.

One result of the Rio Conference and Agenda 21 has been an expansion of international freshwater resource institutions and programs. The World Water Council, a self-described "think tank" for world water-resource issues created in 1996, has hosted World Water Forums every three years (see Biswas, 1995). These are gatherings of government, nongovernment, and private agency representatives to discuss and collectively determine a vision for the management of water resources over the next quarter century. These forums have led to the creation of the World Water Vision, a forward-looking declaration of philosophical and institutional water management needs, as well as the creation of coordinating and implementing agencies, such as the World Commission on Water for the Twenty-First Century and the Global Water Partnership. These forums also led to the Camdesus panel on financing; the results of which were discussed in the G-8 meetings of 2003 (World Water Council, 2003). The Second World Water Forum also served as the venue for a Ministerial Conference in which the leaders of participating countries signed a declaration concerning water security in the twenty-first century. Continued momentum of these recent global water initiatives is supported by a number of interim appraisal meetings to review actions taken since the Earth Summit. In the Johannesburg Declaration on Sustainable Development, for example, delegates at the World Summit on Sustainable Development (WSSD) reaffirmed a commitment to the principles contained in Agenda 21 and called on the United Nations to review, evaluate, and promote further implementation of this global action plan (United Nations, 2002a). Seven Millennium Development Goals (MDGs) have been identified and agreed to by most nations as targets to move toward. As the sidebar on MDG shows, however, water is critical to reaching each one of these beyond goal seven, which refers to water. Collectively they actually argue for more integration across the goals and there several implied uses of water resources (see Sidebar 4.1).

This conference also identified hydropower as a renewable resource and began to reconnect water's traditionally critical role to development within the sustainable development community. And although shared waters were *not* mentioned explicitly in official documents, implementation of the World Water Vision was assessed during the Third World Water Forum (WWFIII) held in Japan in 2003. The WWFIII and the Fourth World Water Forum (WWFIV) have broadened the world water debate to include discussions of floods and

Sidebar 4.1 Water Supply and Sanitation (WSS): Key Ingredient in Millennium Development Goals (MDGs)

Goal 1. Eradicate extreme poverty and hunger
WSS essential for improving quality of life – for health and economic development

Goal 2. Achieve universal primary education
WSS keeps children fit and underpins healthy school environment

Goal 3. Promote gender equality and empower women
WSS saves women's time and provides opportunities for women to lead

Goal 4. Reduce child mortality

Goal 5. Improve maternal health

Goal 6. Combat HIV/AIDS, malaria, and other diseases

Goal 7. Ensure environmental sustainability
2000–2015: Halve proportion of people without sustainable access to safe drinking water
2020: Have achieved a significant improvement in the lives of at least 100 million slum dwellers – access to improved sanitation

Goal 8. Develop a global partnership for development

Source: United Nations Development Programme, 2007.

disasters, hydropower, and navigation. WWFIV has also begun to provide for elected local officials and parliamentarians to discuss water in addition to the traditional fora for diplomats and ministers.

The large meetings have also spawned critiques, mainly by some donors and selected international NGOs, as becoming too large and numerous (see, for example, Varady and Iles Shih, 2005, and Biswas and Tortajada, 2006). At the same time many local NGOs, while also criticizing the meetings, are quick to add that it may be better to err on the side of too many than too few meetings because they have become places where local NGOs have been able to interact with the so-called world water elite and visa versa. Indeed the meetings have seemed to echo what has been happening within countries. There is increasing awareness that "water people" alone will not deal with these problems and that more and different types of stakeholders must be involved. In addition, politicians interact more with professionals. Indeed, for many years, the world water meeting were mostly run and attended by water professionals. The large meetings have begun to change this to reflect the broader trends occurring within countries. The WWFIV emphasized implementing local actions and spawned a movement to identify and monitor these actions. It remains to be seen whether the water professional community can accept this.

Through these meetings, the international community has reinforced its commitment to satisfy the water quality and quantity requirements of the global population and its surrounding environment and has identified attendant tasks and policy measures needed to fulfill its pledge. Although many of the strategies in Agenda 21 and subsequent statements are directed primarily at national water resources, their relevance extends to international transboundary waters. In fact, the Ministerial Declaration at the Second World Water Forum included "sharing water" (between different users and States) as one of its seven major challenges to achieving water security in the twenty-first century. Many of the other six challenges, which are meeting basic needs, securing the food supply, protecting the ecosystem, managing risks, valuing water, and governing water wisely, are also applicable to waters in an international setting. Furthermore, policy measures prescribed by the international community to build greater institutional capacity, such as integrated water resource management, expanded stakeholder participation, and improved monitoring and evaluation schemes, are likewise important components of international watercourse management.

Like Agenda 21, however, none of these post-Rio statements or declarations focuses exclusively on international freshwater sources. Thus, although many of the principles of national water management apply to international waters, the political, social, and economic dynamics associated with waters shared between sovereign States can require special consideration.

Nevertheless, after decades of institutional risk aversion and a general lack of leadership in international waters, the 1990s and 2000s are turning out to be a period of tremendous momentum on the ground as well: the World Bank and UNDP have collaborated to facilitate the Nile Basin Initiative, which looks close to establishing a treaty framework and development plan for the basin, and the Bank is taking the lead on bringing the riparians of the Guaraní Aquifer in Latin America to dialogue. The U.S. State Department, a number of UN agencies, and other parties have established a Global Alliance on Water Security, aimed at identifying priority regions for assistance, which may help countries get ahead of the crisis curve. The Global Environment Facility (GEF) is now active in fifty-five international basins. The UNECE has programs on ten European and Central Asian basins and supports the International Water Assessment Center. The Southern African Development Community and the Economic and Social Commission for Asia and the Pacific have been taking the lead in establishing transboundary dialogues within their respective regions. The International Network of Basin Organizations (INBO) has created a thriving network of those managing international and transboundary rivers, including a "twinning" program that brings together diverse basin managers to share experience and best practices. The International Water Academy has engaged researchers from around the world to address these difficult issues, as has the Universities Partnership for Transboundary Waters. And UNESCO and Green Cross International have teamed up for a broad-based, multiyear project called From Potential Conflict to Co-operation Potential. Both organizations are working with the Organization for Security and Cooperation in Europe on their project on international waters (United Nations Educational, Scientific and Cultural Organization – UNESCO-PCCP, 2007). Moreover, UNESCO is taking the lead in helping to develop a global "Water Cooperation Facility" to help prevent and resolve the world's water disputes.

4.3.2 International water law and treaty development

INTERNATIONAL LAW[3]

There is a vast and growing literature on international water law (see, for example, the excellent summaries by Wescoat, 1996; Salman and de Chazournes, 1998; McCaffrey, 1999; Wouters, 2000; McCaffrey, 2001; and Paisley, 2003). Patricia Wouters and her team at the University of Dundee have created a legal assessment model to help countries develop transboundary institutions (Wouters, 2003). According to Cano (1989, p. 168), international water law was not substantially formulated until after World War I. Since that time, organs of international law have tried to provide a framework for increasingly intensive water use, focusing on general guidelines that could be applied to the world's watersheds. These general principles of customary law, codified and progressively developed by advisory bodies and private organizations, are not intended to be legally binding in and of themselves but can provide evidence of customary law and may help crystallize that law. Wouters (personal communication, 2003) notes that "customary law is not soft law, even though it might be found in codification efforts of nongovernmental organizations (NGOs) or even the ILC rules of customary law are rules of international law and considered as sources." Although it is tempting to look to these principles for clear and binding rules, it is more accurate to think in terms of guidelines for the process of conflict resolution: "(T)he principles (of customary law) themselves derive from the process and the outcomes of the process rather than prescribe either the process or its outcome" (Dellapenna, personal communication, 1997).

The International Court of Justice (ICJ) refers to the following guidelines, in order of precedence, for its rulings (Cano, 1989; Rosenne, 1995):

1. The law of treaties and conventions ratified by governments
2. Customs

3. Generally accepted principles
4. Decisions of the judiciary and doctrines of qualified authors

How this works in practice can be complex. The concept of a "drainage basin," for example, was accepted by the International Law Association (ILA) in the Helsinki Rules of 1966, which also provides guidelines for "reasonable and equitable" sharing of a common waterway (International Law Association, 1966; Caponera, 1985). Article V lists no fewer than eleven factors that must be taken into account in defining what is "reasonable and equitable." The factors include a basin's geography, hydrology, climate, past and existing water utilization; economic and social needs of the riparians; population; comparative costs of alternative sources; availability of other sources; avoidance of waste; practicability of compensation as a means of adjusting conflicts; and the degree to which a state's needs may be satisfied without causing substantial injury to a cobasin state. There is no hierarchy to these components of "reasonable use"; rather, they are to be considered as a whole. One important shift in legal thinking in the Helsinki Rules is that they address the right to "beneficial use" of water rather than to water *per se* (International Law Association, 1966; Housen-Couriel, 1994, p. 10). The Helsinki Rules have been used explicitly only once to help define water use – the Mekong Committee used the definition of "reasonable and equitable use" in the formulation of their Declaration of Principles in 1975, although no specific allocations were determined. Although this is the sole case of the Helsinki Rules definitions being used explicitly in treaty text, the concept of "reasonable and equitable use" is quite common, as is described here.

When the United Nations considered the Helsinki Rules in 1970, objections were raised by some nations as to how inclusive the process of drafting had been. In addition and, according to Biswas (1993), more importantly, some States (including Brazil, Belgium, China, and France) objected to the prominence of the drainage basin approach, which might be interpreted as an infringement on a nation's sovereignty. Finland and the Netherlands argued that a watershed was the most "rational and scientific" unit to be managed. Others argued that, given the complexities and uniqueness of each watershed, general codification should not even be attempted. On December 8, 1970, the General Assembly directed its own legal advisory body, the International Law Commission (ILC) to study "Codification of the Law on Water Courses for Purposes other than Navigation."[4]

It is testimony to the difficulty of marrying legal and hydrologic intricacies that the ILC, despite an additional call for codification at the UN Water Conference at Mar de Plata in

---

[3] This section draws from Wolf (1997) and Giordano and Wolf (2003).

[4] In its reference to the ILC, the General Assembly excised all mention to the Helsinki Rules to allay political concerns over the drainage basin approach (Wescoat 1992, p. 307).

1977, took 21 years to complete its Draft Articles. It took until 1984, for example, for the term *international watercourse* to be adequately defined (a process described in exquisite detail by Wescoat, 1992; see also Teclaff, 1996). Problems both political and hydrological slowed the definition: in a 1974 questionnaire submitted to member states, about half of the 32 that responded by 1982 supported the concept of a drainage basin (e.g., Argentina, Finland, and the Netherlands), while half were strongly negative (e.g., Austria, Brazil, and Spain) or ambivalent (Wescoat, 1992, p. 311). For example, "watercourse system" connoted a basin, which threatened sovereignty issues, and borderline cases, such as glaciers and confined aquifers, both excluded at one point, had to be determined. In 1994, more than two decades after receiving its charge, the ILC adopted a set of thirty-two draft articles (United Nations, 1994). The UN General Assembly adopted the articles, with some revisions, as the Convention on the Law of the Non-Navigational Uses of International Watercourses on May 21, 1997 (Appendix A; United Nations, 2005). The vote was 103 in favor, 3 against (Turkey, China, and Burundi), and 27 abstentions. Much of the debate focused on issues such as the place of environmental sustainability, the degree to which the Convention affected past and future treaties, and the relationship between "reasonable and equitable use" and the "obligation not to commit harm," as will be explored in this section. See Tanzi (1997) for more detail. Wouters (personal communication, 2003) notes that there is marked distinction between the ILA's and the ILC's and the UN Convention's approach to the substantive rules that govern the legitimacy of new and existing uses.

The 1997 Convention includes language very similar to the Helsinki Rules (International Law Association, 1966), requiring riparian states along an international watercourse in general to communicate and cooperate. Provisions are included for exchange of data and information, notification of possible adverse effects, protection of ecosystems, and emergency situations. Allocations are dealt with through equally vague, but positive, language. Much of the discussions leading to the Convention centered on how "reasonable and equitable use" within each watercourse state, "with a view to attaining optimal utilization thereof and benefits therefrom," is balanced with an obligation not to cause significant harm (Tanzi, 1997). Reasonable and equitable use is defined similar to the Helsinki Rules, to be based on a nonexhaustive list of seven relevant factors. These factors are

- Geographic, hydrographic, hydrological, climatic, ecological, and other natural factors
- Social and economic needs of each riparian state
- Population dependent on the watercourse
- Effects of use in one state on the uses of other states

- Existing and potential uses
- Conservation, protection, development, and economy of use and the costs of measures taken to that effect
- The availability of alternatives, of corresponding value, to a particular planned or existing use

The text of the ILC articles does not offer guidelines for prioritizing these factors, suggesting in Article 6 only that "the weight to be given to each factor is to be determined by its importance" and that "all relevant factors are to be considered together." Article 10 says both that "in the absence of agreement or custom to the contrary, no use . . . enjoys inherent priority over other uses" and that "in the event of a conflict between uses . . . [it shall be resolved] with special regard being given to the requirements of vital human needs."

UN CONVENTION

The UN Convention (see Appendix A and United Nations, 2005) codifies many of the principles deemed essential by the international community for the management of shared water resources, such as equitable and reasonable utilization of waters with specific attention to vital human needs; protection of the aquatic environment; and the promotion of cooperative management mechanisms. The document also incorporates provisions concerning data and information exchange and mechanisms for conflict resolution. Once ratified, the UN Convention would provide a legally binding framework, at least upon its signatories, for managing international watercourses. Even without ratification, its guidelines are being increasingly invoked in international forums, and it was cited as evidence of customary law in 1997 by the International Court of Justice in the Danube case, mentioned here, even without technically being in force.

The approval of the Convention by UN member states, however, does not entirely resolve many legal questions concerning the management of internationally shared waters. First, the Convention would technically be binding only on those nations that have ratified or consented to be bound by the agreement. To date, only fourteen countries are party to the UN Convention, well below the requisite thirty-five instruments of ratification, acceptance, accession, or approval needed to bring the Convention into force (United Nations, 2002b). As of January 2006, Finland, Hungary, Iraq, Jordan, Lebanon, Libya, Namibia, the Netherlands, Norway, Portugal, Qatar, South Africa, Sweden, and Syria were party to the Convention. Second, international law only guides conduct between sovereign nations. Thus, grievances of political units or ethnic groups within nations over the domestic management of international waterways would not be addressed. Third, although the Convention offers general guidance to coriparian states,

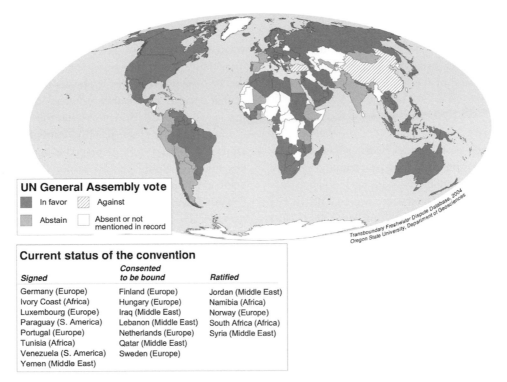

UN General Assembly vote

In favor | Against
Abstain | Absent or not mentioned in record

Current status of the convention

Signed	Consented to be bound	Ratified
Germany (Europe)	Finland (Europe)	Jordan (Middle East)
Ivory Coast (Africa)	Hungary (Europe)	Namibia (Africa)
Luxembourg (Europe)	Iraq (Middle East)	Norway (Europe)
Paraguay (S. America)	Lebanon (Middle East)	South Africa (Africa)
Portugal (Europe)	Netherlands (Europe)	Syria (Middle East)
Tunisia (Africa)	Qatar (Middle East)	
Venezuela (S. America)	Sweden (Europe)	
Yemen (Middle East)		

Figure 4.3 Map of United Nations General Assembly votes on the 1997 Convention on the Law of Non-navigational Uses of International Watercourses (Transboundary Freshwater Dispute Database, 2004).

its vague and occasionally contradictory language can result in varied, and, indeed, conflictive interpretations of the principles contained therein. As stated by Biswas (1999), the "vague, broad, and general terms" incorporated in the UN Convention "can be defined, and in certain cases quantified, in a variety of different ways." Fourth, there is no practical enforcement mechanism with which to back up the Convention's guidance. The International Court of Justice, which came into being in 1946, with the dissolution of its predecessor, the Permanent Court of International Justice, hears cases only with the consent of the parties involved and only on very specific legal points. Although the earlier body did rule on four international water disputes during its existence, from 1922 to 1946, the ICJ has decided only one case in its 55-year history (apart from those related to boundary definitional disputes) pertinent to international waters – that of the Gabçíkovo–Nagymaros Project on the Danube between Hungary and Slovakia in 1997. And in that case the court essentially told the parties that they each had committed errors and they needed to negotiate outside of court among themselves for resolution. Finally, the Convention addresses only those groundwater bodies that are connected to surface water systems (i.e., unconfined aquifers), yet several nations are already beginning to tap into confined groundwater systems, many of which are shared across international boundaries. Nevertheless, and despite the fact that the process

of ratification is moving extremely slowly, the Convention's common acceptance, and the fact that the International Court of Justice referred to it in its decision on the 1997 case on the Gabçikovo Dam, gives the Convention increasing standing as an instrument of customary law (Figure 4.3).

The Bellagio Draft Treaty, an early document focusing on groundwater, was developed as a document of "soft law" in a process described by Hayton and Utton (1989, p. 677). Like the UN Convention, it includes eight factors for consideration in allocations – (1) hydrogeology and meteorology, (2) existing and planned uses, (3) environmental sensitivity, (4) quality control requirements, (5) socioeconomic implications, (7) water conservation practices, (7) artificial recharge potential, and (8) comparative costs and implications of alternative sources of supply – and suggests that "the weight to be given to each factor is to be determined by its importance in comparison with that of the other relevant factors" (Hayton and Utton, 1989, p. 677). In separate comments, Hayton and Utton suggest that a Commission, established under treaty, should also consider the traditional rights of nomadic or tribal peoples of a border region. In 2002, the ILC took up the issue of shared groundwater resources and began a process of codification similar to the 1997 Convention. This process will present its own challenges, not least of which are the many types of aquifers (Eckstein and Eckstein, 2003, 2005, define six challenges, each with its own legal complication), and the scientific uncertainty (often spanning several orders of magnitude) inherent in the medium. Eckstein (2004) describes the issues being considered by the ILC, as well as its UNESCO-arranged advisory panel. (The current version of the ILC rules for groundwater is included in Appendix A.)

SUMMARY

The uniqueness of each basin and its riparian states suggest that any universal set of principles must, by necessity, be fairly general. Problems arise when attempts are made to apply this reasonable but vague language to specific water conflicts. For example, riparian positions and consequent legal rights shift with changing boundaries, many of which are still not recognized by the world community. Furthermore, international law only concerns itself with the rights and responsibilities of nation-states. Some political entities who might claim water rights, therefore, would not be represented, such as the Palestinians along the Jordan or the Kurds along the Euphrates. Dellapenna (personal communication, 1997) points out that there are differences between these two examples, however, in that the Palestinians do have some degree of autonomy and even sovereignty within their territory. He uses the term *national communities* for the riparians of the Jordan River to make this distinction.

Within nation-states, the same type of conflict between geography and jurisdictions and legal procedures exists. In the United States, water use allocation has been set, project by project, through various Congressional legislations. However, the demographic as well as the needs have changed over time. Nevertheless, it is very difficult to change the allocation or to reapportion the uses to fit new realities in the face of old legal precedents. This is especially true because states within the nation-state of the United States have sovereignty over the water.

The process is further complicated in the rare cases of formal litigation or arbitration – there are few specialized institutions for international law making, interpreting, or enforcing. The International Court of Justice in The Hague, for example, hears cases only on specific points of law and only with the consent of the parties involved, and there is no practical enforcement mechanism to back up the Court's findings. A nation-state with pressing national interests can therefore disclaim entirely the Court's jurisdiction or findings (Rosenne, 1995).

4.3.3 Rights-based criteria: Hydrography versus chronology

EXTREME PRINCIPLES

Customary international law has focused on providing general guidelines for the watersheds of the world. In the absence of such guidelines, some principles have been claimed regularly by riparians in negotiations, often depending on where along a watershed a riparian nation-state is situated. Many of the common claims for water rights are based either on hydrography, that is, from where a river or aquifer originates and how much of that territory falls within a certain state, or on chronology, that is, who has been using the water the longest (Molle, 2004).

Initial positions are usually extreme (Matthews, 1984; Housen-Couriel, 1994). As noted in Chapter 2, the "doctrine of absolute sovereignty" or the Harmon Doctrine is often initially claimed by an upstream riparian. According to U.S. Attorney General Harmon in 1895, "The fundamental principle of international law is the absolute sovereignty of every nation, as against all others, within its own Territory" (cited in LeMarquand, 1993, p. 63). Harmon was making the hydrologically preposterous argument that upstream water diversions within the territorial United States would not legally affect downstream navigation on international stretches of the Rio Grande because the diversions were to be carried out by individuals, not states (McCaffrey, 1997). Considering that this doctrine was immediately rejected by Harmon's successor and later officially repudiated by the United States (McCaffrey, 1996a, 1996b), it was never implemented in any water treaty (with the rare exception of some internal tributaries of international waters), was not invoked as a sources for judgment in any international water legal ruling, and was explicitly rejected by the international tribunal over the Lac Lanoux case in 1957 (described in the next section), the Harmon Doctrine is wildly overemphasized as a principle of international law. As far back as 1911, the Institut de Droit International had asserted that the dependence of riparian states on each other precludes the idea of absolute autonomy over shared waters (Laylin and Bianchi, 1959, p. 46).

The downstream extreme claim often depends on climate. In a humid watershed, the extreme principle advanced is "the doctrine of absolute riverain integrity," which suggests that every riparian is entitled to the natural flow of a river system crossing its borders. This principle has reached acceptance in the international setting as rarely as has absolute sovereignty. In an arid or exotic (i.e., humid headwaters region with an arid downstream) watershed, the downstream riparian often has older water infrastructure, which is in its interest to defend. The principle that rights are acquired through older use is referred to as "historic rights" (or "prior appropriation" in the United States), that is, "first in time, first in right."

These conflicting doctrines of hydrography and chronology clash along many international rivers, with positions usually defined by relative riparian positions. The inherent conflict between upstream and downstream riparian occurs in most settings and scales. Crawford (1988, pp. 88–90) describes such disputes along the traditional *acequia* canal systems in New

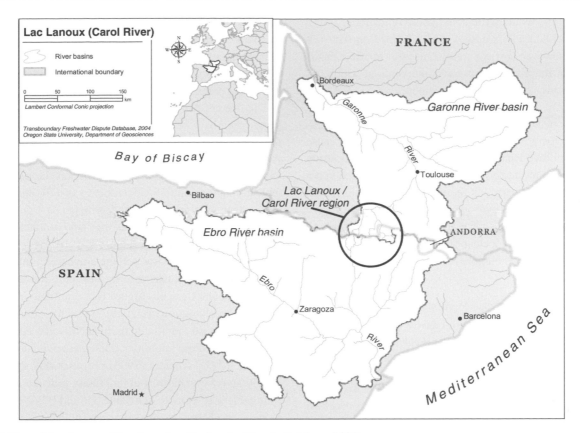

Figure 4.4 Map of Lac Lanoux (Transboundary Freshwater Dispute Database, 2004).

Mexico. Downstream riparians, such as Iraq and Egypt, often receive less rainfall than their upstream neighbors, such as Ethiopia and Turkey, and therefore have historically depended on river water. As a consequence, modern "rights-based" disputes often take the form of upstream riparians arguing in favor of the doctrine of absolute sovereignty, with downstream riparians taking the position of historic rights. For examples of these respective positions, see the exchange between Jovanovic (1985, 1986a, 1986b) and Shahin (1986) about the Nile and the description of political claims along the Euphrates in Kolars and Mitchell (1991) and in Kibaroğlu (2002a).

MODERATED PRINCIPLES

It quickly becomes clear in negotiations that keeping to an extreme position leaves very little room for bargaining. Over time, rights become balanced by responsibility such that most States eventually accept some limitation to both their own sovereignty and to the river's absolute integrity. The process that led to the disavowal of the legal principles of absolute sovereignty and absolute riverain integrity was the Lac Lanoux case (Laylin and Bianchi, 1959; MacChesney, 1959). The Carol River crosses from the French Pyrenees into the Spanish Pyrenees (Figure 4.4). In the early 1950s, France, asserting absolute sovereignty, proposed diverting water from the river

across a divide toward the Font-Vive for hydropower generation, with monetary compensation for Spain. Spain objected, asserting absolute riverain integrity and the existing irrigation needs on its side of the border. Even when France agreed to first divert back the water needed for Spanish irrigation, then *all* of the water being diverted, through a tunnel between watersheds, Spain insisted on absolute riverain integrity, claiming it did not want French hands on its tap. This concern is raised regularly in negotiations, for example, between Egypt and Ethiopia and for a series of proposed canals from Turkey or Lebanon into the Jordan basin. It is primarily this concern that causes Israel to emphasize desalination over possibly less-expensive water import schemes.

In the Lac Lanoux case, both absolute principles were effectively dismissed when a 1957 arbitration tribunal ruled that "territorial sovereignty...must bend before all international obligations," effectively negating the doctrine of absolute sovereignty. Yet the tribunal also admonished the downstream state from the right to veto "reasonable" upstream development, thereby negating the principle of natural flow or absolute riverain integrity. This decision made possible the 1958 Lac Lanoux treaty (revised in 1970), in which it is agreed that water is diverted out of basin for French hydropower generation, and a similar quantity is returned before the stream reaches

Spanish territory. Wouters (personal communications, 2003), notes that the Lac Lanoux case was *not* decided on issues of sovereignty, however, it was a treaty law case. In effect this is an example of the use of third-party intervention, which has links to new forms of negotiated agreements.

The "doctrine of limited territorial sovereignty" reflects rights to reasonably use the waters of an international waterway, yet with the acknowledgment that one should not cause harm to any other riparian state. In fact, the relationship between "reasonable and equitable use" and the obligation not to cause "significant harm" is the more subtle manifestation of the argument between hydrography and chronology. As we have noted, the 1997 Convention includes provisions for both concepts, without setting a clear priority between the two (see Appendix A and United Nations, 2005). The relevant articles are

Article 5: Equitable and reasonable utilization and participation

1. Watercourse States shall in their respective territories utilize an international watercourse in an equitable and reasonable manner. In particular, an international watercourse shall be used and developed by watercourse States with a view to attaining optimal and sustainable utilization thereof and benefits therefrom, taking into account the interests of the watercourse States concerned, consistent with adequate protection of the watercourse.
2. Watercourse States shall participate in the use, development, and protection of an international watercourse in an equitable and reasonable manner. Such participation includes both the right to utilize the watercourse and the duty to cooperate in the protection and development thereof, as provided in the present Convention.

Article 7: Obligation not to cause significant harm

1. Watercourse States shall, in utilizing an international watercourse in their territories, take all appropriate measures to prevent the causing of significant harm to other watercourse States.
2. Where significant harm nevertheless is caused to another watercourse State, the States whose use causes such harm shall, in the absence of agreement to such use, take all appropriate measures, having due regard for the provisions of articles 5 and 6, in consultation with the affected State, to eliminate or mitigate such harm and, where appropriate, to discuss the question of compensation.

Article 10: Relationship between different kinds of uses

1. In the absence of agreement or custom to the contrary, no use of an international watercourse enjoys inherent priority over other uses.
2. In the event of a conflict between uses of an international watercourse, it shall be resolved with reference to the principles and factors set out in articles 5 to 7, with special regard being given to the requirements of vital human needs.

Not surprisingly, upstream riparians have advocated that the emphasis between the two principles be on "equitable utilization," because that principle gives the needs of the present the same weight as those of the past. Likewise, downstream riparians (along with the environmental and development communities) have pushed for emphasis on "no significant harm," effectively the equivalent of the doctrine of historic rights in protecting preexisting use.

The debate over which doctrine, "reasonable use" or "no harm" shall have priority has been intense and was one of the focuses of discussion leading to the Convention (Tanzi, 1997). According to Khassawneh (1995, p. 24), the Special Rapporteurs for the ILC project came down on the side of "equitable utilization" until the incumbency of J. Evensen, the third Rapporteur who argued for the primacy of "no appreciable harm." Commentators have had the same problem reconciling the concepts as the Rapporteurs. Khassawneh (1995, p. 24) suggests that the latter Rapporteurs are correct that "no appreciable harm" should take priority, while, in the same volume, Dellapenna (1995, p. 66) argues for "equitable use" and suggests that the evolution of Article 7 (which, in the Convention, includes a clause to mitigate harm and discuss compensation) is evidence of these intentions (Dellapenna, personal communication, 1997). Wouters (1996) proposes that the ILC Draft clearly favors "no harm" but that treaty practice suggests that "equitable use" is more advisable. Utton (1996) describes the roots of "no harm" more as a water quality issue and advises that the Convention be written as such. The World Bank, which must follow prevailing principles of international law in its funded projects, recognizes the importance of equitable use in theory but, for practical considerations, gives "no appreciable harm" precedent – it is considered easier to define – and will not finance a project that causes harm without the approval of all affected riparians (see World Bank, 1993, p. 120; and Krishna, 1995, pp. 43–45).

This debate actually has parallels to the debate of water professionals over which way is the best to manage the resources in the public interest. Like the equitable case, concept planners would say the best way is to look at the comprehensiveness and integration of the whole system. Others would approach the river primarily through regulations, which use principles and would look at case-by-case at impacts.

Even as the principles for sharing scarce water resources evolve and become more moderate over time, the essential argument still emphasizes the *rights* of each state – the sense that a riparian is entitled to a certain quantity or use of water depending on certain physical or historical constructs – generally resting on the fundamental dispute between hydrography and chronology. In addition, defining concepts that are intentionally vague both for reasons of legal interpretation and for political expediency – "reasonable," "equitable," and "significant" – guarantee continued ambiguity in the principles of customary law. McCaffrey (2001) suggests that there are three

Figure 4.5 Photo of ancient treaty on clay tablet. The history of international water treaties dates as far back as 2500 BC, when the two Sumerian city-states of Lagash and Umma crafted an agreement related to the Tigris River, ending the only true "water war" in history. Photo credit: András Szöllösi-Nagy.

main general principles of the customary law of international watercourses that are widely accepted:

- Equitable and reasonable utilization
- Prevention of significant harm
- Prior notification of potentially harmful planned activities

He also suggests that an emerging principle is the protection of ecosystems of international watercourses from harm through pollution and other human activities.

4.4 INTERNATIONAL WATER TREATIES: PRACTICE[5]

At the heart of water conflict management is the question of "equity." A vague and relative term in any event, criteria for equity are particularly difficult to determine in water conflicts, where international water law can be seen as ambiguous and contradictory and no mechanism exists through which to enforce agreed-on principles (Van der Zaag, Seyam, and Savenije, 2002; Naff and Dellapenna, 2002). However, the application of an "equitable" water-sharing agreement along the volatile waterways of the world is a prerequisite to hydropolitical stability, which could help propel political forces away from conflict in favor of cooperation. In addition to the efforts of the international community, riparian States have themselves developed a rich history of treaties concerning the man-

agement of shared watercourses. In contrast with the vague and sometimes contradictory global declarations and principles, the institutions developed by coriparian nations have been able to focus on specific basin-level conditions and concerns. An evaluation of these institutions over the past half-century, with particular attention to treaties signed since the Rio Conference, offers insights into how appropriately the emphasis areas highlighted in Agenda 21, and subsequent declarations and conventions on freshwater resource management address the needs of international transboundary waters specifically.

The literature includes very little systematic work on the body of international water treaties as a whole, although authors have often used treaty examples to make a point about specific conflicts, areas of cooperation, or larger issues of water law (see, for example, Vlachos, 1990; Eaton and Eaton, 1994; Housen-Couriel, 1994; Dellapenna, 1995; Kliot, 1995; Kliot, Shmueli, and Shamir 1999; Dinar, 2004; Dinar and Dinar, 2003; Fischhendler, 2004, Fischhendler and Feitelson, 2005; and Conca, 2006). In two important exceptions, Dellapenna (1994) describes the evolution of treaty practice dating back to the mid-1800s, and Wescoat (1996) assesses historic trends of water treaties dating from 1648 to 1948 in a global perspective. Furthermore, the reports of the ILC Rapporteurs and related commentaries provide rich assessments of water treaty practice.

The history of international water treaties dates as far back as 2500 BC, when the two Sumerian city-states of Lagash and Umma crafted an agreement related to the Tigris River, ending

[5] This section draws from Giordano and Wolf (2003).

the only true "water war" in history (Figure 4.5). Since then, a large body of water treaties has emerged. The Food and Agriculture Organization of the United Nations has identified more than 3,600 treaties dating from AD 805 to 1984 (United Nations Food and Agriculture Organization, 1978; United Nations Food and Agriculture Organization, 1984; Wolf, 1998). While the majority of these treaties relate to some aspect of navigation, a growing number address nonnavigational issues of water management, including flood control, hydropower projects, or allocations for consumptive or nonconsumptive uses in international basins. Since 1820, more than 400 water treaties and other water-related agreements have been signed, more than half of which were concluded in just the past 50 years (TFDD, 2006).

Despite their growth in numbers, however, a review of treaties from the past half-century reveals an overall lack of robustness. Water allocations, for example, the most conflictive issue area between coriparian states, are seldom clearly delineated in water accords. Moreover, in the treaties that do specify quantities, allocations are often in fixed amounts, thus ignoring hydrologic variation and changing values and needs. Information on 145 of these treaties is summarized in Table 4.1.

Despite increasing sophistication, this survey suggests that the legal management of transboundary rivers is still in its conceptual infancy (Molle, 2004). Almost half of these 145 treaties have no monitoring provisions and, perhaps as a consequence, two-thirds do not delineate specific allocations and four-fifths have no enforcement mechanism. Moreover, the treaties that do specify quantities allocate a fixed amount to all riparian nations but one and that one nation must then accept the balance of the river flow, regardless of fluctuations. Finally, multilateral basins are, almost without exception, governed by bilateral treaties, precluding the integrated basin management long-advocated by water managers.

The treaty record is replete with agreements that do not allow for the vagaries of nature and the scientific unknown, misunderstandings that often lead to tense political standoffs. For example, the waters of the Colorado were already overallocated between the upper and lower U.S. states when a treaty with Mexico was signed in 1944, which also neglected the entire issue of water quality. After legal posturing on both sides as water quality continued to degrade, the United States subsequently built a massive desalination plant at the border so the water delivered would at least be usable. Currently, the fact that shared groundwater is likewise not covered in the treaty is leading to its share of tensions between the two nations (Kenney, 2005).

In December 1996, a treaty between India and Bangladesh was finally signed, allocating their shared Ganges waters after more than 35 years of dispute. In April 1997, however – the

Table 4.1 *International treaty statistics summary sheet*

Signatories	Bilateral 124/145 (86%)
	Multilateral 21/145 (14%)
Principal focus	Water supply 53/145 (37%)
	Hydropower 57/145 (39%)
	Flood control 13/145 (9%)
	Industrial uses 9/145 (6%)
	Navigation 6/145 (4%)
	Pollution 6/145 (4%)
	Fishing 1/145 (<1%)
Monitoring	Provided 78/145 (54%)
	None/not available 67/145 (46%)
Conflict resolution	Council 43/145 (30%)
	Other governmental unit 9/145 (6%)
	United Nations/third party 14/145 (10%)
	None/not available 79/145 (54%)
Enforcement	Council 26/145 (18%)
	Force 2/145 (1%)
	Economic 1/145 (<1%)
	None/not available 116/145 (80%)
Unequal power relationship	Yes 52/145 (36%)
	No/unclear 93/145 (64%)
Information sharing	Yes 93/145 (64%)
	No/not available 52/145 (36%)
Water allocation	Equal portions 15/145 (10%)
	Complex but clear 39/145 (27%)
	Unclear 14/145 (10%)
	None/not available 77/145 (53%)
Nonwater linkages	Money 44/145 (30%)
	Land 6/145 (4%)
	Political concessions 2/145 (1%)
	Other linkages 10/145 (7%)
	No linkages 83/145 (57%)

Source: Hamner and Wolf (1998).

very first season following signing of the treaty – the two countries were involved in their first conflict over cross-boundary flow: water passing through the Farakka Dam dropped below the minimum provided in the treaty, prompting Bangladesh to insist on a full review of the state of the watershed (Bandyopadhayay, 2002; Swain, 2004).

In 1994, Israel and Jordan signed one of the most creative water treaties on record. The treaty has Jordan store winter runoff in the only major surface reservoir in the region – the Sea of Galilee – even though that lake happens to be in Israel; it allows Israel to lease from Jordan, in 50-year increments, the wells and agricultural land on which Israel has come to rely; and it created a joint water committee to manage the

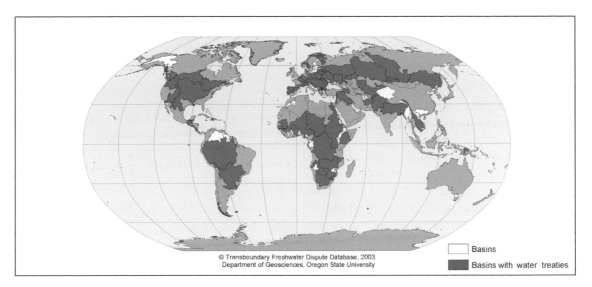

Figure 4.6 Map of international river basins with existing or historical water agreements (Transboundary Freshwater Dispute Database, 2003).

shared resources. But the treaty did not adequately describe what would happen to the prescribed allocations in a drought. In early 1999, this excluded issue roared into prominence with a vengeance, as the worst drought on record caused Israel to threaten to renege on its delivery schedule, which in turn caused protests in the streets of Amman, personal outrage on the part of the King of Jordan, and, according to some, threatened the very stability of peace between the two nations before a resolution was found (Shmueli and Shamir, 2001; Fischhendler, 2008).

Likewise, water quality provisions have played only a minor role in coriparian agreements historically. Enforcement mechanisms are also absent in a large percentage of the treaties. Finally, international basins with water agreements remain in the minority. Formal management institutions have been established in only 117 of the 263 international basins (see Figure 4.6) and even within these, few include all nations riparian to the affected basins. This precludes the integrated basin management advocated by the international community.

More encouraging characteristics are the inclusion of information sharing, monitoring, and conflict management/ resolution provisions in many of the past half-century's treaties. In addition, there has been a broadening in the definition and measurement of basin benefits. Traditionally, coriparians have focused on water as a commodity to be divided – a zero-sum, rights-based approach. Precedents now exist for determining formulas that equitably allocate the benefits derived from the use of water, not the water itself – a positive-sum, integrative approach. For example, as part of the 1961 Columbia River Treaty, the United States paid Canada for the benefits of flood control and Canada was granted rights to divert

water between the Columbia and Kootenai for hydropower purposes. Similarly, a 1975 Mekong River agreement among the four lower riparian states of Laos, Vietnam, Cambodia, and Vietnam defined "equality of right" not as equal shares of water but as equal rights to use water on the basis of each riparian's economic and social needs (Wolf, 1999a). (In the context of navigation, the 1995 Mekong River agreement, which superseded the 1975 agreement, again referenced, but in this case did not define, the concept of "equality of right.") In other words, the parties have created benefits jointly that have not existed before. They have gone beyond fighting over allocating flows to creating joint benefits, which, in the aggregate, will be greater for each than if they had pursued traditional zero-sum approaches. In negotiations language, water allowed them to expand the pie.

A review of treaties signed within the past 10 years also reveals some encouraging developments (Giordano and Wolf, 2003). At least fifty-four new bilateral and multilateral water agreements have been concluded since the Rio Conference, representing basins in Asia, Africa, Europe, North America, and South America. Since the mid-1950s European water accords have continued to dominate; however, agreements from other regions, in particular Asia, have grown disproportionately. The fact that agreements representing European basins dominate the treaty record is not surprising, given that Europe has the largest number of international basins (69) followed by Africa (59), Asia (57), North America (40), and South America (38) (Wolf et al., 1999; UNEP and OSU, 2002). In addition to greater geographic representation, a number of improvements can be seen in this more recent set of treaties, compared with the past half-century as a whole. First, a growing percentage of treaties address some aspect of water quality, a finding consistent with Rio's goal of both managing and protecting freshwater resources. Second, provisions concerning monitoring and evaluation, data exchange, and conflict resolution are included in many of the post-Rio treaties. Third, a number of agreements establish joint water commissions with decision-making and/or enforcement powers, a significant departure from the traditional advisory standing of basin commissions. Fourth, country participation in basin-level accords appears to be expanding. Although few of the agreements incorporate all basin states, a greater proportion of treaties are multilateral and many incorporate all major hydraulic contributors. Finally, although the exception, a 1998 agreement on the Syr Darya Basin, in which water management is exchanged for fossil fuels, provides a post-Rio example of basin states broadly capitalizing on their shared resource interests.

Institutional vulnerabilities still exist in a number of key areas, however. Many treaties, for example, ignore issues of allocation, and of those that do include allocative measures, few possess the flexibility to handle changes in the hydrologic regime or in regional values. References to water quality, related groundwater systems, monitoring and evaluation, and conflict resolution mechanisms, while growing in numbers, are often weak in actual substance. Furthermore, enforcement measures and public participation, two elements that can greatly enhance the resiliency of institutions are largely overlooked.

4.4.1 International water treaties: Allocations/quantity

Wolf (1999) studied the 145 treaties in the Transboundary Freshwater Dispute Database (TFDD, 2006) and culled those that had an explicit mechanism for allocating water between two or more nations. He excluded treaties that established basin authorities or described specific flood control or hydro-electricity projects *if* specific allocations are not described. For example, the 1957 accord that establishes the Mekong Committee was excluded from the study, but a 1975 Declaration of Principles among the same riparians, which describes principles for water allocations, was included. Of the collection of 145 treaties, he found that 49 described allocations for consumptive or nonconsumptive uses. Those treaties with water allocations generally came about in conjunction with boundary waters agreements, river development agreements, and/or single-project agreements (see the summary of treaties in Appendix G).

What is noticeable in reading through the *practice* of water conflict prevention and resolution, as documented in the forty-

Table 4.2 *Unique allocation practice*

Principle	Treaties ($n = 145$)	
	Number	Percentage
Half of flow to each of two riparians	9	6
Absolute sovereignty on tributaries	3	2
Relinquish prior uses	1	<1
Prioritize uses	4	3
Equal allocations of benefits	2	1
Compensation for lost benefits	10	7
Payments for water	4	3

Source: Wolf, 1999.

nine treaties listed in Appendix G, is just how rarely the general *principles* are explicitly invoked, particularly the extreme principles of absolute sovereignty or absolute riverain integrity. Neither of these principles is encoded in a single one of the documents surveyed in the 1999 study (Wolf, 1999). Some have pointed out that the fact that extreme principles are not invoked is *precisely* evidence that "equitable utilization" is the dominant underlying principle. Although it may be true that for an agreement to be reached, both sides have to see some degree of "equity" in an arrangement, its legal definition seems overly vague and relies too heavily on approval by the parties themselves. The argument that a normative principle needs be defined in the application of that principle feels somewhat circular. Furthermore, examination of the process details of our in-depth case studies in Appendix C reveals that these legal principles simply are not invoked in the process leading up to a treaty.[6] Rather than building from the legal principles, technocrats generally enlist lawyers late in the process to help codify water management practices, based primarily on the hydrologic and political landscape.

Each local setting is so diverse, both hydrologically and politically, that one is struck by the creativity of the negotiators in addressing specific code to each very specific situation (Table 4.2). As will be explored below, some divide *waters* equally between riparians and some divide the *benefits* derived from the waters equally – which is not at all the same thing. Most favor existing uses and guarantees to downstream riparians; the upstream riparian is favored only rarely. Each has sections that address the specific setting and concerns of local geography. The trends found in these treaties, as documented in Appendices D, E, and F are described in the following sections.

---

[6] The exception covered in the river basin case studies in Appendix C is the 1995 Mekong Agreement, probably because it is the only case where the mediator/facilitator, George Radosevich, is himself an international lawyer.

4.4.2 International water treaties – quality

Water quality has also been dealt with in treaties, but much less extensively than quantity. Meredith Giordano (2002) reviewed 180 treaties and 49 U.S. compacts in the Transboundary Freshwater Dispute Database, while searching for those mentioning "water quality." She then classified them into one of three groups depending on the amount of detail devoted to the issue of water quality as follows: (1) agreements with either explicit water quality standards or an established framework for water quality management; (2) agreements that reference general water quality objectives or programs; and (3) agreements that include an indefinite commitment to abate, control, or prevent water pollution (see Appendix E).

She found that 63 (28 percent) of the 228 agreements contained references to water quality. Seven were classified as Category One agreements (explicit standards), forty as Category Two (general objectives), and sixteen as Category Three (vague commitments). Her findings are presented here.

CATEGORY ONE: EXPLICIT STANDARDS

Four international treaties and two U.S. interstate compacts comprise the first category of water quality related treaties. Of the four international treaties, the 1978 Great Lakes Water Quality Agreement, is the broadest in terms of scope and provides the greatest detail concerning water quality standards. The 1972 and 1973 agreements between the United States and Mexico, while much narrower in scope, contain specific guidelines to reduce the salinity of Colorado waters entering Mexico. The 1994 Convention on the Cooperation for the Sustainable Use of the Danube River, like the Great Lakes Agreement, covers a range of issues related to water quality and its management and outlines a number of cooperative measures to protect the Danube waters. However, rather than defining specific standards, the Convention provides a general framework from which the signatories can devise appropriate water quality objectives and criteria. Of the three U.S. interstate compacts included in this first category, the 1941 Interstate Sanitation Commission, one of the oldest compacts addressing water quality, provides the most complete set of effluent standards. The other two compacts, 1938 Rio Grande Compact and 1948 Ohio River Valley Water Sanitation Compact, each set standards related to particular substances (e.g., suspended solids and sodium).

CATEGORY TWO: GENERAL OBJECTIVES

The majority of the documents reviewed fall into this second category of agreements, those that reference general objectives or programs related to water quality. Included in this category are the remaining nine interstate compacts and 31 of the 53

international treaties containing water quality provisions. The dates of these agreements span nearly the entire twentieth century and the international treaties relate to basins located in Asia, Africa, the Middle East, and Europe. The signatories to these documents agree to certain water quality goals and in many cases broadly describe measures, to be undertaken individually or jointly, to manage the quality of their shared waters. When mentioned, the details of the water quality measures outlined are entrusted to the contracting parties for further negotiations and consultations, often with the assistance of existing or newly created water commissions.

CATEGORY THREE: INDEFINITE COMMITMENTS

Category Three includes documents containing only vague references to pollution abatement, prevention, and control. Although similarities exist between the category two and three agreements, those placed in the latter category are, in general, less specific in nature and do not describe measures to achieve the stated water quality objectives. Included in this category are sixteen international water treaties drafted throughout the twentieth century and representing a wide range of geographic regions. Although the references to water quality in the Category Three agreements are generally brief, many of the treaties, like those in the previous category, include commitments by the respective signatories to further coordinate water-quality management efforts (Giordano, 2002).

4.4.3 International water treaties: Groundwater

With one-third of the world's fresh water being stored in the underground pore and fracture space known as aquifers (Shiklomanov, 1993), groundwater provides a critical source of potential water supplies, now accounting for all water uses in some parts of the world (Puri et al., 2001). Yet, precisely because it is found at varying depths underground, in rock strata of varying complexity, and in storage for varying periods of time, groundwater has certain characteristics that make its management intrinsically difficult (Puri, 2003). Its depth makes bringing it to the user tremendously expensive – the deeper the costlier. Because of its long contact with surrounding rocks, it is often saline and occasionally unusable without extensive treatment. Its diffusion throughout the substrate makes it tremendously difficult, if not impossible, to clean up once polluted. But mostly, the uncertainties associated with trying to understand flow through complex underground structure, regularly reaching five or ten orders of magnitude, can lead to extreme barriers in efficient management. Add an overlying political boundary into the mix, and it is no wonder that transboundary groundwater management has been called

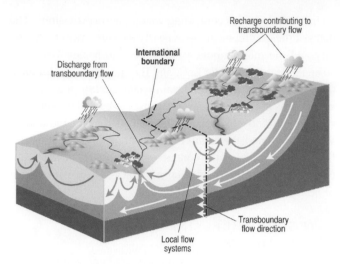

Figure 4.7 Schematic illustration of a transboundary aquifer (Puri et al., 2001).

"as close to witchcraft as we get in the sciences" (Bahr, 1988).

International aquifers can range in surface area from several hundred square kilometers to tens of thousands of square kilometers. As depicted on Figure 4.7, the principal component of transboundary aquifers is subsurface flow that is intersected by an international boundary. Water transfers from one side of the boundary to the other may occur naturally or due to capture by wells located on one side of the boundary. In many cases, the aquifer might receive most of its recharge on one side then most of its discharge would occur on the other side. The subsurface flow system at the international boundary itself can be visualized to include regional, as well as local movement of water (Puri et al., 2001; Matsumoto, 2002; Eckstein and Eckstein, 2003; Jarvis, 2006).

Matsumoto (2002) documented the evolution of the international community's attempts to deal with transboundary groundwater, as shown in Table 4.3. Given the complexities of the subject, it is no wonder that little beyond general principles of cooperation have been spelled out in detail. The most detailed work on the topic is being carried out by the International Shared Aquifer Resources Management (ISARM) program, initiated by UNESCO in 2000. ISARM anticipates publishing an inventory of transboundary aquifer systems in 2006. Transboundary aquifer systems currently being assessed by ISARM include the following:

- Guaraní Aquifer (South America)
- Nubian Sandstone Aquifers (Northern Africa)
- Karoo Aquifers (Southern Africa)
- Vechte Aquifer (Western Europe)
- Slovak Karst—Aggtelek Aquifer (Central Europe)
- Praded Aquifer (Central Europe)

Table 4.3 *Summary of international law related to groundwater*

Helsinki Rules (1966) – International Law Association (ILA)
    Defines a body of underground water as part of an international drainage basin, except confined groundwater

Seoul Rules (1986) – ILA
    Defines international drainage basin, "An aquifer intersected by the boundary between two or more States that does not contribute water to, or receive water from, surface waters of an international drainage basin constitutes an international drainage basin for the purposes of the Helsinki Rules"

Bellagio Draft Treaty (1989) – from United States–Mexico negotiations on transboundary water
    Recognizes hydrologic interdependence between surface water and groundwater
    Recognizes transboundary aquifer as a part of an international basin

Agenda 21 (1992) Chapter 18 – United Nations Conference on Environment and Development
    Suggests the comprehensive action plan for environmental management
    Recognizes groundwater as freshwater bodies and gives parallel status to surface water
    Recommends holistic freshwater management
    Neglects transboundary aspect of freshwater resource management

The Draft of the Law of the Non-Navigational Use of International Watercourses (1994) – International Law Commission (ILC), United Nations General Assembly
    Uses International Watercourse approach
    Does not include confined aquifers

Convention on the Law of the Non-Navigational Uses of International Watercourses – ILC
    Uses the same definition of watercourses as in the draft (1994)

The Resolution of the Law of Non-Navigational Use of International Watercourses (1994) – ILC
    Recognizes that confined aquifers, that is, groundwater not related to an international watercourse, is also substantial
    The Rules regarding water management that are presented in the draft of the Law may be applicable to transboundary confined aquifers

The Protocol on Water and Health to the 1992 Convention on the Protection and Use of Transboundary Watercourses and International Lakes (1999) – Economic and Social Council
    Recommends extending the levels of water resource management to transboundary, State's level in order to protect human health and well-being
    Recommends integrated water resources management, including groundwater

Source: Matsumoto, 2002.

Table 4.4 *Comparison among physical characteristics of transboundary surface water and groundwater and institutional issues*

Transboundary rivers	Transboundary aquifers	Institutional lessons for aquifers
Long linear features	Bulk three-dimensional systems	Responsibility must be "basin /aquifer wide"
Use of resources generally limited to vicinity of the river channel	Resources may be extracted from and used extensively over outcrop and subcrop	As above, but must address diverse users, for example, industry as well as irrigation
Replenishment always from upstream resources	Replenishment may take place from any, or all of three dimensions	The resource planning mandate has to be wide
Rapid and time-constrained gain from replenishment	Replenishment could be slow, net gain can be drawn upon over longer periods	Planning horizon must be relate to the aquifer response time
Abstraction has an immediate downstream impact	Abstraction impact can be much slower — can be 10s of years	As above
Little impact on upstream riparian sites	Could have an equal impact on both upstream and downstream riparian sites	A mandate for multinational linkage of institutions
Pollution transported rapidly downstream	Slow movement of pollution	Relate to the response time
Pollutant transport invariably downstream, upstream source may be unaffected	Pollutant transport controlled by local hydraulics; an operating well may induce "upstream" movement toward itself	Both qualitative and quantitative responsibility needed

Source: Puri and Naser, 2002.

Momentum in attempts to address this difficult topic is growing. UNESCO and the United Nations Food and Agriculture Organization (FAO) recently collaborated on a legislative study of groundwater in international law (Burchi and Mechlem, 2005). The Madrid Workshop on Intensely Exploited Aquifers, held in 2001, provided ideas and suggestions to improve water management where there is intensive use of groundwater. The Valencia International Symposium on Intensive Use of Groundwater held in 2002 built on these issues (Llamas and Custodio, 2002; Puri and El Naser, 2002). There are many contrasts between transboundary rivers and aquifers. Some of these are listed in Table 4.4, and these peculiarities need to be accounted for in developing mechanisms for jointly managing transboundary groundwater.

To get at how transboundary aquifers are actually dealt with in treaties, Matsumoto (2002) culled the 400 (at the time) treaties of the Transboundary Freshwater Dispute Database and categorized them as in Table 4.5. Treaties that mentioned groundwater were also categorized by their inclusion of the following groundwater issues: (1) water quality, including pollution; (2) water quantity, including allocations of groundwater; (3) territory/ boundary concerns; (4) physical relationship with surface water; (5) water rights; (6) and others. She found that 109 of the 400 treaties at least allude in passing to groundwater, dating from an 1864 treaty between Portugal and Spain dealing with shared springs along their common border. Of these, she was able to examine 62 in detail (see Appendix E).

Only six of sixty-two treaties deal specifically with groundwater quality, and just eight refer to groundwater quantity. However, seventeen treaties discuss groundwater in relation to border concerns as well as the physical groundwater and surface water interconnection. Nine of the sixty-two treaties contain explicit groundwater management provisions, which are classified as Level 3 (Table 4.5). These treaties can be additionally divided and placed under the following categories: (1) spring and aquifer extraction limits, (2) water allocations, and (3) incorporating management principles. Out of these nine treaties, only the treaty about Iran and Iraq's national borders and neighborly conduct, clearly cites water allocation. According to the protocol (June 13, 1975), these two countries should share their spring water on an hourly basis.

Table 4.5 *Description of the levels of groundwater resource management*

Level	Description
Level 1	Indirectly mentioned groundwater; no specific provisions of management
Level 2	Briefly mentioned groundwater provisions of management; water rights of groundwater are assigned to a state although specificity of allocation is absent
Level 3	Deals with groundwater regulations specifically, including allocation, quality provisions, and/or protection of land

Source: Matsumoto, 2002.

Five treaties discussing restrictions on groundwater pumping can be placed under the first and second categories. Among these agreements, the Mexico–U.S. Agreement aims to provide long-term solutions for the salinity problem in the Colorado River basin (Minute 242), by limiting groundwater pumping in the basin area. Another agreement, the Convention on Environmental Impact Assessment in a Transboundary Context, Espoo (September 19, 1997), affirms that a given amount of groundwater pumping commonly affects the environment. Furthermore, the Israeli–Palestinian provisional agreement concerning the West Bank and the Gaza Strip (1995; Article 40), provides expansive water management strategies regarding water and sewage concerns. In accordance with this agreement, the two countries are obligated to share groundwater. It also delineates the rates of pumping allowed from the Eastern, Northeastern, and Western Aquifers. Furthermore, the agreement implemented a joint water committee, and delineated the guidelines for the committee's roles. Another treaty in the Middle-East, between Israel and the Hashemite Kingdom of Jordan (1994), is defined in Article IV: Groundwater in Emek Ha'arava/Wadi Arava. This treaty bounds extraction rates from wells to 10 MCM/year. An additional treaty, the Franco–Swiss agreement, discusses the protection, use, and regulation of the Geneva Aquifer. This major agreement also defines groundwater extraction limits.

Four of the agreements, which address principles of groundwater management, fall under Level 3. One such agreement, the Danube River Convention (June 29, 1994), discusses the issues of cooperation, protection, and sustainable use of the river. This convention demonstrates that groundwater regulations should include all aspects of the hydrologic cycle. It incorporates temporal as well as spatial components of water and clearly cites groundwater's allocated use. Furthermore, the treaty specifies "groundwater resources subject to a long-term protection as well as protection zones valuable for existing or future drinking water supply purposes" (Article 6). In addition, the treaty calls for long-term groundwater management needs as well as means to protect the land, which filters the water. Another agreement, the 1910 convention between the bordering territories of Great Britain and the Sultan of Abdali and the water supply of Aden (Yemen), establishes that British water infrastructure should not influence water quality or quantity in the Sultan of Abdali's wells. A different treaty, known as the Johnston negotiations (December 31, 1955), serves as an effort to make an agreement concerning water management and politics. This unratified agreement between Syria, Israel, Jordan, and Lebanon incorporates a stipulation to decrease salinity in Lake Tiberias by diverting water from saline springs. However, its focus is on surface water, not groundwater. The last agreement citing management principles is the Convention on the Geneva Aquifer (1977), which most effectively addresses the complexity of groundwater management. This agreement provides means for a joint groundwater management scheme.

Groundwater is referred to cursorily in thirty-three agreements in Level 2 (Table E.2, Appendix E). Although groundwater is discussed as an extension of surface water in seventeen treaties, the physical interconnection between groundwater and surface water is cited in thirteen agreements. Furthermore, seven agreements address groundwater in connection with boundary issues. Although these treaties are generally related to surface water pollution, they recognize the need to protect groundwater to guard the quality of surface water. The convention instituting a protocol for the Niger basin (1980) addresses the groundwater and surface water connection as it affirms the necessity for "the initiating and monitoring of an orderly and rational regional policy for the utilization of the surface and underground waters in the Basin" (Article 4). Although this protocol cites the need for cooperative management of groundwater and surface water, it is narrow in scope as the agreement only delineates the groundwater in the Niger basin.

A keyword relating to groundwater (i.e., aquifer, groundwater, spring, subsoil, subsurface, underground, or wells) is cited in nineteen treaties in Level 1 (Table E.3 Appendix E). The boundary delineation of transboundary water bodies is central in nine of these treaties. For instance, the agreement between Poland and the German Democratic Republic (July 6, 1950) reveals the State's control over subsoil. Although it defines a boundary as a "[b]orderline [that] also applies in the subsoil," it fails to discuss groundwater explicitly. However, the statement connects the physical interconnection between groundwater and surface water as well as it tacitly entails that groundwater should be protected.

In addition, the agreement between the Federal Republic of Nigeria and the Republic of Niger (July 17, 1986) [7] delineates groundwater under certain conditions [8]:

Groundwater resources shall not be accounted for the purpose of equitable sharing determination unless: (a) such resources are part of shared river basins within the meaning of Article 1, paragraph (3)20; or (b) such resources lie in whole or only in part within the shared

---

[7] An agreement between the Federal Republic of Nigeria and the Republic of Niger Concerning the Equitable Sharing in the Development, Conservation and Use of Their Common Water Resources.

[8] Article 1, 2. "The shared river basins to which this Agreement applies are: a. the Maggia/Lamido River Basin; b. the Gada/Goulbi of Maradi River Basin; c. the Tagwai/El Fadama River Basin; and d. the lower section of the Komadougou-Yobe River Basin, and each River Basin shall be defined by reference to the Maps annexed to, and forming an integral part of, this Agreement."

river basins and are bi-sected by the common frontier between the Contracting Parties. (Article 9)[9]

Matsumoto concludes her thorough survey with these suggestions. Overall, this study quantitatively shows that the treaties of the past do deal with groundwater. However, groundwater is usually treated as a secondary issue to surface water. Thirty-four of the examined agreements are in the categories of territory/boundary issues or physical relationships between surface and groundwater. Nine agreements of sixty-two have specific provisions for groundwater management.

4.4.4 Negotiation process: From rights to needs to benefits to equity[10]

Although there are no "blueprints" for water conflict transformation, there seem to be general patterns in approaches to water conflict that have emerged over time. This section offers observations about one path to the transformation of water disputes from zero-sum, intractable disputes to positive-sum, creative solutions. It centers on the process of transformation in negotiations – the point at which parties move from thinking of themselves as representing countries to perceiving more broadly the needs of all stakeholders within a basin. This is a critical juncture in negotiations, where movement from "rights-based" to "needs-based" to "interest-based" to "equity-based" negotiations suddenly becomes possible. In international basins, this transformation may take years or even decades, during which time political tensions are exacerbated, ecosystems go unprotected, and water is generally managed, at best, inefficiently. However, discussing water often provides a venue for dialogue and to gradually help the parties to envision the possibilities of joint gains and the creation of new benefits.

As we have seen, most international negotiations surveyed begin with parties basing their initial positions in terms of rights – the sense that a riparian is entitled to a certain allocation based on hydrography or chronology of use. Upstream riparians often invoke some variation of the Harmon Doctrine, claiming that water rights originate where the water falls. India claimed absolute sovereignty in the early phases of negotiations over the Indus Waters Treaty, as did France in the Lac Lanoux case, and Palestine over the West Bank aquifer. Downstream riparians often claim absolute river integrity, claiming rights to an undisturbed system or, if on an exotic stream, historic rights based on their history of use. Spain insisted on absolute sovereignty regarding the Lac Lanoux project, while Egypt

Table 4.6 *Examples of needs-based criteria*

Treaty	Criteria for allocations
Egypt/Sudan (1929, 1959, Nile)	"Acquired" rights from existing uses, plus even division of any additional water resulting from development projects
Johnston Accord (1955, Jordan)	Amount of irrigable (by gravity) land within the watershed in each State
India/Pakistan (1960, Indus)	Historic and planned use (for Pakistan) plus geographic allocations (western vs. eastern rivers)
South Africa (Southwest Africa)/Portugal (Angola) (1969, Cunene)	Allocations for human and animal needs, and initial irrigation
Israel–Palestinian Interim Agreement (1995, shared aquifers)	Population patterns and irrigation needs

Source: Wolf, 1999.

claimed historic rights against first Sudan, and later Ethiopia, on the Nile.

In almost all of the resolved disputes, however, particularly on arid or exotic streams, the paradigms used for negotiations have not been "rights-based" at all, neither on relative hydrography nor specifically on chronology of use. Instead, they have been "needs based." This experience argues strongly for the use of third-party and assisted-negotiations techniques. Needs are defined by irrigable land, population, or the requirements of a specific project (Table 4.6). (Here we distinguish between "rights" in terms of a sense of entitlement and legal rights. Obviously, once negotiations lead to allocations, regardless of how they are determined, each riparian has legal "rights" to that water, even if the allocations were determined by "needs.") In agreements between Egypt and Sudan signed in 1929 and in 1959, for example, allocations were determined on the basis of local needs, primarily of those of agriculture. Egypt argued for a greater share of the Nile because of its larger population and extensive irrigation works. In 1959, Sudan and Egypt then divided future water from development equally between the two countries. Current allocations of 55.5 billion cubic meters (BCM) per year for Egypt and 18.5 BCM per year for Sudan reflect these relative needs (Waterbury, 1979). It should be pointed out, however, that not everyone's needs were considered in the Nile Agreements, which included only two of the ten riparian states – Egypt and Sudan, both minor contributors to the river's flow. The notable exception to the treaty, and the one that might argue most adamantly for greater sovereignty,

---

[9] Article 1 Paragraph (3) "Subject to the provisions of Article 9, a reference to the shared river basins shall include a reference to underground waters contributing to the flow of surface waters."
[10] This section draws from Wolf (1999b).

is Ethiopia, which contributes 75 to 85 percent of the Nile's flow.

Likewise, in the Jordan River basin, the Johnston Accord emphasized the needs rather than the inherent rights of each of the riparians. Johnston's approach, based on a report written under the direction of the Tennessee Valley Authority, was to estimate, without regard to political boundaries, the water needs for all irrigable land within the Jordan Valley basin that could be irrigated by gravity flow (Main, 1953). National allocations were then based on these in-basin agricultural needs, with the understanding that each country could then use the water as it wished, including diverting it out of basin. This formula was not only acceptable to the parties at the time, but it allowed for a breakthrough in negotiations when a land survey of Jordan concluded that its future water needs were lower than previously thought. Years later, Israel and Palestine came back to needs in the Interim Agreement of 1995, when Israel first recognized Palestinian water rights on the West Bank – a formula for agriculture and per capita consumption determined future Palestinian water needs at 70–80 MCM per year and Israel agreed to provide 28.6 MCM per year toward those needs.

Needs are the most prevalent criteria for allocations along arid or exotic streams outside of the Middle East as well. Allocations of the Rio Grande/Rio Bravo and the Colorado between Mexico and the United States are based on Mexican irrigation requirements; Bangladeshi requirements determined the allocations of the Ganges; and Indus negotiations deferred to Pakistani projects (although estimates of needs are still disputed and changing, particularly in these latter two examples).

One might speculate as to why negotiations move from rights-based to needs-based criteria for allocation. The first reason may have something to do with the psychology of negotiations. Rothman (1995), among others, points out that negotiations ideally move along three stages: the adversarial stage, where each side defines its positions or rights; the reflexive stage, where the needs of each side bringing them to their positions is addressed; and, finally, the integrative stage, where negotiators brainstorm together to address each side's underlying interests. The negotiations in these examples seem to follow this pattern. Each negotiator may initially see him- or herself as Egyptian or Israeli or Indian, with the rights of his or her own country of paramount importance. Yet, over time, the negotiator begins to empathize, recognizing that even one's enemy, whether Sudanese, Palestinian, or Pakistani, requires the same amount of water for the same use with the same methods as oneself.

The second reason for the shift from rights to needs, may simply be that rights are not quantifiable but needs are.

We have seen the vague guidance that the 1997 Convention provides for allocations: a series of occasionally conflicting parameters to be considered as a whole (see Appendix A). If two nations insist on their respective rights of upstream versus down, for example, there is no spectrum along which to bargain, no common frame of reference. A needs-based criterion – irrigable land or population, for example – is more easily quantified for each nation. Even with differing interpretations, once both sides feel comfortable that their minimum needs are being met, talks eventually turn to straightforward bargaining over numbers along a common spectrum.

Because of its relative success, needs-based allocations have been advocated in recent disputes, notably in and around the Jordan River watershed, where riparian disputes exist not only along the river itself but also over several shared groundwater aquifers. Gleick (1996) defines basic human needs, regardless of climate, as 50 liters per capita per day for personal use alone (18.25 m^3/year), and in earlier work (Gleick, 1994) suggests 75 m^3/year as appropriate minimum levels per capita for the Middle East. Shuval (1992) also argues for a minimum baseline allocation among Israel, West Bank Palestinians, and Jordan, based on a per capita allotment of 100 m^3/year for domestic and industrial use plus 25 m^3/year for agriculture. He adds 65 percent of urban uses for recycled wastewater and advocates a series of water import schemes and desalination plants to provide the difference between regional supply and future demand.

Wolf (1993) likewise advocates a needs-based approach but considers new sources such as recycled wastewater as separate issues. He plans for total urban needs of 100 m^3/year per person and extrapolates to the point in the future where *all* of the basin's 2,500 MCM/year has to be allocated first to these needs, when the regional population reaches 25 million, expected in the early part of the next century.

4.5 RELATIVE HYDROGRAPHY VERSUS CHRONOLOGY OF USE

As described above, generalized legal principles focus on some version of upstream-versus-downstream relations, whether defined in the extreme as absolute sovereignty versus absolute riverain integrity or versus historic rights or, more moderately, as equitable use versus the obligation not to cause harm. In practice, the only situation in which there is still any ambiguity is along humid, underdeveloped rivers. Along arid or exotic streams, where some aspect of consumptive use is involved, there is very little debate – prior uses are *always* protected in the treaties that describe them (with only one exception, described later in the prior uses section) and downstream needs are generally favored. Nine treaties do not address the issue at all, simply basing their allocations equally between two riparians.

4.5.1 Absolute principles

As noted above, the dispute that led to the disavowal of the legal principles of both absolute sovereignty and absolute riverain integrity was the Lac Lanoux case of 1957, which found, in short, that "the upstream State has a right of initiative... provided it takes into consideration in a reasonable manner the interest of the downstream State" (cited in MacChesney 1959, p. 170).

The only situations in which absolute rights are codified in treaties are relating to some tributaries of international waterways in conjunction with broader boundary waters accords, always in a quid pro quo arrangement. Such is the case in only three of our case studies listed in Appendix G. Mexico and the United States each retain absolute sovereignty to some internal tributaries of the Rio Grande/Rio Bravo, for example (Figure 4.8). In a 1950 boundary waters agreement, of five tributaries of the Isar that flow from Austria to Bavaria, one is allowed to flow freely to Bavaria, two can be developed entirely by Austria, and two can be developed by Austria, provided it allows minimum flows during winter months. Interestingly, and perhaps adding incentive to a particularly creative agreement, Austria is an upstream riparian on these tributaries to the Isar, and then becomes a downstream riparian to Bavaria (Germany) after the Isar flows into the Danube, which bends back into Austria. In contrast, a 1925 accord on the streams that form the boundaries between Finland and Norway allocates each State half of the boundary streams, but absolute sovereignty to each State over all the tributaries to those streams in which both banks are within one country.

4.5.2 Prior uses

In contrast to the extreme rarity with which absolute principles are codified, prior uses are regularly protected (with one major exception, described later in this section. The entire focus of some treaties is on protecting existing uses. All of the six existing treaties regarding the Nile, for example, protect Egyptian uses in early years and later those of Egypt and Sudan. More often, a clause that protects existing uses, whether the focus is on boundary demarcations, boundary waters, or water resources development, is included in a broader treaty. Peru continues to supply water to Ecuadorian villages, for example, as part of their 1944 boundary demarcation. The boundary water accords between the United States and Canada and

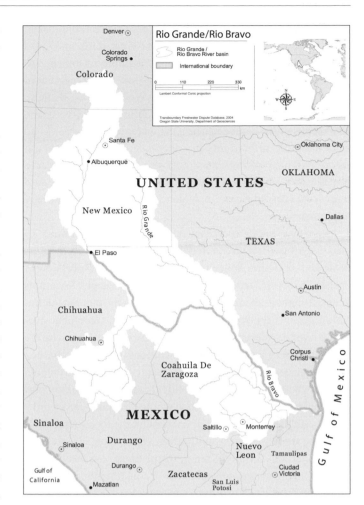

Figure 4.8 Map of the Rio Grande–Rio Bravo basin (Transboundary Freshwater Dispute Database, 2004).

between the United States and Mexico all include prior use clauses. A 1969 accord between Portugal, for Angola, and South Africa, for Southwest Africa, which describes an elaborate river development project, includes "humanitarian" allocations for human and animal requirements in Southwest Africa.

The supremacy of prior uses would not necessarily be surprising in those cases along arid or exotic streams, where investment in irrigation infrastructure has long relied on the knowledge of a stable supply. Even in humid regions, and even when water is divided proportionally, prior uses are generally protected. The boundary agreement between Russia and China along the Horgos River divides the water equally but protects the uses of existing canals and one Chinese outpost. The three boundary waters accords among Austria, Hungary, and Czechoslovakia all allocate each two signatories half of the natural flow of the shared rivers, "without prejudice to acquired (or existing) rights."

The only treaty in which existing uses were relinquished is the 1995 Israel–Palestine accord on West Bank and Gaza aquifers. Israel began tapping into these aquifers as long ago as 1955; before the accord they made up as much as 40 percent of Israel's renewable freshwater supply (Wolf, 1995b). Because two of the three West Bank aquifers naturally flow to Israel, and because they had been using the water longer, Israelis had been claiming prior rights in peace negotiations. By recognizing and quantifying Palestinian needs, and by agreeing to provide 28.6 MCM/year toward those needs, the 1995 accord represents the only case in which prior rights are explicitly relinquished.

Again, we might speculate on the inherent supremacy of prior uses. First, we have noted the shift in thinking from rights to needs – existing water use is a pretty clear expression of "needs." Second, treaties with clauses for water allocations generally come about in conjunction with a boundary delineation, a division of boundary waters, or an agreement over future river development. In each of these cases, those who are using the water are important constituents of the negotiators. In those cases regarding boundary waters, negotiations would probably be carried out in the political arena where the support of those living within a watershed would be vital to an accord's success. In the case of river development, the technocrats who negotiate these treaties, usually from water agencies, are generally extremely aware of the needs of those in a basin. In all cases, existing uses represent existing constituents, in contrast to hypothetical users or future generations.

4.5.3 Upstream–downstream relations

Rights inherent in an upstream or downstream position are not explicitly claimed in any of the treaties in our collection. This should not be understood to suggest that the upstream–downstream relationship is ignored, only that when it is addressed, it is done so implicitly.

In general, the downstream riparian is favored, or at least its allocations are protected, along arid and exotic streams. This is not to say that the downstream riparian receives more water because this is not always the case – Mexico receives less water on both the Colorado and the Rio Grande/Rio Bravo than the United States – only that it is the allocations of the downstream riparian that are generally delineated and protected. Mexico, Egypt, Bangladesh, and Pakistan all have their needs defined and guaranteed in their respective treaties. This precedence probably comes about as a consequence of two earlier observations – that rights give way to needs and that prior uses are generally protected. Because there is more, and generally older, irrigated agriculture downstream on an arid

or exotic stream, and because agricultural practices predate more recent hydroelectric needs – the sites for which are in the headwater uplands – the downstream riparian would have greater claim whether measured by needs or by prior uses of a stream system.

The only treaties in which upstream allocations are delineated (except for the internal tributaries granted absolute sovereignty as we have noted), are on boundary waters agreements in humid regions. The 1956 boundary waters accord between Austria and Hungary grants the upstream state up to one-third of the water of any of the covered river systems. (This is an interesting exception, for which we have no explanation – similar treaties between Austria and Czechoslovakia and between Czechoslovakia and Hungary have no such provision.) Three other humid boundary water agreements simply divide the waters equally – Austria/Hungary, Czechoslovakia/ Hungary, and Finland/Norway. In the only treaty that explicitly favors the upstream riparian, the 1925 accord on the Gash between Italy, for Eritrea, and the United Kingdom, for Sudan, grants upstream Eritrea all of the low flow and half of the moderate flow of the stream; Sudan also agrees to pay Eritrea a share of what was received for agricultural cultivation in the Gash Delta (Figure 4.9).

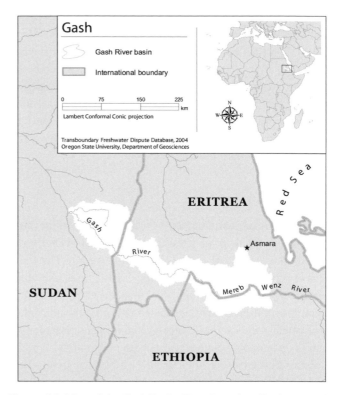

Figure 4.9 Map of the Gash Basin (Transboundary Freshwater Dispute Database, 2004).

Table 4.7 *Prioritizing uses*

Order of priorities	United States/Mexico Boundary Waters (1906, 1944)	United States/Canada Boundary Waters (1910)	Indus Waters Treaty (1960)	Mekong Agreement (1975)
1	Domestic	Domestic and sanitary	Domestic	Domestic and urban uses
2	Agriculture	Navigation	Nonconsumptive	Other criteria from Helsinki Rules without priority
3	Electric power	Power and irrigation	Agriculture	
4	Other industry		Hydropower	
5	Navigation			
6	Fishing			
7	Other beneficial uses			

Source: Wolf, 1999.

4.5.4 Prioritizing use

The Helsinki Rules list eleven hydrographic and sociopolitical factors that ought to be taken into account in water allocations (International Law Association, 1966); the 1997 Convention lists seven but does suggest that the "requirements of vital human needs" be given "special regard" (Appendix A; United Nations, 2005). Neither set of parameters has been explicitly used in any treaty to derive allocations. The Helsinki Rules *are* listed, verbatim, only in the 1975 Mekong Agreement – and the criteria that a benefit–cost ratio for each proposed project be performed is added – but no allocations are derived.

Four treaties do differentiate between types of use (other than existing uses, described above), but they use far less criteria and each list is prioritized (Table 4.7). After listing the criteria from the Helsinki Rules, for example, the Mekong Agreement gives domestic and urban uses a preference (International Law Association, 1966). The two sets of boundary waters agreements between the United States and Canada, and the United States and Mexico prioritize differently, probably due to the amount of water available along each border region: the former prioritizes by domestic and sanitary, navigation, and power and irrigation; the latter gives descending weight to domestic, agriculture, electric power, other industry, navigation, fishing, and other beneficial uses. The 1960 Indus Waters Treaty lists its order of priority as domestic, nonconsumptive, agriculture, and hydropower. Notably absent in all of these lists are any instream or other environmental requirements. This may be changing, however, at a 1997 meeting on international waters of Latin America, a representative of the Global Environmental Facility suggested that watershed needs start with the environmental needs at the delta and work backward.

4.5.5 The unique local setting

Although most of the debate in the realm of customary law has been over trying to accommodate as many concerns as possible in an attempt to find generalized principles for all of the world's international water, riparians of these basins have, in the meantime, been negotiating agreements that focus on specifically local concerns and conditions. Although many of these treaties incorporate particularly local issues, they often include a clause that explicitly disavows the treaty as setting an international precedent. The 1950 accord on Austria–Bavaria boundary waters is typical: "Notwithstanding this agreement," it reads, each State maintains its "respective position regarding the legal principles of international waters." The most-recent agreement in our collection, the 1996 Ganges Agreement, includes the similar provision that the parties are "desirous of finding a fair and just solution without … establishing any general principles of law or precedent" (see Appendix G).

The changes of local needs over time are seen in the boundary waters between Canada and the United States. Even as the boundary waters agreements of 1909 were modified in 1941 to allow for greater hydropower generation in both Canada and the United States along the Niagara to bolster the war effort, the two States nevertheless reaffirmed that protecting the "scenic beauty of this great heritage of the two countries" was their primary obligation (see Appendix G). A 1950 revision continued to allow hydropower generation, but it allows a greater minimum flow over the falls during summer daylight hours, when tourism is at its peak.

Cultural geography can overwhelm the capacity of generalized principles as well. In 1997 discussions among the riparians of the Euphrates basin, Syrians objected strenuously to proposals for water pricing. This led to a temporary impasse, until it was explained by an outside observer that some Islamic legal interpretation forbids charging money for water itself; the term was modified to "tariff," to represent costs only for storage, treatment, and delivery, and discussions were able to proceed (see Appendix G).

In what will no doubt become a classic modification of the tenets of international law, Israelis and Jordanians invented legal terminology to suit particularly local requirements in their 1994 peace treaty. In negotiations leading up to the treaty, Israelis, arguing that the entire region was running out of water, insisted on discussing only water "allocations," that is, the future needs of each riparian. Jordanians, in contrast, refused to discuss the future until past grievances had been addressed – they would not negotiate "allocations" until the historic question of water "rights" had been resolved (see Appendix G).

There is little room to bargain between the past and the future, between "rights" and "allocations." Negotiations reached an impasse until one of the mediators suggested the term "rightful allocations" to describe simultaneously historic claims and future goals for cooperative projects – this new term is now immortalized in the water-related clauses of the Israel–Jordan Treaty of Peace (see Appendix G).

4.6 FROM ALLOCATING WATER TO SHARING BASKETS OF BENEFITS

The next step in the process has been to go from needs to benefits. One productive approach to the development of transboundary waters has been to examine the benefits in the basin from a regional approach. Riparians must get past looking at the water as a commodity to be divided – a zero-sum, rights-based approach – and develop an approach that equitably allocates not the water, but the benefits derived therefrom – a positive-sum, integrative approach.

It requires the parties to envision the real possibilities that they can create jointly benefits that will exceed what they can have by pursing their interests in a zero-sum way – that is to have them jointly increase their own pie for negotiating. This notion clearly illustrates how water negotiations are vitally linked to socioeconomic development and are central to it. Often water negotiations are mistakenly viewed as a type of environmental negotiations over a fixed amount of resources or a resource that is dwindling. If so, this would mean that the only negotiations possible would be over the distribution and redistribution of diminishing resources. This is a recipe for violence. The history of water, however, especially when it is

connected to development, shows other paths. Because water can be used for so many activities, it offers the clear potential for creating more wealth and value rather than providing only a fixed or diminishing amount. Too often the rich countries bring this fixed or diminishing resource assumption – implicitly – to the table. They then say they come to avoid conflict, but could actually end up creating the very conflict they say they seek to avoid. Water is more complex than this notion allows. This is why water's link to development and growth and income generation is so important to the developing world.

4.6.1 Beneficial uses

Economists suggest that water, like any scarce resource, should be allocated to its most efficient use. In practice, economic criteria have influenced water allocations only in the exception. The one topic most affected by economic criteria is when principles of "beneficial" uses are specifically defined, notably in treaties describing hydropower or river development projects. Of the twenty-eight treaties in these two categories, five allocate water equally. Two of the twenty-eight refer not to equal allocations but to equal allocations of benefits – which, as noted above, is not at all the same thing. For example, as previously discussed, the boundary waters agreement between the United States and Canada and the 1975 Mekong accord allocate water according to equal benefits.

While compensation for lost power generation or flooded land is fairly common, appearing in ten of the twenty-eight development treaties, compensation for water itself is not: only four of all forty-nine treaties have such provisions. In the first such accord, a 1910 agreement on Aden groundwater, Great Britain agreed to pay the Sultan of the Abdali 3,000 rupees a month if the proposed wells went unmolested; otherwise, the price dropped to 15 rupees per 100,000 gallons. In a 1926 accord on the Cunene River, no charge was made for water diverted for subsistence, but South Africa would pay unspecified fees to Portugal if the water were used for "purposes of gain." South Africa not only paid much of the development costs of the Lesotho Highlands project, but it pays Lesotho outright for water delivered. In a slight twist, Great Britain agreed in 1925 to pay upstream Eritrea a share of its cultivation in the Gash Delta: 20 percent of any sales over £50,000. Payments were discontinued when Great Britain took control of Eritrea in World War II.

The treaty with the most economic influence is the 1995 groundwater agreement between Israel and Palestine. Although no payments are made outright for water, provisions are included to consider water markets in the future, and the two sides agree not to subsidize marketed water – moves long encouraged by economists to promote efficient use.[11]

4.6.2 "Baskets" of benefits

In most of these treaties, water issues are dealt with alone, separate from any other political or resource issues between countries – water *qua* water. By separating the two realms of "high" (political) and "low" (resource economical) politics, or by ignoring other resources that might be included in an agreement, some have argued, the process is either likely to fail, as in the case of the 1955 Johnston accords on the Jordan, or achieve a suboptimum development arrangement, as is currently the case on the Indus agreement, signed in 1960 (see Lowi, 1993, and Waterbury, 1993). In addition, water negotiations are usually separate from any other resource disputes, which may preclude some creative tradeoffs. In fact, in a quest to generate creative options in water negotiations, the best solution may involve other resources entirely and usually does. Dinar (2004) describes how these "baskets" can help overcome power imbalances between nations – going beyond water alone raises the chances that each State may be able to put forward a benefit for mutual gain. Increasingly, linkages are being made between water and politics and between water and other resources. These multiresource linkages may offer more opportunities for creative solutions to be generated, allowing for greater economic efficiency through a "basket" of benefits. Indeed this is the notion that has driven the classic approaches of multipurpose water planning refined over the 1960s, 1970s, and 1980s in North America and that now appears in many other parts of the world as integrated water resources management. The following are some of the resources that have been included in water negotiations.

FINANCIAL RESOURCES

An offer of financial incentives is occasionally able to circumvent an impasse in negotiations. World Bank financing helped resolve the Indus dispute, while UN-led investments helped achieve the Mekong Agreement. Cooperation-inducing financing has not always come from outside of the region. Thailand helped finance a project in Laos, as did India in Pakistan, in conjunction with their respective watershed agreements. Egypt pays Sudan outright for water as a provision of the Nile Waters Treaty to which they both agreed. Sudan had rights, but it was not able to use them.

It should be noted that financial incentives have often not been sufficient to overcome hostilities. The World Bank has offered to help finance the Unity Dam on the Yarmuk River since the late 1970s, and it is currently offering help with a variety of projects in conjunction with the Middle East multilateral working group. The Bank provision that all riparians agree has so far been enough to preclude any large-scale development project.

ENERGY RESOURCES

One increasingly common linkage being made is that between water and energy resources. As noted above, in conjunction with the Mekong Agreement, Thailand helped to fund a hydroelectric project in Laos in exchange for a portion of the power to be generated. In the particularly elaborate 1986 Lesotho Highlands Treaty, South Africa agreed to help finance a hydroelectric/water diversion facility in Lesotho. South Africa acquired rights to drinking water for Johannesburg and Lesotho receives all of the power generated.[12] Similar arrangements have been suggested in China on the Mekong, Nepal on the Ganges, and between Syria and Jordan on the Yarmuk.

Energy and water resources can be linked in other ways. It has been suggested, for example, that a possible Saudi contribution to the Middle East peace process might come in the form of oil or natural gas to help lower the cost of desalination in the region (Wolf, 1993). Another link might come in the form of energy infrastructure. The Trans-Arabian Pipeline from the Persian Gulf to Lebanon has been unused since the early 1970s. Although the pipe itself is corroded, the parallel access road still exists and, it has been suggested, might be used to reduce the costs of piping water in the opposite direction, from Lebanon toward the Gulf (Wolf, 1993).

One policy question inevitably raised when discussing linkages is whether increased integration of infrastructure between nations leads to increased potential for political conflict or to greater impetus for cooperation. Or does increased interdependency mean increased or decreased vulnerability? Indeed, the interdependency does decrease the vulnerability to the exigencies of nature. Politically, the ability to deal with the perturbations of nature has been a key to developing stable societies. Every Western country, for example, has dealt with the

---

[11] Water subsidies within each party's territory are not covered by the agreement and will probably continue.

[12] Months before the signing of the LHWP treaty, there was a coup in Lesotho and some scholars (e.g., Homer-Dixon, 1994) interpreted the coup as a "water coup." Later assessments (Aline-Baillat, 2004) suggested that the coup was not an outcome of preceding South African blockade and if there was any involvement whatsoever, it was aiming to stop the Lesotho support to ANC. The new government formed by Lekhanya after the coup was more conciliatory to South Africa than the Jonathans government and it stopped support to the ANC, although it did not hand them over to South Africa as it was requested. Signing of the LHWP treaty seems to be another result of the better relations between Lesotho and South Africa. (Thanks to Jakub Landovsky and Olga Zarubova-Pfeffermannova for the research for this note.)

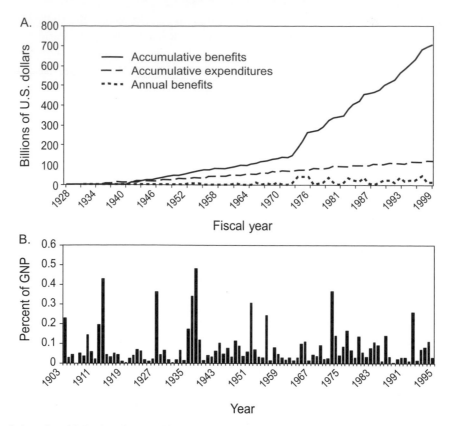

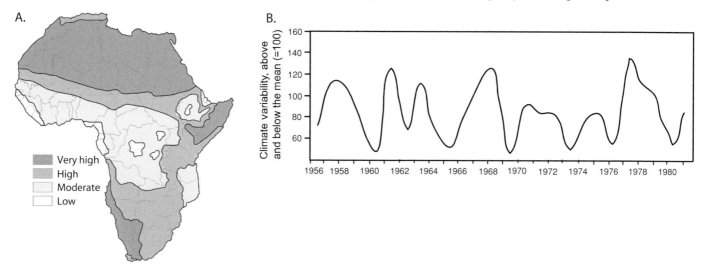

Figure 4.10 U.S. economic benefits of federal projects and flood damage as percent of GNP. (a) Benefits of federal projects (damages prevented) and accumulative Corps expenditures (principle plus O&M; billions of dollars adjusted to 1999 using construction cost index). (b) National flood damages suffered, as percent of GNP (USACE, 2004). USACE projects: 16,000 miles levees, 383 reservoirs, 400 miles – shore protection.

investments necessary in water infrastructure to produce a capability for storing peak flows for when the water is needed most. (See Figure 4.10.) But look what happens with those countries that have not – these are the poor of Africa and Asia. We can see how Mozambique, Ethiopia, Tanzania, and others are vulnerable to the ups and downs of rainfall and that the inability to store heavy flows can produce GDP fluctuations of 25 percent or more (Figure 4.11; Tables 4.8 and 4.9). Such figures, if correct, would render attempts at predictable development and of breaking a cycle of expected periodic disaster

Figure 4.11 Climate variability; risk of recurrent drought (a) Africa's natural legacy: areas of low to very high annual precipitation; (b) Africa's risk of recurrent drought; climate variability 1956–1980 (World Bank; Sadoff and Grey 2005).

Table 4.8 *Kenya variability in climate and GDP*

Climate event	Impact	Billions of dollars (U.S.)
Flood, October 1997 to February 1998	Infrastructure damage	**2.39**
Drought, October 1998 to May 2000	Crop loss	0.24
	Livestock loss	0.14
	Reduction in hydropower	0.64
	Reduced industrial production	1.39
	Total	**2.41**
Both events combined	Cost of climate variability	**4.8**
Approximate (annual) GDP		**22**
Impact as % GDP/annum		**22%**

Source: World Bank.

and recovery almost impossible. However, as we see in the affluent United States, flood damages increase absolutely over time but as a percentage of GPD they continue go down (Figure 4.10).

This shows that the means to mitigate, avoid, and minimize the effects of the perturbations of rainfall and the hydrograph are related to the ability to conduct socioeconomic activity with little interruption. This is why the treaty that helps the parties trades upstream flood control benefits in Canada for downstream hydropower benefits in the United States are real. They

are vital for the U.S. Pacific Northwest region's GDP. Thus, flattening the hydrological cycle becomes important not only for dealing with the primordial fears of floods and droughts but also for creating a platform to break out of the cycle of despair expectations. Interdependency is a way to deal with this traditional demand of populations on their elected leader. Water negotiations are becoming more and more central, as population increases and urbanizing and populations depend on water that crosses borders.

In support of the latter interpretation, it might be noted that the flow of electricity between Laos and Thailand and of water between Lesotho and South Africa was never interrupted, despite dramatic political changes in both regions. Many parts of the world show closer and closer linkages between hydrologic and social interdepedencies. Perhaps Southern Africa is the most dramatic, as shown in Figures 4.12 and 4.13 (Turton and Earle, 2005).

POLITICAL LINKAGES

Political capital, like investment capital, might likewise be linked to water negotiations, although no treaty to date includes such provisions. This linkage might be done implicitly, as, for example, the parallel but interrelated political and resource tracks of the Middle East peace talks, or explicitly, as talks between Turkish acquiescence on water issues have been linked in a quid pro quo with Syrian ties to Kurdish nationalists.

DATA AND TECHNOLOGY

As water-management models become more sophisticated, water data is increasingly vital to management agencies. It

Table 4.9 *Rainfall affects growth: The case of Mozambique's year 2000 floods*

	Actual		Projection			
			Before the floods		After the floods	
	1998	1999	2000	2001	2000	2001
Real GDP (annual growth rate)	12.0	9.0	7.0[a]	7.2	5.4[a]	7.9
Inflation (annual average, %)	0.6	2.0	6.6[b]	5.0	9.5[b]	5.0
External current account						
Before grants	−20.5	−31.7	−23.0	−15.7	−31.5	−18.5
After grants	−12.4	−21.5	−16.3	−9.1	−19.7	−11.0
Fiscal balance						
Before grants	−10.7	−12.1	−12.1	−10.7	−16.0	−11.5
After grants	−2.4	−1.2	−5.2	−4.4	−7.0	−5.1
Memorandum						
GDP (Mt billion)	46,134	52,913	60,177	67,790	61,471	69,673

Source: World Bank estimates, IMF, and Government of Mozambique.
[a] −23%
[b] +44%

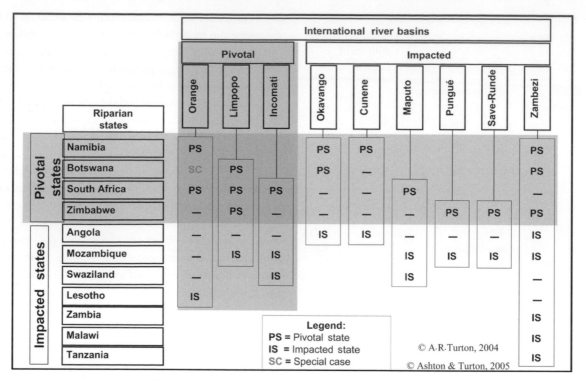

Figure 4.12 Southern African hydropolitical complex. Role of international rivers as an element of a regional security complex is as yet largely unexplored. Threats to economic security derive from the role of water as a foundation for the economic growth and prosperity of a given state. International river basins form an important element of the Southern African regional security complex (Turton, 2004; Ashton and Turton, 2005).

is now possible and inexpensive for parties to jointly create the actual simulation models which will become the algorithms by which tradeoffs are calculated (see, for example, work by Palmer et al., 1999, Cady and Soden, 2001; Nandalal and Simonovic, 2003, and Leitman, 2005). As such, data itself can be used as a form of negotiating capital. Data sharing can lead to breakthroughs in negotiations: for example, an engineering study allowed circumvention of an impasse in the Johnston negotiations when it was found that Jordan's water needs were not as extensive as had been thought, allowing for more bargaining room. In contrast, the lack of agreed-to criteria for data in negotiations on the Ganges has hampered progress over the years (Chakraborty, 2004; Swain, 2004).

Data issues, when managed effectively, can also allow a framework for developing patterns of cooperation in the absence of more contentious issues, particularly water allocations. Data gathering can be delegated to a trusted third party or to a joint fact-finding body made up of representatives from the riparian states. Perhaps the best example of this internationally is on the Mekong, where the Mekong Committee's first 5-year plan consisted almost entirely of data-gathering projects, effectively precluding data disputes in the future and allowing the riparians to get used to cooperation and trust (Le-Huu and Nguyen-Duc, 2003; Swain, 2004).

WATER-RELATED "BASKETS"

Some of the most complete "baskets" were negotiated between India and Nepal in 1959 on the Bagmati and the Gandak and in 1966 on the Kosi (all tributaries of the Ganges). These two treaties include provisions for a variety of water-related projects, including irrigation/hydropower, navigation, fishing, related transportation, and even afforestation: India plants trees in Nepal to contain downstream sedimentation. Although Nepal has expressed recent bitterness to both these accords, the structures of these treaties are good examples of how broader "baskets" can allow for more creative solutions (Sidebar 4.2).

4.7 LESSONS FOR THE INTERNATIONAL COMMUNITY[13]

A review of the provisions contained in the agreements surveyed here highlights a number of positive trends in international river basin management over the past century. First, the hydrologic linkages formed by the world's international basins create shared interests among each basin's coriparian states.

[13] This section draws from Giordano and Wolf (2003).

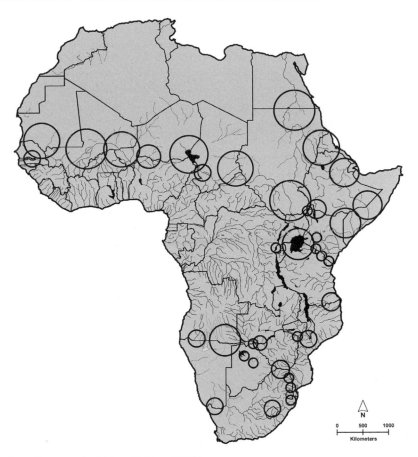

Figure 4.13 Map of potential conflict areas in Africa (Ashton, 2000).

Sidebar 4.2 Kosi and Gandak treaties

The Kosi and the Gandak River treaties have been subject to major controversies. They were signed, respectively, in 1954 and 1959 and are still today in force, but many Nepalese feel cheated by these two treaties. These treaties were subject, indeed, to high criticism within Nepal and the domestic pressure was such that successive Nepali governments had to renegotiate the treaties. India accepted to amend them in 1964 and 1966 (and again slightly in 1971 and 1978) after months and even years of talks. Despite significant modifications in the provisions of both treaties, Nepalese opinion remained that Nepal policy makers were under undue influence by Indian to sell off the water resources of the country. Some scholars present the Kosi and Ganduk projects as a positive undertaking for Nepal, especially if one considers the fact that Nepal "was and remains unable to construct large water projects on its own" (Elhance, 2000).

Agriculture, industry, recreation, hydropower, flood control, environmental integrity, and human health are all connected to some degree within an international basin. Although individual sectors and countries may have exploited their riparian position or dominance at times throughout history, basin states have likewise demonstrated a remarkable ability to cooperatively capitalize on their shared interests and to focus not only on the division of shared water resources themselves but on the broader benefits from their use or control.

Second, basin states have illustrated a great deal of creativity in formulating treaty provisions that meet the unique hydrological, political, and cultural settings of their individual basins.

Third, conditions and priorities within a basin can change considerably over time, necessitating some degree of flexibility in the institutions created to manage shared water systems. Although further progress is needed in this area, precedents exist for incorporating provisions into basin accords to accommodate changing needs and values. The 1987 Agreement on the Action Plan for the Environmentally Sound Management of the Common Zambezi River System, for example, allows for the future accession of additional riparian states to the treaty (Nakayama, 1997). Other examples of treaties with built-in flexibility include water allocation formulas that account for hydrologic fluctuations or changing needs and values, such as in the 1996 Treaty between India and Bangladesh on Sharing of the Ganga–Ganges Waters at Farakka, the 1986 Lesotho Highlands Water Project Agreement, and the 1992 Komati River Basin Treaty between South Africa and Swaziland.

A final notable development in the twentieth-century treaty record has been a use, albeit limited, of multiresource linkages, effectively broadening the "basket of benefits" considered in international water agreements and expanding the possibility for positive-sum solutions to resource problems. Although countries have traditionally treated water separately from other transboundary issues, a number of precedents exist in which water negotiations were explicitly linked to other issues. As noted above, in treaties concluded in 1959 and 1966, India and Nepal, for example, bundled projects related to irrigation, hydropower, navigation, fishing, and afforestation. More far-reaching examples can be found in the Middle East, where the 1994 and 1995 agreements between Israel and Jordan and Israel and the Palestinian Authority, respectively, incorporate water within a broader framework for peace in the region. And, currently, the Nile Basin Initiative is developing a regionwide plan to explore opportunities for maximizing the benefits of the river's waters through cooperative development and management of the basin.

SUMMARY

A review of international water relations and institutional development over the past 50 years provides important insights into water conflict and the role of institutions. The historical record of water conflict and cooperation suggests that although international watercourses can cause tensions between co-riparian states, acute violence is the exception rather than the rule. A much more likely scenario is that a gradual decline in water quantity or quality, or both, affects the internal stability of a nation or region, which may in turn impact the international arena. Early coordination among riparian states, however, can serve to ameliorate these sources of friction.

Too often parties feel that they must include everything in the treaty from the beginning and that it is forever. This is not the case. Demographics and interests and needs will change over time. Often it is not possible to include everything. But what is important is that the parties talk and create a safe space for them to meet and talk. Over time, the agreements and organizations across boundaries will grow. They should not be measured by unrealistic standards of including everything but by standards of how long they last and are they able to keep the parties coming back to talk. They will grow and be able to include new issues and expand their authority as the relationship itself grows.

Sidebar 4.3 Key Factors in the Development of Cooperative Management Networks

- *Adaptable management structure.* Effective institutional management structures incorporate a certain level of flexibility, allowing for public input, changing basin priorities, and new information and monitoring technologies. The adaptability of management structures must also extend to nonsignatory riparians by incorporating provisions addressing their needs, rights, and potential accession.

- *Clear and flexible criteria for water allocations and water quality management.* Allocations, which are at the heart of most water disputes, are a function of water quantity and quality, as well as political fiat. Thus, effective institutions must identify clear allocation schedules and water quality standards that simultaneously provide for extreme hydrological events; new understanding of basin dynamics, including groundwater reserves; and changing societal values. Additionally, riparian states may consider prioritizing uses throughout the basin. Establishing catchmentwide water precedents may not only help to avert interriparian conflicts over water use but also protect the environmental health of the basin as a whole.

- *Equitable distribution of benefits.* Distributing water benefits, a concept that is subtly yet powerfully different than pure water allocation, is at the root of some of the world's most successful institutions. The idea concerns the distribution of benefits from water use – whether from hydropower, agriculture, economic development, aesthetics, or the preservation of healthy aquatic ecosystems – *not* the water itself. Distributing benefits allows for positive-sum agreements, occasionally including even nonwater-related gains in a "basket of benefits," whereas dividing the water only allows for winners and losers.

- *Concrete mechanisms to enforce treaty provisions.* Once a treaty is signed, successful implementation is dependent not only on the actual terms of the agreement but also on an ability to enforce those terms. Appointing oversight bodies with decision-making and enforcement authority is one important step toward maintaining cooperative management institutions.

- *Detailed conflict resolution mechanisms.* Many basins continue to experience disputes even after a treaty is negotiated and signed. Thus, incorporating clear mechanisms for resolving conflicts is a prerequisite for effective, long-term basin management.

The centrality of institutions both in preventive hydrodiplomacy and in effective transboundary water management cannot be overemphasized. Yet, although progress is indeed apparent, the past 50 years of treaty writing suggests that capacity-building opportunities still remain. Many international basins are without any type of cooperative management framework, and even where institutions do exist, the post-Rio treaty record highlights a number of remaining weaknesses. Thus, in combination with its existing efforts, the international community might consider focusing more attention on the specific institutional needs of individual basin communities by assisting riparian states in the development of cooperative management networks (Sidebar 4.3).

5 Public participation, institutional capacity, and river basin organizations for managing conflict

If roads lead to civilization, then water leads to peace.

– Shimon Peres

The river basin has been one of the most persistent examples of how the functional and spatial necessities of water can form civilization. Historically, the river basin concept never seems to die but rather to continually reemerge.

Brittain (1958) notes that although the concept of the river basin may seem modern, it has existed for thousands of years. He sees the myth of Hercules' conquest of the river Achelous as an indicator that men had begun to dream about control over a whole basin. In this myth, Hercules wins his wife, Deianira, by fighting and defeating the god of the river Achelous. Ovid's account of battle is actually a summary of the various steps one might take to establish basinwide management. For example, as soon as the left fork of the river is wrenched off from the main body, it is snatched up into heaven where it is turned into a cornucopia pouring out wealth of fruit and flowers upon the reclaimed valley and enriching the whole kingdom. Ovid pairs this story with another myth, that of Erysichthon. After Erysichthon willfully cuts down a grove of Demeter's, the Greek goddess of agriculture, she calls on Famine to avenge her. Erysichthon's insatiable appetite causes him to strip his whole kingdom barren. The myth appears as a parable against the dangers of deforestation. When Ovid juxtaposes it next to Hercules' fight with Achelous, the symbolic message reflects the most modern of concerns, balanced river basin development (Brittain, 1958, pp. 268–273).

The spatial and functional characteristics of the river basin influenced human settlement and interaction long before the idea of the river basin started to be formalized into legal and administrative terms. The direction of the flow of rivers influenced the movement of civilization. Rivers have been crucial to the communication that led to the formation of political units, especially in desert basins of the fluvial civilizations and in the densely forested regions. Once irrigation canals were adapted to navigation, such canals were built for specific purpose of navigation. The influence of the physical unity of the basin has proved stronger than various political divisions.

Evidence of functional cooperation or unification of states around a river basin can be found in Hammurabi's code on the operations of irrigation trenches, the Chinese Book of the Tang on the operation of waterwheels and private reservoirs, and Herodotus's stories of apportionment of waters in a river basin in Persia. Teclaff notes that the river basin has the most influence on administration where waterways were the best means of communication (Teclaff, 1967, sections IV and IVe). Navigation laid the groundwork for a legal or administrative unity of the river basin in politically divided basins. This sense of unity was built on as the nonnavigation demands and the technological means to meet those demands grew. (This should cause some pause for reflection on the 1997 UN convention on nonnavigable waters. It attempts to set out unifying principles for shared rivers while leaving out the historic legal basis of unity – navigation! See Appendix A; United Nations, 2005.)

Rarely have the political jurisdictions stopped navigation completely. For example, there was considerable freedom of trade and navigation throughout most of ancient Mesopotamia, and so, too, on the Nile (Teclaff, 1967, section IV). In some cases, it has encouraged the opposite: the creation of political entities. For example, facilitating river navigation was a primary motivation for holding the early conventions that led to the Constitutional Convention and eventually the United States federal system in late eighteenth-century North America.

During Roman administration in Europe, navigation was open to the public. Tolls were collected for operations and maintenance. Boatmen's associations exercised considerable influence and should be seen as basinwide attempts at organizing waterborne navigation. Indeed special offices for the arbitration of disputes were created along the Rhone. Basinwide use of the rivers persisted even during Barbarian invasions in Gaul (Teclaff, 1967, section IV).

The river basin has clearly played a major role in unifying communities, stimulating trade, and forging large political–economic organizational units. Historical examples illustrate that communities were integrated through the management of water and land resources for agriculture, river navigation, and settlement networks based on agrarian productivity and transport nodes. River navigation also facilitated the movement of raw materials and manufactured goods from different parts of the basin and among basins.

The use of the rivers as waterways, in effect, has helped form river basins into commercial entities, despite political divisions. This can be traced in the Vistula, Great Lakes, St. Lawrence, Mississippi, and other basins (Teclaff, 1967, section IVc). This commercial unity can be seen in early Supreme Court cases establishing federal power over states – in certain cases and conditions of interstate commerce – in the young United States such as in *Gibbons vs. Ogden* (1824), wherein the court established federal (national) right to assume free flow of commerce along rivers.

The strong sense of commercial necessity tied to the increasing nonnavigational uses at the end of the nineteenth century set the stage for a further evolution of the river-basin idea into multipurpose and basinwide development. As competing uses vied for claims on the water, many began to see the logic of a systemwide integration of the uses to preserve and maximize their use. The functional and spatial exigencies of the river basin began manifesting themselves in higher-order schemes of social organization and administration organized around the river or water.

In the early 1900s, Sir William Willcocks proposed multipurpose plans for the Nile and the Tigris–Euphrates. Theodore Roosevelt, in the United States, stated: "Each river system, from its headwaters in the forest to its mouth on the coast, is a single unit and should be treated as such." His national inland waterways commission confirmed the need for basinwide planning. In England, movement toward basinwide planning began with the 1921 report of the Board of Trade Water Power Resources Committee. Perhaps the best known examples in Europe were the Ruhr basin associations and the Compagnie Nationale du Rhône. These influenced the subsequent development of the French River Basin Committees established in the early 1960s. And these French basin authorities now influence countries in Central and Eastern Europe, Africa, and Asia.

During the 1940s and 1950s, basin authorities emerged throughout the world: in India, Sri Lanka, Brazil, Colombia, Ghana, Australia, and other countries. These took a variety of forms. Some only coordinated planning, whereas others included a broader range of allocation power.

In the mid-1950s the UN Secretary General stated that river basin development would be recognized as an essential feature of economic development. In 1925, Congress authorized the

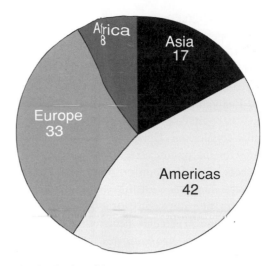

Figure 5.1 Distribution of river basin organizations by region around the world (Mestre, 2004).

U.S. Army Corps of Engineers to complete comprehensive river basin studies, called "308 Reports," throughout the United States. This activity led to a series of commissions from the 1940s to the 1970s. These culminated in a water resources council and a series of river basin commissions added to several existing interstate compacts, the Tennessee Valley Authority (TVA), and a few other river basin commissions.

We should note that the watershed is not always the most effective unit of measurement, even for managing water resources. Omernik has for years urged against the misuse of watersheds as a unit for ecosystem analysis in favor of ecoregions (see, for example, Omernik and Bailey, 1997, and Omernik, 2003) and Allan (2002) has made a strong case for the term *problemsheds* as being more useful, particularly if hydropolitics is the focus of the analysis. Nevertheless, the watershed, as a natural unit within which all aspects of the resource – quality and quantity, surface and groundwater – are connected, has an instinctive appeal in our institutional quest for unity. And, in a sense, the river basin commissions and organizations can be seen as a logical progression of that deeper quest for integration. If increased integration is the next threshold of civilization, then the twentieth century's experiments with river basin organizations are central. Figure 5.1 shows how important the basin approach has become in today's world.

5.1 PUBLIC PARTICIPATION IN WATER RESOURCES

Public participation is increasingly accepted as an important means to reach project ends in development activities (Mostert,

2003; Bruch et al., 2005). This trend has been emerging for years. At the operational, or retail, level, Cernea (1992) and others (see, for example, Nagle and Ghose, 1990) conclude that projects with a participatory approach tend to be more cost-effective and sustainable in the long run. World Bank Operations Evaluation Department (OED) reports and Paul's (1987 and 1991) analyses of institutional capacity building reach similar conclusions. Uphoff (1992) finds a high benefit-to-cost ratio for the participatory components of some irrigation projects in the Philippines and Sri Lanka. For the Sri Lankan projects there was a calculated overall return rate of 24 percent, with participatory components accounting for about half the benefits and 10 percent of the costs. Uphoff (1992) also notes that participatory approach quickly built cooperation in an area with an almost 30-year legacy of conflict.

After scaling up the Philippine participatory project, which was a product of the Ford Foundation and the National Irrigation Administration, the Bank had to learn to scale back to be responsive to demand and capacity (Uphoff, 1992).

Indeed, user participation in irrigation has a long tradition in numerous countries beyond North America, such as in the Mendoza region of Argentina or Chilean water user associations. The condominial sewage systems in northeastern Brazil show how participation can actually create technical options that no one had dreamed of for reducing costs, provide for recoverable user fees in poor areas, and provide service to those previously thought unserviceable (World Development Report, 1992, chapter 5). A World Bank review of participation in the Mexico Hydroelectric Development Project notes that by investing in social infrastructure before construction of the physical infrastructure, stated traditional problems of unsettlement were avoided (Bhatnager, 1992).

But as an end in itself or at the intersectoral, framework, or wholesale levels of water resource development, public participation becomes more controversial. The Swedish International Development Authority (SIDA) notes that participation can be viewed as an objective in itself, as a basic democratic right that should be taken into consideration and promoted in all development projects (Rudquist, 1992). Findings of the World Bank's participation learning process identify reasons why even nonparticipatory governments can and have found net benefits to a participatory approach. However, the experience of participating in decisions that affect their lives can be many times more effective in teaching those "habits and attitudes of governance" that the Bank's governance policy espouses. For hundreds of years the Dutch water boards provided experiences that helped to create a democratic civic culture and eventually the model for modern Dutch democracy. Indeed, the World Bank's Africa Regional Office has established a task force on participation and governance (Bhatnager, 1992).

Managing the physical infrastructure and environment in a participatory way can actually create the civic infrastructure; Europe and Africa provide some recent examples. Grassroots environmental groups have been in the forefront of democratic change and have been some of the principal recruiting grounds for new leadership (Page, 2003). Eastern Europeans grassroots nongovernmental organizations (NGOs) are also now creating sophisticated computer-based information links and data-sharing networks. Kwaku Kyem (2004) describes the use of participatory GIS to manage a conflict over natural resources allocation in southern Ghana.

United States environmental policy legislation in the early 1970s, including the National Environmental Policy Act (NEPA), brought great visibility to participation in water resources. The old agency patterns of decide, implement, and defend began to change to consult, decide, and implement. Delli Priscoli (2005a) points out that in the face of such ethical responsibilities, how could we continue to use the all-too-familiar model: decide, inform the client community and then justify our decision or decide, announce, and defend? This old model must be – and is being – replaced by another model, in which the participants jointly share information, jointly diagnose the problem, jointly reach an agreement about a solution, and jointly implement it. The decide-inform-justify approach usually builds on a paternalistic professional ethic. The professional formulates alternatives or determines options and then, for the good of society, informs the public and thereby justifies those decisions.

We must find new ways to jointly diagnose problems, decide on plans of actions, and implement them. This notion of professionalism is driven by a new ethic of "informed consent" or "consensus seeking," as opposed to paternalism. This informed consent model of professional ethics means that water managers will become balancers and facilitators more than dictators of specific solutions. They must focus not just on the acts but also on the relationships of those who are acting.

As Zilleßen (1991) notes, the same transformation has been occurring in Western Europe over the past decade. For example, representation of various interests conflicting over the use of France's Dordogne River came to consensus on a charter for its waters. It was the culminating event of a participatory process, including officials, professionals, citizens, and others. Although regional, it clearly involved cross-sectoral interests (Ambroise-Rendu, 1992). The Environmental Program for the Danube River, established in 1991, included the first basin-wide international body that actively encouraged public and NGO participation throughout the planning process, which, by diffusing the confrontational setting common in planning helped to preclude conflicts both within countries and, as a consequence, internationally (Bingham, Wolf, and Wohlgenant, 1994).

All this requires a process of collaborative public engineering. The environmental as well as the engineering communities have vital interests in such processes. In short, participation forces us to be more than simply "water customers" or "water clients"; we become "water citizens." Today, participatory processes are doing more than making our democratic institutions perform better. They are becoming catalysts for new civic partnerships and even new governance structures that transcend the old. The Republic of South Africa, based on participation, has written into its constitution a fundamental right to water. It has abolished old riparian systems and created a new system with two reserved rights and all other rights permitted for limited times. Participatory processes in water management have become a fundamental vector for creating a new distribution of civic rights and responsibilities (Creighton and Delli Priscoli, 2004).

The California three-way dialogue designed to produce an agreement among environmental, urban, and agricultural interests is another example of using a participatory process to reach intersectoral water agreement. In the United States, similar efforts are beginning in humid as well as arid areas. These efforts echo pioneering participatory processes of the 1970s, such as on the Susquehanna and Delaware rivers. Similar participants' policy dialogues have been initiated, with mixed success, on national policies such as energy strategy and wetlands use.

Participation in alternative water planning is one of the most interesting collaborative approaches to negotiating long-term cross-sectoral allocation decisions. Such planning encourages representatives of various interests to project various visions for the future, based on their values. Actions that would be taken to achieve their vision are then mapped and compared across future critical actions paths of others. Options that will be foreclosed, as well as a variety of actions that can be taken regardless of the chosen future, are described. The process engages participants in creating options, provides clearer understanding of impacts of options, and often leads to serious tradeoff negotiations. As long ago as the 1980s, the U.S. Bureau of Reclamation used such an approach in four northern California counties to produce a series of critical-path action diagrams and decision trees that provided audit traces of key decision points and assumptions at each point. The study subsequently guided intercounty water management decisions. A similar notion is used in Nicosia, Cyprus, as an aid to getting parties who do not speak with each other to talk about necessary future joint decisions, such as on water supply and sanitation (UNDP, 1987).

The experiences of the industrial West and the reindustrializing East offer some lessons. Because water resources management is likely to move to multipurpose and intersectoral considerations (whether these considerations are handled in administrative, political, planning, or regulatory mode), participation of stakeholders becomes central. Although power among the stakeholders will always be asymmetrical, the number who can stop or stall projects will grow. Without meaningful opportunity to participate in forming positive development goals, negative power will be rewarded and growth will stop. The same pattern is already appearing in the developing world, albeit through different institutional routes. For examples, a water project in Botswana was stopped by a coalition of local people and environmentalists. Mexican and foreign environmentalists and archaeologists have delayed hydroelectric projects on the Usumacinta River (Henry, 1991; Golden, 1992). A similar story is unfolding around World Bank-sponsored projects in Thailand. So participation should be seen as more than an instrumental means for project development, it can become critical to intersectoral dialogues in water policy in the developing world.

In its most elementary form, participation means more than simply giving information to people. It is receiving information from people, listening, and acknowledging how that information is used (Mostert, 2003). It is also recognizing that what may work very well in one culture may not work in another (Creighton and Delli Priscoli, 2004).

Building on Hirschman's notion of voice and project experience, Salman (1989) demonstrates the effectiveness of listening in development projects. Echoing what others find in the industrial world, Cernea (1992) notes that investment priorities made by communities during participatory process are often different from the expert's solutions. As we experience the growing gap between water-resource development needs and capital, investment priority setting will become even more crucial – most likely it will have to become more participatory.

Facilitation techniques are designed to help people listen to one another. UNICEF and often others use the term *animators*, or people who help people explore their situation and build critical awareness of problems and possibilities (Racellis, 1992). By fostering conditions and processes where people learn from each other, facilitation can result in creation of new integrative options. Participation can isolate extremes and create incentives for building new grounds for agreement. Extreme positions will always be present on all sides of water issues, for important ethical and moral reasons. But extreme positions should not be allowed the claim of broadly based constituent support without transparent accountability. Participation can build that transparency. Frequently, the lack of meaningful participation often encourages the very situation most seek to avoid – extreme posturing, little dialogue, and no transparent accountability to constituencies. The level of participation could be viewed as a simple scale: *knowledge about a decision* to *being heard before the decision* to *having an influence on the decision* to *agreeing to the decision*

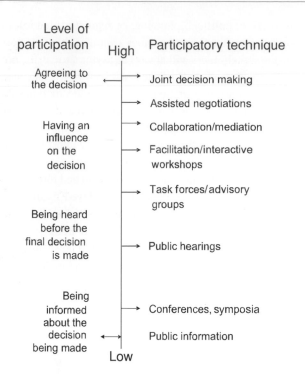

Figure 5.2 Level of institutional participation (Creighton, 1998b, p. 131).

(Figure 5.2). A wealth of practical and theoretical material exists on how to achieve participation at each of these levels (Delli Priscolli, 1983).

Multiparty facilitation and environmental mediation have substantially been products of public participation experience. In the end, participation builds on open access to information and empowerment of people. Participation in water resources seeks to build a sense of shared ownership in alternatives, thus increasing the probability that they will be implemented. Therefore, it must be part of the early design of policy and projects. Kirmani (1990) describes what can happen on the international level when participation and the sense of ownership among riparians are not present, even with external resources. He states that the Mekong Commission is a classic example of external effort, management, and planning, with little involvement of beneficiaries. Even after much engineering study and technical and financial assistance, dreams and hopes have not been realized.

5.2 WATER INSTITUTIONS AND CONFLICT

The fundamental question of water management institutions is: who participates in decision making, and on what basis are the decisions made? Creighton, Delli Priscoli, and Dunning

(1983) offer the suggestion that water managers must think of participants as if the participants were hundreds of decision makers needing objective information if they are going to participate wisely and with confidence in the process. Again, we find a certain consistency across time and scale, which gives us a vast solution set to draw from. The questions of management authority and participation, which Attia (1985) encounters within an oasis community in Tunisia, are essentially the same as those confronted at the international level by Agrawal and Gibson (1999): who allocates the resource and how much input should the public have, and at what levels? Ostrom (1992) has done remarkable work in tying small-scale, local experiences in water management with larger lessons and scales. Wolf (2000) investigates the allocation rules of Berbers and Bedouin, and draws implications from their experiences for international waters.

Another recurring institutional theme is the question of subsidiarity, which suggests that the most efficient management should be at the lowest level consistent with adequate accounting for externalities. If one were to implement this principle, at what level or where? Top down? Bottom up? Something in between? Recent environmental literature, as represented by Milich and Varady (1999), warmly advocates public participation as being more transparent, and more democratic and, through a bit of a leap, as leading to greater environmental sustainability. Agrawal and Gibson (1999) remind us that communities, like nations, are not homogeneous in their interests – that advocates often describe "'mythic communities': small, integrated groups using locally evolved norms to manage resources sustainably and equitably . . . [and] ignore how differences affect processes around conservation, the differential access of actors within communities to various channels of influence, and the possibility of 'layered alliances,' spanning multiple levels of politics."

The United States, like several other nations, is a federalist country. In the United States, states have sovereignty over the water. The federal interest, and thus intervention, occurs only if national interests like flow of interstate commerce are threatened, conflicts among states that paralyze needed action emerge, or national standards are needed. The U.S. system starts from decentralized political systems, not from the top down. Australia and Canada have federal systems, but they are more of a hybrid. But the problem of trying to coordinate among sovereign states, which also control water that crosses their boundaries, sounds familiar to the international transboundary water debates of today, even if they occur within the context of one nation-state. It is interesting to note that the process of balancing sovereignty of the states versus that of the federal government in the United States has been central to nation building in North America.

Nakayama (1997), in a comparative case study of four international basins, suggests that buy-in at the highest possible levels is one of the prerequisites for success in developing institutions across boundaries. Many of the advantages of participatory processes are self-evident. However, it must be remembered that, although it may fit well when the cripparians have democratic roots and warm relations, in many cultural settings consulting with the public is seen as weakness: leaders who turn to the people must, by definition, be ineffectual. In other basins, data are viewed with military secrecy and tied to issues of national security. All negotiation processes are susceptible to the truism that the more people in a room drafting a document, the less it says. Right or wrong, in many settings it can be presumptuous to argue the inherent supremacy of openness, transparency, capacity building, and bottom-up design.

Turton (1999) describes in his account of interaction between NGOs and nations in Southern Africa one final limitation to participation across the borders of international basins: the extreme reluctance of nations to relinquish any degree of sovereignty to outside authority. Despite the tendency of water managers to think in terms of total integration of watersheds, even friendly States often have difficulty relinquishing sovereignty to a supralegal authority, and the obstacles only increase along with the level of suspicion and rancor. At best in some settings, one might strive not for integration but for coordination. Once the appropriate benefits are negotiated, it then becomes an issue of agreeing on a set quantity, quality, and timing of the water that will cross each border. Coordination, when designed correctly, can offer the same benefits as integration and be far superior to unilateral development but does not threaten the sovereignty of a nation.

It is possible to discern convergence on requirements for building water institutions from the fields of international organization, dispute resolution, and recent experience. In talking of regional water cooperation and management, however, three important characteristics should be highlighted. First, water docs not hold still for labeling, fencing, or jurisdictional boundaries (Delli Priscoli, 1983). This makes it difficult to subject water resources to property rights and only the somewhat limited usufructuary right is normally possible. Second, water is highly variable in time and space. Variability compounds the challenges of building cooperative regional management institutions because water flows are uncertain. Third, forming water institutions is almost always done in a broader social context and in light of previous allocation agreements.

The debate over building water organizations can be characterized as a dialectic between two philosophical norms: the rational analytic model, often called the planning norm; and the utilitarian or free market model, often couched in terms of privatization (Beecher, 2000). Each of these caricatured norms implies different visions of how water institutions should change.

The rational analytic view has an explicit holistic notion of the resource and criteria for its use, which should then guide subsequent action. This norm can be driven by grand multiobjective project design, holistic ecological systems theory, or other regional designs, many of which conflict. The norm usually leads to a high degree of explicit or conscious design up front. The market norm sees institutional arrangements emerging from spontaneous interaction of self-interested parties, which reasonably conform in some way to Pareto optimality. This norm usually leads to less conscious design and a more hands-off approach. The rational analytic emphasizes concepts of water scarcity and public participation in technical decision-making processes. The market will emphasize individual freedom and public participation through buying and selling in markets.

Forming water institutions is almost always done in a broad social context and in light of previous allocation agreements. Processes used to solve redistributive issues rarely fit with rational analytic and rational choice models. Water planning is as much flexibility and managing uncertainty as discerning deterministic trends. Therefore, our experience lies between these extremes.

In the United States, numerous presidential commissions have tried unsuccessfully to establish national water policy (Deason, Schad, and Sherk, 2001). During the 1970s, an elaborate institutional and analytical procedure evolved, only to be abandoned as its implementation was beginning. To a great degree, this structure was based on river basins and was fueled by rational analytical notions. It encouraged high-level intersectoral planning and autonomous operating levels. A minianalytical rapprochement among engineers, social scientists, and ecologists was achieved in the form of two planning objectives and four accounts.

In the 1980s, the United States approach moved toward the market norm. National economic development was effectively established again as the prime objective, with environment as a constraint, usually articulated through regulatory policy. New private–public partnerships, called "cost sharing," emerged. Attempts were made to use more realistic pricing (closer to marginal costs) for water through a variety of water market mechanisms. In light of the movement away from planning, recent surrogate-rational analytic planning is emerging through the environmental regulatory structure.

In Europe, the British moved from a public river basin planning model toward privatization. Although the river basins were smaller and were operated for fewer purposes, the system also had national regulatory oversight. Since the 1970s,

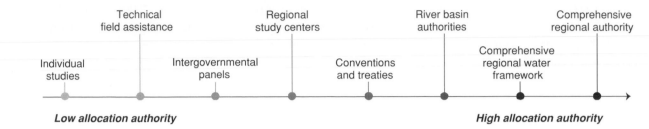

Figure 5.3 Options for water management. Describes a variety of institutional mechanisms and a continuum of options ranging from low allocative power authority to high allocative power authority. To the left of the continuum is represented allocative action based solely on individual national autonomy. To the right, the continuum represents regional, comprehensive authority for decisions in the water resources field. Moving from individual autonomy toward regional authority, various approaches are noted (Delli Priscoli, 1996).

the French have operated a river basin system that falls somewhere closer to the center of these extremes. The major basins have committees that include representation by industry, environment organizations, and the general public. These committees, which formally represent users and are financed through pollution charges, set priorities for users over a period of 20–25 years (Oliver, 1992).

As in the United States, the European Community has begun to move from single to multipurpose orientation of its river basin organizations, such as the Danube and Rhine river basin organizations. However, the focus is far more on planning and coordination and than on allocative authorities. Figure 5.3 describes a variety of institutional mechanisms and a continuum of options ranging from low allocative power/authority to high allocative power/authority.

To the left of the continuum is represented allocative action based solely on individual national autonomy. To the right, the continuum represents regional, comprehensive authority for decisions in the water resources field. Moving from individual autonomy toward regional authority, there are a variety of approaches: individual studies, regional study centers, treaties, conventions, and river basin authorities up to comprehensive regional authority. As water professionals have begun to understand water flows in light of increasing economic development, interdependence, sustainability, and population growth, the realities of the water resource push us from the left to the right of this continuum. However, legitimate and important political realities generally resist such regional notions driven by natural resource conditions.

Few comprehensive regional authorities have come into existence. As we have noted, the TVA is one outstanding example. However, a variety of river basin authorities exist, along with treaties and numerous regional centers. The allocative power/authorities of water resource agencies can also be thought of as moving from low levels of planning to higher levels of allocation operation and revenue generation. Regional and comprehensive water basin authorities, while they exist,

tend to be primarily concerned with planning rather than operations, construction, or legal oversight. Those empowered with higher levels of allocative power/authority tend to focus on single purposes, such as navigation. Few comprehensive authorities that cross jurisdictional boundaries exist for allocation and operating.

Nevertheless, our knowledge of water resources is pushing toward a vision of developing ways and means for comprehensive analysis and operation, so we can better integrate uses. It is also calling us to integrate resources management across jurisdictions. As we begin to reach the limits of use, the flexibility of our organizations to respond to water flow fluctuations and to accommodate future uses becomes crucial. This flexibility is most needed to provide new forums for dealing with political tradeoffs that cross both time and space. Nitze (1991) also notes that flexibility has been central to negotiating international environmental regimes. Indeed, flexibility has been central to recent successful negotiations of international environmental regimes.

5.2.1 Institutional barriers to conflict resolution

Although remaining optimistic, it is worth explicitly noting the difficulties that may present themselves as alternative dispute resolution (ADR) techniques begin to be infused within the government and nongovernment agencies responsible for international resource negotiations. The first barrier that may preclude total reliance on ADR in its current state is that between science and policy analysis. As Ozawa and Susskind (1985) point out, "Scientific advice is [sometimes] reduced to an instrument for legitimating political demands. Scientific analysis, in turn, can distort policy disputes by masking, beneath a veneer of technical rationality, underlying concerns over the distribution of costs and benefits." In addition, scientists seem increasingly to advocate positions with the claim of scientific legitimacy and often oppose one another as scientists claiming objectivity on the same topic. This is viewed as advocacy science. It can actually begin to delegitimize the very enterprise of science. This problem of science's tenuous relationship with policy analysis is exacerbated by the fact that diplomats are often trained in political science or law, whereas those scientists most competent to evaluate resource conflicts are rarely trained specifically in either diplomacy or policy analysis (Faigman, 1999).

The second, somewhat more subtle, barrier that can impede ADR's usefulness in international water disputes is that between ADR practitioners and analysts. Zartman (1992) discusses a common practitioner's approach to environmental disputes either as a case of "problem solving," where the disputants can dissociate themselves emotionally from the problem, considered to be a distinct entity, a "game against nature" or as a case of information dispute, where resolution becomes apparent in the process of clarifying the data. He suggests that these views are incomplete; that they "assume away conflict, rather than explaining and confronting it." He suggests steps, based on the ADR analyst's experience, for recognizing conflicts of nature also as conflicts of interest: "Inherent in the conflict with nature is conflict among different parties' interests; inherent in problem solving is a need for conflict management."

These barriers – between science and policy, between analyst and practitioner – can individually lead to a convoluted and incomplete process of conflict resolution and, together, can preclude arrival at the "best" (Pareto-optimal or win–win) solution to a given problem. In the parlance of game theory, water resources are being treated as a zero-sum commodity and distributive solutions are being emphasized; were the parties to stress the potential *products* of water, however, the conceptual shift to net gain as a positive-sum commodity could take place and the possible solutions could be integrative, or "win–win," in nature. The final result of avoiding these distinctions could be lost opportunities to reduce political tensions in regions of growing hostility.

We should note the hazards of conflating performance criteria. Often, advocates of public participation and collaborative processes define "success" in human terms: people get along better or understand each other's positions more clearly. But in the harsh gauge of measurable performance criteria – water quality parameters, for instance – these processes have a more questionable track record. In his review of Sabatier et al. (2005), Smith (2006) lauds the approach taken by the authors but notes that collaboration is not useful in all cases, especially when clear power disparities exist or when the interests of the powerful are threatened. He also points out that Wolf (1995a) found that, although Wisconsin's rural nonpoint source pollution abatement program was cited regularly as a success in human terms, voluntary actions had no measurable

positive impact on water quality due to limited participation. He poses the difficult question, "Will collaborative efforts, which also have a strong voluntary element to them, meet the same fate?"

The final potential barriers are those inherent in the limitations posed by cross-cultural communication. Shared basins are often defined by crossing political boundaries, but even more profoundly, they cross cultures – those of societies and ethnic groups, of religions and professions, of language and of class. The concept of problem-solving institutions, such as have been described over time in Western academic literature (and, possibly, overly much of the terminology and assumptions in this book draw from this world), but the ideas have deep roots in cultural traditions throughout the world. A facilitator/mediator, however, needs to be acutely aware of, and sensitive to, how cross-cultural dynamics can have an impact on the flow of communication and ideas, as well as their own inherent assumptions.[1]

The whole concept of analytic problem solving, for example, is fraught with cultural assumptions. Abu-Nimer (1996) describes the premises of North American mediators from a Middle Eastern, Muslim perspective, and Lederach (1995, p. 81) describes his experiences acting as a mediator in Central America:

Why is it . . . that in the middle of listening to someone give their side of a problem, I have a natural inclination to make a list, to break their story down into parts such as issues and concerns? But when I ask them about issues, they seem to have a natural inclination to tell me yet another story. The difference . . . lies in the distinction between analytical and holistic thinking. Our North American conflict resolution approaches are driven by analysis; that is the breaking of things down into their component parts. Storytelling . . . keeps the parts together. It understands problems and events as a whole.

Avruch (1998) sums up:

Even while acknowledging that the capacity to reason is a human universal, we face the other fact that the representations of the worlds about which humans bring their reason to bear can differ profoundly from one another. . . . To try to suppress this variance, even in the powerful setting of a conflict resolution problem-solving workshop, seems to be an invitation to failure. (p. 94)

He cites Cohen (1993) for a good model of a culturally aware mediator, not specialists and not globalists:

First, these individuals are aware of the gamut of cultural differences and do not naively 'assume that 'underneath we are all pretty much the same.' Second, they perceive the potency of religious and other cultural resonances. Third, [they] grasp that Western 'rationality' is

---

[1] The Western, academic development of the problem-solving workshop, and culture's impact, can be found in Avruch (1998, pp. 84–100).

based on culture-bound values and assumptions. Finally, they do not take for granted that an expedient (such as face-to-face negotiation) that works for one culture necessarily works for another. (p. 104)

Nevertheless, Zartman (1993) suggests that "culture" is too often used as an excuse for failure, while Lowi and Rothman (1993) use the water negotiations over the Jordan basin to show how cultural differences can actually be harnessed to induce more effective dialogue. Lederach (1995) agrees, "Culture is rooted in social knowledge and represents a vast resource, a rich seedbed for producing a multitude of approaches and models in dealing with conflict" (p. 120).

5.2.2 Summary

Clearly participatory processes are not add-ons to traditional water management. They are central to these processes as they transform water management policies. If some form of participation, whether it is the RBO as participatory means or the RBO using other external participatory means, is not present, integrated water resources management (IWRM) will not be achieved and basin organizations will be minimally effective.

Modern flood management, with its combination of structural and nonstructural measures, will not be achieved. Several RBOs have high formal power but relatively low meaningful participation. To the degree that participation is critical to IWRM, flood management, and RBOs, this gap will narrow. The French basin agencies and the Murray–Darling and the Potomac River commissions probably have the narrowest gap. The point is that high formal power does not necessarily mean high effectiveness in river basin or in flood management. However, high participation without clearly defined performance criteria can result in increased frustration and irresponsible and unimplementable decisions.

At minimum, participation relies on some basics, such as outlined in the Aarhus Convention (United Nations Economic Commission for Europe, 1998): access to information, actual involvement of broad range of stakeholder (meaning right of assembly), rights to access information, and rights or access in environmental matters. Each of these is a fundamental aspect of democratic civic culture. Although they may exist in varying forms and in varying institutional arrangements, some semblance of each is really necessary for participatory processes – and participatory processes are necessary for IWRM and to RBOs. Once begun, the experience of such participation within water management organizations actually becomes a learning ground for building this very democratic culture. Essentially such rights as they are exercised become clearer, more solidified, and indeed expected and codified.

5.3 PRACTICE OF TRANSBOUNDARY INSTITUTION BUILDING

Despite such barriers, people have and will continue to create transboundary water management institutions. Most of these attempts have been in regions within countries. As such, they provide the most fertile source of models for ways to overcome barriers presented by the conflict between human jurisdictional needs and the naturally integrative flow of water. There cannot be a direct transfer of such models from country to country, let alone to an international situation. Water, like politics, is very local. These examples, however, are a rich source of ideas for those negotiating international arrangements. They are sources for formulating purposes or end states of negotiations, for putting immediate event-orientated actions into longer-term evolution of transboundary management arrangements. They can be catalytic to creating possible future visions of shared water management. (Appendix B offers a sampling of such institutions from around the globe.)

Much of the history of water resources management has been a struggle to build institutions that are interjurisdictional (without too much impact on sovereignty) and intersectoral (without too much shock to the real politics of specialized knowledge and interests). This struggle has produced a variety of organizations that have had varying success in fostering collaboration and in allocating water but are rich with lessons for both the water and negotiations fields. We need to start mining this experience for its process and institutional lessons.

Figure 5.4 is a conceptual map to help make sense of this search. The horizontal axis represents various jurisdictions, including primary jurisdictions and subjurisdictions. The vertical axis represents sectors such as agriculture, transportation, and industry. Our water experience has sought to build institutions that fall across the matrix as they seek to allocate and value water, along with establishing and maintaining rights. These institutions are private as well as public and are testimony to great variety in our understanding of what subsidiarity means in water resources.

Much of the professional water-resources literature has examined one sector within a jurisdiction or the vertical space across sectors, but within principle and subsidiary jurisdictions (Figures 5.5 and 5.6). This can be seen in the evolution of water management from single-purpose to multipurpose procedures. California's water banking and the World Bank's call for cross-sectoral stakeholder participation in developing water strategies are two recent examples of these efforts. To varying degrees, this space is characterized by some laws, sanctions, and compliance.

Looking horizontally across sectors, we are often faced with weak laws and little enforcement (Figure 5.7). Early

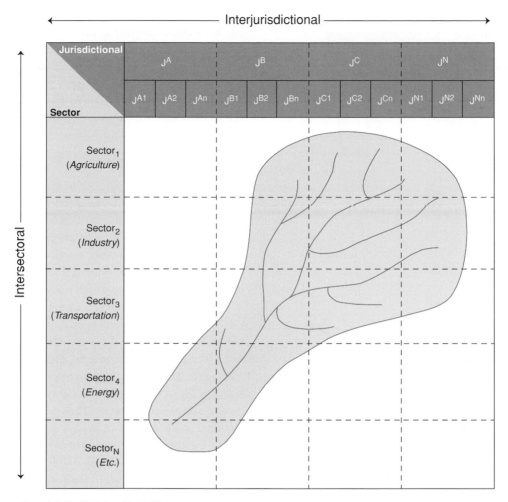

Figure 5.4 Conceptual model (Delli Priscoli, 2000a).

interjurisdictional water institutions grew out of specific sectoral needs, for example, in transportation. Many of these institutions have gradually expanded their authorities to other sectors. We have tended to fund both international and domestic water resources sectorally, however, thus pitting sector (technically defined interests) against jurisdictional logic, which manifests as arguments over what is political and technical.

Water has been treated as an end and as a means. In truth it is both. When water appears plentiful, it is easier to think of it as a means. In arid areas, this is less likely, and water is more likely to become an organizing principle for society. Indeed, there are those who argue, like Wittfogel, that the rise and fall of many civilizations can be traced to their social organization and management of water. If thought of as a means, it is easy to see water as a factor of production and in utilitarian terms. Water as an end often takes on a sanctity and value beyond utilitarian exchange. The West's three main religions – Christianity, Judaism, and Islam – were born in the arid Middle East environment, and water is central to the liturgy of each.

Clearly there is a balance to be reached between viewing water as an end or as a means. But this balance point will differ throughout the world. If left unexamined, value assumptions embedded in models of water institutions of humid areas can be disruptive for arid areas and vice versa.

Techniques and institutions will vary for different sections of the matrix. For example, water markets have long existed in subjurisdictions within one sector (Figure 5.6). But they are modified as they move out to multisector use. Our current need to build new water institutions is being done in the context of increased demands for water even in humid areas.

5.3.1 The international experience

There are many other international examples of regional institutions that cross the continuum. In Asia, the Indus River and its permanent commission has already been mentioned (Alam, 2002). After 1977, a Joint River Committee was established for the Ganges (Bingham, Wolf, and Wohlgenant, 1994; Swain,

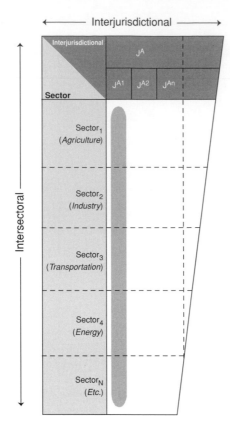

Figure 5.5 Cross-sectoral jurisdiction (Delli Priscoli, 2000a).

2004). Among other mandates it seeks to resolve disputes using joint expert committees. These committees have equal numbers of Indians and Bangladeshis. Unlike other expert commissions, such as now suggested in the current ILA draft, these committees do not include a neutral party from outside the region.

Also in Asia, the Mekong Commission, like the Indus Commission, has continued deliberations, even during periods of conflict (Le-Huu and Nguyen-Duc, 2003). Like many other river basin organizations, it started with a permanent advisory board of professional engineers. About 25 percent of its expenditures ($44 million seed and US$800 million attracted investment) are for data gathering and feasibility studies. Among its achievements are twelve tributary projects providing 210 megawatts of power and supplementary irrigation for 200,000 hectares, flood protection, pump irrigation, agricultural research, and extension, fisheries, and river navigation. However, as Kirmani (1990) notes, the Commission has, until recently, suffered from a weak sense of ownership among the parties of the region because it has been too dependent on external staff and support.

In South America, a Coordinating Intergovernmental Committee (CIC) was established for the La Plata basin, which helped prepare the treaty of La Plata basin. This arrangement can be seen as near the center of the continuum. The CIC responds under a conference of Foreign Ministries. Numerous binational entities and technical commissions have been established for the survey, design, construction, and operation of various water works in the La Plata basin. In practice, the institutional machinery has not worked well.

5.3.2 The intranational experience

Institutions, generally, and for water specifically, are almost by definition individual and culturally specific. Therefore, summaries of institutional lessons, although admirable, often become too general to be meaningful, or sometimes present well-meaning but confusing matrices, which seem to lose important details that actually tell the "on-the-ground story." However, countries and water managers cannot simply lift examples from one country to another.

The French river basins organizations have stood the test of time (see Appendix B). They have institutionalized participation with their water parliaments. They are a prime example of making the RBO define the participatory process. The interests participating in the water parliaments are, in effect, partners in the decision-making process. They generate revenue and affect behavior on the ground. Although somewhat narrow in focus at first, they have recently been trying to expand their planning roles. In doing so, they have also added participatory tools outside the institutional form of the basin parliaments to accommodate a broadening array of stakeholders and interests in the planning process. It remains to be seen whether they can really be successful in this endeavor. It is no surprise that variations of this model are probably the most replicated worldwide.

The Murray–Darling organization is at the forefront of basin organizations today. Like those in the United States it was set within a federalist system whose states are sovereign (McKay, 2005). It provides practical standards against which to design

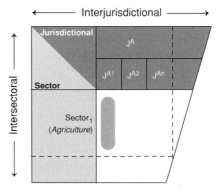

Figure 5.6 Single-sector jurisdiction (Delli Priscoli, 2000a).

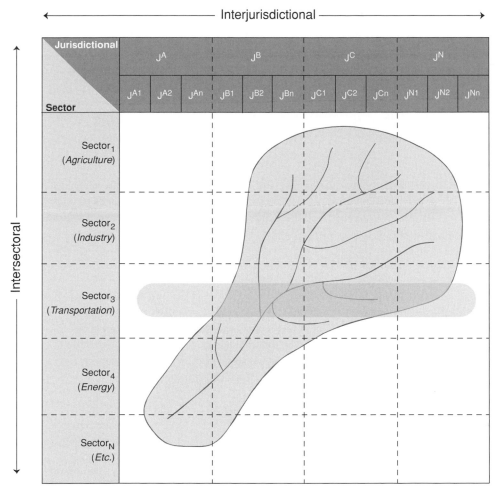

Figure 5.7 Cross-jurisdictional, single-sector authority (Delli Priscoli, 2000a).

RBOs. It can accommodate the full range of water uses, meaning integrating flood management and other uses, land and water, surface water and groundwater, and upstream and downstream activities. It uses citizen advisory committees, but it does so in ways that avoid the major pitfalls of such committees. The committee's influence and reporting requirements were built into the laws establishing the basin organizations from the beginning. The citizen committees have only limited prescribed structure but fit directly into the basin structure. They also have a direct line to the political decision makers. Thus, there is flexibility with clear lines of accountability, expectations, and knowledge of their level of influence. Expectations were clarified and the key actors, although they may have resisted this, knew why such was being done – from the beginning. The citizen committee structure is based on bottom-up knowledge, which feeds directly and meaningfully into the highest ministerial level, which, in effect, sets the agenda for the basin structure. That structure includes

states and the federal government. The committee structure helps set the agenda, which is worked out with the technical expertise of the basin organizations. Thus, the committees also help assure rapid implementation of actions that come out of the process. All of this has resulted in a cap for water uses.

The Mexican experience is important to many countries in Latin America and around the world (Tortajada and Contreras-Moreno, 2005). It represents a practical compromise of the various issues discussed in this chapter. The compromise is borne out of the realities presented by transforming political cultures striving for more open democratic practice, with increasing demands on water, with traditions of technical, bureaucratically driven, paternalistic engineering (most of the water world!), and with a tradition of central federal control over water. It remains to be seen whether the system survives, however, and the degree to which it really continues to include stakeholders and regional perspectives.

Table 5.1 *Socioeconomic progress in the Tennessee Valley and Columbia River basins*

	Tennessee Valley		Columbia[a]	
	Before 1920s–1930s	After +one generation	Before 1920s–1930s	After +one generation
Personal income per capita	$320 in 1930	$2,700 in 1968	$590 ($500–660) in 1930	$3,400 ($2,900–3,800 in 1968
Life expectancy (% of the population aged 65 years and over)	4.3% in 1920 (4.7% nationwide)	9.7% in 1970 (0.7% nationwide)	4.4% in 1920 (4.7% nationwide)	9.6% in 1970 (9.7% nationwide)
Illiteracy	10.3% illiterate among the population of persons 10 years old and over (6% nationwide) Urban = 7% (4.4% nationwide) Rural = 11.6% (7.7% U.S.)	Quasi – no illiteracy by 1920s census definition 94% of persons 14 to 17 years old in school (93% nationwide)	1.6% illiterate among the population of 10 years old and over in 1920 (6% nationwide) Urban = 1.5% (4.4% nationwide) Rural = 1.6% (7.7% nationwide)	Quasi – no illiteracy by 1920s census definition 88% of persons 14 to 17 years old in school (93% nationwide)
Access to water	2% of farms reported water piped in the house in 1920 (10% nationwide)	75% of homes were supplied with water by the public system or a private company in 1970 (82% nationwide)	23% of farms reported water piped in the house in 1920 (10% nationwide)	79% (70–86%) of homes were supplied with water by the public system or a private company in 1970 (82% nationwide)
Units with flush toilet	35% in 1940 (65% nationwide)	72% in 1960 (90% nationwide)	63% (45–74% in 1940 (65% nationwide)	93% (89–95%) in 1960 (90% nationwide)
Units with plumbing facilities	24% in 1940 (55% nationwide)	85% in 1970 (93% nationwide)	55% (38–65% in 1940 (55% nationwide)	96% (95–97% in 1970 (93% nationwide)
Electricity used as house heating fuel	4% in 1950 (0.7% nationwide)	40% in 1970 (8% nationwide)	2.5% (1.6–3.3%) in 1950 (0.7% nationwide)	23% (10–30%) in 1970 (8% U.S.-wide)
Energy source	2% of farms reported gas or electric light in 1920 (7% nationwide)	advanced 100%	13% of farms reported gas or electric light in 1920 (7% nationwide)	More than 50% of homes had central or electric heating in the mid-1950s
Flood damage	High % recurrent	$5.4 billion prevented	High % recurrent	$15.8 billion prevented
Commerce – industry	Little commerce	up 500%++	Light manufacturing	High production of aluminum for war needs

Source: Delli Priscoli, 2005b.

[a] Data for the Columbia River are averages of the states of Washington, Idaho, and Oregon. Data in parentheses show the range for all three states.

The system uses parallel structures of basin agencies (representing the formal public entities) and basin commissions (forums or water parliaments such as the French system). Unlike the French system, the commissions are more consultative rather than the partners in decision. Thus it is important to discern what level of influence on what type of decisions these commissions can continue to exercise. In its current form, the system can accommodate most uses of water, including integrating flood control with other uses.

Although the system may be easy to criticize, it has unleashed new expectations and opened up new avenues for stakeholders. Experience around the world has shown that these genies are almost impossible to put back in the bottle. This is good for IWRM and democratic decision making. This system is a quantum step beyond what still seems to be happening in those countries where the Ministry controls the RBOs (as in several Asian countries). In many countries, the tensions between the recognized need for river basin management and creating sufficiently open and decentralized organizational tools to do it, is growing. The Mexican experience offers some practical guidance to such countries on how to proceed.

The North American experience offers numerous examples of RBOs, including boundary organizations. There have been eight types of approaches to regions or transboundary basin management in the United States: (1) interstate compact commissions; (2) interstate councils; (3) basin interagency committees (ad hoc); (4) interagency–interstate commissions; (5) federal–interstate compact commissions; (6) federal–regional agencies; (7) single federal administrator; and (8) watershed councils/processes.

The Tennessee Valley Authority has clearly been the most successful comprehensive regional development agency built around river basins in the modern era. The TVA is an RBO that, in effect, was as regional social development agency. Its performance, in one generation, is remarkable (Table 5.1).

As a public corporation RBO for regional development, the organizational structure of TVA is still radical. It is, however, probably impossible to replicate anywhere, including the United States. It is too politically difficult. Nevertheless, the means TVA used to integrate revenues across water uses and to integrate social development strategies with water management remain important for other IWRM efforts.

Despite the success of the original organization, attempts at seven more TVA-type organizations spread across the United States, failed in the United States. This was called the "Valley Movement," but it never got off the ground. The authority was not a state and it was not fully the federal government. It was in effect, a new political-administrative organization. As such, existing sovereign entities, such as the states, saw it as a threat.

In terms of starting IWRM and RBOs in the developing world and where jurisdictional boundaries and sovereignty seem overwhelming, the Interstate Commission on the Potomac River Basin is very important. This is true because it is an RBO with little formal authority and high influence on the behavior of sovereign riparian entities. It uses various participatory means, such as joint modeling and stakeholder workshops, enhanced by new software technologies, to position itself as the servant of the entities themselves. As such, it has established itself as the neutral, essential monitor and provider of scenarios and future actions for the entities. Unfortunately, it is the least studied of the North American experiences. Although its focus has been on drought

Sidebar 5.1 Summary of Principles

Whatever models are used for building transboundary or international arrangements, the experience from regions and countries provides some important principles for international water negotiators as they design processes for basin arrangements.

1. Move beyond "impact fixation" to incorporating environmental and other values into creating alternatives, formulating options, and evaluating options and impact mitigation.
2. Bring implementation and operational interests into formulation process.
3. Give preference to operating at the lowest level possible and creating self-sustaining organizations.
4. Explicitly manage the "gray" area between technical and political.
5. Facilitate explicit negotiations among long-term visions and short-term efficiencies.
6. Help place water as driver, or first constraint, in cross-sectoral strategies and negotiations.
7. Use open and transparent rules of behavior.
8. Promote participation of those likely to be impacted, as well as disbursed beneficiaries.
9. Foster norms of collaborative behavior and move beyond reductionist expertise.
10. Better align internal cultural values of water organizations to those external values of collaboration and participation.
11. Facilitate the integration of upstream–downstream and surface water and groundwater uses.
12. Consider political viability the possible and transformative.
13. Let function dictate structure.
14. Create mechanisms that create, disseminate, and foster regional visions and "problemshed," or watershed, or basin-level visions.
15. Use process orientation.
16. Establish mechanisms for resolving disputes.
17. Separate administrative functions and fundamental policy issues and design mechanisms for accountability.
18. Promote flexibility and creativity.
19. Use process such as social learning of each other's interests to create incentives for parties.
20. Focus on creating benefits versus allocating flows.

contingency planning and supply, it has the capacity to work across all water used, from floods to water supply.

Although not an RBO in the formal institutional sense, the organization of the Columbia River Treaty is also instructive. First, it shows paths to negotiating over benefits versus simply fighting over allocation of flows. Second, it shows how to create benefits, the incentives for negotiations, by combining hydropower and flood control efforts across jurisdictional boundaries. In addition, it does this by incorporating U.S. federal agencies, Canadian federal agencies and Canadian provincial government, a parastatal, and the private sector.

Finally, the now-defunct Title II river basin commissions still offer many insights to those seeking to build RBOs. The reasons for their demise are many, but at the root are familiar political issues of cross-sectoral power fighting and cross-jurisdictional fighting, as well as unique situations in the U.S. political landscape at the time. However the commissions are instructive in how they set out levels of planning that accommodated and clarified where the public, federal, state, local, and private roles fit. Second, they included a uniform set of procedures, called principle and standards (P&S) for water planning. These still exist in modified form called principles and guidance. The P&S includes both the content and methods for reaching two goals – environmental quality and economic development – and four objectives: social well-being, economic development, regional development, and environmental quality. It is an accounting system and set of analytic procedures that can be used across agencies with different, even conflicting, missions. The P&S is the probably the closest handbook for line water agencies and organizations on practical analytical planning procedures for what is now called IWRM (Sidebar 5.1).

6 Lessons learned: Patterns and issues

Once I got the sign for water, I got the whole world.
— Hellen Keller, *My Life*

This chapter examines some of the lessons learned through attempts at resolving and managing past water conflicts with the hope that these will help decision makers deal with future water conflicts and enhance their ability to manage them. The chapter is not meant to provide a definitive topology for a generic watershed conflict or a checklist for a hypothetical mediator. Rather, it presents the observations of the authors about a relatively recent approach to the resolution of particularly vital resource conflicts. The chapter is divided into lessons learned in the four general stages of water conflict management: (i) assess the current setting, (ii) take the borders off the map, (iii) enhance the benefits, and (iv) return the borders for institutional capacity building and the equitable distribution of benefits. The focus is on themes and issues that recur throughout our survey of national and international water conflicts. Specific lessons implied by the discussion are presented throughout the chapter.

6.1 FOUR STAGES IN WATER CONFLICT TRANSFORMATION[1]

As the global experience with shared waters becomes more nuanced and sophisticated, a process is beginning to emerge that brings some order to the vast amount of information and disciplinary expertise necessary to move from conflict to cooperation. Imagine a hypothetical basin. Imagine it goes through four stages in its evolution from unilateral development and conflicting interests to coordinated development and shared interests. In a very general sense, the process of building effective transboundary water resources management can be

thought of in four stages of negotiation – adversarial, reflexive, integrative, and action:

Stage I: Initial State: Basins with Boundaries – Scale is interpersonal, focus is on trust building, and analysis is of parties, positions, and interests. Negotiations are often adversarial, with an emphasis on rights.

Stage II: Changing Perceptions: Basins without Boundaries – Scale is intersectoral, focus is on skills building, and analysis is on gap between current and future states. Negotiations move to the *reflexive* stage, and parties define *needs*.

Stage III: Enhancing Benefits – Scale moves beyond the basin, focus is on consensus building, and analysis is on benefits of cooperation. Negotiations are *integrative*, where parties define *benefits*.

Stage IV: Putting It All Together: Institutional and Organizational Capacity and Sharing Benefits – Scale is international, focus is on capacity building, and analysis is on institutional capacity. Negotiations are in the *action* stage, where *equity* is defined and institutionalized.

Although there are no "blueprints" for water conflict transformation, there does seem to be general patterns in approaches to water conflict, which have emerged over time. "Classic" disputes between, for example, developers and environmentalists, rural and urban users, or upstream and downstream riparians suggest zero-sum confrontations, where one party's loss is another's gain and where confrontation seems inevitable. Yet such "intractable" conflicts are regularly and commonly resolved. Over time, creative thinking and human ingenuity allow solutions that draw on a more intricate understanding of both water and conflict to come to the fore, and parties identify joint gains individual and shared interests beyond positions.

This section offers observations about one path to the transformation of water disputes – from zero-sum, intractable disputes to positive-sum, creative solutions – and centers on a migration of thought, generally through four stages. Note that

---

[1] The structure developed in this section draws from a World Bank course skills-building workbook, which was published as Wolf (2008). Len Abrams crafted the "world," and maps of the fictional Sandus River basin.

all stages exist simultaneously and need not be approached in sequence and that no stage must necessarily be achieved for "success." In today's world, many disputes never move beyond the first or second stage, yet are tremendously resilient, whereas a few have achieved the fourth stage and are fraught with tension. Nevertheless, like any skill, it is useful to understand the structure of an "ideal" path, to perfect the tools required for any individual situation. Indeed, much of the negotiations and interactive processes are also social learning and even experimentation.

Although conceptually this can look like a neat linear evolution through four stages, parties within states and between states will go through such stages over long periods of time. For example, successful river basin organizations in the United States have taken more than 50 years to evolve to working primarily in Stage IV. The Danube and Rhine have evolved over most of the twentieth century, changing functions and increasing in salience over time with the experiences of responding to triggering events. As parties go through these stages they may progress to one stage only to return to another. Some attempts may die in one of the stages. Often efforts may progress, seem to regress, and then remain dormant until another triggering event. Much depends on what the parties experience within the basin, how the experience affects their relationship with other parties, and the recurrent triggering events. As the unilateral attempts of parties seem not to solve the problem, parties might try threat, power or (within states) court cases. As these threat and court fightings do not produce implementable and shared approaches, parties often come to see that the price of that type of collective action is likely to be finding ways to work with each other. But this takes time for the parties to reflect on costs of unilateral actions in the face of minimal returns, to think of what the alternative to no agreement may be, and to understand what it means to see the basin as a whole and what basin-level benefits may be. Frequently, external intervention that produces incentives for cooperation combines with triggering events, such as massive floods, to move parties to different stages. The point is that these stages are really not linear but are iterative.

As parties move along in time, they may, however, be progressively spending more time in different stages, as Figure 6.1 shows. For example, in early efforts, a great deal of time may be spent assessing and reassessing the basin, and little time is spent on solutions. However, over time, the parties will presumably be spending more time creating institutions and far less time talking about assessing – even though the assessing and reassessing will continue. The generalized path described here is structured around an understanding of each of the four stages through any of four perspectives, as described in Figure 6.1.

6.1.1 Stage I. Assess the current setting: Basins with boundaries

In Stage I, in its initial, *adversarial*, setting, regional geopolitics often overwhelm the capacity for efficient water resources management. Metaphorically, the political boundaries on a map at this stage are more prevalent than any other boundaries, of interest, sector, or hydrology. Dialogue is often focused on the past, based on the *rights* to which a country or state or province feels it is entitled, and a period of expressing pent-up grievances can be necessary. As a consequence of these initial tensions, the collaborative learning emphasis is on *trust building*, notably on active and transformative listening, and on the process of conflict transformation. By focusing primarily on the rights and interests of countries, states, and/or provinces, inefficiencies and inequities are inevitable (Figure 6.2).

Virtually all negotiations occur within some historical context – they did not just spring to life. Things, including previous negotiations, have been experienced. Indeed, there will inevitably be future negotiations. Thus, it is important to understand this history. Indeed, as Delli Priscoli and Montville (1994) and others note, it can be positive for parties to walk through this history together. But there will also be a future, even if the parties break off and negotiations seem dormant. How the parties interact with each other in the present will affect how they interact in the future. If one party adopts a "blow them out of the water" attitude now, that party might just experience that attitude in the future from the other party.

Previous chapters have outlined numerous methods to help identify stakeholders and bring them to talking. In addition there are numerous training and other skill-building resources to help. The skills developed in alternative dispute resolution (ADR) and conflict management and participation fields will have great payoffs here.

Within countries, water disputes have become multiparty/multi-issue. The stakeholders can be public, private, nongovernmental organizations (NGOs), or other types of groups. This increased range of stakeholders is important to international disputes. In international disputes, we speak of parties as if the parties are unified entities, but that rarely reflects reality. The same multiparty/multi-issue environment that produces multistakeholders is usually bubbling just below the surface within the parties and putting pressure on the states themselves. But more than this, the parties are made of people who themselves represent a variety of constituencies woven within countries. For example, a party (nation-state, state, and/or province) may have the ministries of water, agriculture, land, finance, and others involved on their team. So, too, will the other parties. In fact, some of the interests represented by these diverse groups

Negotiation stage (a)	Common water claims (b)	Collaborative skills (c)	Geographic scope
Adversarial	Rights	Trust building	 Nations
Reflexive	Needs	Skill building	 Watersheds
Integrative	Benefits	Consensus building	 "Benefit-sheds"
Action	Equity	Capacity building	 Region

Figure 6.1 Four stages of water conflict transformation. (a) These stages build primarily on the work of Jay Rothman, who initially described his stages as ARI – adversarial, reflexive, and integrative (Rothman, 1989). When ARI becomes ARIA, adding action, Rothman's terminology (1997) also evolved to antagonism, resonance, invention, and action. We retain the former terms, feeling they are more descriptive for our purposes. (b) These claims stem from an assessment of 145 treaty deliberations described in Wolf (1999). Rothman (1995) too uses the terms *rights*, *interests*, and *needs*, in that order, arguing that *needs* are motivation for *interests*, rather than the other way around, as we use it here. For our purposes, our order feels more intuitive, especially for natural resources. (c) These sets of skills are drawn from Kaufman (2002), who ties each set of dynamics specifically to Rothman's ARIA model in great detail, based on his extensive work conducting "Innovative Problem Solving Workshops" for "partners in conflict" around the world.

may more closely parallel their counterparts within nation-states of the parties in other state ministries than the aggregate of their own state team. For example, construction ministries may have more in common with each other than with environmental ministries within their own nation-state or state or province party. To add to this, each of the parties will be pressured from outside their delegations by various national and international stakeholders, usually NGOs. These NGOs have increasingly begun to form coalitions among themselves across borders and present unified approaches for pressuring the national parties involved in the negotiations. Indeed, this common interest is often a strong incentive for national parties to begin to see the basin as a whole, as it is often such cross-boundary pressures that see beyond traditional national boundaries. At the same time, these cross-boundary pressures might work at cross purposes with more local NGOs within each country that might feel threatened by actions taken at high levels that hurt their more local interests.

Figure 6.3 portrays these relationships in a diagram of two-party negotiations around a table. It is easy to see how management of these negotiations within the parties is often as difficult as those between the parties themselves. When you add more parties to the mix, as is more frequently the case, the management difficulty increases rapidly. Figure 6.4 shows this as five parties around a table. Both figures show that horizontal bargaining is going on within parties. In addition, there is

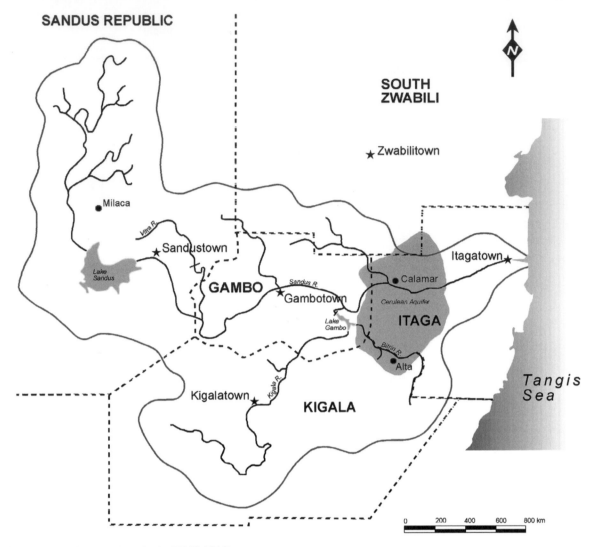

Figure 6.2 Map of the Sandus River basin (Wolf, 2008).

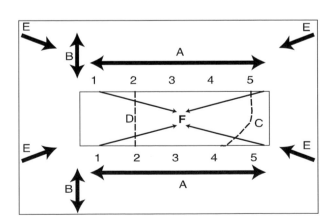

Figure 6.3 Schematic of two-party, multi-interest negotiations (Moore, 1985).

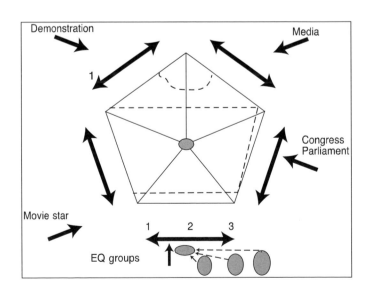

Figure 6.4 Schematic of multiparty, multi-interest negotiations (Moore, 1985).

informal bargaining most likely going on among parties within each party. There is bargaining going on among external parties who are pressuring the country bargaining team and so forth. Consequently, stakeholder identification and management and an understanding of the situation is of major importance and are one reason why outside intervention is also often needed.

Once stakeholders are brought to the table, this stage generally involves classic hydropolitical assessments of the current setting within a basin, including biophysical, socioeconomic, and geopolitical parameters. The processes for assessing many of these aspects are well defined (e.g., hydrologic studies or benefit–cost analyses of development alternatives), whereas many are less quantitative but no less critical (e.g., social impact statements or assessments of indigenous traditions of management).

At this stage, stakeholders often think *nationally*, or provincially are focused on their *rights*, and may be looking disproportionately backward, if only to be able to vent and perhaps address perceived grievances. Although understanding the baseline of any basin may take decades, if it is possible at all, it is not necessary to agree to all data before greater cooperation takes place – these assessments or training workshops can be used in and of themselves as confidence-building measures to move to the next stage, even as greater mutual understanding of the basin is being created.

6.1.2 Stage II. Changing perceptions: Basins without boundaries

As the adversarial stage plays out, occasionally some cracks can be seen in the strict, rights-based, country- (province/state) based positions of each side (although in actual water negotiations, this process can last decades). Eventually, and sometimes painfully, a shift can start to take place when the parties begin to listen a bit more, and when the interests underlying the positions start to become a bit apparent. In this Stage II, a *reflexive stage*, negotiations can shift from *rights* (what a country state/province feels it deserves) to *needs* (what is actually required to fulfill its goals). Conceptually, it is as if we have taken the national, provincial/state boundaries off the map and can, as if for the first time, start to assess the needs of the watershed as a whole. This shift, from speaking to listening, from rights to needs, and from a basin with boundaries to one without, is a huge and crucial conceptual shift on the part of the participants that can be both profoundly difficult to accomplish and absolutely vital to achieve for any movement at all toward sustainable basin management. To help accomplish this shift, the collaborative learning emphasis is on *skills building*, and

we approach the (boundary-less) basin by sector rather than by nation (Figure 6.5).

At this stage, the attention shifts from past to future, as stakeholder examine each others' interests beyond positions. They go beyond seeing negotiation as a competitive sport. A process of social learning sets in. Parties can begin to ask, "What could be?" rather than "what was?" or "what is?" The metaphor for this stage is a basin without borders, where rather than *rights*, there are *needs*; rather than allocating water, we can think about allocating benefits; and rather than thinking of national issues, we might look instead to how different sectors might be developed basinwide.[2]

This shift is transformative – the point at which parties move from thinking of themselves as representing countries or states/provinces to perceiving more broadly the needs of all stakeholders within a basin (whether or not they like these needs) – and this transformation can take years or even decades, during which time political tensions are exacerbated, ecosystems go unprotected, and water is generally managed, at best, inefficiently. Parties begin to understand the needs of the other and thus the requirements that must be met, if binding agreements are to be reached.

6.1.3 Stage III. Enhancing benefits: Beyond the river

Once participants have moved in the first two stages from mostly speaking to mostly listening, and from thinking about rights to needs, the problem-solving capabilities that are inherent to most groups can begin to foster creative, cooperative solutions. In this Stage III, an *integrative stage*, the needs expressed earlier begin to coalesce to form group interests – the "why" underlying the desire for the resource. Conceptually, we start to add *benefits* to the still-boundary-less map, and in fact to think about how to enhance benefits throughout the region, primarily by adding resources other than water and geographic units other than the basin. The collaborative learning emphasis is now on the *consensus building* of the group, and we begin to move in "benefit-shed" rather than being restricted by the basin boundaries. This is often the start of negotiations and it is often called "expanding of the pie." It is the start of moving beyond conceiving of the situation as purely zero-sum gaming into the zone of creating joint gains (Figure 6.6).

Once the shift has been made in thinking about allocating water to allocating benefits, it is a natural progression to think together about how to enhance the benefits within and beyond

---

[2] In the World Bank Course, *sectors* are defined by those in the UNDP "comb" of Integrated Water Resources Management: water supply and sanitation, irrigation and drainage, energy, and environmental services. In addition to these technical sectors, we add social and spiritual needs.

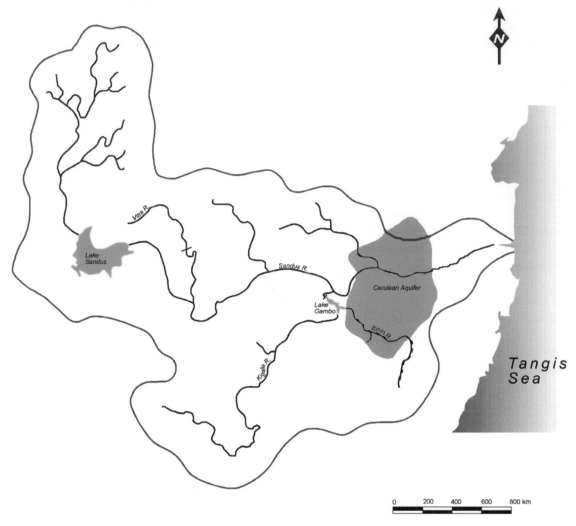

Figure 6.5 Sandus River basin (Changing perceptions: Basins without boundaries).

the basin. This may be done within the realm of water resources alone – a well-designed dam upstream might, for example, both enhance agricultural production downstream and help protect riparian habitat. But it is often helpful to think at this stage about "baskets of benefits" that may go well beyond water or well beyond the basin in question. Indeed, the most successful cases of building regional approaches to water have gone beyond seeing water as the ends to seeing it as a means to achieve other goals, such as socioeconomic development and reduction of fears of floods and drought. Energy production and water development are often linked, for example, as are afforestation programs, transportation networks, and environmental protection. Naturally the transaction costs of including more sectors than water goes up exponentially, but so do the potential benefits. This means bringing in actors beyond the water sector and expanding the basket to be considered.

6.1.4 Stage IV. Putting it all together: Institutional and organizational capacity and sharing benefits

Finally, although tremendous progress has been made over the first three stages, both in terms of group dynamics and in developing cooperative benefits, Stage IV (the last, *action*, stage) helps with tools to guide the sustainable implementation of the plans and to make sure that the benefits are distributed fairly *equitably* among the parties. The scale at this stage is now *regional* where, conceptually, we need to put the political boundaries back on the map, reintroducing the political interest in seeing that the "baskets" that have been developed are to the benefit of all. The collaborative learning emphasis is on *capacity building*, primarily of institutions (Figure 6.7).

Much as water people like to think in terms of basins or watersheds alone, eventually the borders have to come back on

Figure 6.6 Sandus River basin (Growing benefits throughout and beyond the basin).

the map – political entities are primarily responsible for their own benefits and sovereignty, after all, and it is often hard to sell their own constituents on an integrated basin alone. The most critical issues at this stage are "how can the benefits be distributed equitably or perceived as fair" and "how can sustainable and resilient institutions be crafted?" "How are the existing institutions and organizations to be taken care of or compensated for any change?" The first question may require trade or side payments, whereas the second and third questions must evoke the best in institutional design. It is important to remember that conflict potential can actually increase during periods or situations of increased benefits. The increase of benefits alone will not assure the mitigation of conflict. This is because parties may realize benefits they never had, but they may perceive that the other is getting relatively more benefits than they are getting. This problem of relative deprivation that

Gurr identifies is the main cause of violence. Thus the perceptions of fairness, not just the tangible delivery of benefits are critical. This is often hard for the technical engineer to understand.

Parties can undertake a number of actions to create collective actions. Table 6.1 shows a range of actions from simple agreements to sophisticated transboundary organizations.

6.1.5 Analytic framework

Sadoff and Grey (2002, 2005) suggest two spectrums that together define the level of cooperation among riparians on international waterways. The first, described in Section 6.1.4 and in Figure 4.1, delineates increasing cooperative integration, beginning with "dispute" and increasing to total "integration." As we have noted efficient water management

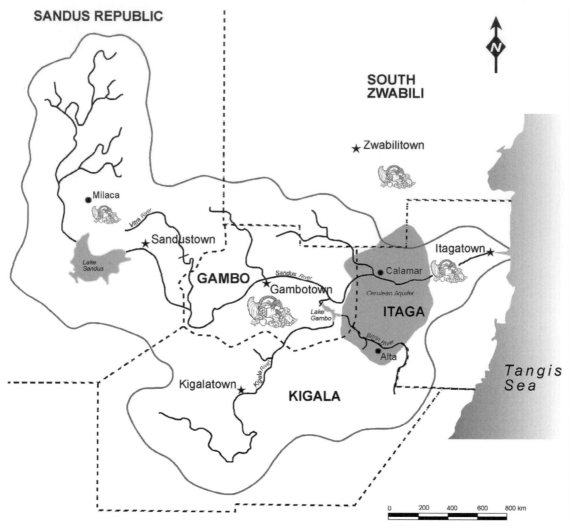

Figure 6.7 Sandus River basin (Putting it all together: Institutional capacity and sharing benefits).

generally trends toward increasing integration, whereas the political needs to protect national sovereignty trend in the opposite direction. The other spectrum is the *type* of benefit that can be gained through cooperation and includes:

Type 1: Benefits to the river – the ecosystem protection that is best gained through cooperative management

Type 2: Benefits from the river – in increased food and energy production, for example

Type 3: Reduction in costs because of the river – primarily the political and economic costs of a conflictive setting, which can be reduced through hydrocooperation

Type 4: Benefits beyond the river – branching out to increase the "basket of benefits" through greater cooperation and even infrastructural and economic integration

The following case studies represent different examples of both the level and integration and the type of benefits, as shown in Figure 4.1.

BOUNDARY WATERS AGREEMENT BETWEEN CANADA AND THE UNITED STATES

This agreement allocates water according to equal benefits, usually defined by hydropower generation. This allocation results in the seemingly odd arrangement in which power may be exported out of basin for gain, but the water itself may not. In the 1964 treaty on the Columbia, an arrangement was worked out where the United States paid Canada for the benefits of flood control and Canada was granted rights to divert water between the Columbia and Kootenai for hydropower. The relative nature of "beneficial" uses is exhibited in a 1950

Table 6.1 *Integration versus transaction costs: Transboundary management structures*

Structure	Number of tasks	Potential for disagreement	Sovereignty infringement	Transaction costs
Unilateral development	Many	High	None	n/a
Watershed monitoring	Single	Low	None	Low
Technical research coordination	Single	Low	None	Low
Resource conservation	Single	Low	None	Low
Training center	Single	Low	None	Low
Apportionment body	Single	High	Limited	Medium
Arbitration body	Single	High	Limited	Medium
Apportionment monitoring	Single	Moderate	None	Low–medium
Investigative advisory body	Few	High	Limited	Medium
Risk management	Few	High	Limited	Medium
Pollution control	Many	Moderate	Significant	High
Joint regulatory bodies	Several	High	Major	Very high
Wastewater utility	Several	Moderate	None	Medium
Water utility	Several	Moderate	None	Medium
Economic development	Several	Moderate–high	Limited	Medium–high
Project management	Several	High	Limited	Medium–high
Water transfers or markets	Several	Moderate	Limited	High
Comprehensive utility	Many	High	Limited	High
Integrated watershed management	Many	Very high	Major	Very high
Centralized joint management	Many	Very high	Major	Very high

Source: Feitelson, (2006).

agreement on the Niagara, flowing between the United States and Canada, which provides a greater flow over the famous falls during "show times" of summer daylight hours, when tourist dollars are worth more per cubic meter than the alternate use in hydropower generation (for further details, see Appendix C).

MEKONG BASIN

In 1957 the creation of the Mekong Committee for Coordination of Investigations of the Lower Mekong Basin was an early attempt in the later part of the twentieth century of UN involvement in a program to develop an international river basin. The 1975 Mekong accord defines "equality of right" not as equal shares of water but as equal rights to use water on the basis of each riparian's economic and social needs. The new Mekong Agreement was signed in 1995 after a relatively short period of negotiation benefiting from a shared data base, long-established relationships, and familiarity of the key players with the provisions of relevant international jurisprudence. The Mekong Agreement clearly states the mutual commitment to cooperate. It established the Mekong River Commission as the international body that implements the Agreement and seeks cooperation on all aspects of water management (see Appendix C).

INDUS BASIN

Despite three wars and numerous skirmishes since 1948, India and Pakistan, with World Bank support, have managed to negotiate and implement a complex treaty on sharing the waters of the Indus River system. The Indus Waters Treaty was finally signed in 1960. During periods of hostility, neither side targeted the water facilities of the other or attempted to disrupt the negotiated arrangements for water management (see Appendix C).

NILE RIVER BASIN

The political will to achieve a basinwide agreement and framework for long-term cooperation on the part of the ten Nile Basin riparian states is gathering pace. In 1992, representatives of all ten states agreed on a Nile River Basin Action Plan, with the task of developing a cooperative scheme for the management of the Nile. In 1995, the World Bank, together with UNDP and the Canadian International Development Agency, accepted the request from the Nile riparian states to give impetus to the project. In 1999 the Nile Basin Initiative was launched, with the membership of all basin states. The international community has facilitated an ongoing dialogue between the riparians of the Nile Basin to a process of dialogue and joint planning (see Appendix C).

DANUBE RIVER BASIN

The Danube Convention is a vital legal continuation of a tradition of regional management along the Danube dating back
140 years. As a document, it provides a legal framework for
integrated watershed management and environmental protection along a waterway with widespread potential for disputes.
The Environmental Program for the Danube River is also a
basinwide international body that actively encourages public and NGO participation throughout the planning process.
This proactive stakeholder participation may help preclude
future disputes both within countries and internationally (see
Appendix C).

JORDAN RIVER BASIN

As we have previously mentioned, even while Israel and Jordan
were legally at war, Israeli and Jordanian water officials met
several times a year at so-called Picnic Table Talks. As a result,
when the Jordan−Israel Peace Treaty was signed in 1994, it
was possible to include a well-developed annex acknowledging
that "water issues along their entire boundary must be dealt
with in their totality" (see Appendix C).

6.2 INTEGRATION VERSUS TRANSACTION COSTS: TRANSBOUNDARY MANAGEMENT STRUCTURES

Creating the ultimate organizations that include all conceivable
issues is not necessary to begin to see success. Far from it.
Most transboundary organizations within countries and among
nations usually start with limited objectives and power. Over
time, because they provide a safe space for nations to negotiate,
they are increasingly used and generally grow. If they do not,
they are unlikely to remain. Thus, cooperation could start with
individual studies of assistance. Or it could include full-blown
regional authorities. However, those organizations lower down
in Table 6.1 usually emerge after a long history of cooperative
attempts and many iterative rounds of the four stages outlined
here.

Nevertheless, research on this growth has begun to show
some lessons. First and foremost, any agreement or transboundary organizations must be built on constituencies from
the bottom up but with buy-in at the highest levels. That is,
the parties on the ground must see something in it for themselves. Those organizations imposed from the top or bottom
alone will fail. That happened with the United States and within
other countries. It has happened in numerous cases in Africa
with donor-driven originations that engendered only minimum
on-the-ground support, where the donor role should have been
primarily to encourage and facilitate the parties to come up
with their own design – not to dictate that design.

6.3 LESSONS LEARNED THROUGHOUT THE FOUR STAGES

6.3.1 Lessons learned: Stage I – basins with boundaries

NATIONAL VERSUS INTERNATIONAL SETTINGS

It should be clear from the cases presented in this study that
national and international experiences with managing conflicts
that cross jurisdictional boundaries have both both inherent
similarities and distinct differences. The differences are more
often stressed, but just how different the two settings are is open
to debate. This is especially true with the experiences of large
federalist countries that are built on smaller sovereign entities
called states or provinces. Assumptions about the differences
between national and international settings that are common
include institutions and authority, law and enforcement, and
presumption of equal power.

INSTITUTIONS AND AUTHORITY

National cases often are played out in relatively sophisticated
institutional settings, particularly in the developed world, while
international conflicts can be hampered by the lack even of an
institutional capacity for conflict resolution.

Even sophisticated institutions, however, have often not
been amenable to relinquishing the traditional, usually legal,
approaches to resolving water conflicts, effectively presenting
the same challenges as the international setting.

LAW AND ENFORCEMENT

The United States and other countries have over the years
established intricate and elaborate legal structures to provide
both guidance in cases of dispute, and a setting for clarifying conflicting interpretations of that guidance. Because the
United States is a federalist system, the states are sovereign
and have sovereign control over the waters. The federal interest has evolved over time due to circumstances where sovereign
states conflict or where national standards are needed. Initially,
the founders of the United States thought that water would be
handled by interstate compacts – the equivalent of interstate
treaties on the international scale. But this proved not to be
enough, for reasons similar to those debates today regarding
international waters. The United States starts from a decentralized political system and thus a decentralized system of
water management. It starts by managing water at the lowest
possible levels. This process has produced a sector with over
300,000 people employed at state and local levels in water,

over 60,000 at the national level, and many more in the private sector. This process has resulted in frustrating duplication and overlap and in conflicting laws and confusion. But it is important to remember that it has resulted from the decentralization that many in the world water community say they now seek. It has also resulted in widespread successful capacity building in water management. International disputes, in contrast, rely on poorly defined water law, a court system in which the disputants themselves have to decide on jurisdiction and frames of reference before a case can be heard, and little in the way of enforcement mechanisms. One result is that international water conflicts are rarely heard in the International Court of Justice. Likewise, of the international cases presented in this volume, only the Mekong Committee and the Southern African Development Community (SADC) protocol have used the legal definition of "reasonable and equitable" use in their agreements.

In the legal realm too, it has been argued that the differences between national and international disputes are more apparent than real. Given the myriad of legal venues open to disputants and the ambiguities of court jurisdiction, creative lawyers can effectively hamstring legal challenges for years, essentially creating a de facto lack of legal authority.

PRESUMPTION OF EQUAL POWER

"All are equal in the eyes of the law" is a common phrase describing national legal frameworks. No such presumption exists in international conflicts, where power inequities define regional relations. Each of the watersheds presented here includes a hegemonic power that brings its power to bear in regional negotiations and that often sees agreements tilt in its favor as a consequence. However, power can be exercised in a variety of ways. Even a nonhegemonic nation, state, or province can exercise certain types of power, such as moral authority. Indeed, the use of interest-based approaches and assisted negotiations are very helpful in bringing such forms of power more clearly into focus before they become destructively used.

Here, too, it has been argued that unequal resources, usually financial or political, result in real-world inequities finding their way into the national settings of conflict resolution as well.

BEST ALTERNATIVE TO A NEGOTIATED AGREEMENT (BATNA)

A difference commonly pointed out between national and international disputes is that, in national water conflicts, war is not usually a realistic BATNA. Although it may be true that intranational "water wars" are not likely, the same is increasingly accepted as being true of the international setting. Although shots have been fired, both nationally and internationally, and

Table 6.2 *Flash points*

International basin	Flash point
Danube	None
Ganges	Farakka Barrage (India)
Indus	Diversion of tributaries (India)
Jordan	Development on border (Israel)
Mekong	None
Multilaterals	None
Nile	Plans for high dam (Egypt)

(Observations taken from case studies in Appendix C.)

troops have been mobilized between countries, no all-out war has ever been caused by water resources alone. Although real differences do exist between the national and international settings for water conflict resolution, these distinctions may not be as great as is often thought. The fortunate corollary to this is that many of the successes of dispute management, ADR, and of building transboundary organizations in nations may be more applicable to the international setting than is commonly argued.

ANTICIPATING CONFLICT AND ACTIVE INVOLVEMENT IN ADVANCE OF CONFLICT

Most of the international water conflicts presented here, with the notable exception of the Mekong and the Danube, are defined by a flash point, a single action on the part of a riparian that led to impending conflict, or recurrent disasters, such as floods or drought, that cannot be managed without cross-jurisdictional cooperation, which only then led to attempts at conflict resolution (Table 6.2). It is worth noting that in the exceptions to this pattern, the Mekong and the Danube, an institutional framework for joint management and dispute resolution was established well in advance of any likely conflict. It is also worth noting the Mekong Committee's impressive record of continuing its work throughout intense political disputes between the riparian countries, as well as that data conflicts, common and contentious in all of the other basins presented, have not been a factor in the Mekong (see Ringler, 2001).

It might be suggested, drawing both from our exceptions, the Mekong and the Danube, as well as from other basins, that when international institutions are established well in advance of water stress these institutions help preclude such dangerous flash points. The single most important lesson that comes out of the global experience in shared waters follows from that pattern.

Lesson: Water conflict mitigation is best attempted before conflicts arise within a watershed. Such an institutional framework for conflict management helps preclude data disputes and provides a pattern of cooperation in the absence of the intense political tensions of a flash point.

Early intervention is also beneficial to the process of conflict resolution, helping to shift the mode of dispute from costly, impasse-oriented, reactive dynamics to less costly, problem-solving planning dynamics. In the heat of some flash points, such as the Nile, the Indus, and the Jordan, as armed conflict seemed imminent, tremendous energy was spent just getting the parties to talk to each other. Hostilities were so pointed that negotiations inevitably began confrontationally, usually resulting in a distributive approach being the only one viable.

In contrast, discussions in the Mekong Committee, the multilateral working group in the Middle East, and on the Danube, have all moved beyond the causes of immediate disputes on to actual, practical projects that may be implemented in an integrative framework.

To be able to entice early cooperation, parties need to perceive incentives. In all of the cases mentioned in this section, not only was there strong third-party involvement in encouraging the parties to come together, but extensive funding was made available on the part of the international community to help finance projects that would come from the process. There is also a history with which all of the activity fits. This suggests the following observation.

Lesson: Not only are outside-party involvement and assisted negotiations vital in bringing about international water conflict resolution, that involvement is most effective if active and backed by both the financial and political support of the international community.

Given that the international community has neither the resources nor the time to help establish a basinwide institution for integrated watershed management as summarized by Bulkley (1995) on each of the world's international rivers and aquifers, patterns do emerge that may be useful in allowing for anticipation of likely conflict.

Lesson: Scarce international resources might be focused most efficiently where the likelihood for intense water conflicts is high, as well as on refining measurements for early-warning indicators and identifying obstacles or where sharp disparities in wealth exist among bordering countries, where water as a resource that can generate wealth has little investment or uneven investment.

Lesson: Mechanisms must be found that provide money to support up-front collaboration and discussion or what can loosely be called "collaborative planning."

It is the key to creating options and shared visions. It is also the key to creating some sense of criteria for discriminating among options. It also maximized the creativity of the engineering community with its impulse to create versus the demand to defend options. (This is also an area of high leverage for small investment often not more than say 1 or 2 percent of large project cost and it is also an activity that will allow more donor coordination and grease the skids for donor and lender funding.)

6.3.2 Lessons learned: Stage II: Basins without boundaries

ASPECTS OF WATER RESOURCES THAT CAN ENCOURAGE COOPERATION

Just as there are difficulties inherent in water-resource conflicts brought on by the qualities particular to the resource, so too does water-resource planning and development offer specific aspects that can encourage cooperation among riparians. A comparatively recent subfield in conflict management, collaborative process and ADR, "dispute systems design," is a process of integrating the potential for ADR in public institutions and other organizations that deal with conflict. Described by Ury, Brett, and Goldberg (1988), and initially instituted by USACE in selected programs, "dispute systems design" may offer lessons to cooperation enhancement in water systems as well. Although most of the work in this field describes incorporating cooperation inducement within organizations, some of the same lessons for "enhancing cooperation capacity" or "design considerations and guidelines" might be applicable to technical or policy systems as well. A water-sharing agreement, or even a regional water development project, for example, might be designed specifically to induce cooperation in ever-increasing integration from the beginning.

In a study of the history of conflict and cooperation over water resources in the Jordan basin, Wolf (1994) described two issues at the heart of resolving water conflict – an equitable allocation of existing resources and control of one's own major water sources. Only when these two issues of equity and control are addressed, it is argued, can the riparians move forward to build increasingly integrated infrastructure. The lessons of that particular basin may be applicable to other contentious watersheds as well. That is, that cooperation-inducing water-resources implementation be pursued along the following general guidelines.

"DIS-INTEGRATING" CONTROL OF WATER RESOURCES

Equitable allocations and control of one's major water sources are of primary concern to each riparian entity, are usually necessary to address past and present grievances, and are prerequisites for market-driven solutions. As such, an initial separation of resources within the basin might be advisable. Because these steps involve a separation of control as a precondition to "integration," the process might be referred to as "dis-integration."

EXAMINING THE DETAILS OF INITIAL POSITIONS FOR
OPTIONS TO INDUCE COOPERATION

Each party to negotiations usually has its own interests upper-most in mind. The initial claims, or "starting points" in the language of conflict management, collaboration, and ADR, often seek to maximize those interests. By closely examining the assumptions and beliefs behind the starting points, one might be able to glean clues for inducing some movement within the "bargaining mix" of each party. Parties, in effect, begin to educate each other on their interests beyond positions. Frequently this dynamic changes the perception of interests within the parties themselves. These underlying beliefs may also provide indications for the creative solutions necessary to move from distributive bargaining over the amount of water each entity should receive, to integrative bargaining – inventing options for mutual gain.

DESIGNING A PLANNING PROCESS OR PROJECT, FROM
SMALL-SCALE IMPLICIT COOPERATION TO EVER-INCREASING
INTEGRATION

Building on the first two steps, riparians who have clear water rights and control of enough water for their immediate needs might begin to work slowly toward increasing cooperation on projects or planning. Even hostile riparians, it has been shown, can cooperate if the scale is small and the cooperation is secret. Building on that small-scale cooperation, and keeping the concerns of equity and control firmly in mind, projects might be developed to increase integration within the watershed, or even between watersheds, over time.

The "cooperation-inducing design" process can be described as moving from small and doable projects to ever-increasing cooperation and integration, remaining always on the cutting edge of political relations. This process has been applied to water rights negotiations, as is currently the case between Palestinians and Israelis, to watershed planning, such as the incremental steps of the Mekong Committee, or to cooperative projects for watershed development, such as the Middle East multilateral working group on water.

Ironically, many of the same aspects of water resources that make them conducive to conflict also allow their management to induce cooperation. These characteristics include:

- Physical parameters and ecological concerns – The fluctuations inherent in the hydrologic cycle result in countries having disparate quantities at differing times, allowing options for trade, as explored earlier.
- "Wheeling" – Water resources, like energy resources, can be traded stepwise over great distances. Any addition to the water budget in the Jordan watershed, for example, can be "wheeled" anywhere else. Litani or Turkish water diverted into the Jordan headwaters in Israel, for instance, can be "credited" for Yarmuk water to Jordan, which in turn might allow more water in the lower Jordan for the West Bank, which might result in surplus West Bank groundwater being diverted to Gaza, and so on. However, this cost-saving practice of "wheeling" can be achieved only when infrastructure is designed for future cooperation from the beginning.
- Structural considerations – Not only can water-resources infrastructure be designed for possible future cooperation, topographic and hydrographic differences between countries can also be taken advantage of for trade between countries. Upstream riparians like China, Nepal, and Ethiopia might have better access to good dam sites, for example, which might be developed cooperatively with downstream riparians. The Sea of Galilee has likewise been suggested as a storage facility for the Jordan riparians in absence of a Unity Dam.
- Economic factors – Water is worth different things to different people, again allowing incentives for trade once, as discussed previously, property rights to the resource have been established.
- Training of water managers – Perhaps more than the managers of any other resource, water managers think regionally, beyond their borders, by training and practice. It is not surprising therefore, that water managers have been able to reach agreements often well in advance of their political counterparts. Indeed, joint training of water decision makers is often an excellent strategy for building the environment for cooperative efforts.
- Water science – Countries within a watershed develop different levels of water technology, often with different emphasis. Although Israel has emphasized drip irrigation and genetic engineering, Gulf States have invested heavily in desalination. Trade of existing technologies and joint research and development projects provide ideal venues to enhance regional cooperation. But a common language of water science is often broadly shared among professionals in states.

OFTEN-IGNORED PHYSICAL PARAMETERS

Although including resources in the bargaining mix other than water may help achieve an agreement, it is perhaps more important to be aware of some aspects particular to water that, if excluded, could impede the durability of understanding. For an agreement to be viable over time, it must incorporate mechanisms for any future misunderstandings to be resolved. This is a final, but crucial, step that has to be taken for a negotiated arrangement to last beyond the signing ceremony. The circumstances that brought about a conflict to begin with are seldom static; neither are the conditions of agreement. This is particularly true for hydrologic conflicts, where supply, demand,

and understanding of existing hydrologic conditions all change from season to season, and from year to year. Experiences in smaller disputes have shown that by discussing eventual disputes – up front – and planning to handle them often has the effect of reducing their likelihood.

Water managers in general are relied on to implement national policy within the limits imposed by:

- normal seasonal and annual variability
- dramatic fluctuations in quantity (droughts and floods)
- groundwater pumping and recharge rates within "safe yield"
- delivery system capability
- adequate water quality for each use
- economic efficiency
- political considerations

Although the international and transboundary agreements that have been reached often include some understanding of these parameters, including mechanisms explicitly dealing with aspects of hydrologic variability, most are weaker in considering other ways in which a basin may change over time. The Nile Waters Agreement, for example, has sections concerning natural variability of the river, as well as guidelines for allocating unanticipated gains and losses between Egypt and Sudan. The Agreement, however, also counts on the gains of implementing a canal through the Sudd wetlands – the negotiators could not have foreseen years of civil strife in Sudan and new concerns about the possible environmental impact precluding such an extensive development.

Some parameters of water resources that are commonly excluded or vague in international agreements include the following.

FLUCTUATIONS IN SEASONAL, ANNUAL, AND LONG-TERM WATER SUPPLY

This aspect of water resources often *is* included in international agreements with varying degrees of success. One method of dealing with quantity fluctuations is to assign one state the "remainder," or "residue" flow, after other states have received a set quantity. This method, used in the Johnston agreement that assigned Israel the "residue" flow, has the drawback of assigning all of the stochastic risk to one riparian. A variation is to allow for fluctuation but to assign each riparian a minimum absolute amount – important in arid and monsoon regions, both of which are particularly susceptible to seasonal fluctuations. Minimum quantities are guaranteed, unofficially, on the Euphrates and the Yarmuk. An alternative is to divide quantity by a percentage of actual flow, which effectively spreads risk among riparians but which puts downstream users at particular risk if changes occur upstream. Such is the case on

the Ganges, where Bangladesh sees decreasing flows due to greater upstream use by India.

The Colorado compact between upper and lower riparians provides an example of the consequences of *not* incorporating quantity fluctuations – the agreement calls for a set amount to each of the two parts of the basin but overestimates the quantity to be divided, as well as initially neglecting Mexico's claims, together resulting in shortfalls in more years than not.

GROUNDWATER

The relationship between groundwater and surface water is rarely codified into law or international agreements as well as with transboundary agreements (Eckstein and Eckstein, 2003). The results of excluding groundwater can include strains on existing relations among riparians – planned deep wells in the West Bank strained relations between Israelis, who undertook the project, and Palestinians, who thought the wells would undercut their own water supplies – or strains on existing agreements – Israel and Jordan got into a brief "pumping war" in competition over two sides of an aquifer that underlies the Yarmuk. An illustration of the interrelationship between groundwater, surface water, and international relations can be found in the Rio Grande basin, on the border of the United States and Mexico.

Fossil aquifers that straddle borders are likewise poorly managed. Fossil aquifers underlie joint borders throughout the Middle East, for example, between Israel and Jordan, Jordan and Saudi Arabia, and Israel and Egypt. As they are increasingly used as alternative sources of water, they may create increased friction between States. A complicating factor is that surface water and groundwater watersheds are not necessarily identical.

WATER QUALITY AND MINIMUM FLOW REQUIREMENTS

Much focus in agreements is often placed on the amount due each riparian, whereas less attention is usually paid to the water quality. The Colorado agreement between the United States and Mexico provides a good example of initially ignoring quality issues, when, after formal Mexican protest, the United States agreed to build one of the most extensive desalination plants in the world to meet Mexican quality needs. In contrast, water quality is explicitly delineated in the Johnston accords, which defines salinity standards, in parts per million, for each branch of the Jordan. The United Nations Food and Agriculture Organization (2003) and Moench (2004) indicate the current that the management of the transboundary groundwater resources must include provisions to protect the quality of the water. Minimum flow agreements can be a powerful mechanism to encourage cooperation in water and in other areas. The Interstate Commission on the Potomac River Basin in the United States is an excellent example of this. It operates with little formal authority – two low flow agreements – yet its ability to influence the decisions of the service suppliers within three jurisdictions along the river continues to increase.

THE PHYSICAL ENVIRONMENT

This vital parameter is almost invariably given perfunctory treatment in international agreements, if it is dealt with at all. Treaties often allocate the entire average flows of river systems between users, leaving no water at all for instream needs. Development projects such as the Jonglei Canal on the Nile tributaries, and the cascade dams on the Mekong, have historically paid little attention to the potential impact on the physical environment. Riparians will need to be more sensitive to the environmental consequences of their water-resource agreements, if only because international agencies increasingly use environmental impact as a measure of development viability rather than because of a "land ethic" in a Leopoldian sense (Leopold, 1949).

CHANGES IN UNDERSTANDING OF THE PHYSICAL SYSTEM

With greater modeling precision and more statistical information, physical systems are better understood over time. This understanding can result in easing negotiations, as was the case when Jordan found it needed less water for its future needs than was thought, allowing for a break in the Johnston negotiations or in strains on an agreement, as is the case in the Colorado compact's allocation of less water than usually exists.

TECHNICAL BREAKTHROUGHS

One interesting question, in light of potential technical breakthroughs, is how each might affect an international and transboundary agreement for water resources development. For example, who would have borne the cost of implementing and maintaining extensive water projects had the early promise of nuclear desalination or cold fusion resulted in dramatically inexpensive water? In addition, inexpensive interactive metasoftware now allows stakeholders to jointly create simulation models using whatever other models they chose. This can build up ownership in the very algorithms that will then be used for tradeoffs.

It is as common to ignore the link between these physical parameters as it is to exclude them separately. This suggests a lesson about approaching a watershed at conflict.

Lesson: The issue is river basin or watershed management, not just water management. It is more than service delivery, it is resources management. This links quality and quantity, surface water and groundwater. Everything is connected to everything else.

6.3.3 Lessons learned: Stage III: Enhancing benefits

Wishart (1989), Sadoff and Grey (2002, 2005), Whittington (2004), and many others demonstrated that cooperative water-resources development within an international river basin or watershed is usually more efficient from a water-resources management perspective, in the economic sense, than conflicting unilateral development. This cooperation has been the story within the United States with the TVA and along the Columbia River, and with the even earlier role of water infrastructure in the United States and nation building on the continent. Although the economic incentives of cooperation alone have rarely been responsible for overcoming other, usually political, obstacles, the paradigm of economics is increasingly being used to help define terms of conflict resolution. As mentioned above, for example, one study (Fisher et al., 2002) is attempting to monetize the water dispute on the Jordan River, while international water markets have been mentioned as a method of increasing regional efficiency.

Some considerations that have been used in the past to enhance the potential for economic cooperation between riparians include:

- Recognizing that although water itself is a finite commodity, and therefore often seen only as zero-sum solutions ("distributive" or "win–lose"), the benefits or welfare derived from water is variable and therefore tradable for non-zero-sum ("integrative" or "win–win") solutions.
- Welfare can be measured basinwide and among all the players participating in cooperation so even when one player's individual welfare is not immediately enhanced by the loss of the resource, the resulting payoffs of trade should result in the region as a whole being better off.
- Infrastructure considerations can enhance the argument for cooperation, especially when considering the variable aspects inherent to water resources. One or another of the riparians may have better resources to deal with fluctuating quantity or quality – more storage potential or better developed water treatment, for example – which can help encourage an alliance. But flattening the hydrograph and thus building more stable expectations into the social system is a prerequisite for development of any kind and to set an atmosphere of stable social expectations.

As economics becomes a more dominant paradigm in conflict resolution, it also worth recognizing the sometimes-overpowering noneconomic values that water users occasionally attribute to their water. These might include:

- political attributes of water, for example, perceived past injustice or national and state and/or province pride

- cooperation per se (e.g., the World Bank and other funding agencies do not include international cooperation as a benefit in benefit–cost analyses)
- physical security
- perceptions of beauty in the environment
- inherent value of noneconomic species and a healthy ecosystem
- food or water security – the psychological value of control
- open space (water is now being subsidized in some countries to help keep agricultural land open against encroaching urban development)
- instream flow

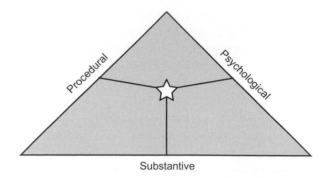

Figure 6.8 Achieving agreement: The satisfaction triangle (Delli Priscoli and Moore, 1985).

BENEFITS OF NATIONAL STRATEGIES AND NATION BUILDING

Although the absence of economic considerations in international agreements can condemn the riparians to "inefficient" resources management or development, hazards exist, too, in relying solely on economics to guide negotiations. There is a difference between economic efficiency at the resources management levels and financial viability at the service delivery level.

Increasingly, linkages are being made between water and other resources. Awareness of options outside the specific water issues being discussed may offer more opportunities for creative solutions to be generated. This suggests the following lesson.

Lesson: Creating incentives for voluntary resolution of water resource conflicts is key.

Although international or transboundary institutions may not have the laws and authorities to enforce solutions, they often have access to other carrots and sticks, which can help induce agreement by capitalizing on differences and creating trades or linkages (Sadoff and Grey, 2002).

6.3.4 Lessons learned: Stage IV: Institutional and organizational capacity and sharing benefits

ELEMENTS OF DURABLE AGREEMENTS

Durable agreements depend on achieving procedural and psychological as well as content satisfaction (Lincoln, 1986; Moore and Delli Priscoli, 1989). By habit, training, and job description, technical professionals and agencies usually focus on the content or substance of discussion. However, technical excellence does not necessarily bestow process credibility. Indeed, professional claims of neutrality based on substantive expertise and objectivity can backfire. Even the best analysis is driven by interests and values. Often, the more these professionals are immersed in the substance, the less aware they

are of the values driving these assumptions. Although data are crucial to agreements, it is not necessary to agree on data to come to an agreement on action. Controversial water projects often uncover equal and opposite expert data and interpretation. The more that water development includes social assessments, environmental, and other concerns, data uncertainties will become more explicit and process more important. Procedural and psychological satisfaction will have to be explicitly managed along with the substantive-content satisfaction (Figure 6.8).

Although we like to distinguish between the technical and political, both are blurred in intranational and international water-allocation decisions. When dominated by the political, poor or even unworkable agreements can result. When dominated by the technical, the results can be too narrow, fall short of satisfying interests, and even exacerbate political and social tensions. The search for new allocation mechanisms is ultimately the search for institutions and processes that facilitate a balance between the technical and the political. The key is to make the technical politically creative and the political technically sound. Waiting until highly adversarial political negotiations emerge can mean acting too late. Building on shared technical language, but driven by different values and interests, early technically sponsored negotiating forums can produce a range of alternatives that could enhance subsequent political negotiations.

SUCCESSFUL INSTITUTIONS

As noted earlier in this chapter, it is important to negotiate around benefits versus allocating around flows. Parties should engage in joint diagnosis, joint creation of options, joint implementation. They should remember that IWRM is more than service delivery or privatization. Parties should recognize social reality of fragmentation and work with it so implementation is more assured. They should be developing, funding, and supporting regional-transboundary organizations for water. The parties should build on indigenous traditions of collaboration and dispute management. Lenders and donors should assess the costs to parties of no water agreements. Parties should support early participation of stakeholders in intersectoral water strategies and assessments. Lenders and donors should go beyond an emphasis on expert panels and encourage early use of facilitation and mediation. All should support the development of technology and interactive decision-support tools. This means they should encourage support for public access databases in countries seeking aid or loans. Donors and lenders need to analyze present and expected water-related investment performance in situations of potential scarcity and transboundary conflict. They should become more like facilitators of – rather than simply evaluators of and designers for – specific solutions. This whole process needs the early support funding of up-front costs for shared visions and strategic views of rivers.

FUNCTIONS AND RESPONSIBILITIES

After examining how transboundary river basin organizations evolve across sovereign states in North America, Kenney and Lord (1994) offer some simple but useful lessons to other parties, even in the international realm, on the practical aspects of designing transboundary and river basin organizations. To do so, parties ought to discuss functions and responsibilities, which include both soft and hard functions. Soft functions include activities like research, monitoring, advising, and advocacy. Regionally focused information and data and information providers accountable to decision makers are needed. Hard functions include activities, such as project development, operation, regulation, assuming oversight or directions of other functions – and power to modify and integrate the policies of others. And there are other functions, like conflict mediation and awareness raising.

MEMBERSHIP AND PARTICIPATION

Membership and participation considerations include questions such as What jurisdictions and agencies and interests must be represented? What are the realistic power sharing and relative balances? What type of actors should lead – technical, political, and administrative? What are the roles for interest groups and NGOs. And what about technical staff? Indeed one of the most pervasive findings about successful transboundary organizations concerns technical staff. To the degree that such organizations can create such staff and to the degree they are used by the parties over time, the organization will prosper and grow.

DECISION MAKING

Operating rules most importantly include decision rules. Will the organization operate by unanimity-consensus, majority rule, or other procedures? Each has implication for how rela-

tionships among the parties will evolve. For example, consensus relies on negotiations. However, majority vote is likely to supports coalition building. In addition, the parties must consider how decision rules will affect political aspects of creating the organization, and authorities and resources are needed to assure the rules stick. They must also ensure that parties have equal access to information.

AUTHORITIES

Consideration on authorities should include these questions: what authority is needed to accomplish functions? Which existing jurisdictions are reluctant to delegate to new organization? This consideration is important because lack of formal authority often means only soft functions will be included. Parties should avoid appeals to negative powers, such as taxing and regulating, although more and more transboundary organizations within countries are actually starting to raise these issues, with some success, in various parts of the world. Parties should appeal to positive powers, such as creating new markets, resolving disputes, implementing agreements, responding to emergencies, and others. Parties need to discuss how much delegation can be done in the political environment?

In considering the legal basis and structures, parties should look from formal to informal, such as agreements, legislation, treaties, and others. Parties need to discuss what the authorities and the membership demands.

FINANCING

Financing is obviously critical. It depends on the situation. It can be from direct appropriations. If so, the parties need to be clear as to what jurisdictions. The reliance on one or few means the organizations will be vulnerable. Contributions from voluntary, mandated, personal, agencies, and others need to be examined. Self-supporting means, such as user fees, bonds, and taxes on users, should be explored. Self-financing is the most stable but is politically difficult. It also tends to move the parties toward creating vendible services. The question then is how much is needed? Overall the question is: how will the funding source decide how money is spent?

The parties also need to look at the *range of issues* before the organization. What are the crosscutting issues over water uses? How do they incorporate new ecological values? Where is the place where land and water and economic interests can meet?

Parties need explicit discussion of how agreements will be implemented. In other words they need to look at:

- What are the social realities of fragmentation and does design account for these?
- Is there solid support within society? Where?

- Is there a system of rights?
- How will trust and credibility of the technical staff be developed?
- Does the design allow for evolution and change?

As the organization provides a secure forum for dialogue it will change to fit the changing realities on the ground, thus it is common to see purposes/uses added as values change.

Lesson: Parties should remember that increased interdependency in basins can be a key to security. Agreements decrease vulnerability to nature, deal directly with primordial anxiety and fear, and enhance wealth generation. In fact, building transboundary water institutions and agreements is the practical work of this view of security. It is a work of integration at the social level (Sidebar 6.1).

POLITICAL PARAMETERS

Many aspects particular to water resources have properties that can both provoke conflict and induce cooperation. The water conflicts presented here suggest that, with early planning, one can help guide riparians along the latter path. To do so, however, takes foresight and awareness of the options throughout the negotiating process. The following lesson is therefore suggested.

Lesson: In planning for implementation, seek mutual understanding of underlying interests of situations where positions may be mutually exclusive but where underlying interests are not. Do it early and iteratively, throughout the process. Awareness and incorporation of links to other issues is vital. Water and politics cannot be separated.

Sidebar 6.1 Successful Cooperative Actions

In successful cooperative actions, parties tend to

- Move beyond "impact fixation to incorporating environmental and other values into creating alternatives, formulating options, and evaluating options and impact mitigation.
- Bring operational and implementation interest into design.
- Create open and transparent rules of behavior.
- Foster norms of collaboration versus experts.
- Establish means for resolving disputes.
- Separate administrative and policy.
- Promote flexibility and creativity.
- Foster regional shared visions of river.
- Seek high political (ministerial) commitment.
- Seek meaningful community input.
- Seek high knowledge levels.
- Achieve clear accountability among participants.
- Be flexible and creative in the river basin organization design.
- Design structures based on functions/missions.
- Foster perceptions of the basin as a whole.
- Use process tools.
- Create means for conflict management.
- Separate administrative and policy and regulating and constructing functions.
- Establish reciprocal incentives ("back to the future") such as with flood control/hydro-storage/navigation-storage/minimum flow pressures from international community.

- Respond to pressures for capital.
- Move beyond allocation to creation/sharing benefits.
- Finding optimal mixes of uses.
- Use words such as *shared* and *common*.
- Bring implementation and operational interests into formulation process.
- Give preference given to operating at the lowest level possible and creating self-sustaining organizations.
- Explicitly manage the "gray" area between technical and political.
- Facilitate explicit negotiations among long-term visions and short-term efficiencies.
- Help place water as driver, or first constraint, in cross-sectoral strategies and negotiations.
- Better align internal cultural values of water organizations to those external values of collaboration and participation.
- Promote participation of those likely to be impacted, as well as disbursed beneficiaries.
- Foster norms of collaborative behavior and move beyond reductionist expertise.
- Facilitate the integration of upstream—downstream and ground and surface uses.
- Consider political viability the possible and transformative.
- Let function dictate structure.
- Create mechanism that create, disseminate, and foster regional visions and "problemshed" or watershed or basin-level visions.

GENERAL HYDROPOLITICS

Although some international and transboundary agreements make some provision for dealing with hydrologic variation, none surveyed here deal explicitly with the possibility of any political variation whatsoever. The Transboundary Freshwater Dispute Database survey (TFDD, 2006) suggests that political change is a major catalyst in either provoking disputes or in bringing about their resolution. Political change has already been mentioned as an indicator of possible water conflict, as many of the conflicts presented here, including those on the Ganges, the Indus, and the Nile, took on international complications as the British Empire gave way to local rule. The Mekong Committee became an "interim" committee when the Khmer Rouge gained control in Cambodia. In contrast, other agreements were hastened when new governments resulted in friendlier relations within basins. Such was the case with Sudan on the Nile and India on the Ganges.

Along with changes in government, other political considerations can be taken into account in international negotiations. These might include changing levels of hostility between riparians changing power relationships, including

- riparian position (e.g., Israel and Pakistan have each shifted riparian relations with their neighbors)
- military power shifts
- legal changes (e.g., clarity of water rights)
- economic growth and stability

- the social environment, for example, population movements (refugees, immigrants, resettlement because of water developments)

ENFORCEMENT MECHANISMS

Most of the agreements reviewed have some description of a feedback mechanism for ongoing conflict resolution. Many of these are innocuous – requiring little more than meetings at progressively higher political levels – and, probably as a consequence, ineffectual. What is notably lacking in all of the agreements is any real mechanism for enforcing the negotiated terms. Although abrogated agreements can be brought before the International Court of Justice, this venue has practical limitations, mentioned earlier, which preclude it as a common method for resolving contracts.

Finally, getting beyond the imperative of "integrated international basin management" has been an important step in some basins. Even friendly states often have difficulty relinquishing sovereignty to a supralegal authority, and the obstacles only increase along with the level of suspicion and rancor. At best in many regions, one might strive for coordination over integration. Once the appropriate benefits are negotiated, it then becomes an issue of agreeing on a set quantity, quality, and timing of water resources that will cross each border. Coordination, when done correctly, can offer many of the same benefits as integration, and be far superior to unilateral development, but does not threaten the one issue all states hold dear – their very sovereignty.

7 Water conflict prevention and resolution: Where to from here?

If there is magic on this planet, it is contained in water.

– Loren Eisely, *The Immense Journey*

7.1 WHY MIGHT THE FUTURE LOOK NOTHING LIKE THE PAST?[1]

This book is partly based on the assumption that we can tell something about the future by looking at the past. This assumption is not unassailable. Thus, it is worth stopping at this point and challenging its foundation. Why might the future look nothing at all like the past? What new approaches or technologies are on the horizon to change or ameliorate the risk to the basins we reviewed or even to the whole approach to basins at risk?

By definition, a discussion of the future cannot have the same empirical backing as a historical study – the data just do not yet exist. Yet there are cutting-edge developments and recent trends that, if one examined them within the context of this study, might suggest some possible changes in store for transboundary waters in the near future. What follows, then, are four possible fundamental changes in the way we approach international and/or transboundary waters.

7.1.1 New technologies for negotiation and management

Most analysis of international waters dates from the mid-1960s onward. In some ways, water management now is very similar to how it was then (or, for that matter, as it was 5,000 years ago). But some fundamental aspects are profoundly different. While global water stresses are increasing, institutions are getting better and more resilient, management and understanding are improving, and these issues are increasingly on the radar screen of global and local decision makers. But what is most important is that the twenty-first century has access to new

technology – including remote sensing and modeling capabilities and technologies and management practices that increase water-use efficiency – which were unheard of in the mid- to late-twentieth century and that add substantially to the ability to both negotiate and manage transboundary waters more effectively. For example, systems can now integrate – in real time – multiple databases, such as weather forecasts and simulation contingences run by in-place models to provide instant "what if" scenarios for decision makers and the publics affected (United States Army Corps of Engineers, 2007). Although new technologies and data cannot replace the political goodwill necessary for creative solutions – and they are not widely available outside the developed world – they can, if appropriately deployed, allow for more robust negotiations and greater flexibility in joint management. This technology includes many information technology (IT) breakthroughs, which have enabled more equal footing in areas where "techies" and agencies previously seemed to dominate; hence, IT has equalized the playing field for many water debates.

7.1.2 Globalization: Private capital, WTO, and circumvented ethics

Very little of the recent attention on globalization and the World Trade Organization (WTO) has centered on water resources, but there is a definite water component to these trends (see, for example, Anderson and Snyder, 1997, and Finger and Allouche, 2002). One of the most profound trends is the shift of development funds from global and regional development banks, such as the World Bank and the Asia Development Bank, to private multinationals, such as Bechtel, Vivendi, and Ondeo (formally Lyonnaise des Eaux). Development banks have, over the years, been susceptible to public pressures and ethical challenges and, as such, have developed procedures for evaluating social and environmental impacts of projects and incorporating them in decision making. On international waters, each development bank has guidelines that generally

---

[1] This section draws from Wolf, Yoffe, and Giordano (2003b).

prohibit development unless all riparians agree to the project, which in and of itself has promoted successful negotiations in the past. Private enterprises have no such restrictions, and nations eager to develop controversial projects have been increasingly turning to private capital to circumvent public ethics. The most controversial projects of the day – Turkey's GAP project, India's Narmada River project, and China's Three Gorges Dam – are all proceeding through the studied avoidance of development banks and their mores. These projects were internally funded through combinations of devices to attract private capital and internal public finding. It is important to note that the vast majority of funding for water investments has always come from within countries.

There is a more subtle effect of globalization, though, that has to do with the WTO and its emphasis on privatization and full-cost recovery of investments. However, water infrastructure is long term and capital intensive. It means that the majority of water will continue to be driven by public funding and decision making. Local and national governments, which have traditionally implemented and subsidized water development systems to keep water prices down, are under increasing pressure from the forces of globalization to develop these systems through private companies. These large, multinational water companies, in turn, manage for profit and, if they use development capital, both push and are pushed to recover the full cost of their investment. This situation can translate not only into immediate and substantial rises in the cost of water, disproportionately affecting the poor, but also to greater eradication of local and indigenous management systems and cultures.

If there is to be water-related violence in the future, it is much more liable to be like the "water riots" against a Bechtel development in Bolivia in 1999 than "water wars" across national boundaries. This raises the issues of where operation and maintenance (O&M) stops and new capital investment begins – and just what is it we are talking about in full-cost recovery. Private markets are unlikely to be funding new, long-term, large-infrastructure needs and capital investment versus O&M. All cost recovery will include, as it always has, some forms of subsidies. The question has become "what is the transparency of the subsidies and where does accountability lie?" For companies this raises issues of financial disclosure and participation in negotiations. It is at the nexus of the conflict between public accountability and private efficiency. In fact, most private companies now seek to partner with strong and clearly accountable elected officials and to sell their expertise as efficient management. But the needs go far beyond the service delivery, which is the heart of the public–private debate.

As WTO rules are elaborated and negotiated, real questions remain as to how much of this process will be *required* of nations in the future, simply to retain membership in the organization. The "commodification" of water as a result of these forces is a case in point. For the past 20 years, no global water policy meeting has neglected to pass a resolution that, among other issues, defined water as an "economic good" – thus setting the stage, at both the 2000 and 2006 World Water Forums, for an unresolved showdown against those who would define water as a human or ecosystem *right*.

The debate looms large over the future of water resources: if water is a commodity, and if WTO rules disallow obstacles to the trade of commodities, will nations be forced to sell their water? Although farfetched now (even as a California company is challenging British Columbia over precisely such an issue under NAFTA rules), the globalization debate between market forces and social forces continues to play out in microcosm in the world of water resources.

7.1.3 The geopolitics of desalination

Twice in the past 50 years – during the 1960s nuclear energy fervor and in the late 1980s, with "discoveries" in cold fusion – much of the world briefly thought it was on the verge of having access to close-to-free energy supplies. "Too cheap to meter" was the phrase during the Atoms for Peace Conference. Although neither the economics nor the technology finally supported these claims, it is not farfetched to picture changes that could profoundly alter the economics of desalination.

The marginal cost of desalinated water (between US$0.55 and US$0.80/m^3) makes it currently cost-effective only in the developed world, where (i) the water will be used for drinking water; (ii) the population to whom the water will be delivered lives along a coast and at low elevations; and (iii) there are no alternatives. The only places not so restricted are where energy costs are especially low, notably the Arabian Peninsula. A fundamental shift either in energy prices or in membrane technology could bring costs down substantially. If either happened to the extent that the marginal cost allowed for agricultural irrigation with seawater (around US$0.08/m^3 on average), a large proportion of the world's water supplies would shift from rivers and shallow aquifers to the sea (an unlikely, but plausible, scenario). And the price of desalinization *is* dropping, dramatically. Recent bid prices for a project in Tampa Bay, Florida, were less than half of the lowest cubic meter prices for desalinated water in the 1990s. This drop in price is important because the trend for desalinization, in many ways, makes it look more competitive with other sources. And we should remember that most of the world's population lives close to the sea.

In addition to the fundamental economic changes that would result, geopolitical thinking about water systems would also need to shift. Currently, there is inherent political power in being an upstream riparian and thus controlling the head-waters. In the scenario for cheap desalination, that spatial position of power would shift from mountains to the valleys and from the headwaters to the sea. Many nations, such as Israel, Egypt, and Iraq, that currently dependent on upstream neighbors for their water supply would, by virtue of their coastlines, suddenly find their roles reversed. Again, this is unlikely but plausible. (Naturally, even if the world addressed the problem of water supply with cheap sources, the water problem is far from solved. Floods, droughts, access problems, and ecosystem degradation would all still need to be addressed.)

7.1.4 The changing sources of water and the changing nature of conflict

Both the worlds of water and of conflict are undergoing slow but steady changes, which may obviate much of the watershed-based thinking in this book. Lack of access to a safe, stable supply of water is reaching unprecedented proportions. Furthermore, as surface water supplies and easy groundwater sources are increasingly exploited throughout the world, two major changes result: quality is steadily becoming a more serious issue to many than quantity, and water use is shifting to less traditional sources. Many of these sources – such as deep fossil aquifers, wastewater reclamation, and interbasin transfers – are not restricted by the confines of watershed boundaries, our fundamental unit of analysis in this study.

Conflict, too, is becoming less about classic nation-state power politics and is increasingly being driven by internal or local pressures or more subtle issues of poverty and stability. The combination of changes in water resources and in conflict suggests that the water disputes of tomorrow may look very different from today.

7.2 WHAT TYPES OF POLICY RECOMMENDATIONS CAN WE MAKE?

Given the lessons of the previous chapter, what can the world water community and those interested and responsible for water do? We divide our suggestions among institutions, funding and development agencies, universities and research agencies, and civil society.

7.2.1 Funding and development agencies

Water-related development needs to be coordinated and focused, relating quality, quantity, groundwater, surface water, and local sociopolitical settings in an integrated fashion. Fund-ing should be commensurate with the responsibility assistance agencies have for alleviating the global water crisis.

Ameliorating the crux of water security – human suffering – often rests with agencies that, given the size of the crisis, are extraordinarily underfunded. One can contrast the resources spent on issues such as global climate change and arms control, laudable for their efforts to protect against potential loss of life in the future, to the millions of people now dying because they lack the resources to access clean, fresh water. Agencies such as United States Agency for International Development (USAID), Canadian International Development Agency (CIDA), and Japan International Cooperation Agency (JICA) have access to technical expertise and experience to help yet are hindered by political and budgetary constraints. Funding agencies often are hamstrung by local politics.

A powerful argument can be made that water-related disease costs the global economy US$125 billion per year, whereas ameliorating the diseases would cost US$7–50 billion in total (Gleick, 1998). Programs such as USAID's Project Forward, which integrates water management with conflict resolution training, offer models for the future. In the end, water must be ranked higher in the budgets and agendas within countries. Governments have to come to see how investing in water is the necessary condition for them to build the platform for efforts to break the cycle of poverty, and that poverty alleviation is, explicitly, a security concern.

In our international system, which lacks a strong compliance structure and needs incentives, international lenders, such as the World Bank, have a comparative advantage in many of the areas experts have identified as critical to forging international and interjurisdictional cooperation.

Although lenders and donors certainly cannot solve all the world's problems, they can assume a leadership role in encouraging and facilitating early collaborative and participatory efforts among parties that would otherwise conflict. If the experience of the industrial world is any indication, this facilitating role could be the key leadership role for these agencies in water resources. Other thoughts include

1. In situations of potential or ripened intersectoral and/or transboundary conflict, the lenders and donors could ask for assessments of the costs to interested parties who do not have water management agreements.
2. Lenders and donors could encourage and support discussions of alternative water futures among interested parties in the early stages of project development and/or intersectoral policy development.
3. Lenders and donors could support early participation of major stakeholders, nongovernmental organizations (NGOs; environmental and other NGOs), those who are

affected by water issues, and others at the intersectoral levels of water assessment. Indeed the World Bank's water policy paper has called for such intersectoral dialogue. This means going beyond public information programs to the active engagement of interests in the formulation of options.

4. Lenders and donors could go beyond the emphasis on expert panels and actively encourage the early use of facilitation and mediation in the formative stages of water projects and water assessments.

5. Lenders and donors could support the development of technology for the use of public access databases in those countries seeking significant water resources loans. As many experts suggest, the ESAs should also encourage and support the use of interactive software as means to describe water futures, tradeoffs, and the best alternative to a negotiated agreement (BATNA).

6. Donors and lenders could do a quantified vulnerability analysis of present and expected water-related investments performance where intersectoral and transboundary conflict and potential water scarcity are involved. One way to accomplish this could be through river basin study groups (Rogers, 1992a).

7. Donors and lenders should start thinking about funding the development of regional and transnational organizations for activities that do not threaten jurisdictional authority.

8. Find and build on indigenous and grassroots traditions of collaboration and dispute management in water resources.

7.2.2 International institutions

Anticipating and ameliorating water disputes are as important, more effective, and less costly than conflict resolution. Watershed and river basin commissions should be developed for those basins that do not have them and strengthened for those that do.

Three characteristics of international and transboundary waters – the fact that conflict is invariably subacute, that tensions can be averted when institutions are established early, and that such institutions are tremendously resilient over time – inform this recommendation. Early intervention can be far less costly than conflict resolution processes. In some cases, such as the Nile, the Indus, and the Jordan, as armed conflict seemed imminent, tremendous energy was spent getting the parties to talk to each other. In contrast, river basin discussions in the Mekong Committee, the multilateral working group in the Middle East, and on the Danube, have all moved beyond the causes of immediate disputes on to actual, practical projects that may be implemented in an integrative framework. The International Network of River Basin Organizations (INBO) has established a registry of international RBOs and instituted

a series of programs for bolstering their capabilities. Similar to what was suggested in the 1967 Water for Peace program, countries could think of building a pool of International water-resources civil servants.

7.2.3 Universities and research agencies

Universities and research agencies can best contribute to alleviation of the water crisis in three major ways: (i) acquire, analyze, and coordinate the primary data necessary for good empirical work; (ii) identify indicators of future water disputes and/or insecurity in regions most at risk; and (iii) train tomorrow's water managers in an integrated fashion.

The Internet's initial mandate is still one of the best: to allow communication among researchers around the world to exchange information and enhance collaboration. The surplus of primary data currently threatens an information overload in the developed world, whereas the most basic information is often lacking in the developing world. Data availability not only allows for greater understanding of the physical world but, by adding information and knowledge from the social, economic, and political realms, indicators showing regions at risk can be identified.

Moreover, universities are best suited to train those who will resolve tomorrow's water disputes, and programs at, for example, UNESCO/IHE-Delft, the University of Dundee, Linkopping University, Tufts University, and Oregon State University are allowing students to focus on *both* conflict transformation *and* in the science and policy of water resources. UNESCO, the World Bank, and the Universities Partnership for Transboundary Waters have been developing and compiling curricula and skills-building manuals to help train the water champions of tomorrow.

In addition, much useful research needs to be done in areas such as the following:

1. Studies of international water resource agreements that analyze how agreements develop and what the internal and external conditions are for their success

2. Studies of the actual operations of dispute clauses and assisted negotiations under current water resources agreements and RBOs

3. Studies of the reasons for past successes and failures of international water resources dispute management

4. Research that relates methods of managing conflicts to the types of water resources decisions we are likely to take. For example, how do regulatory versus planning versus free-market versus assisted negotiation approaches affect water resources decisions such as design, implementation, construction, operations, and maintenance? Who is involved

at what levels in these decisions? How successful have we been in looking at the social utility functions of each? What does each approach tell us about equity, efficiency, and fairness? How does each approach generate options and tradeoffs?

5. Studies that integrate theories from a variety of disciplines, for example, community building, international negotiations, alternative dispute resolution, and multiple-objective planning in water resource management

6. Studies that examine the roles of current international lender and donor institutions — to what degree may they become more facilitators of agreement as opposed to evaluators and/or designers of solutions? In what ways can those institutions that deal with water improve their behavior so as to help prevent conflicts?

7. Research that discerns how our water resources experiences – namely whether we live in humid or arid areas – in turn affect our perceptions, and how such perceptions, in turn, affect both our own policies and those policies that we may recommend for others

8. Research to assess and describe where and how intra- and international-state water issues could threaten political and social security

9. Examination of whether increased integration of infrastructure among hostile neighbors increases or decreases likelihood of conflict

10. Study of what is *minimum* data necessary for informed policy decisions

11. Studies of the impact of globalization, privatization, and commodification of water resources on conflict potential

7.2.4 Private industry

Much of the debate on water will increasingly be around what the appropriate public and private roles for policy formulation are. Broadly speaking, the resource management decisions are public, whereas service delivery can be either private or public under regulation. However, there is much gray area in between.

Private industry has traits that can be harnessed to help ameliorate the world water crisis: its reach transcends national boundaries, its resources are generally greater than those of public institutions, and its strategic planning is generally superb. Historically, private companies, such as Bechtel and Lyonnaise des Eaux, have been involved primarily in large-scale development projects, whereas the smaller-scale projects have been left to development assistance agencies. Recently, a shift in thinking has taken place in some corporate boardrooms. Bank of America, for example, was not involved in the California-wide process of water planning until recently, when its president noticed that practically *all* of the bank's

investments relied on a safe, stable supply of water. This was true whether the investments were in microchip manufacturing, mortgages, or agriculture. When the bank became involved in the "Cal-Fed Plan," it brought along its lawyers, facilitators, planning expertise, and financial resources. Subsequently, progress was made in several areas where previously there had been impasse. Violia has signed 50-year contracts with Shanghai for service deliveries. Increased private sector roles in service delivery imply new and different roles for public sector expertise in water – primarily in regulation. However, none of these possibilities negates the need for an increased role of the public sector in decision tradeoffs on how the whole resources should be used. One consequence is that new forms of private—public partnerships are likely to emerge and, with this, increased need for skills of conflict management and resolution will also grow.

7.2.5 Civil society

Many projects bring significant and critical benefits to large segments of populations throughout the world. However, the distribution of these needed benefits versus who bears their costs is often skewed.

Examples of these include large projects such as dams that have displaced hundreds of thousands of people and eliminated sites of cultural and religious heritage, projects promoting water markets among religious groups for whom the idea is sacrilege, or activities as seemingly minor as cutting down a tree sacred to a village djinn (genie). In recent years, as a consequence, those affected by a project have been increasingly involved in the decision-making process. In this process, a clear distinction between genuinely homegrown, indigenous, civil society groups and those dependent on a few international institutions and NGOs for funding is important. Creating civil society is a key to social and political development and to developing viable democratic political cultures. Water and the creating of the civil infrastructure have much to say about how the civic culture will evolve. Indeed the two have been closely linked throughout history. Doing water management within countries is an enormously important tool for creating the experience base that is vital to the growth of this civic culture. But the civic experiences and the tradeoff processes (i.e., the social-learning processes) must ultimately be indigenous if the social and cultural benefits of viable political cultures are to be achieved. The agendas of the indigenous civil groups will often conflict with those of external aid givers, whether they are environmental NGOs or development-oriented engineers. We are likely to see more clearly the role of water and the role of developing civil society in the foreseeable future.

7.3 A NEW ETHIC FOR WATER MANAGEMENT

It may take a new water ethic to sway public attention toward the critical lessons the river offers civilization: an ethic to help bring a new balance around water decisions; an ethic that helps us guard against "gigantism" and "technological triumphalism" on the one hand but, equally important, against an unwarranted reverence for an overromanticized past and a "technophobia" on the other. Here are three aspects we think are critical as the fulcrum for such a balance.

First, the new ethic we require is not simply one of preservation. It is one that should be built teleologically, on a sense of purpose and on an active codesigning with nature. A recent example will help explain what we mean here.

While releasing Colorado River water to recreate floods in the Grand Canyon, one engineer said, "We are trying to recreate what Mother Nature would have done." Another stated, "This is a test of whether man can do something right with dams rather than always doing something wrong with dams" (Kenworthy, 1996).

But, the reality is complex. Lack of spring floods has changed patterns of sediment flow, the river banks, and the ecology of the canyon. However, the dams have also allowed tamarisk trees to line the banks. These trees provide habitat for endangered western willow flycatchers, which have helped increase the peregrine falcons to the point where they are no longer an endangered species. It is no wonder that these practical engineers revert to incantations to "Mother Nature" paralleling traditional appeals to wisdom goddesses such as Athena. It is no wonder they use the value and emotional language of moral right and wrong. There are no easy answers. Answers depend, to a great degree, on what you want or think the ecology ought to be. They depend on what purpose and value you ascribe to that ecology. It is frightening to have all your scientific knowledge confront you with the reality that you are codesigning the ecology. The norms to guide such decisions really need to appeal to ultimate authority or higher social or environmental goods.

Even wetland restoration and preservation has come to mean conscious intervention, or partnerships, with nature. We are intervening to create or to recreate some preferred state or equilibrium, whether that preference comes from a vision of the future or from romantic notions of the past. But nature is dynamic. Nature's destruction to nature can be greater than anything that humans could dream up. Look at the results of floods or volcanoes and their impact on the atmosphere.

Second, a new ethic must be based on a balance between humans and technology and among structural and nonstructural approaches. Rarely have either worked alone and it is time to stop characterizing them as one versus the other.

Water resources have moved from manipulation of "natural" systems to manipulating the human systems as management tools. Nonstructural measures are the major example. However, what is the safety margin? How far can we ethically reduce what may appear as "excess" structural capacity when such "excess" structural capacity can provide for social adjustment during times of stress on the resources and when our ability to predict extreme hydrologic events is not as refined as the ability to run our systems at their margin. Lack of such capacity could reduce society's buffer to violence or safety net. Such "excess" capacity also can provide more options to deal with stress on the natural and social systems. Determining, creating, and maintaining an appropriate safety net to reduce fear, anxiety, and potential conflict is the noble inspiration of engineering stemming from humans, the toolmakers. It is a serious issue of residual risk.

Third, a new ethic, even in our advanced technological age, should be based on finding a new balance of the sacred and utilitarian in water.

Throughout history, water has been treated as an end and as a means. In truth, it is both. When water appears plentiful, it is easier to think of it as a means. In arid areas this is less likely and water is more likely to become an organizing principle for society. If thought of as a means, it is easy to see water as a factor of production and in utilitarian terms. But as an end, water often takes on a sanctity and value beyond utilitarian exchange.

The sacred refers to those aspects of water through which mystery and unknown or, some would say, the irrational, elements become present to our awareness (Haught, 1996, p. 277). One only has to look at recent history in the United States of introducing the productive uses of wastewater to see current relevance of this concept. But talking of such a balance does not mean returning to a neo-paganism or to pantheism or any other "ism." It does not mean that water should be made a religion. It is, rather, an appreciation of the intrinsic and broad value of water not captured in the traditional utilitarian calculus of transactions. It is to recognize that water is not only a means to other goals, it is also important as an end in itself.

Balancing the sacred and utilitarian in water is not new, although our era's balance point is. From the ancients' respect for the sanctity of water to Thales' and Hippocrates' notions of water as source of life to the Christian fathers' notions of water as producer of life to ancient Egyptians' use of geometry to predict flooding on the Nile to Mayan, Khmer, and other priests who intervened into the uncertainties of planting and harvesting to the Renaissance *fontaineries* (men who combined knowledge of hydraulics, physics, science, and hydromythology) to nineteenth-century technology's "conquest" and democratization of water, which brought water to more people, humans have been constantly rebalancing the sanctity and the utilitarian in water.

Today, our technology tells us that there is enough water – if we cooperate. One of the most important elements for cooperation is something negotiations experts call "superordinate values." These are values beyond immediate utilitarian values to which competing parties can identify. Rekindling the sense of the sacred in water, a superordinate value, is one way to facilitate the escalation of debate on water cooperation to higher levels and thus affect the capacity to reach cooperation and to manage conflict.

Appendices

A 1997 Convention and ILC draft rules on international groundwater

A.1 CONVENTION ON THE LAW OF THE NON-NAVIGATIONAL USES OF INTERNATIONAL WATERCOURSES. ADOPTED BY THE GENERAL ASSEMBLY OF THE UNITED NATIONS ON 21 MAY 1997[1]

The Parties to the present Convention,

Conscious of the importance of international watercourses and the non-navigational uses thereof in many regions of the world,

Having in mind Article 13, paragraph 1 (a), of the Charter of the United Nations, which provides that the General Assembly shall initiate studies and make recommendations for the purpose of encouraging the progressive development of international law and its codification,

Considering that successful codification and progressive development of rules of international law regarding non-navigational uses of international watercourses would assist in promoting and implementing the purposes and principles set forth in Articles 1 and 2 of the Charter of the United Nations,

Taking into account the problems affecting many international watercourses resulting from, among other things, increasing demands and pollution,

Expressing the conviction that a framework convention will ensure the utilization, development, conservation, management and protection of international watercourses and the promotion of the optimal and sustainable utilization thereof for present and future generations,

Affirming the importance of international cooperation and good-neighbourliness in this field,

Aware of the special situation and needs of developing countries,

Recalling the principles and recommendations adopted by the United Nations Conference on Environment and Development of 1992 in the Rio Declaration and Agenda 21,

Recalling also the existing bilateral and multilateral agreements regarding the non-navigational uses of international watercourses,

Mindful of the valuable contribution of international organizations, both governmental and non-governmental, to the codification and progressive development of international law in this field,

Appreciative of the work carried out by the International Law Commission on the law of the non-navigational uses of international watercourses,

Bearing in mind United Nations General Assembly resolution 49/52 of 9 December 1994,

Have agreed as follows:

PART I.
INTRODUCTION
Article 1
Scope of the present Convention

1. The present Convention applies to uses of international watercourses and of their waters for purposes other than navigation and to measures of protection, preservation and management related to the uses of those watercourses and their waters.
2. The uses of international watercourses for navigation is not within the scope of the present Convention except insofar as other uses affect navigation or are affected by navigation.

Article 2
Use of terms

For the purposes of the present Convention:

(a) "Watercourse" means a system of surface waters and groundwaters constituting by virtue of their physical

[1] Adopted by the UN General Assembly in resolution 51/229 of 21 May 1997. Available at http://untreaty.un.org/ilc/texts/instruments/english/conventions/8_3_1997.pdf.

relationship a unitary whole and normally flowing into a common terminus;

(b) "International watercourse" means a watercourse, parts of which are situated in different States;

(c) "Watercourse State" means a State Party to the present Convention in whose territory part of an international watercourse is situated, or a Party that is a regional economic integration organization, in the territory of one or more of whose Member States part of an international watercourse is situated;

(d) "Regional economic integration organization" means an organization constituted by sovereign States of a given region, to which its member States have transferred competence in respect of matters governed by this Convention and which has been duly authorized in accordance with its internal procedures, to sign, ratify, accept, approve or accede to it.

Article 3
Watercourse agreements

1. In the absence of an agreement to the contrary, nothing in the present Convention shall affect the rights or obligations of a watercourse State arising from agreements in force for it on the date on which it became a party to the present Convention.

2. Notwithstanding the provisions of paragraph 1, parties to agreements referred to in paragraph 1 may, where necessary, consider harmonizing such agreements with the basic principles of the present Convention.

3. Watercourse States may enter into one or more agreements, hereinafter referred to as "watercourse agreements", which apply and adjust the provisions of the present Convention to the characteristics and uses of a particular international watercourse or part thereof.

4. Where a watercourse agreement is concluded between two or more watercourse States, it shall define the waters to which it applies. Such an agreement may be entered into with respect to an entire international watercourse or any part thereof or a particular project, programme or use except insofar as the agreement adversely affects, to a significant extent, the use by one or more other watercourse States of the waters of the watercourse, without their express consent.

5. Where a watercourse State considers that adjustment and application of the provisions of the present Convention is required because of the characteristics and uses of a particular international watercourse, watercourse States shall consult with a view to negotiating in good faith for the purpose of concluding a watercourse agreement or agreements.

6. Where some but not all watercourse States to a particular international watercourse are parties to an agreement, noth-ing in such agreement shall affect the rights or obligations under the present Convention of watercourse States that are not parties to such an agreement.

Article 4
Parties to watercourse agreements

1. Every watercourse State is entitled to participate in the negotiation of and to become a party to any watercourse agreement that applies to the entire international watercourse, as well as to participate in any relevant consultations.

2. A watercourse State whose use of an international watercourse may be affected to a significant extent by the implementation of a proposed watercourse agreement that applies only to a part of the watercourse or to a particular project, programme or use is entitled to participate in consultations on such an agreement and, where appropriate, in the negotiation thereof in good faith with a view to becoming a party thereto, to the extent that its use is thereby affected.

PART II.
GENERAL PRINCIPLES
Article 5
Equitable and reasonable utilization and participation

1. Watercourse States shall in their respective territories utilize an international watercourse in an equitable and reasonable manner. In particular, an international watercourse shall be used and developed by watercourse States with a view to attaining optimal and sustainable utilization thereof and benefits therefrom, taking into account the interests of the watercourse States concerned, consistent with adequate protection of the watercourse.

2. Watercourse States shall participate in the use, development and protection of an international watercourse in an equitable and reasonable manner. Such participation includes both the right to utilize the watercourse and the duty to cooperate in the protection and development thereof, as provided in the present Convention.

Article 6
Factors relevant to equitable and reasonable utilization

1. Utilization of an international watercourse in an equitable and reasonable manner within the meaning of article 5 requires taking into account all relevant factors and circumstances, including:

(a) Geographic, hydrographic, hydrological, climatic, ecological and other factors of a natural character;

(b) The social and economic needs of the watercourse States concerned;

(c) The population dependent on the watercourse in each watercourse State;

(d) The effects of the use or uses of the watercourses in one watercourse State on other watercourse States;

(e) Existing and potential uses of the watercourse;

(f) Conservation, protection, development and economy of use of the water resources of the watercourse and the costs of measures taken to that effect;

(g) The availability of alternatives, of comparable value, to a particular planned or existing use.

2. In the application of article 5 or paragraph 1 of this article, watercourse States concerned shall, when the need arises, enter into consultations in a spirit of cooperation.

3. The weight to be given to each factor is to be determined by its importance in comparison with that of other relevant factors. In determining what is a reasonable and equitable use, all relevant factors are to be considered together and a conclusion reached on the basis of the whole.

Article 7
Obligation not to cause significant harm

1. Watercourse States shall, in utilizing an international watercourse in their territories, take all appropriate measures to prevent the causing of significant harm to other watercourse States.

2. Where significant harm nevertheless is caused to another watercourse State, the States whose use causes such harm shall, in the absence of agreement to such use, take all appropriate measures, having due regard for the provisions of articles 5 and 6, in consultation with the affected State, to eliminate or mitigate such harm and, where appropriate, to discuss the question of compensation.

Article 8
General obligation to cooperate

1. Watercourse States shall cooperate on the basis of sovereign equality, territorial integrity, mutual benefit and good faith in order to attain optimal utilization and adequate protection of an international watercourse.

2. In determining the manner of such cooperation, watercourse States may consider the establishment of joint mechanisms or commissions, as deemed necessary by them, to facilitate cooperation on relevant measures and procedures in the light of experience gained through cooperation in existing joint mechanisms and commissions in various regions.

Article 9
Regular exchange of data and information

1. Pursuant to article 8, watercourse States shall on a regular basis exchange readily available data and information on the condition of the watercourse, in particular that of a hydrological, meteorological, hydrogeological and ecolog-

ical nature and related to the water quality as well as related forecasts.

2. If a watercourse State is requested by another watercourse State to provide data or information that is not readily available, it shall employ its best efforts to comply with the request but may condition its compliance upon payment by the requesting State of the reasonable costs of collecting and, where appropriate, processing such data or information.

3. Watercourse States shall employ their best efforts to collect and, where appropriate, to process data and information in a manner which facilitates its utilization by the other watercourse States to which it is communicated.

Article 10
Relationship between different kinds of uses

1. In the absence of agreement or custom to the contrary, no use of an international watercourse enjoys inherent priority over other uses.

2. In the event of a conflict between uses of an international watercourse, it shall be resolved with reference to articles 5 to 7, with special regard being given to the requirements of vital human needs.

PART III.
PLANNED MEASURES

Article 11
Information concerning planned measures

Watercourse States shall exchange information and consult each other and, if necessary, negotiate on the possible effects of planned measures on the condition of an international watercourse.

Article 12
Notification concerning planned measures with possible adverse effects

Before a watercourse State implements or permits the implementation of planned measures which may have a significant adverse effect upon other watercourse States, it shall provide those States with timely notification thereof. Such notification shall be accompanied by available technical data and information, including the results of any environmental impact assessment, in order to enable the notified States to evaluate the possible effects of the planned measures.

Article 13
Period for reply to notification

Unless otherwise agreed:

(a) A watercourse State providing a notification under article 12 shall allow the notified States a period of six months within which to study and evaluate the possible effects of the planned measures and to communicate the findings to it;

(b) This period shall, at the request of a notified State for which the evaluation of the planned measures poses special difficulty, be extended for a period of six months.

Article 14
Obligations of the notifying State during the period for reply

During the period referred to in article 13, the notifying State:

(a) Shall cooperate with the notified States by providing them, on request, with any additional data and information that is available and necessary for an accurate evaluation; and

(b) Shall not implement or permit the implementation of the planned measures without the consent of the notified States.

Article 15
Reply to notification

The notified States shall communicate their findings to the notifying State as early as possible within the period applicable pursuant to article 13. If a notified State finds that implementation of the planned measures would be inconsistent with the provisions of articles 5 or 7, it shall attach to its finding a documented explanation setting forth the reasons for the finding.

Article 16
Absence of reply to notification

1. If, within the period applicable pursuant to article 13, the notifying State receives no communication under article 15, it may, subject to its obligations under articles 5 and 7, proceed with the implementation of the planned measures, in accordance with the notification and any other data and information provided to the notified States.

2. Any claim to compensation by a notified State which has failed to reply within the period applicable pursuant to article 13 may be offset by the costs incurred by the notifying State for action undertaken after the expiration of the time for a reply which would not have been undertaken if the notified State had objected within that period.

Article 17
Consultations and negotiations concerning planned measures

1. If a communication is made under article 15 that implementation of the planned measures would be inconsistent with the provisions of article 5 or 7, the notifying State and the State making the communication shall enter into consultations and, if necessary, negotiations with a view to arriving at an equitable resolution of the situation.

2. The consultations and negotiations shall be conducted on the basis that each State must in good faith pay reasonable regard to the rights and legitimate interests of the other State.

3. During the course of the consultations and negotiations, the notifying State shall, if so requested by the notified State at the time it makes the communication, refrain from implementing or permitting the implementation of the planned measures for a period of six months unless otherwise agreed.

Article 18
Procedures in the absence of notification

1. If a watercourse State has reasonable grounds to believe that another watercourse State is planning measures that may have a significant adverse effect upon it, the former State may request the latter to apply the provisions of article 12. The request shall be accompanied by a documented explanation setting forth its grounds.

2. In the event that the State planning the measures nevertheless finds that it is not under an obligation to provide a notification under article 12, it shall so inform the other State, providing a documented explanation setting forth the reasons for such finding. If this finding does not satisfy the other State, the two States shall, at the request of that other State, promptly enter into consultations and negotiations in the manner indicated in paragraphs 1 and 2 of article 17.

3. During the course of the consultations and negotiations, the State planning the measures shall, if so requested by the other State at the time it requests the initiation of consultations and negotiations, refrain from implementing or permitting the implementation of those measures for a period of six months unless otherwise agreed.

Article 19
Urgent implementation of planned measures

1. In the event that the implementation of planned measures is of the utmost urgency in order to protect public health, public safety or other equally important interests, the State planning the measures may, subject to articles 5 and 7, immediately proceed to implementation, notwithstanding the provisions of article 14 and paragraph 3 of article 17.

2. In such case, a formal declaration of the urgency of the measures shall be communicated without delay to the other watercourse States referred to in article 12 together with the relevant data and information.

3. The State planning the measures shall, at the request of any of the States referred to in paragraph 2, promptly enter

into consultations and negotiations with it in the manner indicated in paragraphs 1 and 2 of article 17.

PART IV.
PROTECTION, PRESERVATION AND MANAGEMENT
Article 20
Protection and preservation of ecosystems

Watercourse States shall, individually and, where appropriate, jointly, protect and preserve the ecosystems of international watercourses.

Article 21
Prevention, reduction and control of pollution

1. For the purpose of this article, "pollution of an international watercourse" means any detrimental alteration in the composition or quality of the waters of an international watercourse which results directly or indirectly from human conduct.
2. Watercourse States shall, individually and, where appropriate, jointly, prevent, reduce and control the pollution of an international watercourse that may cause significant harm to other watercourse States or to their environment, including harm to human health or safety, to the use of the waters for any beneficial purpose or to the living resources of the watercourse. Watercourse States shall take steps to harmonize their policies in this connection.
3. Watercourse States shall, at the request of any of them, consult with a view to arriving at mutually agreeable measures and methods to prevent, reduce and control pollution of an international watercourse, such as:
   (a) Setting joint water quality objectives and criteria;
   (b) Establishing techniques and practices to address pollution from point and non-point sources;
   (c) Establishing lists of substances the introduction of which into the waters of an international watercourse is to be prohibited, limited, investigated or monitored.

Article 22
Introduction of alien or new species

Watercourse States shall take all measures necessary to prevent the introduction of species, alien or new, into an international watercourse which may have effects detrimental to the ecosystem of the watercourse resulting in significant harm to other watercourse States.

Article 23
Protection and preservation of the marine environment

Watercourse States shall, individually and, where appropriate, in cooperation with other States, take all measures with respect to an international watercourse that are necessary to protect and preserve the marine environment, including estuaries, taking into account generally accepted international rules and standards.

Article 24
Management

1. Watercourse States shall, at the request of any of them, enter into consultations concerning the management of an international watercourse, which may include the establishment of a joint management mechanism.
2. For the purposes of this article, "management" refers, in particular, to:
   (a) Planning the sustainable development of an international watercourse and providing for the implementation of any plans adopted; and
   (b) Otherwise promoting the rational and optimal utilization, protection and control of the watercourse.

Article 25
Regulation

1. Watercourse States shall cooperate, where appropriate, to respond to needs or opportunities for regulation of the flow of the waters of an international watercourse.
2. Unless otherwise agreed, watercourse States shall participate on an equitable basis in the construction and maintenance or defrayal of the costs of such regulation works as they may have agreed to undertake.
3. For the purposes of this article, "regulation" means the use of hydraulic works or any other continuing measure to alter, vary or otherwise control the flow of the waters of an international watercourse.

Article 26
Installations

1. Watercourse States shall, within their respective territories, employ their best efforts to maintain and protect installations, facilities and other works related to an international watercourse.
2. Watercourse States shall, at the request of any of them which has reasonable grounds to believe that it may suffer significant adverse effects, enter into consultations with regard to:
   (a) The safe operation and maintenance of installations, facilities or other works related to an international watercourse; and
   (b) The protection of installations, facilities or other works from wilful or negligent acts or the forces of nature.

PART V.
HARMFUL CONDITIONS AND EMERGENCY
SITUATIONS
Article 27
Prevention and mitigation of harmful conditions

Watercourse States shall, individually and, where appropriate, jointly, take all appropriate measures to prevent or mitigate conditions related to an international watercourse that may be harmful to other watercourse States, whether resulting from natural causes or human conduct, such as flood or ice conditions, water-borne diseases, siltation, erosion, salt-water intrusion, drought or desertification.

Article 28
Emergency situations

1. For the purposes of this article, "emergency" means a situation that causes, or poses an imminent threat of causing, serious harm to watercourse States or other States and that results suddenly from natural causes, such as floods, the breaking up of ice, landslides or earthquakes, or from human conduct, such as industrial accidents.
2. A watercourse State shall, without delay and by the most expeditious means available, notify other potentially affected States and competent international organizations of any emergency originating within its territory.
3. A watercourse State within whose territory an emergency originates shall, in cooperation with potentially affected States and, where appropriate, competent international organizations, immediately take all practicable measures necessitated by the circumstances to prevent, mitigate and eliminate harmful effects of the emergency.
4. When necessary, watercourse States shall jointly develop contingency plans for responding to emergencies, in cooperation, where appropriate, with other potentially affected States and competent international organizations.

PART VI.
MISCELLANEOUS PROVISIONS
Article 29
International watercourses and installations in time of armed conflict

International watercourses and related installations, facilities and other works shall enjoy the protection accorded by the principles and rules of international law applicable in international and non-international armed conflict and shall not be used in violation of those principles and rules.

Article 30
Indirect procedures

In cases where there are serious obstacles to direct contacts between watercourse States, the States concerned shall fulfil their obligations of cooperation provided for in the present Convention, including exchange of data and information, notification, communication, consultations and negotiations, through any indirect procedure accepted by them.

Article 31
Data and information vital to national defence or security

Nothing in the present Convention obliges a watercourse State to provide data or information vital to its national defence or security. Nevertheless, that State shall cooperate in good faith with the other watercourse States with a view to providing as much information as possible under the circumstances.

Article 32
Non-discrimination

Unless the watercourse States concerned have agreed otherwise for the protection of the interests of persons, natural or juridical, who have suffered or are under a serious threat of suffering significant transboundary harm as a result of activities related to an international watercourse, a watercourse State shall not discriminate on the basis of nationality or residence or place where the injury occurred, in granting to such persons, in accordance with its legal system, access to judicial or other procedures, or a right to claim compensation or other relief in respect of significant harm caused by such activities carried on in its territory.

Article 33
Settlement of disputes

1. In the event of a dispute between two or more parties concerning the interpretation or application of the present Convention, the parties concerned shall, in the absence of an applicable agreement between them, seek a settlement of the dispute by peaceful means in accordance with the following provisions.
2. If the parties concerned cannot reach agreement by negotiation requested by one of them, they may jointly seek the good offices of, or request mediation or conciliation by, a third party, or make use, as appropriate, of any joint watercourse institutions that may have been established by them or agree to submit the dispute to arbitration or to the International Court of Justice.
3. Subject to the operation of paragraph 10, if after six months from the time of the request for negotiations referred to in paragraph 2, the parties concerned have not been able to

settle their dispute through negotiation or any other means referred to in paragraph 2, the dispute shall be submitted, at the request of any of the parties to the dispute, to impartial fact-finding in accordance with paragraphs 4 to 9, unless the parties otherwise agree.

4. A Fact-finding Commission shall be established, composed of one member nominated by each party concerned and in addition a member not having the nationality of any of the parties concerned chosen by the nominated members who shall serve as Chairman.

5. If the members nominated by the parties are unable to agree on a Chairman within three months of the request for the establishment of the Commission, any party concerned may request the Secretary-General of the United Nations to appoint the Chairman who shall not have the nationality of any of the parties to the dispute or of any riparian State of the watercourse concerned. If one of the parties fails to nominate a member within three months of the initial request pursuant to paragraph 3, any other party concerned may request the Secretary-General of the United Nations to appoint a person who shall not have the nationality of any of the parties to the dispute or of any riparian State of the watercourse concerned. The person so appointed shall constitute a single-member Commission.

6. The Commission shall determine its own procedure.

7. The parties concerned have the obligation to provide the Commission with such information as it may require and, on request, to permit the Commission to have access to their respective territory and to inspect any facilities, plant, equipment, construction or natural feature relevant for the purpose of its inquiry.

8. The Commission shall adopt its report by a majority vote, unless it is a single-member Commission, and shall submit that report to the parties concerned setting forth its findings and the reasons therefor and such recommendations as it deems appropriate for an equitable solution of the dispute, which the parties concerned shall consider in good faith.

9. The expenses of the Commission shall be borne equally by the parties concerned.

10. When ratifying, accepting, approving or acceding to the present Convention, or at any time thereafter, a party which is not a regional economic integration organization may declare in a written instrument submitted to the depositary that, in respect of any dispute not resolved in accordance with paragraph 2, it recognizes as compulsory ipso facto, and without special agreement in relation to any party accepting the same obligation:

   (a) Submission of the dispute to the International Court of Justice; and/or

   (b) Arbitration by an arbitral tribunal established and operating, unless the parties to the dispute otherwise agreed, in accordance with the procedure laid down in the annex to the present Convention.

A party which is a regional economic integration organization may make a declaration with like effect in relation to arbitration in accordance with subparagraph (b).

PART VII.
FINAL CLAUSES
Article 34
Signature

The present Convention shall be open for signature by all States and by regional economic integration organizations from 21 May 1997 until 20 May 2000 at United Nations Headquarters in New York.

Article 35
Ratification, acceptance, approval or accession

1. The present Convention is subject to ratification, acceptance, approval or accession by States and by regional economic integration organizations. The instruments of ratification, acceptance, approval or accession shall be deposited with the Secretary-General of the United Nations.

2. Any regional economic integration organization which becomes a Party to this Convention without any of its member States being a Party shall be bound by all the obligations under the Convention. In the case of such organizations, one or more of whose member States is a Party to this Convention, the organization and its member States shall decide on their respective responsibilities for the performance of their obligations under the Convention. In such cases, the organization and the member States shall not be entitled to exercise rights under the Convention concurrently.

3. In their instruments of ratification, acceptance, approval or accession, the regional economic integration organizations shall declare the extent of their competence with respect to the matters governed by the Convention. These organizations shall also inform the Secretary-General of the United Nations of any substantial modification in the extent of their competence.

Article 36
Entry into force

1. The present Convention shall enter into force on the ninetieth day following the date of deposit of the thirty-fifth instrument of ratification, acceptance, approval or accession with the Secretary-General of the United Nations.

2. For each State or regional economic integration organization that ratifies, accepts or approves the Convention or accedes thereto after the deposit of the thirty-fifth instrument of ratification, acceptance, approval or accession, the Convention shall enter into force on the ninetieth day after the deposit by such State or regional economic integration organization of its instrument of ratification, acceptance, approval or accession.

3. For the purposes of paragraphs 1 and 2, any instrument deposited by a regional economic integration organization shall not be counted as additional to those deposited by States.

Article 37
Authentic texts

The original of the present Convention, of which the Arabic, Chinese, English, French, Russian and Spanish texts are equally authentic, shall be deposited with the Secretary-General of the United Nations.

ANNEX
ARBITRATION
Article 1
Unless the parties to the dispute otherwise agree, the arbitration pursuant to article 33 of the Convention shall take place in accordance with articles 2 to 14 of the present annex.

Article 2
The claimant party shall notify the respondent party that it is referring a dispute to arbitration pursuant to article 33 of the Convention. The notification shall state the subject matter of arbitration and include, in particular, the articles of the Convention, the interpretation or application of which are at issue. If the parties do not agree on the subject matter of the dispute, the arbitral tribunal shall determine the subject matter.

Article 3

1. In disputes between two parties, the arbitral tribunal shall consist of three members. Each of the parties to the dispute shall appoint an arbitrator and the two arbitrators so appointed shall designate by common agreement the third arbitrator, who shall be the Chairman of the tribunal. The latter shall not be a national of one of the parties to the dispute or of any riparian State of the watercourse concerned, nor have his or her usual place of residence in the territory of one of these parties or such riparian State, nor have dealt with the case in any other capacity.

2. In disputes between more than two parties, parties in the same interest shall appoint one arbitrator jointly by agreement.

3. Any vacancy shall be filled in the manner prescribed for the initial appointment.

Article 4

1. If the Chairman of the arbitral tribunal has not been designated within two months of the appointment of the second arbitrator, the President of the International Court of Justice shall, at the request of a party, designate the Chairman within a further two-month period.

2. If one of the parties to the dispute does not appoint an arbitrator within two months of receipt of the request, the other party may inform the President of the International Court of Justice, who shall make the designation within a further two-month period.

Article 5
The arbitral tribunal shall render its decisions in accordance with the provisions of this Convention and international law.

Article 6
Unless the parties to the dispute otherwise agree, the arbitral tribunal shall determine its own rules of procedure.

Article 7
The arbitral tribunal may, at the request of one of the parties, recommend essential interim measures of protection.

Article 8

1. The parties to the dispute shall facilitate the work of the arbitral tribunal and, in particular, using all means at their disposal, shall:
   (a) Provide it with all relevant documents, information and facilities; and
   (b) Enable it, when necessary, to call witnesses or experts and receive their evidence.

2. The parties and the arbitrators are under an obligation to protect the confidentiality of any information they receive in confidence during the proceedings of the arbitral tribunal.

Article 9
Unless the arbitral tribunal determines otherwise because of the particular circumstances of the case, the costs of the tribunal shall be borne by the parties to the dispute in equal shares. The tribunal shall keep a record of all its costs, and shall furnish a final statement thereof to the parties.

Article 10
Any party that has an interest of a legal nature in the subject matter of the dispute which may be affected by the decision in the case, may intervene in the proceedings with the consent of the tribunal.

Article 11

The tribunal may hear and determine counterclaims arising directly out of the subject matter of the dispute.

Article 12

Decisions both on procedure and substance of the arbitral tribunal shall be taken by a majority vote of its members.

Article 13

If one of the parties to the dispute does not appear before the arbitral tribunal or fails to defend its case, the other party may request the tribunal to continue the proceedings and to make its award. Absence of a party or a failure of a party to defend its case shall not constitute a bar to the proceedings. Before rendering its final decision, the arbitral tribunal must satisfy itself that the claim is well founded in fact and law.

Article 14

1. The tribunal shall render its final decision within five months of the date on which it is fully constituted unless it finds it necessary to extend the time limit for a period which should not exceed five more months.
2. The final decision of the arbitral tribunal shall be confined to the subject matter of the dispute and shall state the reasons on which it is based. It shall contain the names of the members who have participated and the date of the final decision. Any member of the tribunal may attach a separate or dissenting opinion to the final decision.
3. The award shall be binding on the parties to the dispute. It shall be without appeal unless the parties to the dispute have agreed in advance to an appellate procedure.
4. Any controversy which may arise between the parties to the dispute as regards the interpretation or manner of implementation of the final decision may be submitted by either party for decision to the arbitral tribunal which rendered it.

A.2 TEXT OF THE DRAFT ARTICLES ON THE LAW OF TRANSBOUNDARY AQUIFERS ADOPTED BY THE COMMISSION ON FIRST READING[2]

PART I
INTRODUCTION
Article 1
Scope. The present draft articles apply to:

(a) utilization of transboundary aquifers and aquifer systems;
(b) other activities that have or are likely to have an impact upon those aquifers and aquifer systems; and

[2] At the time of writing, the report had not been released. It will be at http://untreaty.un.org/ilc/sessions/60/60docs.htm

(c) measures for the protection, preservation and management of those aquifers and aquifer systems.

Article 2
Use of terms
For the purposes of the present draft articles:

(a) "aquifer" means a permeable water-bearing underground geological formation underlain by a less permeable layer and the water contained in the saturated zone of the formation;
(b) "aquifer system" means a series of two or more aquifers that are hydraulically connected;
(c) "transboundary aquifer" or "transboundary aquifer system" means, respectively, an aquifer or aquifer system, parts of which are situated in different States;
(d) "aquifer State" means a State in whose territory any part of a transboundary aquifer or aquifer system is situated;
(e) "recharging aquifer" means an aquifer that receives a non-negligible amount of contemporary water recharge;
(f) "recharge zone" means the zone which contributes water to an aquifer, consisting of the catchment area of rainfall water and the area where such water flows to an aquifer by runoff on the ground and infiltration through soil;
(g) "discharge zone" means the zone where water originating from an aquifer flows to its outlets, such as a watercourse, a lake, an oasis, a wetland or an ocean.

PART II
GENERAL PRINCIPLES
Article 3
Sovereignty of aquifer States

Each aquifer State has sovereignty over the portion of a transboundary aquifer or aquifer system located within its territory. It shall exercise its sovereignty in accordance with the present draft articles.

Article 4
Equitable and reasonable utilization

Aquifer States shall utilize a transboundary aquifer or aquifer system according to the principle of equitable and reasonable utilization, as follows:

(a) they shall utilize the transboundary aquifer or aquifer system in a manner that is consistent with the equitable and reasonable accrual of benefits therefrom to the aquifer States concerned;
(b) they shall aim at maximizing the long-term benefits derived from the use of water contained therein;

(c) they shall establish individually or jointly an overall utilization plan, taking into account present and future needs of, and alternative water sources for, the aquifer States; and

(d) they shall not utilize a recharging transboundary aquifer or aquifer system at a level that would prevent continuance of its effective functioning.

Article 5
Factors relevant to equitable and reasonable utilization

1. Utilization of a transboundary aquifer or aquifer system in an equitable and reasonable manner within the meaning of draft article 4 requires taking into account all relevant factors, including:
   (a) the population dependent on the aquifer or aquifer system in each aquifer State;
   (b) the social, economic and other needs, present and future, of the aquifer States concerned;
   (c) the natural characteristics of the aquifer or aquifer system;
   (d) the contribution to the formation and recharge of the aquifer or aquifer system;
   (e) the existing and potential utilization of the aquifer or aquifer system;
   (f) the effects of the utilization of the aquifer or aquifer system in one aquifer State on other aquifer States concerned;
   (g) the availability of alternatives to a particular existing and planned utilization of the aquifer or aquifer system;
   (h) the development, protection and conservation of the aquifer or aquifer system and the costs of measures to be taken to that effect;
   (i) the role of the aquifer or aquifer system in the related ecosystem.

2. The weight to be given to each factor is to be determined by its importance with regard to a specific transboundary aquifer or aquifer system in comparison with that of other relevant factors. In determining what is equitable and reasonable utilization, all relevant factors are to be considered together and a conclusion reached on the basis of all the factors. However, in weighing different utilizations of a transboundary aquifer or aquifer system, special regard shall be given to vital human needs.

Article 6
Obligation not to cause significant harm to other aquifer States

1. Aquifer States shall, in utilizing a transboundary aquifer or aquifer system in their territories, take all appropriate measures to prevent the causing of significant harm to other aquifer States.

2. Aquifer States shall, in undertaking activities other than utilization of a transboundary aquifer or aquifer system that have, or are likely to have, an impact on that transboundary aquifer or aquifer system, take all appropriate measures to prevent the causing of significant harm through that aquifer or aquifer system to other aquifer States.

3. Where significant harm nevertheless is caused to another aquifer State, the aquifer States whose activities cause such harm shall take, in consultation with the affected State, all appropriate measures to eliminate or mitigate such harm, having due regard for the provisions of draft articles 4 and 5.

Article 7
General obligation to cooperate

1. Aquifer States shall cooperate on the basis of sovereign equality, territorial integrity, sustainable development, mutual benefit and good faith in order to attain equitable and reasonable utilization and appropriate protection of their transboundary aquifer or aquifer system.

2. For the purpose of paragraph 1, aquifer States should establish joint mechanisms of cooperation.

Article 8
Regular exchange of data and information

1. Pursuant to draft article 7, aquifer States shall, on a regular basis, exchange readily available data and information on the condition of the transboundary aquifer or aquifer system, in particular of a geological, hydrogeological, hydrological, meteorological and ecological nature and related to the hydrochemistry of the aquifer or aquifer system, as well as related forecasts.

2. Where knowledge about the nature and extent of some transboundary aquifer or aquifer systems is inadequate, aquifer States concerned shall employ their best efforts to collect and generate more complete data and information relating to such aquifer or aquifer systems, taking into account current practices and standards. They shall take such action individually or jointly and, where appropriate, together with or through international organizations.

3. If an aquifer State is requested by another aquifer State to provide data and information relating to the aquifer or aquifer systems that are not readily available, it shall employ its best efforts to comply with the request. The requested State may condition its compliance upon payment by the requesting State of the reasonable costs of collecting and, where appropriate, processing such data or information.

4. Aquifer States shall, where appropriate, employ their best efforts to collect and process data and information in a manner that facilitates their utilization by the other aquifer

States to which such data and information are communicated.

PART III
PROTECTION, PRESERVATION AND MANAGEMENT
Article 9
Protection and preservation of ecosystems

Aquifer States shall take all appropriate measures to protect and preserve ecosystems within, or dependent upon, their transboundary aquifers or aquifer systems, including measures to ensure that the quality and quantity of water retained in the aquifer or aquifer system, as well as that released in its discharge zones, are sufficient to protect and preserve such ecosystems.

Article 10
Recharge and discharge zones

1. Aquifer States shall identify recharge and discharge zones of their transboundary aquifer or aquifer system and, within these zones, shall take special measures to minimize detrimental impacts on the recharge and discharge processes.
2. All States in whose territory a recharge or discharge zone is located, in whole or in part, and which are not aquifer States with regard to that aquifer or aquifer system, shall cooperate with the aquifer States to protect the aquifer or aquifer system.

Article 11
Prevention, reduction and control of pollution

Aquifer States shall, individually and, where appropriate, jointly, prevent, reduce and control pollution of their transboundary aquifer or aquifer system, including through the recharge process, that may cause significant harm to other aquifer States. In view of uncertainty about the nature and extent of transboundary aquifers or aquifer systems and of their vulnerability to pollution, aquifer States shall take a precautionary approach.

Article 12
Monitoring

1. Aquifer States shall monitor their transboundary aquifer or aquifer system. They shall, wherever possible, carry out these monitoring activities jointly with other aquifer States concerned and, where appropriate, in collaboration with the competent international organizations. Where, however, monitoring activities are not carried out jointly, the aquifer States shall exchange the monitored data among themselves.
2. Aquifer States shall use agreed or harmonized standards and methodology for monitoring their transboundary aquifer or

aquifer system. They should identify key parameters that they will monitor based on an agreed conceptual model of the aquifer or aquifer system. These parameters should include parameters on the condition of the aquifer or aquifer system as listed in draft article 8, paragraph 1, and also on the utilization of the aquifer and aquifer system.

Article 13
Management

Aquifer States shall establish and implement plans for the proper management of their transboundary aquifer or aquifer system in accordance with the provisions of the present draft articles. They shall, at the request by any of them, enter into consultations concerning the management of the transboundary aquifer or aquifer system. A joint management mechanism shall be established, wherever appropriate.

PART IV
ACTIVITIES AFFECTING OTHER STATES
Article 14
Planned activities

1. When a State has reasonable grounds for believing that a particular planned activity in its territory may affect a transboundary aquifer or aquifer system and thereby may have a significant adverse effect upon another State, it shall, as far as practicable, assess the possible effects of such activity.
2. Before a State implements or permits the implementation of planned activities which may affect a transboundary aquifer or aquifer system and thereby may have a significant adverse effect upon another State, it shall provide that State with timely notification thereof. Such notification shall be accompanied by available technical data and information, including any environmental impact assessment, in order to enable the notified State to evaluate the possible effects of the planned activities.
3. If the notifying and the notified States disagree on the possible effect of the planned activities, they shall enter into consultations and, if necessary, negotiations with a view to arriving at an equitable resolution of the situation. They may utilize an independent fact-finding body to make an impartial assessment of the effect of the planned activities.

PART V
MISCELLANEOUS PROVISIONS
Article 15
Scientific and technical cooperation with developing States

States shall, directly or through competent international organizations, promote scientific, educational, technical and other cooperation with developing States for the protection and management of transboundary aquifers or aquifer systems. Such cooperation shall include, *inter alia*:

(a) Training of their scientific and technical personnel;

(b) Facilitating their participation in relevant international programmes;

(c) Supplying them with necessary equipment and facilities;

(d) Enhancing their capacity to manufacture such equipment;

(e) Providing advice on and developing facilities for research, monitoring, educational and other programmes;

(f) Providing advice on and developing facilities for minimizing the detrimental effects of major activities affecting transboundary aquifers or aquifer systems;

(g) Preparing environmental impact assessments.

Article 16
Emergency situations

1. For the purpose of the present draft article, "emergency" means a situation, resulting suddenly from natural causes or from human conduct, that poses an imminent threat of causing serious harm to aquifer States or other States.

2. Where an emergency affects a transboundary aquifer or aquifer system and thereby poses an imminent threat to States, the following shall apply:

(a) The State within whose territory the emergency originates shall:

(i) without delay and by the most expeditious means available, notify other potentially affected States and competent international organizations of the emergency;

(ii) in cooperation with potentially affected States and, where appropriate, competent international organizations, immediately take all practicable measures necessitated by the circumstances to prevent, mitigate and eliminate any harmful effect of the emergency;

(b) States shall provide scientific, technical, logistical and other cooperation to other States experiencing an emergency. Cooperation may include coordination of international emergency actions and communications, making available trained emergency response personnel, emergency response equipments and supplies, scientific and technical expertise and humanitarian assistance.

3. Where an emergency poses a threat to vital human needs, aquifer States, notwithstanding draft articles 4 and 6, may take measures that are strictly necessary to meet such needs.

Article 17
Protection in time of armed conflict

Transboundary aquifers or aquifer systems and related installations, facilities and other works shall enjoy the protection accorded by the principles and rules of international law applicable in international and non-international armed conflicts and shall not be used in violation of those principles and rules.

Article 18
Data and information concerning national defence or security

Nothing in the present draft articles obliges a State to provide data or information the confidentiality of which is essential to its national defence or security. Nevertheless, that State shall cooperate in good faith with other States with a view to providing as much information as possible under the circumstances.

Article 19
Bilateral and regional agreements and arrangements

For the purpose of managing a particular transboundary aquifer or aquifer system, aquifer States are encouraged to enter into a bilateral or regional agreement or arrangement among themselves. Such agreement or arrangement may be entered into with respect to an entire aquifer or aquifer system or any part thereof or a particular project, programme or utilization except insofar as the agreement or arrangement adversely affects, to a significant extent, the utilization, by one or more other aquifer States of the water in that aquifer or aquifer system, without their express consent.

B River basin organizations[1]

Jerome Delli Priscoli

B.1 NORTH AMERICA[2]

B.1.1 The United States

The United States operates under two major systems of water rights: riparian doctrine in the East and prior appropriation in the West. The quantifying of Native American tribal rights and their integration into these systems is becoming more important. The *acequia* system found in the Southwest is one of a few hybrids. It was inherited from the Spanish, who brought it from the Arab world.

The United States of America is a federal system. The states are sovereign entities and they have control over water resources. Like other large countries in the world, river basin operations and organizations revolve first around the alignment of powers among these sovereign entities, which rarely fit river boundaries. Second, they revolve around the exercise of bureaucratic power within the federal and state governments. Multiple agencies work with water usually within their own mandates and sector.

However, there are major federal interests affecting water distribution and use. In fact, one of the United States' earliest court decisions was confining the power of the federal government to regulate commerce involving water navigation. Beyond interstate commerce, federal control over water has been established in a variety of areas, such as for emergencies, flood control, irrigation, public health, environmental issues, and fish and wildlife. Many of these interests have been insti-

tutionalized in numerous federal agencies, which present a formidable coordination task. Complex formulas for the mix of federal and state money in water resources development have evolved for different project purposes and water uses, such as flood control, navigation, recreation, water supply for irrigation, and hydroelectric power. Indeed, the debate around these formulae constitutes one of the principal bargaining arenas for water cooperation.

During the 1980s, the movement has been to reduce the federal role and to enhance the state and private sector roles in water resources development. There has been a reduction in water development and a greater emphasis on the management of existing facilities and projects. The federal regulatory role, especially for environmental purposes, has in many ways become the focal point for regional cooperative planning. However, many observers are now looking again at the need for coordinated water development (Sidebar B.1).

During the twentieth century, several types of arrangements were tried: interstate compact commissions; interstate councils; basin interagency committees (ad hoc); interagency-interstate commissions (Title II); federal interstate compact commissions; federal state agencies; single federal administrators; and watershed councils (see Table 5.1).

Two approaches dominated the early twentieth century: (1) interstate compacts (which can be seen as a parallel to treaties among states) and (2) adversarial court cases. These agreements suffered from the illusion that allocation could and should be permanent. However, as populations shifted, Native American tribal demands grew, and new uses (especially instream) appeared, allocations under compacts have proven too inflexible for management. They are not conducive to taking advantage of the variability in the hydrologic system. In general, the challenges to the compacts have been the impact of upstream developments (and future dreams for such development) on the apportionment to downstream states.

In the United States, numerous presidential commissions have tried unsuccessfully to establish national water policy.

---

[1] The materials for this appendix have been gathered from many sources that include gray literature from inside of several major water organizations as well as published materials. Training sessions on RBOs held around the world and several private communications with individuals have also been important to developing the following descriptions. In this regard communications with Eduardo Mestre, Peter Millington, Evan Chere, and Bernard Barraque were most useful. However, the actual descriptions are the views of the authors. The list should provide a good entrée into further work for interested readers.

[2] See also, Delli Priscoli (1976, 2001); Schad (1964); Wendall and Schwan (1975); U.S. Water Resources Council (1967); and Kenney and Lord (1994).

135

Sidebar B.1 Review of U.S. Coordination Mechanisms (from Kenney and Lord, 1994)

One of the most common themes permeating the literature reviewing coordination mechanisms for U.S. interstate water resources is that the track record of these institutional innovations is generally poor. This should not discourage further innovation, however. Addressing the factors that fragment regional water institution is an extremely difficult task and a task normally attempted with a new coordination mechanism only after more established approaches have failed. And in those basins where a coordination mechanism has not successfully resolved the major water resources problems, the mechanisms have generally not been a step backward – but just a discouragingly small step forward. Thus, most mechanisms have proven to be unsuccessful only in the sense that they have failed to satisfy lofty expectations. If a more tempered enthusiasm for such efforts is utilized, then the track record of coordination mechanism is significantly improved, and proposal for further experimentation will be evaluated in a more forgiving and welcoming political atmosphere.

During the 1970s, an elaborate institutional and analytical procedure evolved, only to be abandoned as its implementation was beginning. To a great degree, this structure was based on river basins and was fueled by rational analytical notions. It encouraged high-level intersectoral planning and autonomous operating levels. A minianalytical rapprochement among engineers, social scientists, and ecologists was achieved in the form of two planning objectives and four accounts.

In the 1980s, the United States' approach became more market oriented. National economic development was effectively established again as the prime objective, with environment quality as a constraint, usually articulated through regulatory policy. New private–public partnerships and cost-sharing formulas emerged. Attempts were made to use more realistic pricing – closer to marginal costs – for water through a variety of water market mechanisms. At the same time, recognition of the importance of environmental restoration and wetland management also grew. During the 1990s, the need for new modes of interstate cooperation increased, in both humid and arid areas. Reliance on court judgments proved to be too expensive, inflexible, time-consuming, and locked into precedent to realistically meet new needs. Indeed, even the U.S. Supreme Court noted the importance of planning for future water uses and information sharing as a prerequisite to adjudication. Various basins and regions, such as Apalachiola-Chattahoochee-Flint (ACF) and Alabama-Coosa-Tallapoosa (ACT) areas of Georgia, Alabama, and Florida, are turning to assisted negoti-

ation techniques such as facilitation and mediation. In response to drought, riparian states on the Missouri River are seeking new forms of coordination and some are calling for a return to a river basin commission. Other areas, such as the Southwest and California, are turning to water banking, marketing, and new forms of pricing.

SINGLE FEDERAL ADMINISTRATOR: (1) THE COLORADO AND DEPARTMENT OF THE INTERIOR (COLORADO LAW OF THE RIVER)[3]

The Colorado River serves about 20 million people in two countries (Figure B.1). Its high variability with severe flooding and drought periods inhibited economic development in the early part of the twentieth century. From the 1920s to the 1960s, considerable structural regulation was added to the river. Floods are mostly controlled and water storage has prevented drought. However, the story of the Colorado is fraught with the political problems of trying to achieve river basin management (RBM) and create a river basin organization (RBO). Starting in the 1920s with a basic apportionment, subsequent federal and state statutes, interstate compacts, court decisions and decrees, international treaties, operating criteria and administrative decisions for the Colorado have together come to be called the "Law of the River." The Law of the River remains the means for the administration of the river; there is no RBO for the whole river.

Like the Delaware River basin, the seven states around the Colorado River basin attempted to use the interstate compacts process. From the 1920s to the early 1980s, the federal government acted as a catalyst to agreements around the Colorado. It made development funds conditional on apportionment agreements. Since the 1970s, a new era, which deemphasizes structural solutions and emphasizes wise use and conservation, has emerged. The absence of an RBO has led to a new emphasis on interstate marketing. That too is fraught with problems, however. Upstream states are concerned that an agreement to use of their allocated water could eventually lead to the arguments that they do not need their allocation. Also the basin lacks an equivalent trusted information provider such as the Delaware River Basin Commission (DRBC) technical staff.

In 1922, after considerable acrimony between the upstream states and California, the Colorado River Compact was signed. The upstream states feared that California had an insatiable appetite for water. Herbert Hoover, who at the time was Secretary of Commerce, tried to mediate and build consensus on the entire river. The compromise, which was forged in lieu of full consensus, split the river and allocations into an upper

---

[3] Johnson (1999); Delli Priscoli (2001, 2004); National Academy of Sciences (1968).

Figure B.1 Map of the Colorado River basin (*Source:* Transboundary Freshwater Dispute Database, 2004).

basin (Colorado, Utah, Wyoming, New Mexico) and lower basin (Arizona, California, Nevada). The split allocated 9,255 million cubic meters (MCM) to each; at the time, the thinking was that average annual flow was over 22,000 MCM. Hydrologists have since discovered that the estimate was high: the average flow is closer to 18,500 MCM. Consequently, over the years the river has seen increased demands and less than average flows. Obviously, this has led to overallocation of the

water, competition, and conflict. This is a classic illustration of the importance of data and trusted information providers.

Arizona refused to ratify the compromise because it disagreed with California over allocation. Thus, the U.S. Congress passed the Boulder Canyon Project Act in 1928. This ratified the Compact despite Arizona's objections and placed the U.S. Secretary of the Interior in the position of implementing the water service to the entities in the lower basin. Because the parties did not agree, the Secretary, using contracting authority, implemented it. This essentially meant a federalizing of the lower basin. Congress through the Secretary of the Interior allocated 5,400 MCM to California and 3,455 MCM to Arizona and 370 MCM to Nevada.

In 1929, California legislature passed the California Limitation Act making it unlawful for various California entities to use more than their share of Colorado River water. In 1931, the California entities entered into the Seven Party Agreement, which allocated the water among themselves. The entities could not agree on specific allocations, however, so they agreed on priority rights to users. Four agricultural entities got the first priority of water shares. The Agreement assigned water rights beyond the basic 5,400 MCM. The excess diversions have been allowed because the upstream states have not been fully developed to date; however, this is changing.

The upper states were able to form an Upper Basin Commission in a 1948 compact. In 1956 Congress ratified this with the California Storage Act. This act authorized various storage facilities, among them the Glen Canyon Dam project, which essentially helped launch the modern era of environmental protest against dams.

In the 1940s and 1950s, California blocked Arizona's attempts at moving forward with the Central Arizona Project (CAP), a 365-mile long canal that would eventually stretch from Lake Havasu City to Tucson. In 1951, Arizona filed suit against California. The U.S. Surpreme Court appointed a Special Master, which is a frequently used conflict-management tool in water disputes in the western United States. The Special Master collected data for more than 10 years and the court found in favor of Arizona. The court thus strengthened the role of the Secretary of the Interior as water master on the lower part of the river. The court also allowed California to continue to use more than its mainstream entitlement because the upper States were not fully using theirs – but it also specified that this was not a long-term right.

Native American rights further complicate the situation. The 1964 court decree reserved rights for five Indian tribes located along the river of 1 million acre-feet (MAF). These rights have come to be very important in the implementation and design of the CAP. In 1968, the Colorado River Basin Project Act was passed.

The Colorado is also an international river. In 1944, the United States and Mexico signed the Mexican Water Treaty. It calls for a delivery of 1,851 MCM/year to Mexico, with some additional water during years of surplus. In 1973, Mexico announced that it was suffering from increased salinity due to irrigation return flows and it filed diplomatic protests. This resulted in the 1973 Mexican Treaty and the 1974 Salinity Control Act, which defined water crossing the border in terms beyond quantity. It also authorized treatment at the border to assure the quality of the water.

Today, increased demand and uses of water in rapidly growing areas and the Arizona CAP are putting pressure on the river. The Secretary of the Interior may have to enforce the basic limits as outlined in the Court decree of 1964. Upper basin development has not fully materialized. California has asked the Secretary of the Interior for more detailed guidance on allocation rules of the surplus water.

The lessons of this history are important to river basin organizations. First, the "Law of the River," like most legal systems, is inflexible. It does not easily accommodate changing needs stemming from demographic and other changes, as would a more flexible management system such as an RBO. Thus the failure to achieve an RBO has really increased the transaction costs of trying to do integrated management of the river. Still, the flood control and drought problems have been solved.

Second, data are critical, even in situations of plentiful data. Much of the problems can be traced to the use of figures that were too high because they were estimated at times of high flow. This is another lesson pointing to the importance of having some form of trusted technical expertise for the river.

Third, because clear water entitlements could not be reached among the California parities, there is difficulty in using marketable rights and trading. Fourth, the use of what is called the "surplus" is likely to cause more problems. States have feared, with some justification, that California would claim in effect a preemptive right to water. With the river overallocated, the Secretary of the Interior will need to develop guidance on how to deal with the surplus (which may not really be a surplus). This means that once again the seven states will have to try and reach consensus. Thus, like other examples in the United States, failure to reach consensus will not make problems go away. Court cases will not solve the problem of operating with flexible rules in an equitable ways. Fifth, the same happened on the Delaware for more than 50 years until the parties finally did develop a consensus on an RBO to provide a safe ground for negotiating needs for integrated management.

Sixth, special new rules for interstate water banking must be developed. Indeed a technical committee was formed to do so. It has resulted in federal regulations for transferring water from upper to lower basin states. Seventh, due in major part to new environmental needs, Glen Canyon Dam has adopted an expensive adaptive management plan. This too is fraught with conflict and carries high transaction costs absent some organizational forum to deal with these new ecological needs. In addition, there are now several endangered species in the lower basin. Although the cooperation with Mexico has generally been good, salinity problems remain, and the impact of

Figure B.2 Map of the Columbia River basin (*Source:* Transboundary Freshwater Dispute Database, 2004).

all this activity on the delta is becoming apparent. All of this points to the need for some form of formal cooperation.

SINGLE FEDERAL ADMINISTRATOR-VARIATION: THE COLUMBIA RIVER TREATY ORGANIZATION[4]

The operations of the Columbia River (Figure B.2) are a variation of the single federal agency, from the perspective of the United States. Essentially the U.S. Army Corps of Engineers is the main U.S. manager of actions under the treaty. Like other cases around the world, the government reacted to precipitating events. In this case, it was floods, particularly the flood of 1948. Canada has 15 percent of the basin area, but 30 percent of the 134 MAF average annual flow. Half of the flow in the 1894 Columbia flood came from Canada. Flow at the border ranges

from 14,000 to 555,000 cubic feet per second (cfs), which is a much wider variation than the Mississippi or St. Lawrence. The idea thus was to optimize U.S. operations to realize the benefits of the Canadian storage.

In 1944, the government asked the International Joint Commission (IJC) to study the development of the Columbia. After many studies between 1945 and 1959, the IJC reported with alternative plans and principles for apportioning the downstream benefits. Negotiations began in 1960 and the treaty was signed in 1961. It was ratified in the U.S. Senate in March 1961. However, the Canadians were not prepared to go forward. The Government of British Columbia (B.C.) wanted to sell the downstream power benefits within the United States and the federal government was opposed.

In response, joint engineering studies were done to determine long-term estimates of power benefits. Negotiations between Canada, B.C., the United States, and mid-Columbia

---

[4] Delli Priscoli (2005b); U.S. Army Corps of Engineers (2000).

utilities agreed on a sale price. Negotiations clarified the treaty and allowed the sale of the Canadian entitlement to downstream power benefits and led to a protocol and Canadian ratification in 1964. The exchange of diplomatic notes implementing the treaty and entitlement sale was completed in September 1964.

The treaty and protocol defined dams, operations, and benefit computations for treaty storage. The Canada–B.C. agreement gave construction and operation obligations and benefits to B.C. and allowed sale of the Canadian entitlement to the United States. The Canadian entitlement sold to Columbia Storage Power Exchange (CSPE) for US$254 million for a period of 30 years following the completion of each project. British Columbia used the funds to construct their dams. The allocation agreements allocated the Canadian Entitlement obligation among the downstream U.S. Columbia River project owners. The Pacific Northwest coordination agreement insured coordination operation of the U.S. project for optimum power to create entitlement. The powerhouse expansion on the mainstream Columbia River projects was justified by increased fall–winter flows from treaty storage operations. The ties between allocation and coordination agreements were justified by PNW power surplus resulting from the U.S. entitlement and the purchase of the Canadian entitlement.

The treaty required Canada to construct and operate 115.5 MAF of storage on the Columbia River and a tributary in Canada for optimum power generation and flood control downstream in the United States and Canada. The treaty allowed the United States and Canada to build Libby Dam, with 5 MAF of storage, on the Kootenai River in Montana. The 90-mile long Lake Koocanusa, the reservoir behind the dam, backs up 42 miles into Canada.

In addition to Libby Dam, the treaty also included Mica, Arrow, and Duncan dams. Storage of 8.45 million acre-feet at Arrow, Duncan, and Mica were assured for flood control operation for 69 years. An additional 7 MAF of treaty storage and 5 MAF of nontreaty storage were available "on call" for large floods at a cost of $1.875 million at each of the first four requests and lost operating cost. Cash payments of US$64.4 million were made to Canada by the U.S. government at the completion of the treaty projects for half of the estimated worth of future flood damage prevented.

For hydropower, 15.5 MAF of Canadian storage is operated for optimum power generation downstream in the United States and Canada. Power benefits from the treaty storage include dependable capacity and average annual usable energy. Canada receives half of the increased power generated downstream in the United States due to the operation of Canadian treaty storage. Actual operation and magnitude of water year do not affect the downstream power benefits. Downstream power benefits resulting from Libby storage operation remain in the country where they are generated. The hydroelectric operation plains provide a monthly reservoir balance relationship for the whole of Canadian storage, allowing Canadians the flexibility to operate individual projects for maximum Canadian benefit.

The assured operation plan (AOP) for Canadian treaty storage is developed for the sixth succeeding operating year from a hydroregulation study designed to achieve optimum power and flood control benefits in Canada and the United States. The AOP defined operation criteria for Mica and the rule curves for Mica, Duncan, and Arrow dams that will be used in actual operation unless otherwise agreed by the entities. The assured operation plan determines the downstream benefits that will be sent to Canada. Once the benefits have been calculated they do not change, no matter how much actual energy is generated.

The treaty allows the entities to prepare the detailed operating plan (DOP) for the upcoming year. The DOP may be fine-tuned from the AOP, which was developed 6 years earlier, to produce Canadian project operations that are more advantageous to both countries. The DOP includes the process for determining real-time project operation, which is not in the AOP. The DOP authorizes the operating committee to agree on mutually beneficial changes to the DOP for power generation and on power proposed. Figure B.3 outlines the structure for coordination. The permanent Engineering Board is appointed by the United States and Canadian governments and assures that treaty provisions are carried out. The permanent Engineering Board meets at least once each year and prepares an annual report for the respective governments. The permanent Engineering Board Committee performs technical and administrative duties for the permanent Engineering Board.

The U.S. entity consists of the Division Commander of the U.S. Army Corps of Engineers for flood control, and the CEO of Bonneville Power Administration for power. The Canadian entity is the CEO of BC Hydro. The entities meet once a year. The treaty coordinators act as liaisons among the entities to the Columbia River Treaty Operating Committee. The entity Secretariat performs administrative duties for the entities.

The operating committee performs all the technical work to implement the treaty. The operation committee develops annual operating plans, calculates annual downstream benefit payments, and assures delivery of the Canadian entitlement. The operating committee meets every other month. Working committees may meet more often than other month.

The U.S. entity makes a weekly treaty flow request for treaty storage based on the annual operating plan and additional operation agreements. Canada operates the three treaty projects as one pool, considering the total 15.5 MAF as one project. Individual project operations may differ from the annual plan, but the total, composite Canadian storage will match the plan.

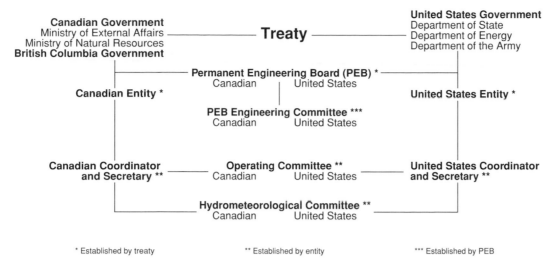

Figure B.3 Columbia River Treaty Organization flowchart (Delli Priscoli, 2005b).

The Columbia shows how hydropower and flood control can be integrated to create benefits, which then are used for negotiations rather than being mired in fighting over the allocation of flows.

COMPREHENSIVE REGIONAL AUTHORITY: THE TVA[5]

Upon signing the Tennessee Valley Authority (TVA) into law in 1933, President Franklin D. Roosevelt said it would bring the best of the private and public sectors together. The Tennessee Valley Authority is a product of the Depression Era and the movement called "valley authorities" in the United States. The authorization of TVA itself did not go through any of the traditional congressional natural resources committee to get to the floor of the U.S. Congress. It went through military-related committees, as it was dealing with old munitions sites of Muscle Shoals, Alabama.

The TVA is one of the few comprehensive regional development organizations based on watershed and river basins. It has sufficient authority to integrate planning, development, and management in one agency, with the coordination focused around the Tennessee River. It has a high level of formal authority: in a sense, it is one of the most integrated and powerful examples of what today we call integrated water resources management (IWRM). TVA has inspired some other regional authorities around the world, for example, in Indonesia, Jordan, and other areas.

Although it is federal and public, TVA is a regional agency and a federal corporation. This is important. The move to create seven more TVAs based on its success was defeated in the U.S.

Congress in the late 1940s, primarily by the power of other federal bureaucracies and states who saw their power, both in the cross sector and geographic sense, being lessened if more regional organizations were added. Private power companies also led the attacks on TVA.

At its height, the TVA used revenues derived from services such as hydropower to subsidize other services such as community development programs. TVA was, at its root, concerned with the total social and economic development of the people of the region. It sought to integrate water uses, from hydropower to flood control, to generate wealth in the region. It brought affordable and reliable basic services to the poor and rural people: electricity, stable water supply, and protection from floods. As such, in one generation, it took one of the poorest and most poverty-stricken parts of the United States into the twentieth century. In fact, the socioeconomic statistics of the region before the TVA are still very familiar to many poor parts of today's world.

Today, the TVA has become focused primarily on power generation and receives little to no money from Congress. It has now also become one of the largest taxpayers in the southern United States. It does maintain an extensive watershed program and some first-class laboratories. During its days as a social development organization, major congressional debates took place over the bonding authority of the TVA. The extent of such authority gave the TVA financial autonomy and setting limits was one of the only ways to put control on the TVA. The TVA reflected the prevailing perspective of water engineering of the time and focused on structures, or what we call "supply side" today. As have others, TVA has changed these perspectives with changing notions of water value.

It is a mistake to call the TVA only a centralized national agency. It was as close as any organization has come in the

---

[5] There is much historical literature on TVA; Philip Selznick (1953); Hubbard (1961); Martin (1956); Delli Priscoli (2005b); Leuchtenburg (1952); Pritchett (1943).

United States to being a regional government. In fact, much of the anti-TVA rhetoric revolved around this very structure, which included a wide range of service departments far beyond what one normally would expect to see in a water agency. It was this comprehensiveness, pervasiveness, and real cross-sectoral organization that made it easy to paint as a huge government bureaucracy.

Using Millington's criteria for success; TVA had high-level support at first; it was oriented to the community; there was clear accountability among the actors; and building a knowledge base and technical expertise was central to its operations.

WATERSHED COUNCILS[6]

Since the demise of the Title II system in the United States in the 1980s, the major river basin management (RBM) innovation has been a renewed focus on watershed management. The U.S. Environmental Protection Agency (EPA), the Association of Metropolitan Sewerage Agencies, the American Planning Association, the U.S. Geological Survey, and Water Quality 2000, have all been advocating this approach. The focus has been on small watershed organizations built from the grassroots. The notion is that watershed councils should be *sui generis* and grassroots in origin, emerging to form shared visions on managing watersheds. Indeed, the watershed focus is far more manageable than the river basin focus.

The central concept has been on consensus. It has been built on the concept of "nested watersheds." This approach is based on the idea that river basin institutions can be composed of interrelated but discrete arrangements organized around nested hydrologic units (i.e., from a large river basin to regional subbasins to local watershed). In this way, the concept is meant to employ both the top-down and bottom-up approaches. Indeed, the success of watershed management has come to be consensus, almost to the exclusion of technical analysis. In addition, it is becoming clear that consensus on numerous small watersheds does not necessarily add up to managing a river basin. However, hundreds of watershed councils have emerged in the United States. They are testimony to a high degree of citizen activism and participation. One of the most visible of these, which has grown to actually bring established agencies into a new partnership structure, is the Chesapeake Bay Council.

The watershed processes, as might be expected, have been highly varied and have focused on water activities. They do address problems versus jurisdictions. In many cases, they have achieved active participation of all local interests (private and nonpublic). They employ highly collaborative design processes. They are consensus driven. They try to use a more

---

[6] Natural Resources Law Center (1997).

	Watersheds	RBOs
Geographic Scope	RBOs are watersheds	Not all watersheds are RBOs
Level of authority	Less	More
Participants	Broader mix public and private	More focused on formal public
Legal basis	Informal	Formal
Issues/services addressed	Multi-issue more local	Multi-issue more regional
Catalyst events	Droughts and floods Fragmentation Demographic mismatching	

Figure B.4 Watersheds versus RBOs summarizes some distinctions between watershed and river basin focus (Delli Priscoli, 2004).

holistic and adaptive systems approach. Information exchange has increased and they have provided good forums for public education. They are nonthreatening venues for dispute management and do force coordination among resources managers and the reduction of duplications. Figure B.4 summarizes some distinctions between the river basin and watershed focus.

INTERAGENCY–INTERSTATE COMMISSIONS: (TITLE II)[7]

Under the Water Resources Planning Act of 1965, the so-called Title II commissions were established. Their purpose was to improve interagency coordination, federal–state coordination, and to complete what was called "comprehensive coordinated joint plans" for river basins. Figure B.5 shows the organizational arrangements for the Title IIs and how they fit into an overall national scheme.

This organizational structure was the result of a long process started in the 1950s and pushed forward by President John F. Kennedy. In many ways, it involved the best thinking available in the United State on how to achieve coordination and planning in the U.S. federal system.

The Title II RBOs had a formal legal status and permanent staffs and treated the states more as equals than did earlier

---

[7] There is a rich literature from the late 1950s to the late 1970s, in the United States on the Water Resources Council and the National Water Commission of the United States and from academia on the following topics. This material is rarely cited in the current debates on water but it is highly relevant, some examples are: Kalter (1971); United States Water Resources Council (1967); Guidelines for Planning Coordination, 1971; Principles and Standards for Water and Land Planning, Washington, DC, 1972; Eckstein (1958); Eckstein and Krutilla (1958); White (1969); Ingram (1971); United States National Water Commission (1974); Ostrom (1971); Arnold (1988).

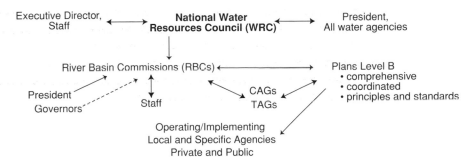

Figure B.5 Structure of the Interagency–Interstate Commission Title II (no longer exists) (Delli Priscoli, 1976, 2004).

commissions. Each member of each commission had one vote. Either consensus or unanimity was used to reach decisions, but the actual process was unclear. Indeed the decisions of Title II were not really enforceable.

President Reagan terminated this whole structure in 1981. Since then there has effectively been no coordinating mechanism among the key U.S. water agencies. Insofar as such occurs, it tends to be achieved through the budget processes. Many water professionals have come to realize the need for such a coordinating institution. When it existed, the National Water Resources Council under the Title II arrangements became mired in political battles. Although the system was good at fostering communications, it was not effective at management conflicts.

In addition, the Title IIs came along when the policy world in the United States was changing. With the environmental movement, the focus became water quality. Under the 1972 Clean Water Act, a large program in water treatment emerged. The focus was on this grants program and on regulation for water quality rather than on traditional planning. The result has been that the United States lost a structure for comprehensive planning along rivers. In addition, some commentators have said that the Title II structure was a structure developed for a time gone past. This is now debated because it is apparent that the structure offered many coordinating means that are much needed in the United States today.

The system also devised an elaborate set of participation procedures and structures. It used citizen advisory committees and technical advisory committees. As such, it went a long way to draw attention to the need for structured and real citizen participation. These committees, however, were subject to many debates. Their effectiveness was highly variable. They appeared at a time when the United States was just beginning to experiment with citizen participation in planning.

The distinction between the citizen and the technical committee was clear on paper. It really built on the idea of separating the technical and political, or nontechnical, in water planning. Subsequent experience has shown that the best route to

participation is in blending the technical and political as much as possible and not keeping them separated. In fact, empirical research on these committees showed that, demographically, the technical committees were often far more reflective of the general citizen population than the citizen advisory groups (Delli Priscoli, 1974). This was primarily because citizen groups came to mean environmental groups at that time. Subsequent experience in the United States has also shown that participation is and must be far more than just participation of environmental groups. It also must move beyond only advisory committees to public workshops and other such interactive means. However, all of this happened at a time when environmental values were just beginning to find a voice in traditional water-resource planning.

In addition to the structure portrayed in Figure B.5, this system of the WRC and the RBCs, under the 1965 act, produced principles and standards (P&S) for planning and a tiered, three-level national system for water-resources planning. The P&S is a remarkable document that tried to set up a system of planning that would be used by all agencies regardless of their mission. It originally did so by using four objectives and four accounts under those objectives: economic development, social well-being, regional development, and environmental quality. In this way, it put forward a uniform national accounting system for water-resources planning and development. The P&S were modified to two objectives (economic development and environmental quality) and four accounts and came to be called principles and guidance (P&G). The P&G still exists and it is one of the best practical guides to doing what is now called "IWRM" for on-the-ground water mangers worldwide (Sidebar B.2). It has been used only by the traditional water agencies, however. Environmental regulators have not adopted the analytical procedures, such as benefit–cost analysis and trade-off analysis. Along with the lack of coordinating organizations for water agencies, this lack of analytical procedures for environmental regulation is also becoming recognized as a serious problem in the United States (Sidebar B.3).

DELAWARE RIVER BASIN COMMISSION (FEDERAL INTERSTATE COMMISSION)[8]

In the early 1920s, drought in the Delaware basin produced allocation conflicts. States initially tried to solve these through judicial remedies. However, judicial formulae were too inflexible and technically inadequate. States began to recognize that enhanced technical capacity, such as information generation and sharing and analysis, was necessary in the hydrological system, if they were to move to a positive-sum negotiating environment.

---

[8] Martin (1960); Delaware River Basin Commission (2006); Delli Priscoli (2001b, 2004, 2005a).

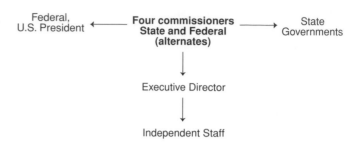

Figure B.6 Structure of the Federal Interstate Compact Commission: Delaware, Susquehanna (Delli Priscoli, 1976, 2004).

Droughts in the 1940s resulted in more judicial rulings, which established equity principles, but still were inadequate for management during droughts. This led to the formation of the Delaware River Basin Commission (DRBC) in the 1960s, which provided a decentralized institution within which to negotiate (Figure B.6). It also enabled the states to draw on its newly instituted technical staff. In subsequent droughts, the experience of negotiating within the DRBC framework and equity principles increased the legitimacy of this technical staff. As a result, the quality of contingency plans improved and a good faith agreement among states was signed in the early 1980s.

The Susquehanna River Basin Commission followed the model of the Delaware River Basin Commission in the 1980s. It too was spawned by continual drought and the need for better management across purposes of water supply, flood control, and drought.

The Delaware (and later the Susquehanna) commissions had sufficient independent authority to act in a management capacity. This includes the ability to block proposed actions that are inconsistent with the regional plans developed by these commissions. In addition, they posses independent and technically competent staff; they have multipurpose mandates and multivalue mandates; they have a large problem-shed geographic scope; and they rely on state political leaders (i.e., governors) rather than bureaucrats in guiding policy decisions. They also have achieved a relatively equal balancing of state autonomy with federal supremacy.

POTOMAC RIVER (INTERSTATE COMPACT COMMISSION)[9]

The Interstate Commission for the Potomac River Basin (ICPRB) was also formed, in large part, because of needs for drought contingency planning. Over a long period, numerous dams were proposed; however, only one major new dam was

---

[9] Hoffman (2001); Steiner, Hagen, and Ducnuigeen (2000); Interstate Commission on the Potomac River Basin (2005); Delli Priscoli (2001b, 2004, 2005a).

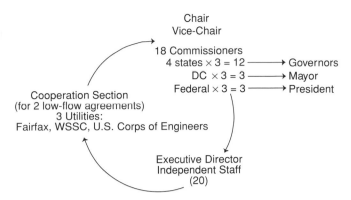

Figure B.7 Structure of the Interstate Compact Commission (Delli Priscoli, 1976, 2004).

built. The remainder of the water supply is provided through negotiated agreements among the states and the federal district (Figure B.7). The ICPRB demonstrates the influence of data and technical analysis in facilitating cooperation. It has little formal power other than to gather data and convene discussions among basin states. Through the use of professional staff and interactive computer approaches such as STELLA,[10] ICPRB has built its technical credibility. Now it manages a real-time river monitoring process, which provides hourly flow projection data and a structure for the riparian states to discuss their responses to that data. Once a year, it facilitates a series of drought contingency simulations for the river. In generating information and analyzing data in this way, it has become the key agent facilitating flexible agreements among the states. And it does this with little mandate other than to help gather and disseminate information.

Kenney and Lord (1994) note that no basin has realized the potential of "nonstructural" innovations better than the Potomac basin. In this basin, the reservoir operations scheme developed and implemented by the ICPRB has increased the overall system yield by 50 percent, while satisfying instream flow and water quality objectives. In contrast, the "structural" solutions proposed earlier by the U.S. Army Corps of Engineers promised an increased yield of 43 percent through the construction of as many as sixteen major projects, with the cost estimates ranging from US$200 million to US$1 billion. It has done this with one major and one minor new reservoir in place of the original sixteen proposed projects.

RBOS AS BOUNDARY ORGANIZATIONS[11]

The two North American border commissions, the International Joint Commission (IJC) and the International Boundary

Waters Commission (IBWC), emphasize their technical objectivity in their attempts to facilitate dispute resolution. Both deal with all waters that either form or cross boundaries between the United States and Canada. Both began with a narrow technical focus – IBWC more so than IJC – but have been under steady pressure to expand their scope as values changed and needs grew. Much of this pressure now comes from transboundary groups advocating new environmental claims on the waters. These are clear cases that participation will go beyond the formal public agencies.

The IJC is more of an appellate, review, and regulatory board because it is mandated to resolve differences. It also has more of a public-access orientation than does the IBWC. The IJC is made up of commissioners from each country. The offices or secretariat for the commission exist in both Ottawa and Washington, DC. Unlike with any other such treaty in the world, these commissioners have no orders from their countries. Instead they must take an oath to uphold the treaty. This enhances the independence of the IJC. It also enhances the IJC's capacity to deal with disputes.

The IJC can deal with any issue around water. Recently the organization has been involved in issues concerning major flooding along border rivers and with water levels of the Great Lakes. From its beginning in 1909, IJC has been mandated to deal with the public health and environmental aspects of boundary waters. Most recently, it has begun a process of initiating cross-boundary watershed committees or councils. If successful, these will be councils of Canadian provinces and U.S. states (thus, subnational sovereign entities) concerned with certain rivers – which will also be unique in the world. Because issues must be forwarded to the IJC by both countries, a certain amount of preliminary negotiations has already occurred by the time IJC gets a case.

The IBWC is made up of two commissioners who must be licensed engineers. Each has an executive staff. IBWC has adopted a low-key mediating approach and nurtures a reputation for neutrality and expertise. It is now under pressure to become more activist, increase its attention to urban and environmental issues, and broaden public access to its deliberations.

B.1.2 Some Canadian experiences[12]

The Canadian Prairie Water Board (PWB) is an example of institutional collaboration falling to the right of the center of the continuum. It monitors flows, provides oversight on water quality, advises on disputes, and uses fact-finding and technical

[10] USACE (2006); United States Army Corps of Engineers Institute for Water Resources (2007).

[11] Delli Priscoli (2005a); International Boundary and Water Commission (2006); International Joint Committee (2006); Utton (1992).

[12] From materials prepared for the U.S. Delegation to the Middle East Multilateral Peace Talks on Water, U.S. Department of State (1992).

committees. It is built on a master agreement among the Canadian Prairie provinces of Alberta, Saskatchewan, and Manitoba. Within the context of this master allocation agreement, provinces have reached bilateral agreements. Each jurisdiction manages its own water inside that jurisdiction, and PWB monitors flow at the borders.

The PWB offers some important lessons: it operates by consensus; maintains strong, technically credible support; and is flexible; and its rules can be redefined as it grows. Requirements are defined at the borders of jurisdictions. It starts with a master agreement on apportionment and then moves to bilateral agreements. Dispute resolution mechanisms are defined. It facilitates information exchange. Many of these lessons are echoed in other basin initiatives. Indeed, a similar process was undertaken on the Mackenzie River.

B.2 EUROPE[13]

There has been great diversity in water management institutions and river basin management in Europe. The diversity reflects the differences in hydrology and geography. The Netherlands, Great Britain, and Portugal have focused more on flood management than have others. France has built the most enduring modern RBO system based on an arbiter role between the polluter and users and on investment incentives. Germany has relatively little river basin planning except the long tradition in the Ruhr Basin, where large industrial demands caused early focus on water quality and multiple uses. Portugal and Spain have long traditions with RBOs, which traditionally focused on supply-side aspects of water management. The new European Union (EU) directive on water requires river basin planning and organization.

The British moved from a public river basin planning model to far more privatization. While the river basins were smaller and were operated for fewer purposes, the system also has national regulatory oversight. Since the 1970s, the French have operated a river basin system that falls somewhere between these approaches. The major basins have committees that include representation by industry, environment organizations, and the general public. These committees, which formally represent users and are financed through pollution charges, set priorities for users over a period of 20–25 years. As in the United States, the European Union has begun to move from single to multipurpose orientation of its river basin organization, such as the Danube and Rhine.

New EU directives now require river basin organizations and planning.

B.2.1 France[14]

The French Water Agencies were initially begun as a funding mechanism and actually were called funding agencies in their initial stages. They are designed as financial intermediaries between polluters and users and water treatment operators. The basin agencies collect revenues from polluters and reallocate these by subsidizing investments in water quality improvement projects. Although water management is financed independently from the general central government budget, the agencies can influence this management – and thus integration of flood management – through this reallocation function (Figure B.8; Sidebar B.4).

From the participatory viewpoint, the basin committees, or water parliaments are the most interesting. These consist of local government, users, and the state. Elected officials from local government and users make up two-thirds of the seats. Representatives for the regions and departments are elected by respective councils and from the communes from the French mayors' associations. Users choose members themselves and various ministries choose other members. All of this constitutes what has been called a "water parliament." The committee is a consultative body. It advises on the levying of taxes and the allocation of revenues and it also approves the tax rates.

In addition, the committees are tasked with preparing the river basin master plans, called "schémas directeur d'aménagement et de gestion des eaux" (SDAGEs). In reality the bulk of this planning is done by the river basin agency. Apparently the users do not always see this activity as useful or have the means and time to pursue the effort. However, as difficult problems are emerging, such as low-flow agreements for basins like the Adour-Garonne, the committees are beginning to see the important role for conflict resolution. These SDAGEs are supplemented by river basin management plans called "schémas d'aménagement et de gestion des eaux" (SAGEs). Local water committees are specifically instituted for this function. They consist of local government members (50 percent), representatives of users, riparian owners, professional organizations, and concerned associations (25 percent), and state representatives (25 percent). These plans continue the older practice of river contract planning. Although the performance of the planning is mixed, there is an indication that the quality of the consensus directly affects the implementation of the plans.

---

[13] From private discussions with Dr. Bernard Barraque, Paris and Evon Chere, Paris; see also Mostert, et al. (1999); Alearts and LeMogine (2002).

[14] Delli Priscoli (2001a); Kaczmark (2002).

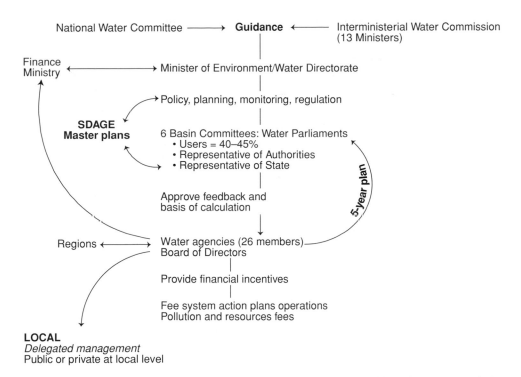

Figure B.8 French system (Delli Priscoli, 2001b, and extensive discussions with several principles of French agencies).

This system is remarkable for its longevity, in how it tries to formalize representation and provide some clear authority lines, and, perhaps most important, how it has attempted local and decentralized water management within the centralist state tradition in France. Indeed the direct participation of users in the bottom-up approach seems to contrast with the centralist government administration.

B.2.2 Germany[15]

As in most federal systems, the states have the primary responsibility for water in Germany. The only state with river basin institution units is in Northern Westphalia (NRW). These consist of water associations responsible for water quality and quantity in the Ruhr region. They began during the periods of rapid population and industrial growth at the turn of the century. Such growth threatened water supply and sewage systems in the regions. Thus water associations were established and membership of local authorities, industry, and supply and treatment companies was mandatory. Each association focused on different functions or task depending on the basin. However, the main concern was with pollution with some concern over water transfer. The associations were established by a special law and are an expression of the German tradition of relegat-

ing water supply management to states. And this is really its primary means for public participation.

The water supply and waste treatment system is self-financing through a system of charges. There is little river basin planning outside this region and the NRW associations. Indeed water planning is just beginning to grow throughout Germany as a result of recent legislation.

B.2.3 The Netherlands[16]

Water is intimately linked to national self-perception in the Netherlands. Indeed the Dutch water boards are among the oldest operating democratic institutions in Europe. Because of its geography, the Dutch concern has traditionally been with flooding. This started with a focus on the sea; recently the focus has moved to the rivers, however, along with concerns over pollution. The water boards are not organized along river basins or watersheds. Integration among the highly fragmented Dutch system is achieved through a system of planning.

The water boards are functional public bodies for management and the provinces are bodies for administration. The board is made up of representatives of interested stakeholders, meaning landowners, and inhabitants of the areas. The right to vote is derived from the duty to pay taxes to the board. There

[15] Huisman, de Jung and Wieriks (2000).

[16] Note articles on Dutch water history in Reuss (2002); TeBrake (2002); and Kaijser (2002).

Sidebar B.4 RBO Participation – Europe

River Basins and Participation in Europe (Barraque, 2000)

It isn't by chance that the only three historically centralized monarchies have their territory now fully covered by river basin authorities: *Confederaciones hidrografica* in Spain, initiated in 1926 and generalized in the 1940s; ten regional water authorities from 1974 to 1989 in England and Wales, which were seen as a centralizing move against the alleged inefficiency of local authorities and are replaced now by even more centralized resource management policy, keeping river basins as a basic unit for water policies; and six French Agencies de L'eau that were initially seen as a central government tool to reinforce its domination over local authorities and are now clearly understood as subsidiary institutions playing a role in postdecentralization governance.

In the subsidiarity countries having kept decentralization and communitarian management traditions, there are practically no river basin institutions as such: in the Netherlands, the water boards in charge of drainage, flood protection, sewage treatment, and sometimes other tasks do not follow catchments, largely because the country is flat and catchment limits are not obvious. But they are clearly subsidiary institutions, having a community type form of representation. In Germany, the *Ruhr Genossenschaften*, which pioneered modern river planning, are quite exceptional and linked to specific geographic and political history. In the rest of the country, river basin water sharing is made by ad hoc working parties between concerned existing institutions, and there are no basin authorities. In Denmark, river basin institutions have been discussed and abandoned. In Sweden and in Finland there are water tribunals operating per groups of basins, but only to settle disputes from hydroelectricity and subsequent river flow modification. In general water management remains at the level of administrative regions. Integrated catchment planning is, however, studied and experimented with.

In the Mediterranean countries, one finds a general hesitation between growing decentralisation of all waters, and correlative development of central or regional government roles, and the river basin approach. In Italy, the administrative region was first preferred, but a subsequent traditional regional planning approach was poorly developed. In 1989 the Autorita di Bacino were created for the nine or ten largest rivers to coordinate interregional planning. They are gaining momentum thanks to their cumulative expertise, but they remain weak until they obtain some direct financing mechanism. The 1944 law increased the public character of all categories of water. In Portugal, the strong legitimacy recovered by the three hundred local authorities at the fall of the dictatorial regime placed the country in a central versus local type of confrontation. This indirectly led to discarding the project in the 1980s to create five river basin institutions and an equivalent number of regions. The country remains in a typical center periphery confrontation. However, there are now five large intergovernmental boards for pollution control around the largest urban areas and a new participative catchment planning approach breaking the country down into fifteen river basin units. In Greece, there are water policy responsibilities at all traditional levels, plus river basin management (fourteen basins) where seemingly the hydroelectricity problem is dominant. In Spain, the new constitution creating the autonomous regions quickly resulted in the Autonomias being the locus of a certain challenge of the traditionally hypercentralized water planning in the Confederaciones. Moreover, the 1985 law legitimized the role of the very ancient water user communities at more local levels and proposed to develop a bottom-up planning approach at their level. As a matter of fact, it is a community of users of a polluted and threatened aquifer, in Prat del Llobregat next to Barcelona, which gave a successful model to the water communities developed in the law.

A very complex situation is developing in most of the member States, because of experiments of public participation. In the United Kingdom, for instance, even though water policy may now appear as the most highly centralized in Europe, and even the most inspired by the "Statist-liberal" paradigm (i.e., privatization and technocratic regulation), there remains river consultative councils that are widely consulted about catchment planning, even though in an informal and on statutory manner, typically in the British style of consensual policy. In Belgium, the extensive federalization of the country, the Walloon region is developing a formal public participation process in river contracts. In the Netherlands, the provinces are now placed at the heart of a complex procedure for integrated river basin planning. In France, the Agencies de L'eau were supplemented as soon as the end of 1970s by an apparently modest approach of river contracts, more informal than Belgium. There are now more than 160 river contracts going on, and then the Ministry of the Environment tried to make things more formal and more binding with the local catchment plans called the SAGE. Although these plans are progressing very slowly, contractual and bottom-up river management is generalizing, including through the changing vision and methods of the approximately sixty river institutions earlier developed for mastering the river only with hydraulic projects.

are managing boards, which reflect the similar composition of stakeholders. Because the Dutch system is so fragmented, consensus is primarily sought through the countrywide planning process. This is actually designed to obtain consensus within the administrative structure. That process includes a national policy document, operational plans for implementing it, and provincial strategic plans for nonstate waters. Accountability is assumed to take place through normal elected process in provincial councils and parliament. Interested stakeholders, such as environmental groups, are also involved in the process of planning. Consensus processes focus on consensus within the administration and not on direct participation of the users. Agricultural interests have traditionally dominated the water boards. There is discussion on separating the policy and operation functions more clearly and better integrating environmental values into the overall process. This discussion has in large part occurred as a result of flooding and a move to rethink the basic approach to flood management, including returning some land to the water.

B.2.4 Portugal

Portugal established four river basin organizations early in the twentieth century. They were focused on hydropower and later on irrigation. Hydropower and irrigation were merged into the Ministry of Public Works. Thus Portugal looks like several other countries in that the river basin organizations were biased toward large infrastructure, supply-side investments and were creatures of the central government. This is similar to China and Nigeria and others. During the 1970s, environmental quality and large fish kills led to questioning this structure. Like in other areas, there was a growing call for more bottom-up approaches, where the local and regional institutions could take more responsibility.

In the 1980s reorganization, five river basin authorities were created. They included environmental aspects of water, licensing of all water uses, planning, and collection of funds based on the user-pays principle. On water supply and waste treatment, the idea was to establish water users' associations of local authorities and other users. These were to be a means for direct participation in planning and management. The plan never really got moving, however, as government changed. But new legislation created planning means for fifteen river basins, which retained the user- and polluter-pays principles. It also established river basin councils, which have the purpose, like the national water council, for developing river basin plans. These plans and planning process are now in process. Although the planning is being done on a river basin basis, it is administered by national and regional authorities that do not correspond to hydrologic units. The participatory processes and involvement processes are not as clearly spelled out as pre-

vious plans. Because water management does not always have a high political profile, it is also hard to get direct participation in consultation processes.

B.2.5 Great Britain

There has been a long gradual evolution of the river basin organizations in Great Britain starting in the Middle Ages. In the modern era, with the land drainage act and the River Boards Act of 1948, organization by river basins emerged. In 1973, the Water Resources Act (RWA) divided the country into ten catchment-based water authorities. These were public bodies responsible for all water functions throughout the water cycle. Although successful in many aspects of integration, the RWA came under criticism because they were both the regulator and polluter. In the late 1980s, the utility functions of water were privatized, with the only asset sell-off seen to date in the water world. However, the National Rivers Authority (NRA) was created to regulate water management, thus carrying on the idea of catchment-based water management. The NRA calls itself guardian of the environment, or custodian of the common waters. It has a fifteen-member board and is organized around eight principal regions based on river basin catchments and subdivisions. It has three statutory committees in each region: a fisheries committee, advisory committee, and flood committee. These committees serve as a type of public participation – a combination of what has been seen elsewhere as technical and citizen advisory committees. They include major stakeholders and users, such as farming, fisheries, conservation, recreation, navigation, water utilities, and industrial users. The NRA's primary focus is to protect and improve the environment and protect against flood.

The NRA tries to use the planning process to integrate its duties into comprehensive plans; however, it has little control over land use change on a catchment basis. It thus relies on partnership, compromise, and negotiations approaches with other relevant planning bodies. In the end, however, it cannot veto a development project. Public and stakeholder involvement lacks a statutory basis and the direct involvement of elected officials under this system. It looks a little like a corporate board. Public input is sought through the planning process and the preparation phase of the catchment plans. Extensive consultation has been held in the various regions as guidelines and standards were formed. In the end, the NRA is not bound by the outcome of such processes, they are all advisory.

B.2.6 Spain[17]

River basin management has been present in Spain, in various forms, for more than 100 years. Participation has both

[17] Hera et al. (2002).

been encouraged and discouraged at various times. At the turn of the twentieth century, the hydrographic basin concept was used for the first national planning efforts. This was followed in the 1920s with the formation for hydrographic unions. These unions were primarily established for river basin planning. They include users and management along with administration. They disappeared in the 1930s due to lack of resources, but the river basin concept remained in the form of hydrographic confederations. Participation of the user continued with the representatives of users chosen by the users themselves.

During the late 1950s, water commissions were established. The commissions followed the same territorial and river basin jurisdictions as the earlier hydrographic basin concepts; however, they did not include user representation or any way for users to participate. The commissions became creatures of the national water administration under the general management of the hydraulic works department of the Ministry of Public Works.

During the 1980s, the reform government produced a new water act. The old commissions and confederations were united into new hydrographic confederations. These had management units called "river basin authorities." River basin plans and a national river plan were called for. The new river basin authorities included a water commissioner, a technical directorate, and secretary general and planning office. Once again, they include representation of users. The steering committee of a river basin authority is composed of user representatives. In various exploitation boards, the users are a majority and have proportional representation on the regulation committees. Thus, the river basin authorities have been organized in accordance with new democratic principles, in accordance with the broader politics of the country.

The river basin authorities attempt to use the concept of "polluter pays." Levies, fees for regulation, irrigation, discharging, and use of public land are all under their purview. Conflicts occur during drought within the context of a "first in use" rights system, established under the Roman law custom. This makes the regulatory committee of the river basin authority very important. New environmental conflicts have also risen, with new laws recognizing ecological claims along with the traditional approaches to water regulation.

The basin concept has been central throughout the previous century in Spain. Conflicts have been managed by collaboration among the users and administration and, for the most part, this has been achieved within the context of basin-management units. This practice has remained remarkably constant, despite the changing governments over this time.

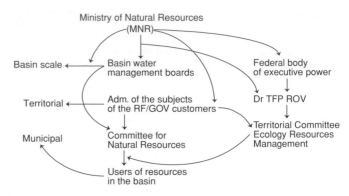

Figure B.9 Organization of the Revival of the Volga (ROV) program (Dukhovny and Ruziev, 1999; Dukhovny and Sokolov, 2002).

B.2.7 Russia[18]

The 1883 Water Law and the 1995 Water Code set the institutional framework for water resources management in Russia, which is built on the concept of State ownership of all water and management structures. It is also based on "user pays" and "polluter pays" principles. The Ministry of Natural Resources (MNR) is the key coordinating body for water at the federal level. Under the MNR there are seventeen river basin agencies (RBA). Five of these are on the Volga River. With the RBAs, the MNR is responsible for the preparation of river basin plans. The RBAs are also responsible for the preparation of the basin-wide water management agreements on uses and protection of water. Legislatively, most of the pieces for IWRM management exist; however, the implementation is weak, especially concerning enforcing regulations management systems.

In 1998, the government established the Revival of the Volga (ROV) program. Its aim was to improve the environment and enhance the quality of life in the basin. It is implemented by seven federal ministries, thirty-nine constituents of the Russian Federation, and numerous other governmental and nongovernmental organizations, all under the coordination of the MNR. The overall coordination is provided by the MNR. At the federal level, the MNR uses a steering committee made up of senior ministers of implementing agencies, state committees, and other institutions; at the regional level, the MNR works with each of the existing Territorial Associations for Economic Cooperation, located in the basin. These consist of the governors of the oblates and republics (Figure B.9).

The ROV program follows four principles and has specific targets. It is supposed to coordinate the complex of institutions and provide for participation of the public. The major water-related functions of the MNR are interbasin and interregional coordination and settlement of disputes; development of laws,

[18] Shevchenko, Rodionov, and Kindler (2002).

regulations, and standards; support of related research and development programs; international cooperation and coordination with state committees; and environmental monitoring. The MNR is responsible for development of federal plans and plans for all the river basins. These plans then are followed by local water agencies, which develop guidelines for the smaller rivers and basins.

The basinwide plans set up water intake limits, forms of payment for water consumption and use, wastewater discharge standards, water quality control measures, and river rehabilitation and use-control programs. The water management agreements are considered and adopted by special commissions that include the federal and regional authorities, water users, and public organizations. Nevertheless, there is a lack of state policy for sustainable water use, a weak capacity to enforce regulations, and a weak management system.

The ROV could impact this whole system, as the Volga River is critical to Russia. Since the 1930s, almost 90 percent of the river has been controlled through a series of mainstem reservoirs. There are twenty-one dams (beyond many small dams) for hydroelectric power and the entire length is navigable to the Black Sea. The basin includes 61 million people, almost 45 percent of the Russian industrial output, and more than 50 percent of its agricultural output. Although these developments have greatly benefited Russia, the environmental costs have been very high. Thus the ROV, in addition to the costs for maintaining the sources of cheap electricity, were not calculated into energy prices. The vast profits were centralized and distributed by the government, with some allocated to reservoir maintenance. All of this changed with the reforms. The river also has very high variability.

The ROV program itself is meant to be an instrument for public participation. It provides for basinwide consulting institutions, monitoring agencies, expert committees, specialized information, and scientific and technological centers. Most of the financing for the ROV comes from the regions (Figure B.9).

The ROV has identified one thousand projects consistent with program objectives. The estimated costs are around 140 billion rubles. These projects clearly recognize that a major shift to more efficient uses, conservation, and demand management are needed. Cost-effective criteria are needed for selection of projects. These projects, like similar projects in other parts of the world, call for new management. Like in many Asian examples, the need for RBOs has been recognized as an outcome of similar forces. Also, the attempt is being made to meet this demand through an essentially traditional top-down bureaucracy and embedded with traditional engineering approaches. Participation is recognized, but its impact on key decisions is uncertain. The power of the MNR is central, so

decentralization may have real limits. Also, like other places, more people are aware of the problem; NGOs have been born and it is hard to ignore or not truly include them.

B.2.8 The Danube[19]

Although there is great diversity along the Danube, there is considerable shared culture, values, and principles along the river. The Danube has a long tradition of cooperation around navigation. This cooperation has provided a forum for expanding into new areas of need concerning integrated management and sustainable environment. Since the mid-1970s, several conventions and treaties have formed an overall environmental program and organizational structure along the river. Figure B.10 summarizes these conventions.

The Danube illustrates how a river basin organization and its experience can provide a basis or safe ground for negotiations on broadening needs. It also has shown how the broader forces of regional integration (in this case, the EU) can encourage cooperation along the river through the promulgation of regulations.

The basin countries and transitions follow these shared principles: using best available technology; control of pollution at the source; the polluter pays; regional cooperation; shared information, and striving for more integration in water management.

B.3 AFRICA[20]

There is a history of river basin or international basin organizations in Africa during the twentieth century. Many of these have been discussed already in the context of general theory. They have often been plagued by donor dependence and in some cases donor competitions, which has resulted in hydrologic models on the same basin that disagree and compete with each other. They have for the most part not been owned by the in-country stakeholders. Nevertheless, the importance of river basin organizations, both internationally and within countries, is growing. This is especially true in Southern Africa. The United Nations Food and Agriculture Organization (FAO) summarized RBO efforts in Africa under four policy approaches or coordinating mechanisms.

Type 1 – No overall coordinating body for water with fragmentation among various ministries and under various acts

---

[19] Nachtnebel (1999, 2002).

[20] Turton (2001); OAC (1999); Vaz (1999); Vaz and Pereira (1999); Delli Priscoli (2004); Shela (1999); Conley and van Niekerk (1999); Heyns (2005).

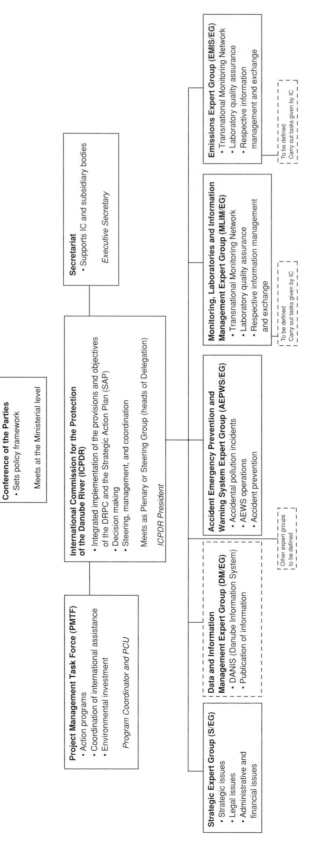

Figure B.10 Structure of the International Commission for the Protection of the Danube River (Conference of the Parties) (Nachtnebel, 1999–2002; Natchkov, 2002).

and little chance for good IWRM. They saw Ghana and Sierra Leone as examples.

Type 2 – An overall coordinating body established. Other specialist agencies are subordinated in one form or another to the Commission or Board. Ethiopia is an example.

Type 3 – A coordinating role within a water ministry that has other responsibilities such as soils, forests, fish, and so forth. Uganda, Zambia, and Nigeria (which has eleven basin authorities) are examples.

Type 4 – A ministry for water resource that controls all aspects of planning and management of water resources. Kenya is an example. Kenya in fact has three basin authorities.

B.3.1 Nigeria

Nigeria is a federation of states. The country is well drained by the Niger River system and its tributary the Benue. Organized water resources development for agriculture started in 1955. Up to that time, water resources development had been in the hands of small-scale and subsistence farmers. The first national development plan of Nigeria, issued in 1962, included agriculture and urban and rural water resources. The drought of the Sahel in the 1970s, like in many other parts of the world, spawned the creation of River Basin Development Authorities (RBDA). This joined the Niger Delta Development authority, which had already been charged with the Niger Delta. The RBDAs were soon increased to encompass the whole nation. Their purposes were to assure systematic use of groundwater and surface water, multipurpose development, supply irrigation water, flood control, erosion control, and general water resources management. As in other countries, they were also seen as tools to stem the tide of rural–urban migration. After the Federal Ministry of Water was set up in 1976, they called for eleven RBDAs with the following functions:

- Undertake comprehensive development, with focus on groundwater and surface water, irrigation, multiobjectives, and flood control.
- Construct, operate, and maintain dams, dykes, and polders wells, irrigation and drainage systems, and other works.
- Supply water from completed schemes for a fee to be determined by RBDA under approval of the ministry.
- Construct, operate, and maintain infrastructure services.
- Develop, maintain, and keep a water resources master plan.
- Undertake a scheme for erosion and flood control for watershed management.
- Allocate water among users and sectors.
- Operate water legislation and control measures in the basin.

In 1979, the RBDA functions were broadened to include fisheries, livestock, and a variety of other activities. In 1984, they were split into eighteen RBDAs, which covered each state of the Federation; in 1986, however, the RDBAs went back to eleven corresponding to river basins. Their functions were modified again and reduced. Their focus became more on development and less on direct production and engagement of extension services. In 1994 the RBDAs were renamed the River Basin and Rural Development Authorities (RBRDAs). Today there are twelve, although there has been movement to reduce them to six.

Although the RBDAs constructed many dams, water use is still not well coordinated. Almost half of the 12 billion cubic meters of stored water are dormant. Of the planned 1 million hectares of irrigated land to be serviced by this infrastructure, only about 100,000 hectares are irrigated. The performance has been poor.

On the whole the RBDAs have not taken ecology and ecosystems into account. They have not dealt well with the conflicts emerging around land acquisition and development because there has been little involvement of the farmers in the process. Government policies have been inconsistent and there is a lack of data and knowledge. The policy on commercialization of the RBDA actually worsened the data collection situation. There has also been a high degree of politicization. Funding has been inadequate and the RBDAs have a low revenue-generating capacity.

Despite much effort and investments made by the federal government through the RBDAs, the achievements are modest compared to the objectives set at the beginning. The unsatisfied water demand has increased twice within a context of tension between users and a constant risk of conflict. The performance of the RBDAs has been mixed, because few of them were either financially or economically sound. In their tasks of monitoring and managing the water resources, all RBDAs have failed to do a satisfactory job. Although their functions are rational, the financial constraints have reduced the programs of several RBDAs to a size that is inconsistent with their relatively large staff and related facilities.

The primary problems with the RBRDAs include (i) RBDAs have no clear role and responsibility, (ii) lack of communication between RBDAs, (iii) top-down approach adopted in dealing with stakeholders, and (iv) lack of autonomy with regard to the federal and the state levels, as well as a lack of continuity in government policies.

Like many public enterprises, the RBDAs have not realized the high expectations to be a vital instrument for the attainment of self-sufficiency in production and harmonious water allocation. Some further reasons for their weakness include (i) the combination of regulatory and management functions, (ii) the frequent changes in policies and the interference in operating decisions, and (iii) poor resources allocation and management.

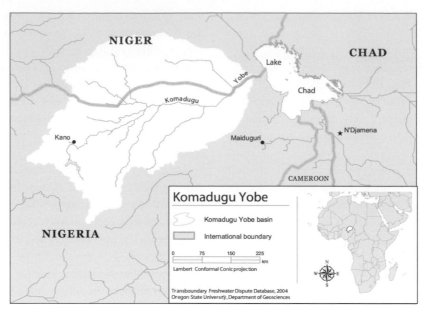

Figure B.11 Map of the Komadugu Yobe basin (*Source:* Transboundary Freshwater Dispute Database, 2004).

In addition, stakeholders' nonparticipation and inappropriate planning have contributed to weaken the RBDAs, leading to increased conflicts among the water users.

To remedy these observed shortcomings, in 1988, the government promulgated Decree 25, which set up the reform package of the RBDAs within the privatization and commercialization program of the public enterprises. The broad objectives of the partial commercialization of the RBDAs were (i) to reorient the RBDAs toward strict commercial principles and practices, (ii) to reduce the constraints on decision-making procedures and administration, and (iii) to move from their dependence on the treasury for funding to a more independent capital structure that would enable them to approach the capital market to fund their operations, without government support. These global objectives have not been fully achieved because the RBDAs were unable to survive without the government subsidies.

Under Decree 101, of 1993, the Federal Ministry of Water Resources (FMWR) embarked on a water-resources management strategy to address water problems including consideration of all the proposed water sector reforms. The prevalence of potential conflict situations, especially in water scarce basins such as the Komadugu Yobe, has led to the formation of six steering committees, each addressing a watershed basin. The Act of 1995 established these steering committees based on the hydrological watershed. Their representatives include river basin commissioners, stakeholders, and relevant federal departments, as well as state representatives. The objective of the steering committees is to facilitate joint and participatory basin management, but they have not yet actually met.

The preparation of the national Water Resources Management Strategy is currently under way. It will address issues such as institutional reforms, legislative reforms, assets, and asset management. The result of this work will serve as the basis for national water resources management reform.

B.3.2 An Example: River Basin Management in the Komadugu–Yobe RBDA[21]

The Komadugu–Yobe basin is located in the semiarid northern part of Nigeria (Figure B.11). It covers about 188.000 km^2 with a population estimated at about 30 million inhabitants. The annual yield of the basin's water resources has been estimated at 13.7 10^6 m^3. The surface water resources, consisting of 8.200 10^6 m^3 annual water yield availability, are more available in the upstream part of the basin. The downstream portion holds the main groundwater resources, estimated at 5.5 10^6 m^3 yield availability, mainly located in the Chad geological formation.

The middle part of the basin contains one of the country's most significant wetlands. The Hadeja wetlands are composed of swamp, grassland, and woodland created by the passage of the Hadeja and Jama'are Rivers. The area flooded annually by river discharge supports various socioeconomic activities and provides a favorable environment for migratory species.

The HJRBDA and CBDA RBDAs have overall responsibility for water-resource management in the Komadugu Yobe basin. The HJRBDA has responsibility for the operational management of the upstream basin in the states of Kano, Bauchi, and Jigawa. The CBDA is responsible for the lower part of the basin from the Nguru wetlands to the river mouth at Lake Chad,

---

[21] Material based on personal communications with O. Dione at the World Bank.

including the states of Borno and Yobe. The two RBDAs are defined by political boundaries rather than hydrological basin limits. In the upper basin, the HJRBDA has constructed fourteen dams, including three major ones: the Tiga, Challawa, and Rwankanya dams. The total water storage of these three dams is estimated about 3.7 billion m³. These assets were developed in order to promote irrigation schemes and provide water supply for rural and urban areas in the RBDA's domain of operation. The water allocation has been developed on an ad hoc basis, taking into account the needs of public irrigation, the water demand to supply the Kano district, and various downstream needs.

The CBDA has to deal with the basin management from the Nguru wetlands to Lake Chad. This river basin management organization does not have dams and water availability and is mainly reliant on upstream water release.

The water-sharing agreement is not clearly defined. In addition to the two RBDAs, several institutions are involved in the basin's institutional management framework. These include the following: (i) the Federal Environment Protection Agency (FEPA); (ii) the governments of the five states that have domains in the basin; (iii) the local government authorities (LGA); (iv) the North East Arid Zone Development Program (NEAZDP); (v) and several water users associations such as the Hadeja Nguru Wetlands Conservation (HNWC), the stakeholders consultative forum, the Dagona Joint Area Development Association and the movement for the survival of the Yobe basin. Nevertheless, the separation of the basin management by two RBDAs has led to a misunderstanding between both RBDAs and various users from upstream to downstream.

Legal considerations in the basin rest mainly based on customary laws; however, these laws are inadequate to address interstate water arrangements and the growing management demands. The promulgation of the Water Law Act of 1993 empowers the Federal Ministry of Water Resources through Decree 101 to regulate the water management in the basin. Similarly, Decree 86 empowers FEPA to ensure that water resources management is not undertaken in a manner that results in negative impacts. To date none of these legal considerations is working efficiently within the basin.

The top-down approach adopted by the two RBDAs in planning and managing the basin has prevented the involvement of users in decision making. The creation of the basin-level Steering Committee for the Komadugu was supposed to bridge the gap between the RBDAs and the stakeholders. There is a need to define and implement a clear framework, with clear roles and mandates for all actors within the basin, so as to implement a clear management process.

For many years, IUCN has been involved in the conservation of the Nguru wetlands. A considerable effort has been made to organize communities and develop a comprehensive and integrated use of the wetlands; however, all these efforts have not led to sustainable successes. The water shortages faced by downstream users and the competing demands with other sectors, such as sugar cane irrigation schemes and water supply, have prevented the achievement of the program objectives set by IUCN at the beginning, with regard to water flow releases from upstream.

Recognition has been reached that sectoral approaches cannot meet the current basin problems. Solutions to the current unbalanced water situation must come through deep institutional reforms that include the involvement of both the RBDAs and stakeholders at all levels.

Although Nigeria has put great emphasis on river basin management and its RBDAs it falls short on Millington's conditions for success. The high-level commitment varies and is actually confusing. There remains a low knowledge base. There is little community participation. Lines of accountability are unclear. The RBDAs, as in many other parts of the world, appear to be handmaidens of the federal government. They have operated in a top-down fashion and have often been highly politicized. They have focused on supply-side solutions with the result that performance has been very poor. They have been unable to accommodate the changing values of water management into their organizations. At the same time there have been highly successful small-scale irrigation operations right next to the large-scale irrigation system failures.

B.4 ASIA[22]

In Asia, the Mekong Commission, roughly at the same point on the continuum as the Indus Commission, has continued deliberations even during periods of conflict. Like many other river basin organizations, it started with a permanent advisory board of professional engineers. About 25 percent of its expenditures (US$44 million seed and US$800 million attracted investment) are for data-gathering and feasibility studies. Among its achievements are twelve tributary projects providing 210 megawatts of power and supplementary irrigation for 200,000 hectares, flood protection, pump irrigation, agricultural research and extension, fisheries, and river navigation. However, as Kirmnani notes, the Commission suffers from weak sense of ownership among the parties of the region. It has been too dependent on external staff and support.

As Frederiksen and many others note, there is little question that the agreement made more than 40 years ago on the Indus was critical to regional security and subsequent economic

---

[22] Delli Priscoli (2001a); Dukhovny and Ruziev (1999); Dukhovny and Sokolov (2002); Li (1999); Mekong River Commission (1999); Niem (2000); Radosevich and Olson (2002); Iyer (1999); Ramu and Herman (2002); Xia, et al. (2001).

development for Pakistan and India in the region. After partition, Pakistan was left as a downstream state. What had been an intranation transboundary water issue became an international transboundary water issue, just as in the Aral Sea and Central Asia in the 1990s. Sharing waters of the Indus, Jhelum, Chenab, Ravi, Beas, and Sutlej in Punjab and Sind was a leading cause of tension between India and Pakistan. It is highly likely that it would have led to major conflict. The treaty and the process of negotiation averted conflict. The treaty has held even during periods of conflict between Pakistan and India. The lessons from this conflict are critical to transboundary conflict management today.

With the help of outside parties, India and Pakistan moved from positional posturing to more interest-based negotiations. Initially, Pakistan called for arbitration and India refused and called for a special court. Both are classic approaches for positional bargaining. Instead, at the initial suggestion of David Lilienthal of the TVA, a World Bank mediation process was initiated. The initial idea was that engineering studies could define optimal uses and stimulate shared operation of the rivers. After more stalemate, this succumbed to creative solutions for dividing the waters proposed by the bank at the request of the parties.

Although the optimal technical solution did not carry the day, talks were begun on the basis of shared epistemic technical and engineering values of integrated assessment. Creative options that expanded the pie were generated by the interplay of such expertise and discussion did begin on sharing benefits and not only allocating waters. The World Bank, as a third party, brought resources and the ability to generate resources to the table. They also brought expertise in development and water. Thus the negotiations moved from positional arbitration to mediated joint problem solving and back to a cross between mediation and arbitration. The well-known solution was to divide the eastern and western waters, provide for a transition period where link canals could be built, provide for India to fund some of the construction, and the generation of international capital to finance other parts of the project, including reservoir storage for Pakistan. In addition to the financial resources available, the World Bank efforts succeeded because it was possible to increase the amount of water available.

B.4.1 Vietnam and Mekong

The new water resources law (1998) in Vietnam calls for a national water resources council (NWRC) and river basin organizations. The NWRC has been established. It is advisory and its job is to recommend strategies to the government. It is in the process of developing a national strategy and action plan. There is still debate on what form the river basin organizations should

take and how much authority and power they should exercise. Whatever the form, there is need for better coordination across various interests in the Mekong Delta, more community awareness of problems and solutions, more stakeholder involvement, and better use of available technical expertise at the local levels.

To date, one such RBO has been formed, the Red River Basin Organization. This RBO and presumably others will be located under the Ministry of Agriculture rural development (MARD). The RBO is supposed to advise the MARD, assure adequate data collection, coordinate planning, and promote public participation in the planning process. The RBO consists of a commission and a support office. It includes representatives of twenty-five provinces and seven central ministries. The chair is appointed by the MARD. It appears that it will be funded from hydropower revenues, fees, and taxes for other water services, along with funds from the central government. A standing committee of fifteen members will carry on the major work. It will reach decision by two-thirds majority vote.

Stakeholder participation is mixed in Vietnam. On the one hand, the system of decentralization and local democracy is strong. On the other hand, political groups other than the communist party are inhibited, thus NGOs and broad civil society groups so important for participatory processes in RBOs do not really exist. Ministry officials have been used to the top-down central planning model.

At the international level there is a long history of basin cooperation under the Mekong River Committee (MC) and now (since 1995) the Mekong River Commission (MRC). This is a history that has spawned the well-known "Mekong spirit."

The Mekong Commission was established in 1957 with headquarters in Bangkok. The MC was under the umbrella of ESACP. It role was to promote, coordinate, supervise, and control planning and investigation of water-resources development projects in the lower Mekong basin. Despite the 40-year history, through all conflict, much of the potential remains untapped. Indeed the Mekong was a main focus of the U.S. President Johnson's Water for Peace program in the early 1960s, diverted eventually by the war.

The MRC resulted from negotiations in the Paris Peace talks of 1991 and the willingness of Cambodia to rejoin the MC. A working group was formed to produce a draft agreement on cooperation for sustainable development of the Mekong River basin. The agreement seeks to promote environmental conservation and sustainable development on a regional and cooperative basis, as well as on an equitable basis that considers the socioeconomic factors of the people. It includes the four lower Mekong Basin States: Cambodia, Laos, Thailand, and Vietnam. It is a three-tiered organization: the Council, which

Figure B.12 Map of the Yellow River basin (*Source:* Transboundary Freshwater Dispute Database, 2004).

is the ministerial level; the Joint Committee at the director general level and the Secretariat at the technical level. The Secretariat has more than 120 people. Each country has it own national Mekong Committee. In Vietnam's case, this committee has been primarily a liaison for coordinating donor-funded activities related to the Delta. The MRC follows a program approach, that is, programs in five work areas.

The MRC sets a type of political space or safe area for meeting and talking. It specifiess general principles, procedures for decisions, and other policies. As such it also can exercise a normative function on countries and be catalytic leader to those countries in areas such as participation and new flood-management tools. The MRC has shown an ability to address common interests and differences quickly. It has shown the values of using outside parties to help negotiate or mediate agreements. It also can help national governments progress in building national institutions. By advocating public awareness and participation in the MC, it encourages each member country to do so also.

Overall, within Vietnam and MARD there is no clear mechanism to bring stakeholders and affected parties together on a regular basis. Thus there is little joint discussion of impacts and solutions and problems. Most of this discussion has been done on a bi-party basis between the government and individual provinces. This process also inhibits transparency, as clear definitions of who plays what role is lacking. Although there has been an attempt at separating functions such as policy setting, management, and operations, it is still important to openly define the relationship between these functions. However, experience has shown that when farmers are given clear responsibilities for managing the parts of an irrigation system close to their areas of interests, they commit funds and effort to maintain and operate systems efficiently.

In effect, the current RBO model seems to be heavily influenced by the traditional top-down bureaucratic experiences using advisory committees with voting procedures and controlled participation in a hierarchical context. However, there is a considerable basis of local and decentralization experience to build on in Vietnam.

B.4.2 China

Flooding is a major concern on many of China's rivers. Flood control is vital to the social and economic development in China. In the modern era, comprehensive planning, with flood control as a central focus, was begun in the 1950s. Until recently the major approaches have involved levees, reservoirs, flood storage, and retention. Protection along China's rivers varies from the 10 to 20-year flood to in excess of the 100-year flood along some parts of the Yellow River (Figure B.12). Flood monitoring, warning, and forecasting systems have become quite sophisticated. In 1997, China passed a law regarding flood control. This law builds on previous laws requiring flood control to be part of river basin planning, which is ultimately accountable to the Central Ministry. The law recognized the need for measures beyond traditional structural solutions, such as population control, policy of economic development, safety in buildings, systems of post flood relief, and compensation and flood insurance programs. It also calls for stakeholder participation.

There are seven major river basin commissions (agencies of the Ministry of Water Resources, MWR) that perform the functions of water administration in the river basins. These functions are quite broad and range from planning to administration, coordination, conflict resolution, operations, and regulations. This institutional arrangement reflects traditional top-down and essentially administrative or technical bureaucratic approaches to flood control. The local water-resources management is really the only element that gets grassroots interests of users into the planning process. These include the provincial, prefecture, country, and village levels. In addition, China puts emphasis on the actual teams and mobilization of people for flood fighting. This process is an important element of participation in the picture of flood management. In the end, the level of flood protection is directly linked to the levels of and changes in socioeconomic development.

Participation looms as a major issue for flood policy and the river basin commissions. New laws, with their calls for measures beyond traditional structural measures, as well as increased pressures on land from population and economic growth, will force calls for direct access to these processes. It is not clear how this will occur. The current structure does not appear to offer such access beyond the interagency and intergovernmental coordination routes. The RBCs are really offices of the MWR; they are commissions without members. There is discussion of legally authorizing them to include representatives of various sectors and users.

In the Tarim basin, the incompatibility of the production objectives and policies set by the central and regional government and those of prefectures around the watersheds of tributaries has become clear. It has inhibited the ability to recover dried-up flow areas of the lower reaches of the basin. Consequently, the reorganization legislation for the Tarim Basin Water Commission (TBWC) includes the five prefectures, along with traditional regional government authorities in decision makers. In addition, under a new basin project, it calls for more community involvement and water user participation. Thus, TBWC is beginning to forge direct linkages with water supply corporations and water user associations. It is seeking to engage such groups in workshops.

China illustrates the complex interaction of political culture, traditional technocratic power, and new pressures for empowerment, along with new demands on water. Although most seem to realize that the river basin is the best way to organize all these divergent forces, it is not clear how willing the MWR and the river basin commissions will be to make what amounts to fundamental organizational change in their management culture to accommodate these forces.

B.4.3 The Yellow River example

The Yellow River Conservancy Commission (YRCC) is an agency of the Ministry of Water Resources. It is responsible for unified management of the basin and resources. This includes flood protection and most of the water uses, coordination, planning supervision, and services. It formulates basinwide policies, strategic plans, and comprehensive plans. It supervises and coordinates tasks and tries to resolves conflict among sectors and users. It is a comprehensive authority with a large and highly competent technical staff. Flood control is one of its most important tasks. Unlike most of China, where local governments perform most of the flood protection, the YRCC does so for the Yellow River.

Although the YRCC is comprehensive and integrative, it is built on the traditional hierarchical, technically driven, top-down model. To this degree it is supply oriented in its approaches. It does the planning, constructing, and operating along the river. These include plans for water and soil conservation, navigation, water utilization, protection, and flood prevention. It carries out a sophisticated process of scenario-based contingency planning using decision support systems. Advanced networks of technology, computer hardware, and cooperative efforts among various departments and levels of government support this. It also prepares and carries out an allocation plan. It is the agency for permitting water abstractions.

Like other states with central control experience, stakeholder participation seems to be done through the various coordinating mechanisms among the state and local and regional organizations. There does not appear to be a developed civil society, as in other countries with one-party domination. In addition, there appears to be little formal stakeholder participation in the formulation of the plans, scenario exercise, decision support systems, and overall planning. The YRCC has achieved a high technical level and some excellent performance. However, it is not clear how and whether it can handle newer ecological demands and accommodate empowerment aspirations and a growing number of interested stakeholders. It is also not clear whether allocation processes and planning respond to higher-value uses versus established political/bureaucratic interests.

Indeed, in 1996, changes were instituted that may be a window into the future of RBO and management in China. The YRCC was authorized to take on overall management of the river. It was tasked to operate on the basis of demand management, and certain amounts of water were required to be reserved for environmental purposes. A consultative mechanism was established that included all users along the main stem. This consisted of regularly held meetings along the river on important issues in planning operations. The plans were also tasked to develop programs that included typical nonstructural measures, such as resettlement, population control, agricultural production restructuring, land-use planning, and safety buildings.

B.4.4 Indonesia

Like other countries in Asia, Indonesia is starting to recognize the importance of river basins and to reorganize. However, it is also having trouble accommodating stakeholders outside the traditional state governmental and corporate entities. Indonesia has two RBOs: the Jatiluhur Water Authority and the Brantas Water Management Corporation. It is planning for six new corporations. At the provincial levels, recent legislation has called for Basin Water Operating Units (BWU) under provincial Water Resources Services, which will have regulatory and operational functions. It also calls for coordination institutions called Provincial Water Management Committees (PWMC) and Basin Management Water Committee (BMWC). The Jatiluhur and the Brantas are centrally owned and managed. The former began on the model of the TVA, but gradually it shifted into the Water Ministry.

Like elsewhere in Asia, these RBOs and the BMWCs really lack stakeholder participation. Stakeholder participation has generally been limited to participation through water user associations (WUAs) in irrigation. There is little direct involvement of stakeholders in planning, program development, or implementation, or in the regulation of basinwide water management. Like other areas in Asia, the idea has been to use line agencies, *government agencies, and corporations* to provide inputs. Since the late 1990s, however, NGO activity has increased along with the broader range of interests needing to be considered on the systemwide basis. However, NGOs and civil society groups are still not well organized and lack financing means. There is little such participation even in the legislative process. The town hall meetings that have been held tend to reflect the paternalistic and hierarchical attitudes of government. The BMWC and provincial committees lack formal stakeholder partition. The various boards also lack such participation. Thus the basinwide work really reflects the voice of the Central Ministry. One result of all this is low general public awareness of water issues such as flooding. Indeed, the Jatiluhur depends on the provincial government for public awareness, while the Brantas has initiated a few awareness programs on its own.

B.5 LATIN AMERICA[23]

There has been dramatic water-resources reform throughout Latin America. Mexico is among the most important. Brazil, Argentina, Peru, Colombia, and Venezuela have each looked

[23] Mestre (2001, 2002); Delli Priscoli (2004); Economic Commission for Latin America and the Caribbean (1997); Johnson (2001).

to the Mexican model; however, the Brazilian and Colombian national laws give stronger emphasis to river basin councils.

In South America, a Coordinating Intergovernmental Committee (CIC) was established for the La Plata basin, which helped prepare the treaty of La Plata basin. This arrangement can be seen as near the center of the continuum. The CIC responds under a conference of Foreign Ministries. Numerous binational entities and technical commissions have been established for the survey, design, construction, and operation of various waterworks in the basin. In practice, the institutional machinery has not worked well (Sidebar B.5).

B.5.1 Brazil

New water-resources management in Brazil now emphasizes the river basin as the main management and planning unit. Water pricing of bulk water is being reformed. Basin committees with state and local water users and civil society groups are being formed. Basin agencies are often becoming the executive arm of the basin committees. Federal and state water-resources councils are the regulators with government, municipalities, water users, and civil society representatives. In general, there is a strengthening of the water rights systems that goes along with all this.

As an example, in the Paraib do sul River basin, the Committee for the Integration of Paraib do Sul basin (CEIVAP), which is an umbrella committee for the whole basin, has technical and financial functions. These include resolving water conflicts; approving basin water plans; setting guidelines for water allocation and water quality; setting water pricing criteria, charges, and investment needs to be approved by the National Water Resource Council; collecting water charges; verifying revenue collections; and helping integrate among all agencies and other entities involved in river basin management.

CEIVAP is a streamlined operation. It will not be the owner of infrastructure nor carry out construction or implementation of investments defined in the basin plans. It will not be responsible for operations and maintenance functions. In this transition period, the expected revenues will be about $6 million per year: 3.6 from domestic users and 2.4 from industrial users. Nevertheless, projected revenues will probably only finance about 10 percent of the projected needs of around $1 billion. It remains to be seen whether the new basin setup will help integrate water quality and quantity. The link between the water pricing and permit systems must be strengthened. In addition, agreement between the national government and the three states involved is needed to provide incentives.

Like other parts of the world, states fear losing their power and control with basin entities. The interaction between the

Sidebar B.5 Perspectives on RBOs and Participation in Latin America

Argentina has attempted to create several river basin organizations (RBOs), triggered either by national government (in lesser numbers) or by provincial efforts (most of them). With few exceptions, national and provincial water institutions are weak. Public participation was ample long ago in Provinces such as Mendoza. However, gradually, centralism reduced some participatory niches, although important efforts have been made in the past few years to revert to such tendencies. (See Mestre, 2000.)

Bolivia is steadily leading the way in public participation with a national law dedicated to this subject. Bolivians are now heavily participating at department and local levels. No relevant RBOs exist, but local authorities have provided space, time, and resources for public participation to take place. However, action derived from public participation is still relatively scarce.

Brazil has fostered the creation of hundreds of river basin organizations of many sorts (the second attempt in 30 years) with the support provided by the Water Law enacted in 1997. Public participation is increasing; independent water movements supported by NGOs are strengthening and today represent one of the best examples of public involvement in water management. The central government is opposing some of these movements and the newly created Agencia Nacional das Aguas will complicate the scenario for a while. The Brazil model was inspired by the Mexican and French models of basin organizations. However, Brazilian Consorcios are a new breed of watershed organization that fosters enthusiastic participation and support (including financial resources).

In Costa Rica, both Tárcoles and Tempisque basins have been operated by pseudo-NGOs with modest but interesting results (both in terms of water management improvement as well as for public participation positive experiences). Government has a light approach on these matters in general terms (the sole exceptions are both basins).

Via pertinent legal supports (i.e., a specific law with valuable concepts throughout), Colombia has created interesting mechanisms (Corporaciones Autónomas) with local entities playing important roles and challenging strong central entities. However, direct public participation is still scarce.

Although the *corporaciones* are numerous and at least three of them are very powerful, the rest of the institutional arrangements are weak.

Chile, with a mature vision in terms of water management and its governing institutions, has made sporadic attempts to create ample public participation schemes with mixed results, heavily related to specific regions and existing socioeconomic issues. RBOs have not been successful in Chile for a number of reasons. Furthermore, the Water Law, with all of its supporters and harsh critics, tends to point to difficult possible decentralized integrated water resources schemes.

Ecuador has weak national and regional institutions to foster public participation; the Consejo Nacional de Recursos Hídricos is still struggling to survive within a sectoral ministry: Agriculture. Decentralization is under way with mixed results. Opposition comes from present strongholds, such as the Corporación del Guayas. However, in specific cases, such as Cuenca, efforts are currently made to incorporate results from systematic public hearings on programs dealing with water treatment, allocation, and protection. Some of the best practices are being supported or directed by entrepreneurs, in a philanthropic manner.

In Guatemala, little advance has been accomplished as all actions are held almost with no public knowledge (i.e., Lake Amatitlán and Guatemala City). Institutions are weak and water laws or acts are inexistent.

Paraguay has struggled to create RBOs and to foster public participation. Lack of legal and institutional frameworks, political willpower, and financial resources has long impeded such actions from taking place. Today, the Paraguayan water sector is heavily in demand of capital investment and a new deal in terms of government reforms of many sorts, especially in those tasks related to possible public involvement.

Uruguay has a twofold approach: on one hand, local authorities and individuals in certain departments tend to participate in considerable numbers, but with very little achievements in the long run; and on the other hand, central government – which struggles against any sort of modernization approaches – has little belief that public participation may be a good idea to improve water management without involving large costs, both economic and political.

Figure B.13 Map of the São Francisco basin (*Source:* Transboundary Freshwater Dispute Database, 2004).

CEIVAP and other subbasin entities also can engender the same fears. This is true with the federal government. There are new partnerships, so parties are adjusting to a new environment.

Similar actions and reactions are taking place in the Curu River basin. The voluntary reforms of the state of Ceara have strong World Bank support. The basin includes local organizations built around reservoirs, called social catchments, as well as basin committees. All of this provides a forum for permanent negotiation for the allocation of water.

These cases are beginning to show what others have learned about building river basin organizations. There is a need to strengthen the intermediate levels of decision making and to encourage more stakeholder participation. However, in the Paraib do Sul, the basin is very large, making it hard for the general public to conceive of the whole. In the Curu, they are relying on strong links among the users of reservoirs and other water bodies. It is clear that the culture and the socioeconomic environment influence the structure of organizations. In other words, one model does not fit everywhere. For example in Curu, there is a more centralized system: basin committees thus do not control their own resources. It remains to be seen how the strong centralized tradition will fare – whether it will essentially defeat the regional idea of the river basin unit, as it has in other parts of the world.

In the PDS, there is more tradition of decentralization. They are using an approach similar to that in the Chapala basin in Mexico: the paired basin committee and basin agency approach. Basin entities are expected to become self-sufficient based on bulk charges. Thus the state and federal government are expected to exert a weaker influence on decision making; whether this holds true remains to be seen, however. Also in the charges are the engines for reform. In the Curu, the stakeholders already have been working together and have been operating in an environment of almost continuous rationing.

There has also been a long history of river basin management concerns over the São Francisco River, one of Brazil's most important rivers. Like with many rivers worldwide, navigation uses first prompted concern for the river as a whole. As in other areas, navigation was a means for bringing agricultural goods to market and for opening up new territories and settlements. Over time, irrigation and hydroelectric power uses grew. All of these produced the familiar pattern of a structural focus, an economic development focus, and then growing sectoral rivalries.

The São Francisco has a major estuary and delta, which exhibit the problems of upstream development and altered flood flows (Figure B.13). Over time, river regulation has favored hydroelectric power and flood control. Numerous government initiatives have emerged to deal with the river. A senate committee has been set to investigate the basin and set the stage for the creation of a river basin development corporation and an interstate liaison committee. The basin issues are familiar: little knowledge of environmental impacts, low traditional stakeholder participation, poorly regulated development with large single-purpose projects, limited human and institutional capacity, lack of a holistic vision to drive management, problems with land regulation, fisheries, urbanization, and water quality.

The principal federal agencies are the CODEVASF (Rio São Francisco Development Agency), CHEF (São Francisco Power Company), and SUDENE (an organization for planning and development in the Northeast). In the early 1980s, a special executive committee, Integrated Studies of the Basin (CEEIVASF), was created under the Commission for Integrated River Basin Studies in Brazil for planning studies on the river. This committee lacked independence, however, like many other such efforts in the world. There is also the Interstate Parliamentary Commission for the Development of the Rio São Francisco (CIPE). This group includes leaders of legislative assemblies of the five riparian states and the union of municipal authorities in the basin. In the late 1980s, there were important planning studies for development efforts in the whole region.

Against this backdrop of long concern, the states of the Northeast in cooperation with the National Water Secretariat formed a group from the water sector of each state in 1995. A special commission for the Development of the São Francisco Valley was also created in 1995 by the federal senate to promote discussion about river basin planning. The states themselves have varying degrees of expertise. In 1997, as we have mentioned, the federal government passed a national water law and a system of public institutions or basin committees for issuing water rights and implementing a charging scheme.

All of this has moved the various entities closer to river basin management. CODEVASF has been increasing user participation through user-controlled water districts. However, CHESF continues to develop hydropower in a single-purpose mode and with little regard to comprehensive management. The agency has encouraged cross-sectional planning and has emphasized strengthening the basin committee as a way to do this. The idea is to find ways for entities to build relationships and work out joint gains, for example, through the creation of more basin committees and user associations. There is a major watershed program geared toward changing the paradigm for water planning.

The São Francisco, like the Colorado in the United States, shows how difficult it is to engage a large basin in river basin management, even when there has been a strong focus on the river basin and numerous attempts and knowledge of the experiments around them. The political motivations of various entities do not always agree. The supply management tradition and top-down tradition continue and are very hard to change.

B.5.2 Mexico

Socioeconomic development in Mexico has always been tightly linked to water. The 1917 constitution established that water is owned by the federal government and cannot be privately owned. The National Irrigation Commission was created in 1926, followed in 1945 by the creation of the Hydraulic Resources Secretariat (HRS), which was a federal ministry responsible for irrigation, river management, and control, along with municipal water and wastewater. From 1947 to 1960, HRS put several river basin executive commissions (RBECs) in place, primarily to promote hydraulic development. These were run by the federal government with little interaction with water users associations other state and local entities. They became powerful regional entities that altered state and regional politics; however, by 1977, they were dismantled. In 1976, a new federal ministry called the Agriculture and Hydraulic Secretariat (AHRS) emerged. In 1989, the National Water Commission (NWC) was the sole federal water authority. It was placed inside the AHRS and had broad responsibilities dealing with water rights, allocation, use, effluent charges, infrastructure, and operations.

In December 1992, the National Water Law (NWL) was enacted, which strengthened the NWC. Through NWL, the roles of regional actors were better defined and participation of state and local water users and general public was encouraged. However, environmental deterioration continued, as did the gap between supply and demand. NWC was transferred into the Environment, Natural Resources and Fishing Secretariat in 1994. Since that time, twenty-four river basin councils have been established, water planning readopted, and thousands of water rights issued. Although efforts have been made to decentralize, especially in the Lerma Chapala basin, they have not moved far (Figure B.14). Water is still primarily a federal activity. The river basin councils are the means to coordinate government institutions and to negotiate with users and social organizations.

Mexico currently has twenty-five river basin councils, six basin commissions, two basin committees, and thirty-eight groundwater technical committees. Users are increasingly jnterested in learning more about management and taking on roles formerly played by government entities. The knowledge base, one of Millington's criteria for success, is growing. The river basin councils are beginning to play a greater role in planning. They do respond to and provide a forum for stakeholder participation. They still need to gain in legitimacy and credibility, along with the NWC and regional councils. The RBCs complement rather than replace the governmental organizations,

Figure B.14 Map of the Lerma Chapala basin (*Source:* Transboundary Freshwater Dispute Database, 2004).

although they are likely to take over even more functions of the NWC in the future. They work through consensus; however, they do need a stronger legal personality. In short they still need further refinement in the definition of lines of accountability and authority.

The Lerma Chapala River Basin Council (LCRBC) has been the forerunner in this trend; it illustrates much of this development and the tensions among federal and other interests when trying to establish river basin organizations (Figure B.15).

B.5.3 Lerma Chapala example

This is one of the richest areas in Latin America and the most important in Mexico. Twenty percent of all national commerce and services activities occur within the basin. The river basin needs coordination across all uses, including flooding. Fre-

quent conflicts over water quality occur. Growing crises in use, allocation, environmental deterioration, and efficiencies spawned social reactions in the early 1990s.

In 1989, a consultative council was formed, which was precursor of the river basin council. It also created a technical working group made of public servants from across sectors. This group became an important engine for keeping the council going. The primary tasks of the council were negotiating resources, coordinating efforts, conciliating different positions, forging consensus, creating legal instruments, and generally facilitating planning and allocation polices. The council met yearly while the technical committee met regularly. The council made decisions through consensus. In difficult cases where such consensus could not be reached, it referred issues to selected high-level members of the technical working group. This mechanism became a critical means for resolving

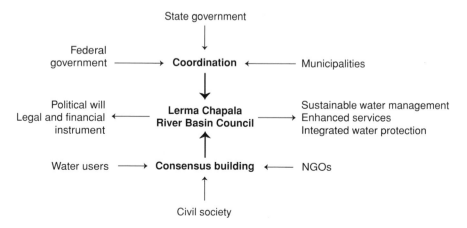

Figure B.15 State government: Lerma Chapala River basin coordination (Mestre, 2001, 2002).

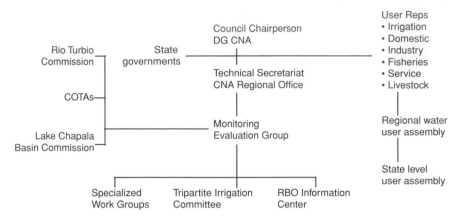

Figure B.16 Structure of the Lerma Chapala River Basin Council (Mestre, 2001, 2002).

disputes. It is very much like the disputes review panels used in North America (Figure B.16).

Like other successful river basin ventures, the council was persistently if not in a low-key way opposed by federal interests. This is much like attempts to create river basin organizations in the United States: the states preferred bilateral agreements and the federal government saw its power as diminished.

The Lerma Chapala River Basin Council was born in 1993, after much negotiation, under the National Water Law, out of the national water council. The first council included more than ten key agency representatives, all five state governors, and six water users of different sectors. The water users could thus interact with the government across sectors and at the various levels. The users eventually came to be chosen by the general assembly; they would come from the state water user organizations within the Lerma Chapala. In effect, the LCRBC is akin to the idea of a water parliament found in the French system, but with less formal power to influence allocations.

At its root, it seems to be an institutional approach to broaden stakeholder participation.

LCRBC is a mixed organization: it is not an authority, public or private. It is legally supported by the NWL. It is consultative; however, it can officially question, propose, and approve decisions. It does not have regulatory responsibility and is not a service provider. It seems to work well to raise issues early, to engender a sense of ownership and participation in decisions, and perhaps forestall or mitigate the intensity of conflicts (Figure B.17).

The Lerma Santiago Pacifico Regional Agency (LSPRA) complements the LCRBC. This is a government organization subordinate to the NWC. It can suggest basin-level tariffs, establish collections schemes for revenues, use financial resources, issue water rights licensing, pollution control, efficiency schemes, and conservation and intervene in disputes and provide water rights.

There is still lack of clarity between the roles of the LCRBC and the LSPRA, however. This makes the system susceptible to creeping centralization attempts by the federal government. This is especially true because of the dependence (to varying degrees) of both LCRBC and the LSPRA on the NWC. The NWC developed the river basin master plan drafts in a top-down manner. It was, however, eventually evaluated by the

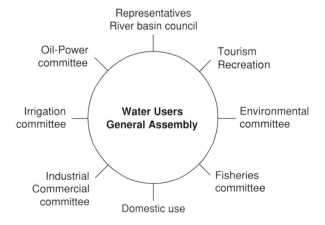

Figure B.17 Members of the Lerma Chapala River Basin Council Water Users Assembly (Mestre, 2001, 2002).

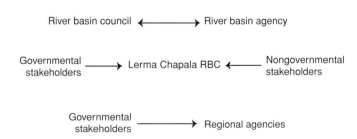

Figure B.18 Approach to regional water management (Mestre, 2001, 2002).

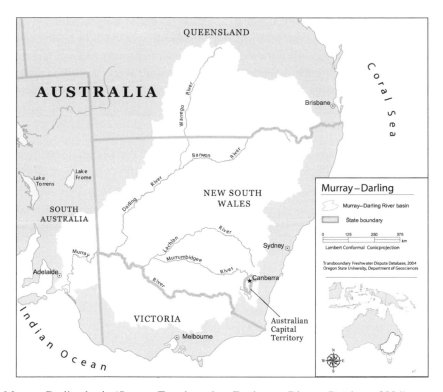

Figure B.19 Map of the Murray–Darling basin (*Source:* Transboundary Freshwater Dispute Database, 2004).

technical working group. Water licensing and rights management remains the exclusive province of the federal government. Many new titles have been issued and this is a major accomplishment. But this also brings some problems where illegal uses have now actually become legal. The Regional Agency of the Government has the key role in water licensing, and the LCRBC plays only a marginal role (Figure B.18).

Although progress has been made, water rights need to be revised. Much more water efficiency is needed; water markets need improvement, as do regulation means. Better water balances are needed. Sanctions must be improved and allocation is not really working properly. Success in these areas will, to a great deal, depend on how well the LCRBC and LSPRA do.

B.6 AUSTRALIA[24]

B.6.1 The Murray–Darling basin

Murray–Darling basin covers more the 1 million square kilometers: almost all of Southeast Australia and a great deal of the continent (Figure B.19). The basin is larger than Spain

and France combined. Attempts to coordinate along the river began as in many places with navigation. Navigation was crucial to opening the interior for development and settlement. By the turn of the century, it was clear that agreement was also needed on water supply because of increasing demands for irrigation. Agreements at the turn of century focused on building storage and structures, as the focus was on development. A Murray–Darling Commission was instituted. Its mission was revised several times during the century, but for the most part remained focused on quantity. More recently, the environmental costs of successful development became apparent. These included collapsing river banks, salinity problems, declining fish, more effluent discharge into the rivers, deteriorating water quality, outbreaks of blue green algae, and reduced flow in the rivers.

In the 1990s, there was the recognition that the states and the commission were not integrating the management of the water or basin resources in ways that were sustainable. There was a need for stable institutional frameworks, better knowledge of problems, better integration across all aspects of natural resources, and much stronger community participation in identifying the problems and solutions. In the 1980s, each state had to refer any development proposals that would impact quantity and quality to the commission. This referral role is similar to that of the Mekong Agreement and to the Delaware

---

[24] Millington (2002); D. Blackmore (2002); private conversations with Australian water officials, 2004 and 2005; Murray–Darling Basin Commission, www.mdbc.gov.au.

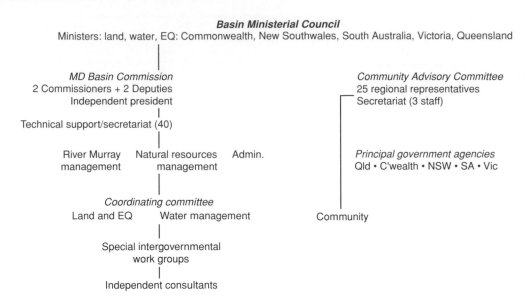

Figure B.20 Murray–Darling Basin Ministerial Council structure (Millington, 2002; Blackmore, 2002; Murray–Darling Basin Commission, 2007; and several personal conversations with principles of Murray–Darling).

River Basin Commission. This expansion of power led to the creation of the Murray–Darling Basin Ministerial Council and a new supporting commission in 1985 (Figures B.20 and B.21).

The objective of the organization is to promote and coordinate effective planning and management for equitable, efficient, and sustainable use of the land, water, and environmental resources. Among its most important innovations was the establishment of the Community Advisory Committee (CAC) to provide independent advice on the Basin communities regarding water and resources issues. This committee represents regional groups from throughout the basin.

The staff at the commission is responsible for managing the Murray–Darling system but not the tributaries in each. It is the engine that drives "on the ground actions." It is strong because it is a community-driven strategy. The action strategy rests on what is called a Community of Common Concern (CCC). This is a flexible concept meant to include a range of communities and types of local action groups. To date there are more than fifty areas with detailed community action programs and more than one hundred land care groups working. There is a comprehensive program called "stream watch and river care" introduced into the schools. This program is meant to get the school communities educated and involved through participation. These programs are successful and are in turn having a major influence on the parents of the school children. The programs encourage communities to develop plans for management of the resources. Indeed, local land and management

Queensland
- Dept. of Primary Industries
- Dept. of EQ Heritage
- Water Resources Commissions

New South Wales
- Dept. of Water Resources
- Dept. of Conservation/Land
- EQ Protection
- NSW Fisheries
- NSW Agriculture

Commonwealth
- Dept. of Primary Industries
- Dept. of Arts, Sports EQ
- Territories

South Australia
- Dept. of Primary Industry
- Dept. of EQ/Land
- Engineering/Water Supply

Victoria
- Dept. of Agriculture
- Dept. of Conservation/Natural Resources
- Rural Water Corps

Figure B.21 Murray–Darling: Principal government agencies (Millington, 2002; Blackmore, 2002; Murray–Darling Basin Commission, 2007; and several personal conversations with principals of Murray–Darling).

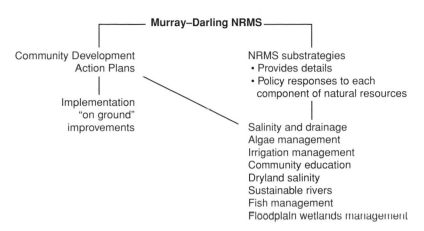

Figure B.22 Murray–Darling natural resources management strategy (NRMS) (Millington, 2002; Blackmore, 2002; Murray–Darling Basin Commission, 2007; and several personal conversations with principles of Murray–Darling).

plans developed with high participation are considered essential precursors to any eventual private leasing, under public authority, of water supply systems or irrigation.

The community actions programs include the identification and definition of local problems, the investigation of options, and the drafting of plans with implementation strategies. The adopted plans require landholders, industry, and government to agree on some long and medium term commitment. The plan will include initial benchmarking, performance targets, cost-sharing arrangements, and a monitoring system (Figure B.22).

Perhaps the most important bottom line to the Murray–Darling experience is the broad and deep recognition across society of the national, regional, and local economic importance of proper river basin planning. A clear understanding up and down is necessary to avoid bad decisions and to capture potential benefits. It is also essential for building stable water rights and entitlement systems, which are necessary for legitimate and accepted allocation decisions and polices. As an example of monitoring the council commissioned a water audit report on water use. This resulted in a cap on increased diversions while detailed discussions take place on appropriate policy for off-stream diversions and management regimes. None of this could occur without the Murray–Darling Basin Council (MDBC) structures, which brings a wide variety of stakeholders into the debate in ways that allow them to feel responsible and that they are participating. For this, the local implementers must be involved, together with the policy makers, so the knowledge base must include local as well as national and technical knowledge.

The MDBC organization has the key attributes for good basin organization: high-level ministerial commitment, meaningful community input, high knowledge levels, and clear accountability among participating members.

B.6.2 The heart of the Murray–Darling organizations: participatory processes

The CAC was constituted because many in the national government were concerned that unless this concept was expressly stated, the state governments, over time, might reject the idea of such a high level "voice" affecting them. State governments with either a strong development or a strong environmental attitude have not necessarily liked such a strong high-level CAC.

Usually the CAC gets a three-year term, but reappointment can occur and often does. The Ministerial Council appoints the Chair after consensus agreement. This is not an easy choice; choices have ranged from former bureaucrats to regional and farmer representatives. The second Chair was a farmer/regional government representative, who lifted the profile of CAC, focused its debates, and generally got the basin community to see how the CAC could be useful as a means to get ideas directly to the political leadership. There are twenty-six members, twenty-one of whom are state representatives chosen on a catchment or regional basis. They are spread among each state in about the same proportion as the main subbasins in each state, plus representatives from four "special interest" organizations and an Aboriginal representative. The latter five are nominated by the twenty-one groups. The states choose the catchment and regional representatives. In practice, these representatives usually come from the CCCs in each state. For example in New South Wales, there are seven major subbasins, each with a catchment management committee (or CMC, but generically a CCC) and a representative of each of these goes onto the CAC – mostly with no interference from the state minister. This means that there is a good connection from the CMC, or CCC, upward to the CAC and a good connection of issues and transference downward of experiences in one part of the subbasin to another. (In each subbasin the CMC is made up of a wide cross-section of the community, and there are good links downward to the many land care, river care, stream watch, and other such groups that are achieving real change on the ground.)

The CAC deals first with issues that are referred to it by the MC or the Commissioner, and concerns of the basin community that needs to be referred to MC or the Commission. These must be major issues. Discerning issue salience is a key role for the CAC Chair and it is crucial to keeping the CAC relevant to the basin and the MDC objectives. In effect, this means that the CAC can actually be assessing community opinion before the MC determines policy options rather than exploring reactions to new policies.

There is no fixed approach to the CAS operations. Generally it does some strategic planning when a 3-year term starts. It usually holds a series of workshops or briefings at key locations throughout the basin on a particular issues, but this depends on the understanding in the basin and on the responses of the twenty-one major CCC groups in the basin represented on the CAC. The CAC has a budget from the MC that allows it to commission limited studies and obtain expert briefings. The CAC usually reports to each MC meeting, of which there are generally two per year.

Some of the Murray–Darling terminology can be confusing. The CCC is really a generic term. Some states use total catchment management committees (TCM), others use integrated catchment management, and still others use catchment management committees. At lower levels, there are the land, river care, and other local action committees.

In sum, the participatory Murray–Darling process seeks to combine a bottom-up knowledge with top-down action. The participatory processes through the CAC and the CCCs have teeth and directly affect the ministerial council. The focus is really the twenty-one subbasin CCCs. Each of these is supposed to create a good natural resource strategic plan for its area, which goes upward into the CAC deliberations and downward to smaller groups within each subbasin. The idea is for these plans to take the basinwide and particular state policy frameworks and put them into strategies and local actions. These actions then link downward to the various action groups within a subbasin such as land care, river care, and stream watch within the school structure. This linkage aims to ensure that these smaller community groups, where the real action occurs at property or subcatchment level, do put their efforts into a common strategic direction that benefits the region as a whole.

The MDBC and state bureaucracies must develop the "technical" story on emerging resource problems and help these groups to develop responses. Each key agency has representatives on the twenty-one senior CCCs; the chairs of these twenty-one CCCs are thus key people. They can make or break to enthusiasm and output at this level.

It does appear that the MDBC is developing resource knowledge and injecting this into the community, even if there is residual reluctance in some states for getting out and selling ideas. There is also a need to look more carefully at the economic impact of reforms being instituted. As could be expected, the quality of the CCC chairs varies. The strategic support also varies. Most farmers still need some convincing to participate. This makes the role and enthusiasm of the state agency CEOs and the CCC chairs so important.

C   Case studies of transboundary dispute resolution

Aaron T. Wolf and Joshua T. Newton

A clear understanding of the details of how water conflicts have been resolved historically is vital in discerning patterns that may be useful in resolving or, better, preventing, future conflict. A total of eighteen systems are presented in a unified format to allow comparative study (see Table C.1). These systems include twelve watersheds (Danube, Euphrates–Tigris, Ganges, Indus, Jordan, Kura–Araks, La Plata, Mekong, Middle East, Nile, Salween, and Senegal); three aquifer systems (Guaraní, U.S.–Mexico, and West Bank); three sets of lakes (Aral Sea, Great Lakes, and Lake Titicaca); and one engineering works (Lesotho Highlands Project). Each case study includes a summary sheet, description of the process and issues of conflict management, a chronology, and lessons learned.

Table C.1 *Features of case study watersheds*

| Name | Riparian states (with % of national available water being utilized)[a,b] | Riparian relations (with dates of most recent agreements) | Watershed features[a] | | | |
			Average annual flow (km³/yr)[c]	Size (km²)	Climate	Special features
Aral	Afghanistan (47.7), China (n/a), Kazakhstan (n/a), Kyrgyzstan (n/a), Pakistan (n/a), Tajikistan (n/a), Turkmenistan (n/a), Uzbekistan (n/a)	Cool to warm (1993 and 1995 Agreements on Aral Action Plans)	10,201	1,231,400	Dry to humid continental	Case of lake management exacerbated by internationalization of basin
Danube	Albania (1.6), Austria (6.1), Bulgaria (7.1), Bosnia and Herzegovina (n/a) Croatia (n/a), the Czech Republic (n/a), Germany (43.8), Hungary (35.5), Italy (26.6), Moldova (n/a), Poland (42.9), Romania (22.0), Slovakia (n/a), Slovenia (n/a), Switzerland (9.8), Ukraine (n/a), Serbia and Montenegro (14.4)	Cold to warm (1994 Danube River Protection Convention)	206	790,100	Dry to humid	1994 Convention is first treaty developed through process of public participation

(continued)

Table C.1 *(continued)*

Name	Riparian states (with % of national available water being utilized)[a,b]	Riparian relations (with dates of most recent agreements)	Watershed features[a]			Special features
			Average annual flow (km³/yr)[c]	Size (km²)	Climate	
Ganges–Brahmaputra–Meghna	China (19.3), Bangladesh (1.0), Bhutan (0.1), India (57.1), India, claimed by China (n/a), India control, claimed by China (n/a) Myanmar (Burma) (n/a), Nepal (14.8)	Cold to warm (1985 Agreement between India and Pakistan lapsed in 1988; new treaty in 1996.)	971	1,634,900	Humid to tropical	Scheduled to be model/workshop case – limited riparians; ongoing dispute
Guaraní Aquifer	Argentina, Brazil, Paraguay, Uruguay		–	1,200,00	–	Groundwater management should be integrated into regional water management strategies.
Indus	Afghanistan (47.7), China (19.3), Chinese control, claimed by India (n/a), India (57.1), Indian control, claimed by China (n/a), Nepal (n/a), Pakistan (53.8)	Cool (1960 Indus Water Treaty between India and Pakistan)	238	1,138,800	Dry to humid sub-tropical	Scheduled as case to be "back-modeled"
International Joint Commission (Great Lakes)	Canada (1.4), United States (21.7)	Warm	225,001	509,200	Humid continental	Case of small number of riparians with good relations
Jordan	Israel (95.6), Jordan (67.6), Lebanon (20.6), Palestine (100.0), Syria (102.0), West Bank (n/a), Egypt (n/a), Golan Heights (n/a)	Cool to warm (1994 Treaty of Peace – Israel/Jordan;1995 Interim Agreement – Israel/Palestine)	1.4	42,800	Dry to Mediter-ranean	Complex conflict and attempts at conflict resolution since 1919
Kura–Araks	Azerbaijan, Iran, Armenia, Georgia, Turkey, Russia			193,200	Dry to Cold	
La Plata	Argentina (3.5), Bolivia (0.7), Brazil (0.5), Paraguay (0.2), Uruguay (0.6)	Warm (1995 Mercosur – Southern Common Market – adds impetus to "hydrovia" canal project)	470	2,954,500	Tropical	Good example of intersectoral, plus international, dispute
Lake Titicaca–Poopo System	Bolivia, Peru, Chile			111,800.00	Dry to montane	
Lesotho Highlands	Lesotho (1.5), South Africa (28.4)	Warm	n/a	n/a	Humid marine	Interesting institutional arrangement exchanging water, financial considerations, and energy resources

| Name | Riparian states (with % of national available water being utilized)[a,b] | Riparian relations (with dates of most recent agreements) | Watershed features[a] | | | |
			Average annual flow (km³/yr)[c]	Size (km²)	Climate	Special features
Mekong	Cambodia (Kampuchea) (0.1), China (19.3) Laos, People's Democratic Republic of (0.8), Myanmar (Burma) (0.4), Thailand (32.1), Vietnam (2.8)	Cool to warm (1957 Mekong Committee reratified as 1995 Mekong Commission)	470	787,800	Humid to tropical	Good example of resilience of agreement

Multilateral Working Group (Middle East)

Name	Riparian states	Riparian relations	Average annual flow	Size (km²)	Climate	Special features
Nile	Burundi (3.1), Central African Republic, Congo, Democratic Rebulic of (Kinshasa), Egypt (111.5), Egypt, administered by Sudan (n/a), Eritrea (n/a), Ethiopia (7.5), Kenya (8.1), Rwanda (2.6), Sudan (37.3), Sudan, administered by Egypt, Tanzania, United Republic of (1.3), Uganda (0.6), Zaire (0.2)	Cold to warm (1959 Nile Water Agreement only includes Egypt and Sudan)	84	3,038,100	Dry to tropical	Scheduled as complex model/ workshop
Salween	China (19.3), Myanmar (Burma) (0.4), Thailand (32.1)	Cool to warm	122	244,000	Humid to tropical	Scheduled as conflict preclusion model/ workshop
Senegal	Mauritania, Mali, Senegal, Guinea			436,000	Dry to tropical	
Tigris–Euphrates	Iran (n/a), Iraq (86.3), Jordan (n/a), Saudi Arabia (n/a), Syria (102.0), Turkey (12.1)	Cool	46	789,000	Dry to Mediterranean	Ongoing tripartite dialog but no international agreement
United States–Mexico Aquifers (Groundwater)	Mexico (22.3), United States (21.7)	Warm (1944 Water Treaty, modified in 1979)	n/a	n/a	Dry	Groundwater not included in original treaty, leading to uncertainty in relations
West Bank Aquifers	Israel (95.6), Palestine (100.0)	Cool (1995 Interim Agreement)	n/a	n/a	Dry	Interim Agreement relegates groundwater allocations to future negotiations

[a] Values for lakes under "Annual Flow" are for storage volume.
[b] Source: Kulshreshtha (1993)
[c] Sources: Gleick (1993); UN Register of International Rivers (1978).
Other data: TFDD.

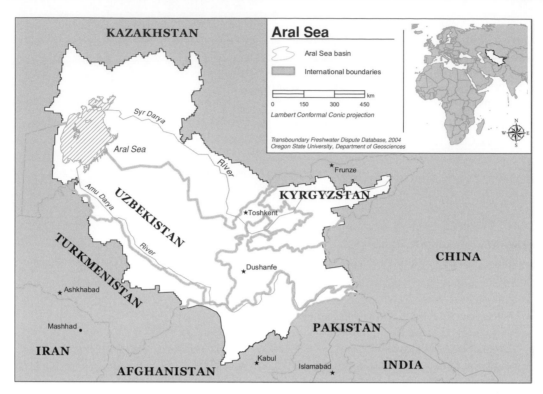

Figure C.1 Map of the Aral Sea and its tributaries, notably the Syr Darya and Amu Darya (*Source:* Transboundary Freshwater Dispute Database, 2004).

C.1 ARAL SEA

C.1.1 Case summary

River basin: Aral Sea and its tributaries, notably the Syr Darya and the Amu Darya (Figure C.1).

Dates of negotiation: Numerous agreements between 1992 and 2002 among various riparian states.

Relevant parties: Kazakhstan, Kyrgyzstan, Tajikistan, Turkmenistan, and Uzbekistan (directly); Afghanistan, Iran, and China (indirectly); Russia has been active observer.

Flash point: None.

Issues: Stated objectives: stabilize and rehabilitate watershed, improve management and build capacity of regional institutions.

Additional issues: Water-related: interstate and intersectoral allocation; Nonwater: general political relations between riparians.

Excluded issues: Transboundary oil pipelines.

Criteria for water allocations: Initially based on central planning for cotton self-sufficiency within the Soviet Union, now moving to "equitable use."

Incentives/linkage: Financial: Extensive funding from international community; Political: Facilitated relations between riparians.

Breakthroughs: Breakup of Soviet Union required development of international cooperation among newly independent Central Asian States.

Status: Initial agreements reached in 1992, 1993, with Program implementation beginning in 1995. Subsequent agreements were reached in 1997 and 2002. Some concerns about funding, legal overlap, priorities. Magnitude of environmental, health, and welfare problems remains extremely large.

C.1.2 Background

The Aral Sea was, until recently, the fourth largest inland body of water in the world. Its basin covers 1.8 million km^2, primarily in the independent republics of Kazakhstan, Kyrgyzstan, Tajikistan, Turkmenistan, and Uzbekistan, formerly part of the Soviet Union. Small portions of the basin headwaters are also located in Afghanistan, Iran, and China. The major sources of the Sea, the Amu Darya and the Syr Darya, are fed from glacial meltwater from the high mountain ranges of the Pamir and Tien Shan in Tajikistan and Kyrgyzstan.

Irrigation in the fertile lands between the Amu Darya and the Syr Darya dates back millennia, although the Sea remained in relative equilibrium until the early 1960s. At that time, the central planning authority of the Soviet Union devised the "Aral

Sea plan" to transform the region into the cotton belt of the USSR. Vast irrigation projects were undertaken in subsequent years, with irrigated area doubling between 1960 and 1990 (Aslov, 2003).

Such intensive cotton monoculture has resulted in extreme environmental degradation. Pesticide use and salinization, along with the region's industrial pollution, have decreased water quality, resulting in high rates of disease and infant mortality. Water diversions, sometimes totaling more than the natural flow of the rivers, have reduced the Amu Darya and the Syr Darya to relative trickles – the Sea itself has lost three-quarters of its volume and half its surface area, and salinity has tripled, all since 1960. The exposed seabeds are thick with salts and agricultural chemical residue, which are carried aloft by the winds as far as the Atlantic and Pacific oceans.

C.1.3 The problem

The environmental problems of the Aral Sea basin are among the worst in the world. Water diversions, agricultural practices, and industrial waste have resulted in a disappearing sea, salinization, and organic and inorganic pollution. The problems of the Aral, which previously had been an internal issue of the Soviet Union, became international problems in 1991. The five new major riparians – Kazakhstan, Kyrgyzstan, Tajikistan, Turkmenistan, and Uzbekistan – have been struggling since that time to help stabilize, and eventually to rehabilitate, the watershed.

C.1.4 Attempts at conflict management

The intensive problems of the Aral basin were internationalized with the breakup of the Soviet Union. Prior to 1988, both use and conservation of natural resources often fell under the jurisdiction of the same Soviet agency, but each often acted as powerful independent entities. In January 1988, a state committee for the protection of nature was formed, elevated as the Ministry for Natural Resources and Environmental Protection in 1990. The Ministry, in collaboration with the Republics, had authority over all aspects of the environment and the use of natural resources. This centralization came to an end with the collapse of the Soviet Union in 1991. Shortly after, the Interstate Coordination Water Commission (ICWC) was formed by the newly independent states to fill the regional planning void that accompanied the loss of Soviet central control.

In February 1992, the five republics negotiated an agreement to coordinate policies on their transboundary waters. Subsequent agreements in the 1990s and in 2002 have updated policies and reorganized transboundary water management institutions.

C.1.5 Outcome

The Agreement on Cooperation in the Management, Utilization, and Protection of Interstate Water Resources was signed on February 18, 1992 by representatives from Kazakhstan, Kyrgyzstan, Tajikistan, Turkmenistan, and Uzbekistan. The agreement calls on the riparians, in general terms, to coordinate efforts to "solve the Aral Sea crisis," including exchanging information, carrying out joint research, and adhering to agreed-to regulations for water use and protection. The agreement also establishes the Interstate Commission for Water Management Coordination to manage, monitor, and facilitate the agreement. Since its inception, the commission has prepared annual plans for water allocations and use, and defined water use limits for each riparian state.

In a parallel development, the Agreement on Joint Actions for Addressing the Problems of the Aral Sea and its Coastal Area, Improving of the Environment and Ensuring the Social and Economic Development of the Aral Sea Region was signed by the same five riparians on March 26, 1993. This agreement also established a coordinating body, the Interstate Council for the Aral Sea (ICAS), which has primary responsibility for "formulating policies and preparing and implementing programs for addressing the crisis." Each State's minister of water management is a member of the council. In order to mobilize and coordinate funding for the Council's activities, the International Fund for the Aral Sea (IFAS) was created in January 1993.

A long-term "Concept" and a short-term "Program" for the Aral Sea were adopted at a meeting of the Heads of Central Asian States in January 1994. The Concept describes a new approach to development of the Aral Sea basin, including a strict policy of water conservation. Allocation of water for preservation of the Aral Sea was recognized as a legitimate water use for the first time. The Program has four major objectives

1. Stabilize the environment of the Aral Sea
2. Rehabilitate the disaster zone around the Sea
3. Improve the management of international waters of the basin
4. Build the capacity of regional institutions to plan and implement these programs

These regional activities are supported and supplemented by a variety of governmental and nongovernmental agencies, including the European Union, the World Bank, UNEP, and UNDP.

In 1995 the Nukus Declaration was signed by heads of state of the Aral Sea basin nations and indicated the need for a "unified multi-sectoral approach and the development of cooperation amongst the states and with the international community" (McKinney, 1997). Despite this forward momentum, some concerns were raised about the potential effectiveness of these plans and institutions. Some have noted that not all promised funding has been forthcoming. Others (e.g., Dante Caponera, 1995) have noted duplication and inconsistencies in the agreements and warn that they seem to accept the concept of "maximum utilization" of the waters of the basin. Vinagradov (1996) has noted the legal problems inherent in these agreements, including some confusion between regulatory and development functions, especially between the commission and the council.

In 1998 the ICAS and IFAS were merged into a reorganized International Fund for the Aral Sea. The principal project goals and components of the IFAS were defined and to be implemented starting in 1998 as follows (Aral, 2001):

Component A "Water and Salt Management" prepares the integrated regional water and salt management strategy on the basis of national strategies.

Subcomponent A2 "Water Conservation Competition" disseminates the experience of farms, water users' associations, and rayon water management organizations in water conservation.

Component B "Public Awareness" educates the general public to conserve water and to accept burdensome political decisions.

Component C "Dam and Reservoir Management" raises reliability of operation and sustainability of dams.

Component D "Transboundary Water Monitoring" creates the basic physical capacity to monitor transboundary water flows and quality.

Component E "Wetlands Restoration" rehabilitates a wetland area near the Amu Darya delta (Lake Sudoche) and contributes to global biodiversity conservation.

Ever since its formation in 1998, IFAS has been under severe constraints and has had difficulties with its credibility and dealing with multisectoral issues. The organization was not very successful with its mandate at developing regional water management strategies. Because of this, the board of the IFAS did not meet until 2002, after a 3-year hiatus, when it came together to propose a new agenda (McKinney, 2003). Operational agreements and working sessions occurred frequently in the late 1990s among the riparians, and in 2002, Kazakhstan, Kyrgyzstan, Tajikistan, and Uzbekistan created the Central Asian Cooperation Organization (CACO) with a broad mandate to promote cooperation among member states on water, energy, and the environment. As of this writing, a secretariat still not been established, but one is being planned.

C.1.6 Lessons learned

A strong regional economic entity can provide support when issues arise between basin states.

The Central Asian Economic Community, now the Central Asian Cooperation Organization, played a key role in mediating between the Aral Sea basin states when there were difficulties within the International Fund for the Aral Sea. Even though regional economic entities sometimes may be too narrow in their interests, they can provide a stability that basin states may otherwise not have.

Lack of trust and credibility can hinder the process of cooperation.

It was apparent during the years of "dormancy" of the International Fund for the Aral Sea that issues of trust and credibility were having a severe effect on the functioning of the organization.

C.1.7 Creative outcomes resulting from resolution process

As a result the atmosphere gained from the heads of state of the Aral Sea basin nations and their recognition that the benefits of cooperation are much higher than that of competition, interstate water management has been coupled with broader economic agreements including trade of hydroelectric energy and fossil fuel to promote regional goals.

C.1.8 Time line

1960–1990 Soviet policy lead to large-scale environmental degradation in the region. Independence of Central Asian States and Interstate Coordination Water Commission formed. Agreement on Cooperation in the Management, Utilization, and Protection of Interstate Water Resources.

January 1993 International Fund for the Aral Sea established. Interstate Council for the Aral Sea formed.

March 1993 Agreement on Joint Actions for Addressing the Problems of the Aral Sea and its Coastal Area, Improving of the Environment and Ensuring the Social and Economic Development of the Aral Sea Region was signed by Kazakhstan, Kyrgyzstan, Tajikistan, Turkmenistan, and Uzbekistan.

January 1994 Long-term "Concept" and short-term "Program" were adopted in a meeting of the Heads of Central Asian States. Nukus Declaration signed by heads of states of the Aral Sea basin. Interstate Council for the Aral Sea and the International Fund for the Aral Sea merge under the name International Fund for the Aral Sea.

2002 Heads of state of the Aral Sea basin nations meet for first time after 3-year hiatus. Central Asian Cooperation Organization (CACO) established.

C.2 THE ENVIRONMENTAL PROGRAM FOR THE DANUBE RIVER

C.2.1 Case summary

River basin: Danube (Figure C.2).

Dates of negotiation: 1985–1994.

Relevant parties: All riparian states of the Danube: Albania, Austria, Bulgaria, Bosnia and Herzegovina, Croatia, the Czech Republic, Germany, Hungary, Italy, Moldova, Poland, Romania, Slovakia, Slovenia, Switzerland, Ukraine, Serbia. Convention is the first designed through the process of public participation, including nongovernmental organizations (NGOs), journalists, and local authorities

Flash point: None – good example of "conflict preclusion."

Issues: Stated objectives: to provide an integrated, basinwide framework for protecting Danube water quality.

Additional issues: Water-related: encourage communication among water-related agencies, NGOs, and individuals; Non-water: none.

Excluded issues: Strong enforcement mechanism.

Criteria for water allocations: None determined.

Incentives/linkage: World Bank/donor help with quality control.

Breakthroughs: No untoward barriers to overcome; creation of the Danube River Basin Strategy on Public Participation.

Status: Convention signed in 1994. Cooperation has continued to be fruitful and well managed.

C.2.2 Background

The Danube River basin is the heart of central Europe and is Europe's second longest river, at a length of 2,857 km. The drainage basin drains 817,000 km^2 including all of Hungary, most of Romania, Austria, Slovenia, Croatia, and Slovakia; and significant parts of Bulgaria, Germany, the Czech Republic, Moldova, Serbia, and Ukraine. Bosnia, and Herzegovina, and small parts of Italy, Switzerland, Albania, and Poland are also included in the basin. The Danube River discharges into the Black Sea through a delta, which is the second largest wetland area in Europe.

The river is shared by a large and ever-growing number of riparian states that for decades were allied with hostile political blocs, some of which are currently locked in intense national disputes. As a consequence, conflicts in the basin tended to be both frequent and intricate and their resolution especially formidable.

Nevertheless, in recent years, the riparian states of the Danube River have established an integrated program for the basinwide control of water quality, which, if not the first such program, has claims to probably being the most active and the most successful of its scale. The Environmental Program for the Danube River is also the first basinwide international body that actively encourages public and NGO participation throughout the planning process, which, by diffusing the confrontational setting common in planning, may help preclude future conflicts both within countries and internationally.

As an example of international basinwide watershed management, the process that led to the development of the Environmental Program for the Danube River merits a detailed description.

C.2.3 The problem

Prior to World War II, the European Commission of the Danube – with roots dating back to the 1856 Treaty of Paris and made up of representatives from each of the riparian countries – was responsible for administration of the Danube River. The primary consideration at the time was navigation, and the Commission was successful at establishing free navigation along the Danube for all European countries. World War II resulted in new political alliances for the riparians, resulting in a new management approach. At a 1948 conference in Belgrade, the East Bloc riparians – a majority of the delegates – shifted control over navigation to the exclusive control of each riparian. This Belgrade Convention also gave the Commission semi-legislative powers but only regarding navigation and inspection.

The main task of the Danube Commission has historically been to assure navigation conditions along the river. In addition, the Commission has developed regional plans for river projects, dissemination of country proposals to all the riparians for comment, and developing unified systems for regulations, channel marking, and data collection. The Commission meets once a year or in special session and, though a majority vote is sufficient to pass a proposal, in practice unanimity is solicited. The Commission has no sovereign powers and its decisions take the form of recommendations to the governments of its members.

By the mid-1980s, it became clear that issues other than navigation were gaining in importance within the Danube basin, notably problems with water quality. The Danube passes by numerous large cities, including four national capitals (Vienna, Bratislava, Budapest, and Belgrade), receiving the attendant waste of millions of individuals and their agriculture and industry. In addition, thirty significant tributaries have been identified as "highly polluted." The breakup of the USSR has also contributed to water quality deterioration, with nascent

Figure C.2 Map of the Danube River basin (*Source*: Transboundary Freshwater Dispute Database, 2004).

economies finding few resources for environmental problems, and national management issues being internationalized with redrawn borders. Recognizing the increasing degradation of water quality, the eight (at that time) riparians of the Danube signed the "Declaration of the Danube Countries to Cooperate on Questions Concerning the Water Management of the Danube," commonly called the Bucharest Declaration, in 1985. This would lead, in turn, to the 1994 Danube River Protection Convention.

C.2.4 Attempts at conflict management

World War II resulted in new political alliances for the riparians, which resulted in a new management approach. At a 1948 conference in Belgrade, the East Bloc riparians – a majority of the delegates – shifted control over navigation to the exclusive control of each riparian. By the 1980s, though, quality considerations brought the Bucharest Declaration, which reinforced the principle that the environmental quality of the river depends on the environment of the basin as a whole, and committed the riparians to a regional and integrated approach to water basin management, beginning with the establishment of a basinwide unified monitoring network. Basinwide coordina-

tion was strengthened at meetings in Sofia in September 1991, in which the riparians elaborated on a plan for protecting the water quality of the Danube. At that meeting, the countries and interested international institutions met to draw up an initiative to support and reinforce national action for the restoration and protection of the Danube River. With this initiative, named the Environmental Program for the Danube River Basin, the participants agreed that each riparian would

- Adopt the same monitoring systems and methods of assessing environmental impact
- Address the issue of liability for cross-border pollution
- Define rules for the protection of wetland habitats
- Define guidelines for development so that areas of ecological importance or aesthetic value are conserved

The meeting also agreed to create an interim Task Force to coordinate efforts, while a convention to steer the program was being negotiated. Members of the Task Force include the Danube countries of Austria, Bulgaria, Croatia, the Czech Republic, Germany, Hungary, Moldova, Romania, Slovakia, Slovenia, and Ukraine; the European Commission (EC), European Bank for Reconstruction and Development (EBRD), European Investment Bank (EIB), Nordic Investment Bank, the

United Nations Development Programme (UNDP), the United Nations Environment Programme (UNEP), the World Bank, the Netherlands, the United States; and NGOs: World Conservation Union (WCU), World Wide Fund for Nature (WWF), the Regional Environmental Centre, and the Barbara Guntlett Foundation.

The interim Task Force first met in Brussels in February 1992. At that meeting, a Program Work Plan was adopted that listed a series of actions and activities necessary to strengthen coordination between the governments and NGOs involved. Although the Commission of European Communities (G-24 Coordinator) has overall responsibility for coordinating the plan, a Program Coordination Unit was established and given the task of supporting the Task Force, monitoring and coordinating Program Work Plan action, and providing support to the financing partners to implement funds made available. Two "expert subgroups" were also established – one responsible for establishing an early warning system for environmental accidents, and one for data management.

Along with the institutional details, the Environmental Program also established several key principles for coordination and participation, which make it unique in integrated planning on this scale. Although the Program's work plan describes its overall strategy in terms fairly common in watershed management – "to provide an operational basis for strategic and integrated management of the Danube River basin environment while focusing initially on priority environmental issues" – specific strategic principles add a new dimension: "The approach should protect and enhance environmental values and promote a mix of actions in the public and private sectors. In addition, the strategy should be integrated, participatory, and coordinated."

In establishing the principles of "integration" and "coordination," the Plan started along the same approach as the Mekong Committee 40 years earlier – that internal issues within each nation are not particularly amenable to international management and that the most important contribution a unit responsible for integrated planning can make is to coordinate between the national representatives and between nations and donor organizations. The Danube Environmental Program went one crucial step further, though, by including the principle of "participation." This inclusion explicitly recognizes the vital link between internal politics among different sectors and political constituents within a nation on the one hand, and the strength and resilience of an agreement reached in the international realm, on the other.

C.2.5 Outcome

The principle of "participation" has been taken seriously in the work of the Environmental Program and the Coordination Unit.

Initially, each riparian country was responsible for identifying two individuals to help coordinate activity within the basin. The first, a "country coordinator," usually a senior official, would act as liaison between the work of the program and the country's political hierarchy. The second, a "country focal point," would coordinate the actual work plan being carried out.

In July 1992, the coordination unit held a workshop in Brussels to help facilitate communication between the coordinators, the focal points, and the donor institutions. Representatives from each of the (by then) eleven riparians and fifteen donor and nongovernmental organizations attended. An important outcome of the workshop was that the participants themselves designed a plan for each issue covered. One issue, for example, was an agreement to produce National Reviews of data availability and priority issues within each country. The information would be used by prefeasibility teams funded by donors who were to identify priority investments in the basin. During the workshop, participants developed the criteria for the National Reviews and agreed on a schedule for completion.

The principle of participation was carried one level deeper at the third Task Force meeting in October 1993 in Bratislava. At that meeting, the Task Force agreed to prepare a "Strategic Action Plan" (SAP) for the Danube basin, with the provision that, "consultation procedures should be strengthened." In moving from planning to implementation, it was determined, the proposed Strategic Action Plan should include the following concerns, raised during informal consultations between members of the Coordination Unit and riparian countries:

- Measures detailed must be "concrete" and aim to achieve results in the short term.
- Major environmental threats to the basin must be clearly addressed with realistically costed actions and constraints for problem solving together with proposals for overcoming them.
- The SAP should be updated regularly to allow amendments and additions as circumstances develop.
- Wide consultation during preparation of the SAP is desirable, in particular with parties who would be responsible for its implementation.

This last point is particularly noteworthy because it is the first time public participation has been required during the development of an international management plan. This concept rejects the principle that internal politics within nations ought to be treated as a geopolitical "black box," whose workings are of little relevance to international agreements, and instead embraces the vital need for input at all levels in order for a plan to ensure that the plan has the support of the people who will be affected by its implementation.

The eleven-member drafting group that was identified to prepare the Strategic Action Plan included representatives of four riparian countries, Austria, Hungary, Bulgaria, and Romania, each of whom were also to represent bordering nations. The World Bank, UNDP, and the Danube Environmental Coordination Unit also provided individuals to work on the drafting group.

During late 1993 and early 1994, another major Danube River activity was being carried out in the basin. At the same time that the Danube Environmental Program was developing the Strategic Action Plan for the Danube River basin, the riparian countries were developing the Convention on Cooperation for the Protection and Sustainable Use of the River Danube (the Danube River Protection Convention), which is aimed at achieving sustainable and equitable water management in the basin.

When the drafting group for the Strategic Action Plan held its first meeting in Vienna in January 1994, members agreed that the SAP should be designed as a tool to support implementation of the new Danube Convention that the riparian countries were planning to ratify in June 1994. During the first drafting group meeting, a schedule was drawn up for the drafting and adoption of the Danube Strategic Action Plan. Public consultation was built into the process from the beginning.

The public consultation process consisted of two steps:

1. Each of the nine downstream riparian countries was requested to designate a "country facilitator," whose task would be to facilitate a public consultation meeting. This individual was to ensure that public input was solicited and then fed back to the drafting group for possible incorporation into the SAP.
2. In order to guarantee a level of uniformity in the process, a "training of trainers" workshop was held in Vienna in February 1994.

The proposed audience to each of these consultation meetings consisted of thirty to thirty-five people, including representatives from the following institutions (with the ideal number from each in parentheses): government ministries, including environment (3), water (1), forestry (1), tourism (1), agriculture (1), industry (1), finance (1), health (1), transportation/navigation (1); mayors of municipalities and managers of public utilities involved in basin studies (2 from each basin study area); consultants from private sector firms who have worked on basin studies or other Danube-related activities (2); managers of research institutions or organizations responsible for monitoring laboratories and data collection (3); managers of large enterprises that have a stake in the results of the Strategic Action Plan (3); Danube-focused NGO representatives, to be coordinated with the NGO Danube

Forum (3); and environmental journalists – representatives of the mass media who have reported on Danube issues in the past (3).

In principle, the individuals who participated in the workshops would form a nucleus, which would not only have input in the drafting of a SAP, but would be involved in reviewing future activities that would be implemented as part of the Plan.

By July 1994, two consultation meetings were held in each of the nine countries. The first round of meetings, held in March 1994, described the purpose of the proposed Strategic Action Plan and sought input on major issues facing the basin. The second round, held during June 1994, solicited comments on the first draft of the SAP. A training-of-trainers workshop also preceded the second round of consultation workshops. Following the public consultation meetings, the country facilitators each prepared a workshop report containing recommendations for the drafting group. A number of revisions have been incorporated into the SAP in response to recommendations from the consultation process.

On June 29, 1994, in Sofia, the Danube River basin countries and the European Union signed the Convention on Cooperation for the Protection and Sustainable Use of the Danube River (the Danube River Protection Convention). The Convention notes that the riparians of the Danube, "concerned over the occurrence and threats of adverse effects, in the short or long term, of changes in conditions of watercourses within the Danube River Basin on the environment, economies, and well-being of the Danubian States," agree to a series of actions, including

- striving to achieve the goals of a sustainable and equitable water management, including the conservation, improvement and the rational use of surface waters and groundwater in the catchment area as far as possible
- cooperating on fundamental water management issues and take all appropriate legal, administrative, and technical measures, to at least maintain and improve the current environmental and water quality conditions of Danube River and of the waters in its catchment area and to prevent and reduce as far as possible adverse impacts and changes occurring or likely to be caused
- setting priorities as appropriate and strengthening, harmonizing, and coordinating measures taken and planned to be taken at the national and international level throughout the Danube basin aiming at sustainable development and environmental protection of the Danube River

The Danube Convention was a vital legal continuation of a tradition of regional management along the Danube, which dated back 140 years. As a political document, it provided a legal framework for integrated watershed management and environmental protection along a waterway with tremendous potential for conflict.

The Strategic Action Plan of the Environmental Program for the Danube River Basin provided the direction and a framework for achieving the goals of regional integrated water management and riverine environmental management expressed in the Danube River Protection Convention. It also aimed to provide a framework in support of the transition from central management to a decentralized and balanced strategy of regulation and market-based incentives. The SAP laid out strategies for overcoming the water-environment-related problems in the Danube River basin. It set targets to be met within 10 years and defined a series of actions to meet them.

The Action Plan addressed the officials of national, regional, and local levels of government who share responsibility for implementing the Convention and the national environmental action programs under the Lucerne Environmental Action Plan for Central and Eastern Europe. Industry, agriculture, citizen-based organizations, and the public also had important roles to play. The regional strategies set out in the Action Plan were intended to support national decision making on water management and on the restoration and protection of vulnerable and valuable areas in the Danube River basin. The Action Plan supports the process of cooperation and collaboration set out in the Convention to address transboundary problems. It will be revised and developed to take into account changing environmental, social, and economic conditions in the basin.

The Task Force formally adopted the Strategic Action Plan on October 28, 1994, and Ministers of Environment or Water or the Ministers' designates signed a Ministerial Declaration supporting the Strategic Action Plan in Bucharest on December 6, 1994.

The SAP describes a framework for regional action, which has been implemented through National Action Plans. It contains three goals for the environment of the Danube River basins (1) strategic directions, including priority sectors and policies; (2) a series of targets within a time frame; and (3) a phased program of actions to meet these targets. These goals concern the improvement of aquatic ecosystems and biodiversity in the Danube River basin and the reduction of pollution loads entering the Black Sea; maintaining and improving the quantity and quality of water in the Danube River basin; control of damage from accidental spills; and the development of regional cooperation in water management. These goals can only be achieved by means of intergrated and sustainable management of the waters of the Danube River basin. A number of short- and medium-term targets have been identified in the National Action Plans to reach the four goals of the SAP to be achieved within three and ten years, respectively.

The public participation and collaborative problem solving approach used in the development of the Strategic Action Plan (SAP) significantly shortened the time of preparation and approval. The SAP was addressed to the officials of national, regional, and local levels of government who share responsibility for implementing the Danube River Protection Convention and the national environmental action programs under the Environmental Action Programme for Central and Eastern Europe. Industry, agriculture, nongovernment organizations, and the public will play important roles. The regional strategies set out in the SAP were intended to support national decision making on water management, and on the restoration and protection of vulnerable and valuable areas in the Danube River basin.

The degree of cooperation among representatives of participating governments and the importance given to public participation in developing the SAP mark significant achievements in promoting regional cooperation in water resources management. Ultimately, the success of this process would be revealed by the degree to which the goals, strategies, and targets set in the agreement are implemented "on the ground." It is one thing to agree to goals and targets in time frames; it is another thing to, for example, agree to shut down a polluting factory, or to create and enforce industrial wastewater pretreatment standards, or develop rigorous monitoring and enforcement regimes. Additionally, because agreement signatories are at the Ministerial level in the water sector (versus at the level of the Foreign Minster), it is not clear if the agreement has the force of an international treaty behind it.

In the years just before the ratification of the Danube River Convention, the riparian states of the Danube River extended the principle of integrated management, and established a program for the basinwide control of water quality, which, if not the first such program, has claims to probably being the most active and the most successful of its scale. The Environmental Program for the Danube River was also the first basinwide international body that actively encouraged public and NGO participation throughout the planning process, which, by diffusing the confrontational setting common in planning, may help preclude future conflicts both within countries and, as a consequence, internationally.

In 1996, the Task Force and the basin countries approved the concept of a Strategic Action Plan Implementation Programme (SIP). The SIP marked the end of SAP activities per se by collecting, evaluating, and analyzing information collected by the SAP, and its activities are seen as the implementation of SAP findings. Its activities focused on six fields: Contamination and Human Health, Sustainable Land Use, Wetlands and Nature Conservation, Sustainable Use of Water Resources, Institutional Capacity Building and Basinwide Projects. It is considered to have exponentially increased the level of international cooperation on the Danube River basin.

The Danube Pollution Reduction Programme (DPRP) was created in 1997 with the support of the UNDP Global Environmental Fund. The goal of the DPRP was to define transboundary measures and actions and to develop an investment program for national, regional, and international cooperation to control and reduce water pollution and nutrient loads in the Danube River and its tributaries with effects to Black Sea ecosystems.

The International Commission for the Protection of the Danube River (ICPDR), mandated by the Danube River Convention, is the overarching management group for cooperation over the basin. Two of its committees, the International Planning Steering Group (IC/STG) and the International Commission Plenary (IC/PLN), met seven times between 1998 (when the Convention entered into force) and 2002 (ICPDR). Examples of the meetings include

1999, Second Meeting IC/PLN adopts the Danube Black Sea Memorandum of Understanding.

2000, Third Meeting IC/PLN approves the Joint Action Programme for the Danube River Basin, January 2001–December 2005; Water Framework Directive becomes highest priority; Danube Watch magazine begins publication.

2001, Sixth Meeting IC/STG establishes an Expert Sub-Group for "Cartography and GIS."

2001, Fourth IC/PLN Meeting agrees to revoke the contributions of Moldova for 1999, 2000, and 2001 due to its very difficult economic situation. Moldova agrees to begin payments in 2002.

In 2003, the ICPDR set out to define the Danube River Basin Strategy for Public Participation in accordance with the 2000 EU Water Framework Directive (WFD). This move is a breakthrough in cooperation over international river basins. The importance of public participation in river basin development decisions is well understood by water resource management bodies, but the ICPDR's attempt at formulating a detailed strategy is the first of its kind.

The Strategy's objectives are

• to ensure public participation (PP) in WFD implementation in the Danube River basin (DRB), especially in the first instance concerning the development of the Danube River Basin Management Plan (RBMP)
• to facilitate the establishment of effective structures and mechanisms for PP in the DRB that will continue operating beyond the first cycle of RBM planning
• to provide guidance to national governments on how to comply with their obligations under the WFD by providing them with practical support and guidance in addressing PP in RBM planning

• to inform other key stakeholders about appropriate PP activities and structures at the different levels (Danube River Basin Strategy for Public Participation in River Basin Management Planning 2003–2009, 2003, p. 3)

The strategy emphasized that public participation must to start immediately (2003), so that future management plans could be based on commonly supported initiatives. This meant that it was a work in progress, but a good model on which other large, diverse river basins' management teams could base its own public participation strategies.

It is structured according to the Water Framework Directive requirement of four levels of public participation that are necessary to obtain valuable comprehensive input

International level: among the basin countries
National level: deals with the implementation strategies and management plans
Subbasin level: various pilot projects at different parts of the basin
Local level: where the WFD is actually implemented

Each phase of the strategy contains activities at each level of participation. For example, in the Preparatory Phase (2003–2004), activities at the international level concentrate on cooperation and organizational analysis of ICPDR with regard to public participation. Activities at the national level focus on the establishment of government structures to coordinate public participation. At each level potential stakeholders are defined by subbasin, village, and/or economic group, and trainings on the theory, implementation, and responsibility for engaging in public participation will be held for management officials from high level, ministerial conferences to trainings for local water providers.

At the international level, Phase One (2004–onward) of the strategy emphasizes the dissemination of information about public participation to all stakeholders through the improvement of Web pages dealing with the Danube, the organization of hearings for all interested parties and the declaration of June 29 as "Danube Day," as well as the creation of a structure within the ICPDR to facilitate public participation. Activities at the national, regional, and local levels in Phase One involve analysis of the local environmental situation, development of action plans and the creation of monitoring and evaluation mechanisms.

Phase Two (2004–onward) is designed to assess activities in Phase One and make adjustments to the original strategy. Phase Three (2004–onward) activities focus on implementing the adjustments needed (as defined in Phase Two), such as developing regional frameworks for water councils, the integration of key stakeholders into discussions on program objectives. In Phase Four (2005–onward), the revision of dissemination materials will continue, evaluations of public participation will be made and feedback mechanisms created.

C.2.6 Lessons learned

Public participation within the management of an international river basin can facilitate greater cooperation between nations with regard to its water resources.

The use of public participation within the Strategic Action Plan of the International Commission for the Protection of the Danube River (ICPDR) since its inception in 1994 has permitted the basin states of the Danube to move forward rather quickly with several initiatives.

C.2.7 Creative outcomes resulting from resolution process

Public participation included early in the decision-making processes can help facilitate cooperation and prevent conflict over the management of international waters.

C.2.8 Time line

1856 Treaty of Paris establishes European Commission of the Danube.

1948 Belgrade Convention signed giving control of navigation exclusively to each riparian nation. Convention also gives European Commission of the Danube semilegislative powers but only with regards to navigation and inspection.

1985 Declaration of the Danube Countries to Cooperate on Questions Concerning the Water Management of the Danube signed, more commonly known as the Bucharest Declaration.

1991 Environmental Program for the Danube River Basin and its Task Force, created after meetings of the riparian states in Sofia, Bulgaria. First use of participation to assist cooperation.

February 1992 First Interim Task Force meeting in Brussels; Program Work Plan developed and Program Coordination Unit created.

July 1992 Coordination Unit holds workshops held in Brussels to help facilitate communication among coordinators, focal points, and donor institutions.

1993 Third Task Force meeting in Bratislava creates the Strategic Action Plan and requires community participation in developing the international management plan.

1993–1994 Development of the Convention on Cooperation for the Protection and Sustainable Use of the River Danube (the Danube River Protection Convention).

January 1994 First Strategic Action Plan meeting.

March–June 1994 Two consultation meetings held in each country describing the purpose of the Strategic Action Plan and to solicit comments on the first draft of the SAP.

June 1994 Danube River Protection Convention signed.

October 1994 Task Force officially adopts Strategic Action Plan.

December 1994 Ministers of Environment or Water or their designees or each riparian state sign Ministerial Declaration supporting the Strategic Action Plan.

1996 Basin states approve the Strategic Action Plan Implementation Programme.

1997 Danube Pollution Reduction Programme created.

1998 Danube River Protection Convention comes into force.

2000 European Union Water Framework Directive requires public participation in river basin management planning.

2003 Danube River Basin Strategy for Public Participation is implemented.

2004 First annual Danube Day.

C.3 GANGES RIVER CONTROVERSY

C.3.1 Case summary

River basin: Ganges River (Figure C.3).
Dates of negotiation: 1960–Present.
Relevant parties: Pre-1971, India, Pakistan; Post-1971, India, Bangladesh.
Flash point: India builds and operates Farakka Barrage diversion of Ganges water without long-term agreement with downstream Bangladesh.
Issues: Stated objectives: negotiate an equitable allocation of the flow of the Ganges River and its tributaries among the riparian states; develop a rational plan for integrated watershed development, including supplementing Ganges flow.
Additional issues: Water-related: appropriate source for supplementing Ganges flow; amount of data necessary for decision making; Indian upstream water development; flood hazards mitigation; management of coastal ecosystems; Nonwater: appropriate diplomatic level for negotiations.
Excluded issues: Other riparians, notably Nepal, until recently.
Criteria for water allocations: Percentage of flow during dry season.
Incentives/linkage: Financial: none; Political: none.
Breakthroughs: Minor agreements reached, but no long-term solution.
Status: Short-term agreements reached in 1977, 1982, and 1985. Treaty signed in 1996.

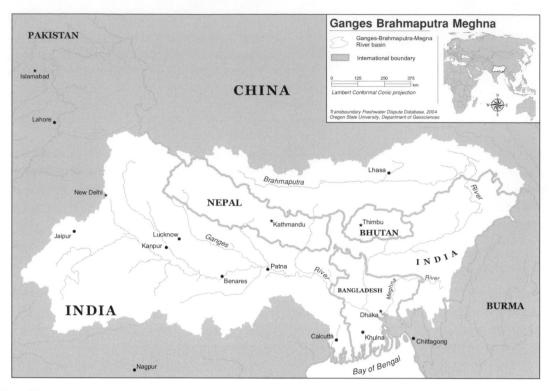

Figure C.3 Map of the Ganges–Brahmaputra–Meghna basins (*Source:* Transboundary Freshwater Dispute Database, 2004).

C.3.2 Background

Although blessed with an abundance of water resources, much of the management problems of the Indian subcontinent come about from the dramatic seasonal variations in rainfall. This management problem is compounded with the creation of new national borders throughout the region. So, too, the problems that have developed between India and Bangladesh, initially India and Pakistan, over the waters of the Ganges River.

The headwaters of the Ganges and its tributaries lie primarily in Nepal and India, where snow and rainfall are heaviest. Flow increases downstream even as annual precipitation drops, as the river flows into Bangladesh, pre-1971 the eastern provinces of the Federation of Pakistan, and on to the Bay of Bengal.

On October 29, 1951, Pakistan officially called Indian attention to reports of Indian plans to build a barrage at Farakka, about 17 kilometers from the border. The barrage would reportedly divert 40,000 cusecs or cubic feet per second (cubic feet per second = 0.0283 cubic meters per second; because all negotiations were in English units, that is what is reported here) out of a dry season average flow of 50,000 cusecs from the Ganges into the Bhagirathi–Hooghly tributary, to provide silt-free flow into Calcutta Bay, which would improve navigability for the city's port during dry months and keep saltwater from the city's water supply. On March 8,

1952, the Indian government responded that the project was only under preliminary investigation, and that concern was "hypothetical."

Over the next years, Pakistan occasionally responded to reports of Indian plans for diversion projects of the Ganges, with little Indian response. In 1957, and again in 1958, Pakistan proposed that

- the advisory and technical services of a United Nations body be secured to assist in planning for the cooperative development of the eastern river systems
- the projects in the two countries be examined jointly by experts of the two countries before their implementation
- the Secretary-General of the UN be requested for the appointment of an engineer or engineers to participate in the meetings at experts level
- India turned down these proposals, although it was agreed that water resources experts of the two countries should, "exchange data on projects of mutual interests." These expert-level meetings commenced June 28, 1960.

C.3.3 The problem

The problem over the Ganges is typical of conflicting interests of upstream and downstream riparians. India, the upper riparian, developed plans for water diversions for its own irrigation, navigability, and water-supply interests. Initially Pakistan, and later Bangladesh, had interests in protecting the historic flow of the river for its own downstream uses. The potential clash between upstream development and downstream historic use set the stage for attempts at conflict management.

C.3.4 Attempts at conflict management

The first round of expert-level meetings between India and Pakistan was held in New Delhi from June 28 to July 3, 1960, with three more to follow by 1962. While the meetings were still in progress, India informed Pakistan on January 30, 1961, that construction had begun on the Farakka Barrage. A series of attempts by Pakistan to arrange a meeting at the level of minister was rebuffed with the Indian claim that such a meeting would not be useful, "until full data are available." In 1963, the two sides agreed to have one more expert-level meeting to determine what data was relevant and necessary for the convening of a minister-level meeting. The meeting at which data needs were to be determined, the fifth round at the level of expert, was not held until May 13, 1968. After that meeting, the Pakistanis concluded that agreement on data, and on the conclusions that could be drawn, was not possible, but that enough data were nevertheless available for substantive talks at the level of minister. India agreed only to a series of meetings at the level of secretary, in advance of a minister-level meeting.

These meetings, at the level of secretary, commenced on December 9, 1968, and a total of five were held in alternating capitals through July 1970. Throughout these meetings, the different strategies became apparent. As the lower riparian, the Pakistani sense of urgency was greater, and their goal was "substantive talks on the framework for a settlement for equitable sharing of the Ganges waters between the two countries." India in contrast, whether actually, or as a stalling tactic, professed concern at data accuracy and adequacy, arguing that a comprehensive agreement was not possible until the data available was complete and accurate.

At the third secretaries' level meeting, Pakistan proposed that an agreement should provide for

- guarantee to Pakistan of fixed minimum deliveries of the Ganges waters on a monthly basis at an agreed point
- construction and maintenance of such works, if any, in India as may be necessary in connection with the construction of the Ganges Barrage in Pakistan
- setting up of a permanent Ganges Commission to implement the agreement
- machinery and procedure for settlement of differences and disputes consistent with international usages

India again argued that such an agreement could only take place after the two sides had agreed to "basic technical facts."

The fifth and final secretaries-level meeting was held in New Delhi from July 16–21, 1970, resulting in three recommendations (1) the point of delivery of supplies to Pakistan of such quantum of water as may be agreed on will be at Farakka; (2) constitution of a body consisting of one representative from each of the two countries for ensuring delivery of agreed supplies at Farakka is acceptable in principle; (3) a meeting would be held in 3 to 6 months time at a level to be agreed to by the two governments to consider the quantum of water to be supplied to Pakistan at Farakka and other unresolved issues relating thereto and to eastern rivers, which have been subject matter of discussions in these series of talks. Little of practicality came out of these talks, and India completed construction of the Farakka Barrage in 1970. Water was not diverted at the time, though, because the feeder canal to the Bhagirathi–Hooghly system was not yet completed.

Bangladesh came into being in 1971, and by March 1972, the governments of India and Bangladesh had agreed to establish the Indo-Bangladesh Joint Rivers Commission, "to develop the waters of the rivers common to the two countries on a cooperative basis." The question of the Ganges, however, was specifically excluded, and would be handled only between the two prime ministers.

Leading up to a meeting between prime ministers was a meeting at the level of minister from July 16–17, 1973, where the two sides agreed that a mutually acceptable solution to issues around the Ganges would be reached before operating the Farakka Barrage, and a meeting between foreign ministers from February 13–15, 1974, at which this agreement was confirmed. The prime ministers of India and Bangladesh met in New Delhi from May 12–16, 1974 and, in a declaration on May 16, 1974, they observed that during the periods of minimum flow in the Ganges, there may not be enough water for both an Indian diversion and Bangladeshi needs; agreed that during low flow months, the Ganges would have to be augmented to meet the requirements of the two countries; agreed that determining the optimum method of augmenting Ganges flow should be turned over to the Joint Rivers Commission; and expressed their determination that a mutually acceptable allocation of the water available during the periods of minimum flow in the Ganges would be determined before the Farakka project is commissioned.

There were two general approaches to augmenting Ganges flow presented to the Commission, which defined the negotiating stance for years: (1) augmentation through storage facilities within the Ganges basin, proposed by Bangladesh, and (2) augmentation through diversion of water from the Brahmaputra to the Ganges at Farakka by a link canal, proposed by India.

Through a series of five Commission meetings between June 1974 and January 1975, and one minister-level meeting in April 1975, the two sides developed the positions detailed below.

BANGLADESH POSITION

- There is adequate storage potential of monsoon flow in the Ganges basin for Indian needs.
- There is additional storage along the headwaters of the Ganges tributaries in Nepal, and that country might be approached for participation.
- A feeder canal from the Brahmaputra to the Ganges is both unnecessary and would have detrimental effects within Bangladesh, not least of which would be massive population resettlement.
- Indian needs would be better met through amending the pattern of diversion of Ganges water into the Bhagirathi–Hooghly, and constructing a navigation link from Calcutta to the sea via Sunderban.

INDIA POSITION

- Additional storage possibilities in India are limited, and not sufficient to meet Indian development needs.
- The most viable option both to supplement the low flow of the Ganges, and for regional development, is a link canal and storage facilities on the Brahmaputra, to be developed in stages for mutual benefit.
- Approaching Nepal or other third countries is beyond the scope of the Commission, as is discussing amending the pattern of diversion into the Bhagirathi–Hooghly.
- Constructing a separate navigation canal is not connected to the question of optimum development of water resources in the region.

At a minister-level meeting in Dhaka from April 16–18, 1975, India asked that, while discussions continue, the feeder canal at Farakka be run during that current period of low flow. The two sides agreed to a limited trial operation of the barrage, with discharges varying between 11,000 and 16,000 cusecs in 10-day periods from April 21 to May 31, 1975, with the remainder of the flow guaranteed to reach Bangladesh. Without renewing or negotiating a new agreement with Bangladesh, India continued to divert the Ganges waters at Farakka after the trial run, throughout the 1975–1976 dry season, at the full capacity of the diversion – 40,000 cusecs. There were serious consequences in Bangladesh resulting from these diversions, including desiccation of tributaries, salination along the coast, and setbacks to agriculture, fisheries, navigation, and industry.

Four more meetings were held between the two states between June 1975 and June 1976, with little result. In Jan-uary 1976, Bangladesh lodged a formal protest against India with the General Assembly of the United Nations which, on November 26, 1976, adopted a consensus statement encouraging the parties to meet urgently at the ministerial level for negotiations, "with a view to arriving at a fair and expeditious settlement." Spurred on by international consensus, the negotiations recommenced on December 16, 1976. At an April 18, 1977, meeting, an understanding was reached on fundamental issues, which culminated in the signing of the Ganges Waters Agreement on November 5, 1977.

C.3.5 Outcome

In principle, the Ganges Water Agreement covers

1. Sharing the waters of the Ganges at Farakka
2. Finding a long-term solution for augmentation of the dry season flows of the Ganges

Specific provisions, described as *not* establishing any general principles of law or precedent include (paraphrased)

Article I. The quantum of waters agreed to be released would be at Farakka.

Article II. The dry season availability of the historical flows was established from the recorded flows of the Ganges from 1948 to 1973 on the basis of 75 percent availabilities. The shares of India and Bangladesh of the Ganges flows at 10-day periods are fixed, the shares in the last 10-day period of April (the leanest) being 20,500 and 34,500 cusec respectively out of 55,000 cusec availability at that period.

In order to ensure Bangladesh's share in the event of any lower availability at Farakka, Bangladesh's share should not fall below 80 percent of the stated share in a particular period shown in a schedule annexed to the agreement.

Article III. Only minimum water would be withdrawn between Farakka and the Bangladesh border.

Article IV–VI. Provision was made for a Joint Committee to supervise the sharing of water, provide data to the two governments, and submit an annual report.

Article VII. Provisions were made for the process of conflict resolution: The Joint Committee would be responsible for examining any difficulty arising out of the implementation of the arrangements of the Agreement.

Any dispute not resolved by the Committee would be referred to a panel of an equal number of Indian and Bangladeshi experts nominated by the two governments.

If the dispute is still not resolved, it would be referred to the two governments which would, "meet urgently at the appropriate level to resolve it by mutual discussion and failing that by such other arrangements as they may mutually agree upon."

Article VIII. The two sides would find out a long-term solution of the problem of augmentation of the dry season flows of the Ganges.

The Agreement would initially cover a period of 5 years. It could be extended further by mutual agreement. The Joint Rivers Commission was again vested with the task of developing a feasibility study for a long-term solution to the problems of the basin, with both sides reintroducing plans along the lines described above. By the end of the 5-year life of the agreement, no solution had been worked out.

In the years since, both sides and, more recently, Nepal, have had years of greater and less success at reaching toward an agreement. Since the 1977 accord:

- A joint communiqué was issued in October 1982, in which both sides agreed not to extend the 1977 agreement, but would rather initiate fresh attempts to achieve a solution within 18 months – a task not accomplished.
- An Indo-Bangladesh Memorandum of Understanding was signed on November 22, 1985, on the sharing of the Ganges dry season flow through 1988, and establishing a Joint Committee of Experts to help resolve development issues. India's proposals focused on linking the Brahmaputra with the Ganges, while Bangladesh's centered on a series of dams along the Ganges headwaters in Nepal. Although both the Joint Committee of Experts and the Joint Rivers Commission met regularly throughout 1986, and although Nepal was approached for possible cooperation, the work ended inconclusively.
- The prime ministers of Bangladesh and India discussed the issue of river water-sharing on the Ganges and other rivers in May, 1992, in New Delhi. Each directed their ministers to renew their efforts to achieve a long-term agreement on the Ganges, with particular attention to low flows during the dry season. Subsequent to that meeting, there has been one minister-level and one secretary-level meeting, at which little progress was reportedly made.

Between 1988, when the last agreement lapsed, and 1996, no agreement was in place between India and Bangladesh. During this time, India granted Bangladesh only a portion of the flow of the Ganges, with no minimum flow guaranteed, and no special provisions for drought years. Each side kept roughly to its positions as stated above, with little room for compromise. Regional schemes were proposed, often providing benefits not

Table C.2 *Ganges River allocations*

Flow amount	India	Bangladesh
< 70,000 cusecs	50%	50%
70,000–75,000 cusecs	Balance of flow	35,000 cusecs
> 75,000 cusecs	40,000 cusecs	Balance of flow

Source: Ganges River Allocations

only to India and Bangladesh, but also to Nepal, landlocked but with tremendous hydropower potential that might be traded for access to the sea. In December 1996, a new treaty was signed between the two riparians, based generally on the 1985 accord, which delineates a flow regime under varying conditions.

The most notable change in the 1996 Ganges River Treaty is the establishment of a new formula for the distribution of Ganges waters from January 1 to May 31, the region's dry season, at Farraka Barrage. The following schedule is to be respected with regard to 10-day period flows (Table C.2).

If flows at Farakka Barrage should fall below 50,000 cusecs, the two governments will meet together to consult as to the appropriate actions taking into consideration "principles of equity, fair play and no harm to either party." The two governments are required by the treaty to review the sharing arrangements at 5-year intervals. If the parties are not able to come to agreement, India is to release no less than 90 percent of Bangladesh's flow at Farraka as stated by this schedule until a solution can mutually agreed on.

Although this agreement should help reduce regional tensions, issues such as extreme events and upstream uses are not covered in detail. Notably, Nepal, China, and Bhutan, not party to the treaty, have their own development plans that could impact the agreement. In addition, the treaty does not contain any arbitration clause to ensure that the parties uphold its provision.

The 1996 treaty was based on data about water discharges at the Farakka dam between 1949 and 1988. Since that time, however, increased upstream draws have significantly lowered the discharges and statistical analysis indicates that neither Bangladesh nor India will be able to withdraw their respective allocations (Mirza, 2002). The very first season following signing of the treaty, in April 1997, India and Bangladesh were involved in their first dispute over cross-boundary flow: water passing through the Farakka dam dropped below the minimum provided in the treaty, prompting Bangladesh to request a review of the state of the watershed. A study that simulated water availability under the 1977 and 1996 treaties concluded that the newer treaty is unlikely to make any substantial contribution to alleviate water scarcity during the dry season in southwestern Bangladesh (Tanzeema, 2001).

In addition to the issue over low flows to Bangladesh during the dry season has been added that of India's Mega River Linking Project, a plan to link dozens of rivers throughout India by way of aqueducts and pumping stations to transport water from the Ganges River to parts of southern and eastern India that are prone to water scarcity. This project would exacerbate the issue of flows to Bangladesh and has the country very worried. India, acting unilaterally has up to this point not agreed to speak with Bangladesh regarding the topic (Pearce, 2003).

C.3.6 Lessons learned

Unequal power relationships, without strong third-party involvement, create strong disincentives for cooperation.

India, the stronger party both geostrategically and hydrostrategically, has little incentive to reach agreement with Bangladesh. Without strong third-party involvement, such as that of the World Bank between India and Pakistan on the Indus, the dispute has gone on for years.

Requests for increasingly detailed data clarifications can be an effective delaying tactic.

Agreeing on the minimum data necessary for a solution, or delegating the task of data-gathering to a third party may speed the pace of negotiations. India used the veracity and detail of data as an effective tactic in postponing a long-term solution with Bangladesh. Interestingly, India was able to surmount this problem on the Indus by stipulating that data could be used in an agreement, *without* agreeing to its accuracy.

Likewise, insisting on bilateral negotiations, as opposed to watershed-wide negotiations, favors the party with greater power.

India has insisted on separate negotiations with each of the riparians of its international rivers. It was thus able to come to arrangements with Nepal on Ganges tributaries without considering Bangladeshi needs.

Agreeing early on the appropriate diplomatic level for negotiations is an important step in the prenegotiation phase.

Much of the negotiations between India and Pakistan and, later, India and Bangladesh, were spent trying to resolve the question of what was the appropriate diplomatic level for negotiations.

Short-term agreements which stipulate that the terms are not permanent can be useful steps in long-term solutions. However, a mechanism for continuation of the temporary agreement in the absence of a long-term agreement is crucial.

Agreements on the distribution of Ganges waters have been short in duration, providing initial impetus for signing, but providing difficulties when they lapse.

C.3.7 Creative outcomes resulting from resolution process

The 1977 Ganges Waters Agreement was reached perhaps more quickly specifically as a short-term agreement, and specifying that it was not establishing any precedents.

C.3.8 Time line

October 29, 1951 Pakistan first calls Indian attention to reports of Indian plans to build a barrage at Farakka to divert Ganges water to Calcutta Bay. India responds that the project was only under preliminary investigation.

June 28, 1960 Meetings commence at level of "expert" between Pakistan and India to exchange data on regional projects.

1960–1968 Experts level meetings continue; there are five in all, most focusing on data issues.

January 30, 1961 India informs Pakistan that construction had begun on the Farakka Barrage.

1968–1970 Five meetings continue at the level of secretary. Fundamental disagreements over approaches to Ganges development and the data required to make policy decisions.

1970 India completes construction of Farakka Barrage.

1971 Bangladesh comes into being, replacing eastern Pakistan.

March 1972 India and Bangladesh establish Indo-Bangladesh Joint Rivers Commission, specifically excluding issues of Ganges development.

May 16, 1974 Prime ministers of India and Bangladesh sign a declaration agreeing to find a mutually acceptable solution to Ganges development, and to turn the question of the best way of supplementing Ganges flow over to the Joint Rivers Commission.

April 16, 1975 The two sides agree to a limited trial operation of the Farakka Barrage. India continues to divert Ganges water after the trial run, without renewing or negotiating a new agreement with Bangladesh.

June 1975–June 1976 Meetings continue, with little result.

January 1976 Bangladesh lodges a formal protest against India with the United Nations, which adopts a consensus statement encouraging the parties to meet urgently, at the level of minister, to arrive at a settlement.

November 5, 1977 Ganges Waters Agreement signed, covering allocation of Ganges water between the two riparians for a period of 5 years. No long-term solution was found within that time frame.

October 1982 Joint communiqué issued, pledging to resolve Ganges issues within 18 months, a task not accomplished.

November 22, 1985 Memorandum of understanding issued, on the sharing of Ganges dry season flow through 1988. When accord lapses, no new agreement is signed.

September 29, 1988 Summit in New Delhi between heads of government: Bangladesh Secretary of Irrigation and India's Secretary of Water Resources were given the task to work on an integrated formula for the sharing of common rivers between India and Bangladesh.

April 1990–February 1992 Secretaries' Committee met six times alternatively between Dhaka and New Delhi.

December 12, 1996 Ganges Water Treaty signed by the Prime Ministers of India and Bangladesh

1996–2004 Bangladesh's attempts to talk with India over agreements concerning seven rivers is met with noninterest.

December 2002 India announces plans for river linking project connecting rivers from north to those in the south and east.

C.4 GUARANÍ AQUIFER

C.4.1 Case summary

River basin: Guaraní Aquifer (Figure C.4).
Dates of negotiation: 2000–present.
Relevant parties: Argentina, Brazil, Paraguay, Uruguay.
Flash point: None.
Issues: Stated Objectives: Relevant parties to design and implement a coordinated management program for preserving and monitoring the Guaraní Aquifer for current and future use.
Criteria for water allocations: None determined
Incentives/linkage: Financial: Protection of aquifer is significantly less costly than remediating a polluted aquifer in the future; Political: None.
Breakthroughs: Four countries, along with support from the Global Environment Facility (GEF), agreed on Project for the Environmental Protection and Sustainable Development of the Guaraní Aquifer System in 2000.
Status: Ongoing design of international Water Management Framework for the Guaraní Aquifer.

C.4.2 Background

The Guaraní Aquifer is the largest groundwater resource in the world, with 45,000 km^3 of water and a surface area of 1.2 million km^2 (Organization of American States, 2004, p. 1; Valente, 2002, pp. 1–2. The transboundary aquifer is shared by Argentina, Brazil, Paraguay, and Uruguay. Table C.3 illustrates the distribution of the aquifer across the four nations and the relevant uses, environmental issues, and information relating to each nation and the aquifer.

C.4.3 The problem

The economic and social importance of the Guaraní Aquifer to the four riparians has spurred concern over the pollution and overexploitation of its groundwater, especially in the context of growing demand for freshwater resources in all four states. Although the level of pollution and use has not yet reached critical levels, the potential for future problems in these areas has led to immediate action and cooperation among the four states to develop an aquifer management strategy.

Additionally, the hydrothermal character of certain areas of the aquifer represents a resource for tourism as well as "clean energy" production. Considering that all four countries are in the process of economic development, and have also signed the Kyoto Protocol, access to the aquifer for these purposes could also be a source of conflict.

C.4.4 Attempts at conflict management

Considering the coordinated efforts of the four nations to implement an aquifer management program before significant problems with pollution and overuse could occur, there has not been any significant conflict over the shared groundwater resource to date. Additionally, these four states have a history of collaboration (for example, the Intergovernmental Committee for the La Plata River Basin and the MERCOSUR trade mechanism) rather than conflict in recent decades.

In order to prevent conflict in the future over the Guaraní aquifer, the four states have been involved in the GEF-funded Project for the Environmental Protection and Sustainable Development of the Guaraní Aquifer System. The US$27 million project (US$13M from the GEF and US$14M from participating countries, the Organization of American States, and other donors) includes five major areas to address the sustainable management of the aquifer: (a) expansion and consolidation of the current knowledge base, (b) joint development and implementation of a Guaraní Aquifer Management Framework, (c) public participation through an appropriate information and institutional framework, (d) implementation of measures to deal with nonpoint source pollution, and (e) monitoring and evaluation (Valente, 2002, p. 2; GEF, 2000, p. 7).

The Guaraní Aquifer Management Framework is one subsection of the much larger La Plata River Basin Integrated Water Resource Management Program. This program was

Figure C.4 Map of the Guaraní Aquifer (*Source:* Transboundary Freshwater Dispute Database, 2004).

originally developed to manage the surface waters contained in the La Plata watershed; however, the original program largely ignored the management of any groundwater resources, including the Guaraní Aquifer.

C.4.5 Outcome

Experts working on the GEF project have until 2007 to develop the plan for all four states to share management of the aquifer. All four states have signed on to the project and participated thus far in the design of the management and monitoring program for the Guaraní Aquifer. Other institutions who have participated in the process include: the World Bank (WB), the Global Environmental Facility (GEF), the Organization of American States (OAS), United Nations Environmental Programme (UNEP), United Nations Education, Scientific and Cultural Organization (UNESCO), the International Atomic Energy Association (IAEA), and the German Government (GEF, 2001, p. 7).

The current GEF/WB project will involve three major sectors: sustainable water resource management, trans-

boundary water management, and energy use. Sustainable water resource management will largely include institutional arrangements between stakeholders, investments in water infrastructure (for use and monitoring) construction and maintenance, as well as measures for pollution control and prevention (GEF, 2001, p. 2). Transboundary water management will be institutionalized by integrating the management framework for the Guaraní Aquifer into the existing framework for the management of the La Plata River basin (Mejia et al., 2004, slide 8). Finally, an initial assessment of the potential energy generation capacity (and tourism potential) of the hydrothermal sections of the basin will allow for the creation of a management strategy for energy use (GEF, 2001, p. 2).

So far, initial surveys of the aquifer have given more detailed information relating to the quantity of water, and the geography, distribution and use of the water, giving stakeholders and policymakers a better understanding of how the aquifer will need to be managed. In fact, some "hot spots" of pollution or overuse have been identified, and new management practices have been initiated in these areas. Additionally, the project has succeeded in raising awareness about the aquifer, which

Table C.3 *Current knowledge and importance of the Guaraní Aquifer in Argentina, Brazil, Uruguay, and Paraguay (Global Environment Facility, 2000, p. 4)*

	Argentina	Brazil	Paraguay	Uruguay
Approximate extension of the Guaraní Aquifer (km^2)	225,500	839,800	71,700	45,000
Surface of territory occupied by the aquifer (%)	5.9	9.8	17.6	25.3
Characteristics	Supply source	Recharge and supply area	Recharge and supply area	Recharge and supply area
Extent of exploitation	6 deep wells for thermal use; about 100 wells for drinking and irrigation	Between 300 to 500 cities partially or entirely supplied by the Guaraní Aquifer	About 200 wells	347 wells for public supply (250), irrigation (90), and thermal tourism (7)
Main environmental issue	1. Potentially uncontrolled drilling and extraction 2. Subject to pollution effects from other countries	1. Point and nonpoint source pollution 2. Uncontrolled drilling and extraction	1. Point and nonpoint source pollution 2. Uncontrolled drilling and extraction 3. Subject to pollution impact from other countries	1. Point and nonpoint source pollution 2. Uncontrolled drilling and extraction 3. Subject to pollution impact from other countries
Level of information	Limited information available	Considerable information available but dispersed in different states and institutions	Limited structured information available	Considerable information available

has resulted in increased international interest, forums for dialogue and the engagement of universities and NGOs (Mejia et al., 2004, slide 14).

C.4.6 Lessons learned

Groundwater management needs to be integrated into regional water management strategies and programs.

Most of the Integrated Water Resource Management (IWRM) program in the region had been devoted to surface waters, largely ignoring one of the largest underground freshwater resources in the world.

Managing a transboundary aquifer effectively requires coordinated collaboration, cooperation, and communication between national and subnational governments as well as the private sector, international organizations, and local civil society.

With an integrated management strategy that affects international politics, economics, the environment, and social wellbeing, it is necessary to include all stakeholders in the process from design to implementation to maintenance, in order for

the program to be effective and sustainable. There needs to be a broad understanding of a common goal and a clear strategy and methodology to achieve that goal.

C.4.7 Creative outcomes resulting from resolution process

The foresight with which the four basin states are using to plan the use of the Guaraní Aquifer System has lead to holistic, sustainable management plans that include public participation and education and are based on preventative actions.

C.4.8 Time line

1969 Plata Basin Framework Agreement.

1981 MERCOSUR Common Market Agreement.

1991 GEF gives birth to an experimental task force aimed at the preservation and sustainable management of the Guaraní Aquifer.

December 1999 São Paulo Workshop discuss use and protection of aquifer by the state.

Figure C.5 Map of the Indus River basin (*Source:* Transboundary Freshwater Dispute Database, 2004).

January–February 2000 Stakeholder workshop at Foz de Iguazu for an endorsement of a project concept note by central and state government representatives, university researchers, NGOs, municipalities, and international organizations (OAS, IICA).

2002–2004 Regular meetings between stakeholders relating to the ongoing design of the integrated management and monitoring framework for the Guaraní Aquifer. Final plan is scheduled to be finished by 2007.

May 2003 All four basin states signed the "Environmental Protection and Sustainable Development of the Guaraní Aquifer System" agreement.

C.5 INDUS WATER TREATY

C.5.1 Case summary

River basin: Indus River and tributaries (Figure C.5).
Dates of negotiation: 1951–1960.
Relevant parties: India, Pakistan.
Flash point: Lack of water-sharing agreement leads India to stem flow of tributaries to Pakistan on April 1, 1948.

Issues: Stated Objectives: negotiate an equitable allocation of the flow of the Indus River and its tributaries between the riparian states; develop a rational plan for integrated watershed development.

Additional issues: Water-related: financing for development plans, whether storage facilities are "replacement" or "development" (tied to who is financially responsible); *Nonwater*: General India–Pakistan relations.

Excluded issues: Future opportunities for regional management; Issues concerning drainage.

Criteria for water allocations: Historic and planned use (for Pakistan) plus geographic allocations (western rivers vs. eastern rivers).

Incentives/linkage: Financial: World Bank organized International Fund Agreement; Political: None.

Breakthroughs: Bank put own proposal forward after 1953 deadlock; International funding raised for final agreement.

Status: Ratified in 1960, with provisions for ongoing conflict resolution. Some suggest that recent meetings have been lukewarm. Physical separation of tributaries may preclude efficient integrated basin management. Renewed attempts to resolve Wuller Barrage and Baglihar dam conflicts begin to take place in July 2004.

C.5.2 Background

Irrigation in the Indus River basin dates back centuries; by the late 1940s the irrigation works along the river were the most extensive in the world. These irrigation projects had been developed over the years under one political authority, British India, and any water conflict could be resolved by executive order. The Government of India Act of 1935, however, put water under provincial jurisdiction, and some disputes did begin to crop up at the sites of the more-extensive works, notably between the provinces of Punjab and Sind.

In 1942, a judicial commission was appointed by the British government to study Sind's concern over planned Punjabi development. The Commission recognized the claims of Sind and called for the integrated management of the basin as a whole. The Commission's report was found unacceptable by both sides, and the chief engineers of the two sides met informally between 1943 and 1945 to try to reconcile their differences. Although a draft agreement was produced, neither of the two provinces accepted the terms, and the dispute was referred to London for a final decision in 1947.

Before a decision could be reached, however, the Indian Independence Act of August 15, 1947, internationalized the dispute between the new states of India and Pakistan. Partition was to be carried out in 73 days, and the full implications of dividing the Indus basin seem not to have been fully considered, although Sir Cyril Radcliffe, who was responsible for the boundary delineation, did express his hope that, "some joint control and management of the irrigation system may be found" (Mehta, 1986, p. 4). Heightened political tensions, population displacements, and unresolved territorial issues, all served to exacerbate hostilities over the water dispute.

As the monsoon flows receded in the fall of 1947, the chief engineers of Pakistan and India met and agreed to a "Standstill Agreement," which froze water allocations at two points on the river until March 31, 1948, allowing discharges from headworks in India to continue to flow into Pakistan.

On April 1, 1948, the day that the "Standstill Agreement" expired, in the absence of a new agreement, India discontinued the delivery of water to the Dipalpur Canal and the main branches of the Upper Bari Daab Anal. Several motives have been suggested for India's actions. The first is legalistic – that of an upper riparian establishing its sovereign water rights. Others include an Indian maneuver to pressure Pakistan on the volatile Kashmir issue, to demonstrate Pakistan's dependence on India in the hope of forcing reconciliation, or to retaliate against a Pakistani levy of an export duty on raw jute leaving East Bengal. Another interpretation is that the action was taken by the provincial government of East Punjab, without the approval of the central government.

C.5.3 The problem

Even before the partition of India and Pakistan, the Indus posed problems between the states of British India. The problem became international only after partition, though, and the attendant increased hostility and lack of supralegal authority only exacerbated the issue. Pakistani territory, which had relied on Indus water for centuries, now found the water sources originating in another country, one with whom geopolitical relations were increasing in hostility.

The question over the flow of the Indus is a classic case of the conflicting claims of up- and downstream riparians. The conflict can be exemplified in the terms for the resumption of water delivery to Pakistan from the Indian headworks, worked out at an Inter-Dominian conference held in Delhi from May 3 to May 4, 1948. India agreed to the resumption of flow, but maintained that Pakistan could not claim any share of those waters as a matter of right (Caponera, 1987, p. 511). This position was reinforced by the Indian claim that, because Pakistan had agreed to pay for water under the Standstill Agreement of 1947, Pakistan had recognized India's water rights. Pakistan countered that they had the rights of prior appropriation and that payments to India were only to cover operation and maintenance costs (Biswas, 1992, p. 204).

Although these conflicting claims were not resolved, an agreement was signed, later referred to as the Delhi Agreement, in which India assured Pakistan that India would not withdraw water delivery without allowing time for Pakistan to develop alternate sources. Pakistan later expressed its displeasure with the agreement in a note dated June 16, 1949, calling for the "equitable apportionment of all common waters" and suggesting turning jurisdiction of the case over to the World Court. India suggested rather that a commission of judges from each side try to resolve their differences before turning the problem over to a third party. This stalemate lasted through 1950.

C.5.4 Attempts at conflict management

In 1951, Indian Prime Minister Nehru, whose interest in integrated river management along the lines of the Tennessee Valley Authority had been piqued, invited David Lilienthal, former chairman of the TVA, to visit India. Lilienthal also visited Pakistan and, on his return to the United States, wrote an article outlining his impressions and recommendations (the trip had been commissioned by Collier's Magazine – international water was not the initial aim of the visit). These included steps from the psychological – a call to allay Pakistani suspicions of Indian intentions for the Indus headwaters to the practical – a proposal for greater storage facilities and cooperative

Table C.4 *Water allocations from Indus negotiations, in MAF/year*[a]

Plan	India	Pakistan
Initial Indian	29.0	90.0
Initial Pakistani	15.5	102.5
Revised Indian	All of the eastern rivers and 7% of the western rivers	None of the eastern rivers and 93% of the western rivers
Revised Pakistani	30% of the eastern rivers and none of the western rivers	70% of the eastern rivers and all of the western rivers
World Bank Proposal	Entire flow of the eastern rivers[b]	Entire flow of the western rivers[c]

[a] Initial estimates of supplies available differed only slightly, with the Indian Plan totaling 119 MAF and the Pakistani Plan arriving at 118 MAF. The "eastern rivers" consist of the Ravi, Beas, and Sutlej tributaries; the "western rivers" refer to the Indus, Jhelum, and Chenab.

[b] India would agree to continue to supply Pakistan with its historic withdrawals from these rivers for a transition period to be agreed on, which would be based on the time necessary to complete Pakistani link canals to replace supplies from India.

[c] The only exception would be an "insignificant" amount of flow from the Jhelum, used at the time in Kashmir.

management. Lilienthal also suggests that international financing be arranged, perhaps by the World Bank, to fund the workings and findings of an "Indus Engineering Corporation" to include representatives from both states, as well as from the World Bank.

The article was read by Lilienthal's friend, David Black, president of the World Bank, who contacted Lilienthal for recommendations on helping to resolve the dispute. As a result, Black contacted the prime ministers of Pakistan and India, inviting both countries to accept the Bank's good offices. In a subsequent letter, Black outlined "essential principles" that might be followed for conflict resolution. These principles included the following: that water resources of the Indus basin should be managed cooperatively and that problems of the basin should be solved on a functional and not on a political plane, without relation to past negotiations and past claims. Black suggested that India and Pakistan each appoint a senior engineer to work on a plan for development of the Indus basin. A Bank engineer would be made available as an ongoing consultant.

Both sides accepted Black's initiative. The first meeting of the Working Party included Indian and Pakistani engineers, along with a team from the Bank, as envisioned by Black, and met for the first time in Washington, DC, in May 1952. The stated agenda was to prepare an outline for a program, including a list of possible technical measures to increase the available supplies of Indus water for economic development. After three weeks of discussions, an outline was agreed to. The outline's points included:

- determination of total water supplies, divided by catchment and use;
- determination of the water requirements of cultivable irrigable areas in each country;

- calculation of data and surveys necessary, as requested by either side; and
- preparation of cost estimates and a construction schedule of new engineering works which might be included in a comprehensive plan.

In a creative avoidance of a potential and common conflict, the parties agreed that any data requested by either side would be collected and verified when possible but that the acceptance of the data, or the inclusion of any topic for study, would not commit either side to its "relevance or materiality."

When the two sides were unable to agree on a common development plan for the basin in subsequent meetings in Karachi, November 1952, and Delhi, January 1953, the Bank suggested that each side submit its own plan. Both sides did submit plans on October 6, 1953, each of which mostly agreed on the supplies available for irrigation but varied extremely on how these supplies should be allocated (Table C.4). The Indian proposal allocated 29 million acre-feet (MAF) per year to India and 90 MAF to Pakistan, totaling 119 MAF (MAF = 1233.48 million cubic meters; because all negotiations were in English units, that is what is reported here). The Pakistani proposal, in contrast, allocated India 15.5 MAF and Pakistan 102.5 MAF, for a total of 118 MAF.

The two sides were persuaded to adjust somewhat their initial proposals, but the modified proposals of each side still left too much difference to overcome. The modified Indian plan called for all of the eastern rivers (Ravi, Beas, and Sutlej) and 7 percent of the western rivers (Indus, Jhelum, and Chenab) to be allocated to India, while Pakistan would be allocated the remainder, or 93 percent of the western rivers. The modified Pakistani plan called for 30 percent of the eastern rivers to be allocated to India, while 70 percent of the eastern rivers and all of the western rivers would go to Pakistan.

The Bank concluded that not only was the stalemate likely to continue, but that the ideal goal of integrated watershed development for the benefit of both riparians was probably too elusive a goal at this stage of political relations. On February 5, 1954, the Bank issued its own proposal, abandoning the strategy of integrated development in favor of one of separation. The Bank proposal called for the entire flow of the eastern rivers to be allocated to India, and all of the western rivers, with the exception of a small amount from the Jhelum, to be allocated to Pakistan. According to the proposal, the two sides would agree to a transition period while Pakistan would complete link canals dividing the watershed, during which India would continue to allow Pakistan's historic use to continue to flow from the eastern rivers.

The Bank proposal was given to both parties simultaneously. On March 25, 1954, India accepted the proposal as the basis for agreement. Pakistan viewed the proposal with more trepidation and gave only qualified acceptance on July 28, 1954; they considered the flow of the western rivers to be insufficient to replace their existing supplies from the eastern rivers, particularly given limited available storage capacity. To help facilitate an agreement, the Bank issued an aide memoir, calling for more storage on the western rivers and suggesting India's financial liability for "replacement facilities" – increased storage facilities and enlarged link canals in Pakistan that could be recognized as the cost replacement of prepartition canals.

Little progress was made until representatives from the two countries met in May 1958. Main points in contention included whether the main replacement storage facility ought to be on the Jhelum or Indus rivers – Pakistan preferred the latter but the Bank argued that the former was more cost-effective; and what the total cost of new development would be and who would pay for it – India's position was that it would only pay for "replacement" and not "development" facilities.

In 1958, Pakistan proposed a plan including two major storage facilities: one each on the Jhelum and the Indus; three smaller dams on both tributaries; and expanded link canals. India, objecting both to the extent and the cost of the Pakistani proposal, approximately $1.12 billion, proposed an alternative plan that was smaller in scale but which Pakistan rejected because it necessitated continued reliance on Indian water deliveries.

By 1959, the Bank evaluated the principal issue to be resolved as follows: which works would be considered "replacement" and which "development," in other words, for which works would India be financially responsible. To circumvent the question, Black suggested an alternative approach in a visit to India and Pakistan in May. Perhaps one might settle on a specific amount for which India is responsible rather than argue over individual works. The Bank might then help raise additional funds among the international community development for watershed development. India was offered help with construction of its Beas Dam, and Pakistan's plan, including both the proposed dams would be looked at favorably. With these conditions, both sides agreed to a fixed payment settlement and to a 10-year transition period during which India would continue to provide Pakistan's historic flows to continue.

In August 1959, Black organized a consortium of donors to support development in the Indus basin that raised close to $900 million, in addition to India's commitment of $174 million. The Indus Water Treaty was signed in Karachi on September 19, 1960, and government ratifications were exchanged in Delhi in January 1961.

C.5.5 Outcome

The Indus Water Treaty addressed both the technical and financial concerns of each side and included a time line for transition. The main points of the treaty included:

- an agreement that Pakistan would receive unrestricted use of the western rivers, which India would allow to flow unimpeded, with minor exceptions;
- provisions for three dams, eight link canals, three barrages, and 2,500 tube wells to be built in Pakistan;
- a 10-year transition period, from April 1, 1960, to March 31, 1970, during which water would continue to be supplied to Pakistan according to a detailed schedule;
- a schedule for India to provide its fixed financial contribution of US$62 million, in ten annual installments during the transition period; and
- additional provisions for data exchange and future cooperation.

The treaty also established the Permanent Indus Commission, made up of one Commissioner of Indus Waters from each country. The two Commissioners would meet annually in order to establish and promote cooperative arrangements for the treaty implementation; promote cooperation between the Parties in the development of the waters of the Indus system; examine and resolve by agreement any question that may arise between the Parties concerning interpretation or implementation of the Treaty; submit an annual report to the two governments.

In case of a dispute, provisions were made to appoint a "neutral expert." If the neutral expert fails to resolve the dispute, negotiators can be appointed by each side to meet with one or more mutually agreed-upon mediators. If either side (or the mediator) views mediated agreement as unlikely, provisions are included for the convening of a Court of Arbitration. In addition, the treaty calls for either party, if it undertakes any engineering works on any of the tributaries, to notify the other of its plans and to provide any data which may be requested.

Since 1960, no projects have been submitted under the provisions for "future cooperation," nor have any issues of water quality been submitted at all. Other disputes have arisen, and been handled in a variety of ways. The first issues arose from Indian nondelivery of some waters during 1965–1966, but instead became a question of procedure and the legality of commission decisions. Negotiators resolved that each commissioner acted as government representatives and that their decisions were legally binding.

One controversy surrounding the design and construction of the Salal Dam was resolved through bilateral negotiations between the two governments. Other disputes, over new hydroelectric projects and the Wuller Barrage on the Jhelum tributary and the Baglihar dam on the Chenab River in Kashmir, have yet to be resolved.

C.5.6 Lessons learned

Shifting political boundaries can turn intranational disputes into international conflicts, exacerbating tensions over existing issues.

Shifting borders and partition exacerbated what was, initially, an intranational Indian issue. After partition, political tensions, particularly over Kashmir territory, contributed to tensions of this newly international conflict.

Power inequities may delay the pace of negotiations.

Power inequities may have delayed pace of negotiations. India had both a superior riparian position, as well as a relatively stronger central government, than Pakistan. The combination may have acted as disincentive to reach agreement.

Positive, active, and continuous involvement of a third party is vital in helping to overcome conflict.

The active participation of Eugene Black and the World Bank were crucial to the success of the Indus Water Treaty. The Bank offered not only their good offices, but a strong leadership role as well. The Bank provided support staff, funding, and, perhaps most important, its own proposals when negotiations reached a stalemate.

Coming to the table with financial assistance can provide sufficient incentive for a breakthrough in agreement.

The Bank helped raise almost US$900 million from the international community, allowing for Pakistan's final objections to be addressed.

Some points may be agreed to more quickly, if it is explicitly agreed that a precedent is not being set.

In the 1948 agreement, Pakistan agreed to pay India for water deliveries. This point was later used by India to argue that, by paying for the water, Pakistan recognized India's water rights. Pakistan, in contrast, argued that they were paying only for operation and maintenance. In an early meeting (May 1952), both sides agreed that any data may be used without committing either side to its "relevance or materiality," thereby precluding delays over data discrepancies.

Sensitivity to each party's particular hydrologic concerns is crucial in determining the bargaining mix.

Early negotiations focused on quantity allocations, while one of Pakistan's main concerns was storage – the timing of the delivery was seen to be as crucial as the amount.

In particularly hot conflicts, when political concerns override, a suboptimal solution may be the best one can achieve.

The plan pointedly disregards the principle of integrated water management, recognizing that between these particular riparians, the most important issue was control by each state of its own resource. Structural division of the basin, although crucial for political reasons, effectively precludes the possibility of increased integrated management.

C.5.7 Creative outcomes resulting from resolution process

In a creative avoidance of a potential and common conflict, the parties agreed that any data requested by either side would be collected and verified when possible, but that the acceptance of the data, or the inclusion of any topic for study, would not commit either side to its "relevance or materiality."

Water was separated out from other contentious issues between India and Pakistan. This allowed negotiations to continue, even in light of tensions over other topics. Water problems were to be viewed as "functional" rather than political.

When both sides were unable to agree on a common development plan in 1953, the Bank suggested that each prepare its own plan, which the Bank would then inspect for commonalities. This active strategy to breaking impasses is currently being attempted with the riparians of the Jordan River watershed in conjunction with the multilateral working group on water.

C.5.8 Time line

Pre-1935 British India has authority to resolve interstate water conflicts by executive order.

1935 Government of India Act makes water a subject of provincial jurisdiction, unless asked to intervene by states.

October 1939 Province of Sind formally requests Governor-General to review new Punjabi irrigation project and potential detriment to Sind.

September 1941 Indus Commission established.

July 1942 Commission submits its report suggesting that withdrawals by Punjab would cause "material injury" to inundation canals in Sind, particularly during the month of September. Incidentally, it called for management of the river system as a whole. Report found unacceptable to both sides.

1943–1945 Chief engineers of both states meet informally, finally producing a draft agreement – provinces refuse to sign. Dispute referred to secretary of state for India in London early 1947.

August 15, 1947 Independent states of India and Pakistan established. Eastern Punjab becomes part of India; western Punjab and Sind become part of Pakistan. Conflict becomes international; British role now irrelevant. Chair of Punjab Boundary Commission suggests that Punjab water system be run as joint venture – declined by both sides.

December 10, 1947 "Standstill Agreement" negotiated by chief engineers of west and east Punjab, freezing allocations at two points until March 31,1948.

April 1, 1948 Without a new agreement, India discontinues delivery of water to Dipalpur Canal and main branches of Upper Bari Daab Canal.

April 30, 1948 India resumes water delivery as negotiations undertaken.

May 3–4, 1948 Inter-Dominion conference, and an agreement is signed. India assures Pakistan that India will not withdraw water delivery without allowing time for Pakistan to develop alternate sources. Other issues remain unresolved.

June 16, 1949 Pakistan sends a note to India expressing displeasure with agreement. The note calls for a conference to resolve the "equitable apportionment of all common waters" and suggesting giving the World Court jurisdiction on the application of either party. India objects to third-party involvement, suggests judges from each side might narrow dispute first. Stalemate results through 1950.

1951 David Lilienthal, past chairman of the Tennessee Valley Authority, invited to India as Prime Minister Nehru's guest. He later publishes an article with his suggestions, which captures the attention of Eugene Black, president of the World Bank.

August 1951 Black invites both prime ministers to meeting in Washington. Both accept and agree on outline of essential principles.

January–February 1952 Meetings continue, Black finds "common understanding," at least that neither side will diminish supplies for existing uses.

May 1952 First meeting of working party in Washington, DC comprised of engineers of Bank engineers and engineers from India and Pakistan. Agreement to: determine future supply and demand; calculate available and desired data; prepare cost estimates and construction schedule of necessary infrastructure.

November 1952 and January 1953 Meetings continue in Karachi and Delhi without agreement. Bank suggests each side submit its own plan.

October 6, 1953 Plans submitted with proposed allocations and sources for each state. Agreement on available supplies but not on allocations.

February 5, 1954 Bank puts forth own proposal, essentially suggesting dividing the western tributaries to Pakistan, and the eastern tributaries to India. The proposal also provided for continued deliveries to Pakistan during transition period.

March 25, 1954 India accepts proposal. Pakistan is less enthusiastic – it would have to replace existing facilities.

July 28, 1954 Pakistan delivers a qualified acceptance of proposal.

May 21, 1956 Bank Aide Memoire suggests that replacement facilities be financed by India.

May–November 1958 Disagreements over which storage facilities are "replacement," for which India would pay, and which are "development" for which Pakistan would be responsible.

May 1959 Black visits India and Pakistan. Suggests that India's share be a fixed cost, rather than by facility, and that the Bank would arrange for additional financing. India agrees, and accepts a 10-year transition period.

September 1960 Bank arranges an international Indus Basin Development Fund Agreement; raises US$893.5 million.

September 19, 1960 Indus Water Treaty signed in Karachi. Provisions call for an Indian and Pakistani engineer to constitute the Permanent Indus Commission, which will meet at least once a year to: establish and promote cooperative arrangements.

July 29, 2004 Talks about the Wuller barrage and Baglihar dam begin in Lahore. Pakistan indicates that it might seek World Bank arbitration if the matter is not sorted out through bilateral talks.

Figure C.6 Map of all transboundary waters along the Canada–United States border (*Source:* Transboundary Freshwater Dispute Database, 2004).

C.6 THE INTERNATIONAL JOINT COMMISSION: CANADA AND THE UNITED STATES

C.6.1 Case summary

River basin: All transboundary waters along the United States–Canada boundary (Figure C.6).

Dates of negotiation: 1905–1909.

Relevant parties: Canada (originally negotiating through UK), United States.

Flash point: Water quality concerns of early twentieth century.

Issues: Stated objectives: to provide an institutional framework to deal with issues related to boundary waters.

Additional issues: Water-related: water quality issues were re-emphasized in 1978; Nonwater: 1987 Protocol and 1991 Agreement added air pollution.

Excluded issues: Tributaries to transboundary waters; some sovereignty issues.

Criteria for water allocations: "Equal and similar rights."

Incentives/linkage: None.

Breakthroughs: Canada accepted sovereignty argument; United States accepted arbitration function.

Status: More than 130 disputes have been averted or reconciled.

C.6.2 Background

Canada and the United States share a 4,000-mile boundary between the main portions of their States, and an additional 1,500 miles between the Canadian Northwest Territories and Alaska. Crossing these boundaries are some of the richest waterways in the world, not least of which are the vast water resources of the five Great Lakes. The ad hoc commissions, which until then had been established to resolve water-related issues, were not sufficient to handle the growing issues. Even the International Waterways Commission, established in 1905, only dealt with issues on a case-by-case basis.

C.6.3 The problem

Canada and the United States share one of the longest boundaries in the world. Industrial development in both countries, which in the humid eastern border region relied on water resources primarily for waste disposal, had led to decreasing water quality along their shared border to the point where, by the early years of the twentieth century, it was in the interest of both countries to seriously address the matter. Prior to 1905, only ad hoc commissions had been established to deal with issues relating to shared water resources as they arose. Both States considered it within their interests to establish a more-permanent body for the joint management of their shared water resources.

C.6.4 Attempts at conflict management

As Canada and the United States entered into negotiations to establish a permanent body to replace the International Waterways Commission, both countries entered talks with their own interests mind. For the United States, the overriding issue was sovereignty. Although it was interested in the practical necessity of an agreement to manage transboundary waters, it did not want to relinquish political independence in the process. This concern was expressed by United States position that absolute territorial sovereignty be retained by each state for the waters within its territory – tributaries should not be included in the Commission's authority. The new body might retain some of the ad hoc nature of prior bodies, so as not to acquire undue authority. Canada was interested in establishing an egalitarian relation with the United States. It was hampered not only because of the relative size and level of development of the two states at the time but also because Canadian foreign policy was still the purview of the United Kingdom – negotiations had to be carried out among Ottawa, Washington, and London. Canada wanted a comprehensive agreement, which would include tributaries, and a Commission with greater authority than the bodies of the past.

C.6.5 Outcome

The "Treaty Relating to Boundary Waters between the United States and Canada," signed between the United Kingdom and the United States in 1909, reflects the interests of each negotiating body. The Treaty establishes the International Joint Commission with six commissioners, three appointed by the governments of each State. Canada accepted U.S. sovereignty concerns to some extent – tributary waters are excluded. The United States in turn accepted the arbitration function of the Commission and allowed it greater authority than it would have liked.

The Treaty calls for open and free navigation along boundary waters, allowing Canadian transportation also on Lake Michigan, the only one of the Great Lakes not defined as boundary water. Although it allows each State unilateral control over all of the waters within its territory, the Treaty does provide for redress by anyone affected downstream. Furthermore, the Commission has "quasijudicial" authority: any project that would affect the "natural" flow of boundary waters has to be approved by both governments. Although the Commission has the mandate to arbitrate agreements, it has never been called to do so. The Commission also has investigative authority – it may have development projects submitted for approval or be asked to investigate an issue by one or another of the governments. Commissioners act independently, not as representatives of their respective governments.

Water quality has been a focal concern of the Commission, particularly in the waterways of the Great Lakes. The Great Lakes–St. Lawrence River system contains one-fifth of the world's surface fresh water and includes the industrial lifelines of each State. Perhaps as a consequence, the antipollution provisions of the Treaty met little opposition on either side. A 1972 "Great Lakes Water Quality Agreement" calls for the States both to control pollution and to clean up waste waters from municipal and industrial sources. This led to the signing of a new Agreement in 1978, and a comprehensive Protocol in 1987, each of which expanded the Commission's authorities and activities with respect to water quality.

These agreements define specific water quality objectives – the 1987 Protocol called on the Commission to review "Remedial Action Plans," prepared by governments and communities, in forty-three "Areas of Concern" – yet allow the appropriate level of government of each side to develop its own plan to meet the objectives. The 1987 Protocol implemented an "ecosystem" approach to pollution control and called for the development of "lakewide management plans" to combat some critical pollutants. It also included new emphasis on non-point source pollution, groundwater contamination, contaminated sediment, and airborne toxics. In 1991, the two States signed an "Agreement of Air Quality" under which the Commission was given limited authority over joint air resources.

The International Joint Commission has met some criticism over the years; most recently some have questioned whether the limited authority of the Commission – politically necessary when the Commission was established – is really conducive to the "ecosystem" approach called for in the 1987 Protocol or whether greater supralegal powers are necessary. Others have questioned the commitment of the Commission to the process of public participation. Nevertheless, given the vast amount of water resources under its authority, and the myriad layers of government to which it must be responsible, the Commission stands out as an institution which has effectively and peacefully managed the boundary waters of two nations over some

90 years, reconciling or averting more than 130 disputes in the process.

C.6.6 Lessons learned

Even with an established binational management organization with significant experience can have difficulty with certain initiatives.

After talks about pollution controlled failed in 1920, more than 50 years went by when the issue was addressed again before creating the Great Lakes Water Quality Agreement in 1972. Both countries had anti-pollution programs domestically, but an international agreement proved complicated to work out even though relations were good between the two States.

An international agreement can bring together a community to work together for greater ends.

Since the inception of the 1909 Boundary Waters Treaty, Canada and the United States, and all stakeholders within the Great Lakes basin, have worked together and have been brought together as a community as a result of the commitment in preserving the shared waters of the two countries.

C.6.7 Creative outcomes resulting from resolution process

A mutual acceptance of the difference in political and cultural systems between the two countries has transcended the gap into allowing the International Joint Commission to arrive at mutually beneficial agreements where this may be an impediment to similar situations elsewhere.

Flexibility within the agreement permits the IJC to adapt to new situations as a result of new information and a change in circumstances. As technology, politics and knowledge of the shared waters changes, the IJC is better prepared than if it were not able to adjust thereby making it an organization with periodic development.

C.6.8 Time line (Dworsky and Allee, 1997)

1909 United Kingdom and the United States sign Boundary Waters Treaty. Creation of the International Joint Commission (IJC).

1912 First meeting of the IJC.

1918 IJC reports on the terrible pollution conditions within the Great Lakes.

1919 Canada and United States ask IJC to create legislation to address the pollution problem.

1920 Canada expresses interest in a treaty to control pollution, but United States declines. Topic left unaddressed.

1972 Canada and the United States sign Great Lakes Water Quality Agreement.

1978 Canada and the United States sign New Great Lakes Water Quality Agreement building on experience that was gained from under the previous Agreement with respect to water quality and pollution.

1987 The two nations sign the Great Lakes Water Quality Agreement Protocol in which more importance was placed on ecosystem well-being.

C.7 JORDAN RIVER: JOHNSTON NEGOTIATIONS, 1953–1955; YARMUK MEDIATIONS, 1980s

C.7.1 Case summary

River basin: Jordan River and tributaries, directly; Litani, indirectly (Figure C.7).

Dates of negotiation: 1953–1955; 1980s through the present.

Relevant parties: United States (initially sponsoring); United States and Russia (sponsoring multilateral negotiations); riparian entities: Israel, Jordan, Lebanon, Palestine, and Syria.

Flash point: 1951 and 1953 Syrian–Israeli exchanges of fire over water development in demilitarized zone; 1964–1966 water diversions.

Issues: Stated objectives: negotiate an equitable allocation of the flow of the Jordan River and its tributaries between the riparian states; develop a rational plan for integrated watershed development.

Additional issues: Water-related: Out-of-basin transfers; level of international control ("water master"); location and control of storage facilities; inclusion or exclusion of the Litani River. Nonwater: political recognition of adversaries.

Excluded issues: Groundwater; Palestinians as political entity (initially).

Criteria for water allocations: Amount of irrigable land within watershed for each state (in Johnston negotiations); "needs-based" criteria developed in current peace talks.

Incentives/linkage: Financial: United States and donor communities have agreed to cost-share regional water projects. Political: Multilateral talks work in conjunction with bilateral negotiations.

Breakthroughs: Harza study of Jordan's water needs (in Johnston talks); question of water rights successfully regulated to bilateral talks; creation of Palestinian Water Authority accepted by all parties.

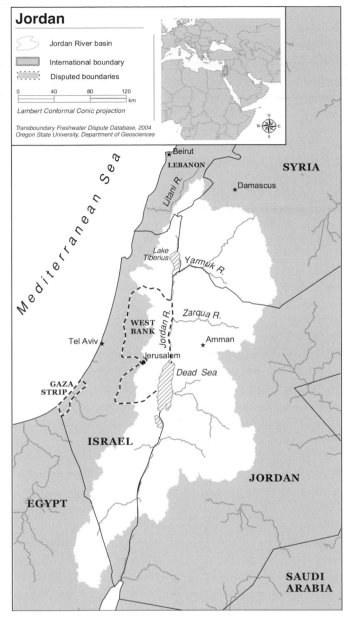

Jordan

- Jordan River basin
- International boundary
- Disputed boundaries

0   40   80   120
km

Lambert Conformal Conic projection

Transboundary Freshwater Dispute Database, 2004
Oregon State University, Department of Geosciences

Mediterranean Sea

Beirut
LEBANON
Litani R.

SYRIA

Damascus

Lake Tiberius
Yarmuk R.

WEST BANK
Jordan R.
Zarqua R.

Tel Aviv
Jerusalem
Amman

Dead Sea

GAZA STRIP

ISRAEL

JORDAN

EGYPT

SAUDI ARABIA

Figure C.7 Map of the Jordan River and tributaries (directly and indirectly, including Litani) (*Source:* Transboundary Freshwater Dispute Database, 2004).

Status: Israel–Jordan Peace Treaty (1994); Israel–Palestine Interim Agreement (1993, 1995) each have major water components.

C.7.2 Background

In 1951, several states announced unilateral plans for the Jordan watershed. Arab states began to discuss organized exploitation of two northern sources of the Jordan – the Hasbani and the Banias. The Israelis made public their "All Israel

Plan," which included the draining of Huleh Lake and swamps, diversion of the northern Jordan River, and construction of a carrier to the coastal plain and Negev Desert – the first out-of-basin transfer for the watershed in the region.

Jordan announced a plan to irrigate the East Ghor of the Jordan Valley by tapping the Yarmuk. At Jordan's announcement, Israel closed the gates of an existing dam south of the Sea of Galilee and began draining the Huleh swamps, which infringed on the demilitarized zone with Syria. This action led to a series of border skirmishes between Israel and Syria that escalated over the summer of 1951.

In March 1953, Jordan and the UN Relief and Works Agency for Palestine Refugees (UNRWA) signed an agreement to begin implementing the "Bunger Plan," which called for a dam at Maqarin on the Yarmuk River with a storage capacity of 480 MCM and a diversion dam at Addassiyah that would direct gravity flow along the East Ghor of the Jordan Valley. The water would both open land for irrigation and provide power for Syria and Jordan and offer resettlement for 100,000 refugees. In June 1953, Jordan and Syria agreed to share the Yarmuk but Israel protested that its riparian rights were not being recognized.

In July 1953, Israel began construction on the intake of its National Water Carrier at the Bridge of Jacob's Daughters, north of the Sea of Galilee and in the demilitarized zone. Syria deployed its armed forces along the border and artillery units opened fire on the construction and engineering sites. Syria also protested to the UN and, though a 1954 resolution allowed Israel to resume work, the USSR vetoed the resolution. The Israelis then moved the intake to its current site at Eshed Kinrot on the northwestern shore of the Sea of Galilee.

Against this tense background, President Dwight Eisenhower sent his special envoy Eric Johnston to the Mideast in October 1953 to try to mediate a comprehensive settlement of the Jordan River system allocations and design a plan for its regional development.

C.7.3 The problem

The Jordan River flows between five particularly contentious riparians, two of which rely on the river as the primary water supply. By the early 1950s, there was little room for any unilateral development without affecting other riparian states. The initial issue was an equitable allocation of the annual flow of the Jordan watershed between its riparian states – Israel, Jordan, Lebanon, and Syria. Egypt also was included, given its preeminence in the Arab world. This is because water was (and is) deeply related to other contentious issues of land, refugees, and political sovereignty. The Johnston negotiations, named after U.S. special envoy Eric Johnston, attempted to mediate the dispute over water rights among all the riparians in the mid-1950s.

Until the current Arab–Israeli peace negotiations, which began in 1991, political or resource problems were always handled separately. Some experts have argued that by separating the two realms of "high" and "low" politics, each process was doomed to fail. The initiatives, which were addressed as strictly water resource issues, namely the Johnston Negotiations of the mid-1950s, attempts at "water-for-peace" through nuclear desalination in the late 1960s, negotiations over the Yarmuk River in the 1970s and 1980s, and the Global Water Summit Initiative of 1991, all failed to one degree or another, because they were handled separately from overall political discussions. The resolution of water resources issues then had to await the Arab–Israeli peace talks to meet with any tangible progress.

C.7.4 Attempts at conflict management

Johnston's initial proposals were based on a study carried out by Charles Main and the Tennessee Valley Authority at the request of UNRWA to develop the area's water resources and to provide for refugee resettlement. The TVA addressed the problem with a regional approach, pointedly ignoring political boundaries in their study. In the words of the introduction, "the report describes the elements of an efficient arrangement of water supply within the watershed of the Jordan River System. It does not consider political factors or attempt to set this system into the national boundaries now prevailing."

The major features of the Main Plan included small dams on the Hasbani, Dan, and Banias, a medium-size (175 MCM storage) dam at Maqarin, additional storage at the Sea of Galilee, and gravity flow canals down both sides of the Jordan Valley. Preliminary allocations gave Israel 394 MCM/year, Jordan 774 MCM/year, and Syria 45 MCM/year (see Table C.5). In addition, the Main Plan described only in-basin use of the Jordan River water, although it conceded that "it is recognized that each of these countries may have different ideas about the specific areas within their boundaries to which these waters might be directed" and excluded the Litani River.

Israel responded to the "Main Plan" with the "Cotton Plan," which allocated Israel 1,290 MCM/year, including 400 MCM/year from the Litani, Jordan 575 MCM/year, Syria 30 MCM/year, and Lebanon 450 MCM/year. In contrast to the Main Plan, the Cotton Plan called for out-of-basin transfers to the coastal plain and the Negev; included the Litani River; and recommended the Sea of Galilee as the main storage facility, thereby diluting its salinity.

In 1954, representatives from Lebanon, Syria, Jordan, and Egypt established the Arab League Technical Committee under

Table C.5 *Water allocations from the Johnston Negotiations, in MCM/year*

Plan	Israel	Jordan	Lebanon	Syria
Main	393	774	–	45
Cotton (Israel)[a]	1290	575	450	30
Arab	182	698	35	132
Unified	400[b]	720[c]	35	132

[a] Cotton Plan included integration of the Litani River into the Jordan basin.

[b] Unified Plan allocated Israel the "residue" flow, what remained after the Arab States withdrew their allocations, estimated at an average of 409 MCM/year

[c] Two different summaries were distributed after the negotiations, with a difference of 15 MCM/year on allocations between Israel and Jordan on the Yarmuk River. This difference was never resolved and was the focus of Yarmuk negotiations in the late 1980s.

Egyptian leadership and formulated the "Arab Plan." Its principal difference from the Johnston Plan was in the water allocated to each state. Israel was to receive 182 MCM/year, Jordan 698 MCM/year, Syria 132 MCM/year, and Lebanon 35 MCM/year, in addition to keeping all of the Litani. The Arab Plan reaffirmed in-basin use, excluded the Litani, and rejected storage in the Galilee, which lies wholly in Israel.

Johnston worked until the end of 1955 to reconcile United States, Arab, and Israeli proposals in a Unified Plan amenable to all of the states involved. His dealings were bolstered by a U.S. offer to fund two-thirds of the development costs. His plan addressed the objections of both sides, and accomplished no small degree of compromise although his neglect of groundwater issues would later prove an important oversight. Though they had not met face to face for these negotiations, all states agreed on the need for a regional approach. Israel gave up on integration of the Litani and the Arabs agreed to allow out-of-basin transfer. The Arabs objected, but finally agreed, to storage at both the (unbuilt) Maqarin Dam and the Sea of Galilee, so long as neither side would have physical control over the share available to the other. Israel objected, but finally agreed, to international supervision of withdrawals and construction. Allocations under the Unified Plan, later known as the Johnston Plan, included 400 MCM/year to Israel, 720 MCM/year to Jordan, 132 MCM/year to Syria, and 35 MCM/year to Lebanon (Table C.5).

Although the agreement was never ratified, both sides have generally adhered to the technical details and allocations, even while proceeding with unilateral development. Agreement was encouraged by the United States, which promised funding for future water development projects only as long as the

Johnston Plans allocations were adhered to. Since that time to the present, Israeli and Jordanian water officials have met several times a year, as often as every 2 weeks during the critical summer months, at so-called Picnic Table Talks at the confluence of the Jordan and Yarmuk Rivers to discuss flow rates and allocations.

C.7.5 Outcome

The technical committees from both sides accepted the Unified Plan, and the Israeli Cabinet approved it without vote in July 1955. President Nasser of Egypt became an active advocate because Johnston's proposals seemed to deal with the Arab–Israeli conflict and the Palestinian problem simultaneously. Among other proposals, Johnston envisioned the diversion of Nile water to the western Sinai Desert to resettle two million Palestinian refugees.

Despite the forward momentum, the Arab League Council decided not to accept the plan in October 1955 because of the political implications of accepting, and the momentum died out. The agreement was never ratified, but both sides have generally adhered to the allocations.

C.7.6 Negotiations over the Yarmuk River

Although the watershed-wide scope of the Johnston negotiations has not been taken advantage of, the allocations that resulted have been at the heart of ongoing attempts at water conflict resolution, particularly along the Yarmuk River, where a dam for storage and hydroelectric power generation has been suggested since the early 1950s.

In 1952, Miles Bunger, an American attached to the Technical Cooperation Agency in Amman, first suggested the construction of a dam at Maqarin to help even the flow of the Yarmuk River and to tap its hydroelectric potential. The following year, Jordan and UNRWA signed an agreement to implement the Bunger plan the following year, including a dam at Maqarin with a storage capacity of 480 MCM and a diversion dam at Addassiyah, and Syria and Jordan agreed that Syria would receive two-thirds of the hydropower generated, in exchange for Jordan's receiving seven-eighths of the natural flow of the river. Dams along the Yarmuk were also included in the Johnston negotiations – the Main Plan included a small dam, 47 meters high with a storage capacity of only 47 MCM, because initial planning called for the Sea of Galilee to be the central storage facility. As Arab resistance to Israeli control over Galilee storage became clear in the course of the negotiations, a larger dam, 126 meters high with a storage capacity of 300 MCM, was included.

Although the idea faded with the Johnston negotiations, the idea of a dam on the Yarmuk was raised again in 1957, in a Soviet–Syrian Aid Agreement, and at the First Arab Summit in Cairo in 1964, as part of the All-Arab Diversion Project. Construction of the diversion dam at Mukheiba was actually begun but was abandoned when the borders shifted after the 1967 war – one side of the projected dam in the Golan Heights shifted from Syrian to Israeli territory.

The Maqarin Dam was resurrected as an idea in Jordan's Seven Year Plan in 1975, and Jordanian water officials approached their Israeli counterparts about the low dam at Mukheiba in 1977. Although the Israelis proved amenable at a ministerial-level meeting in Zurich – a more-even flow of the river would benefit all of the riparians – the Israeli government shifted that year to one less interested in the project.

This stalemate might have continued except for strong U.S. involvement in 1980, when President Carter pledged a $9 million loan toward the Maqarin project and Congress approved an additional $150 million – provided that *all* of the riparians agree. Philip Habib was sent to the region to help mediate an agreement. Although Habib was able to gain consensus on the concept of the dam, on separating the question of the Yarmuk from that of West Bank allocations, and on the difficult question of summer flow allocations – 25 MCM/year would flow to Israel during the summer months – negotiations were hung up winter flow allocations, and final ratification was never reached.

Syria and Jordan reaffirmed mutual commitment to a dam at Maqarin in 1987, whereby Jordan would receive 75 percent of the water stored in the proposed dam, and Syria would receive all of the hydropower generated. The agreement called for funding from the World Bank, which insists that all riparians agree to a project before it can be funded. Israel refused until its concerns about the winter flow of the river were addressed.

Against this backdrop, Jordan in 1989 approached the U.S. Department of State for help in resolving the dispute. Ambassador Richard Armitage was dispatched to the region in September 1989 to resume indirect mediation between Jordan and Israel where Philip Habib had left off a decade earlier. The points raised during the following year were as follows:

- Both sides agreed that 25 MCM/year would be made available to Israel during the summer months, but disagreed as to whether any additional water would be specifically earmarked for Israel during the winter months.
- The overall viability of a dam was also open to question – the Israelis still thought that the Sea of Galilee ought to be used as a regional reservoir, and both sides questioned what effects ongoing development by Syria at the headwaters of the Yarmuk would have on the dam's viability. Because the

U.S. State Department had no mandate to approach Syria, their input was missing from the mediation.

- Israel eventually wanted a formal agreement with Jordan, a step that would have been politically difficult for the Jordanians at the time.

By fall of 1990, agreement seemed to be taking shape, by which Israel agreed to the concept of the dam, and discussions on a formal document and winter flow allocations could continue during construction, estimated for more than 5 years. Two issues held up any agreement. First, the lack of Syrian input left questions of the future of the river unresolved, a point noted by both sides during the mediations. Second, the outbreak of the Gulf War in 1991 overwhelmed other regional issues, finally preempting talks on the Yarmuk. The issue has not been brought up again until recently in the context of the Arab–Israeli peace negotiations.

In the absence of an agreement, both Syria and Israel are currently able to exceed their allocations from the Johnston accords, the former because of a series of small storage dams and the latter because of its downstream riparian position. Syria began building a series of small impoundment dams upstream from both Jordan and Israel in the mid-1980s., while Israel has been taking advantage of the lack of a storage facility to increase its withdrawals from the river. Syria currently has twenty-seven dams in place on the upper Yarmuk, with a combined storage capacity of approximately 250 MCM (its Johnston allocations are 90 MCM/year from the Yarmuk), and Israel currently uses 70–100 MCM/year (its Johnston allocation are 25–40 MCM/year). This leaves Jordan approximately 150 MCM/year for the East Ghor Canal (as compared to its Johnston allocations of 377 MCM/year).

By 1991, several events combined to shift the emphasis on the potential for "hydroconflict" in the Middle East to the potential for "hydrocooperation." The Gulf War in 1990 and the collapse of the Soviet Union caused a realignment of political alliances in the Mideast that finally made possible the first public face-to-face peace talks between Arabs and Israelis, in Madrid on October 30, 1991. During the bilateral negotiations between Israel and each of its neighbors, it was agreed that a second track be established for multilateral negotiations on five subjects deemed "regional," including water resources.

Since the opening session of the multilateral talks in Moscow in January 1992, the Working Group on Water Resources, with the United States as "gavel-holder," has been the venue by which problems of water supply, demand and institutions has been raised among the parties to the bilateral talks, with the exception of Lebanon and Syria. The two tracks of the current negotiations, the bilateral and the multilateral, are designed explicitly not only to close the gap between issues of politics and issues of regional development but perhaps to use progress on each to help catalyze the pace of the other, in a positive feedback loop toward "a just and lasting peace in the Middle East." The idea is that the multilateral working groups would provide forums for relatively free dialogue on the future of the region and, in the process, allow for personal ice-breaking and confidence building to take place. Given the role of the Working Group on Water Resources in this context, the objectives have been more on the order of fact-finding and workshops rather than on tackling the difficult political issues of water rights and allocations or the development of specific projects. Likewise, decisions are made through consensus only.

The pace of success of each round of talks has vacillated but, in general, has been increasing. By this third meeting in 1992, it became clear that regional water-sharing agreements, or any political agreements surrounding water resources, would not be dealt with in the multilaterals, but that the role of these talks was to deal with nonpolitical issues of mutual concern, thereby strengthening the bilateral track. The goal in the Working Group on Water Resources became to plan for a future region at peace and to leave the pace of implementation to the bilaterals. This distinction between "planning" and "implementation" became crucial, with progress being made only as the boundary between the two was continuously pushed and blurred by the mediators.

The multilateral activities have helped set the stage for agreements formalized in bilateral negotiations: the Israel–Jordan Treaty of Peace of 1994 and the Interim Agreements between Israel and the Palestinians (1993 and 1995). For the first time since the states came into being, the Israel–Jordan peace treaty legally spells out mutually recognized water allocations. Acknowledging that, "water issues along their entire boundary must be dealt with in their totality," the treaty spells out allocations for both the Yarmuk and Jordan rivers, as well as regarding Arava/Araba groundwater, and calls for joint efforts to prevent water pollution. Also, "[recognizing] that their water resources are not sufficient to meet their needs," the treaty calls for ways of alleviating the water shortage through cooperative projects, both regional and international. The Interim Agreement also recognizes the water rights of both Israelis and Palestinians but defers their quantification until the final round of negotiations.

C.7.7 Lessons learned

In highly conflictual settings, separating resource issues from political interests may not be a productive strategy.

Eric Johnston took the approach that the process of reaching a rational, watershed-management plan: (1) may, itself, act as a confidence-building catalyst for increased cooperation in the political realm and (2) may help alleviate the burning political issues of refugees and land rights. By approaching peace through water, however, several overriding interests remained unmet in the process. The plan finally remained unratified mainly for political reasons.

Issues of national sovereignty that were unmet during the process included:

- The Arab states saw a final agreement with Israel as recognition of Israel, a step they were not willing to make at the time.
- Some Arabs may have felt that the plan was devised by Israel for its own benefit and was "put over" on the United States

The plan allowed the countries to use their allotted water for whatever purpose they saw fit. The Arabs worried that if Israel used their water to irrigate the Negev (outside the Jordan Valley), the increased amount of agriculture would allow more food production, which would allow for increased immigration, which might encourage greater territorial desires on the part of Israel.

Issues of national sovereignty can manifest itself through the need for each state to control its own water source and/or storage facilities.

The Johnston Plan provided that some winter flood waters be stored in the Sea of Galilee, which is entirely in Israeli territory. The Arab side was reluctant to relinquish too much control of the main storage facility. Likewise, Israel had the same kinds of control reservations about a water master.

Ignoring a riparian party, even one without political standing, can hamper agreement.

There was some concern over whether the Plan was designed to "liquidate the Palestinian refugee problem rather than to give the refugees their right of return." In fact, Palestinians were not addressed as a separate political entity.

Along with political entities, many interests affected by river management were not included in the process. These included NGO's, public interest groups, and environmental groups. Perhaps as a consequence, the *entire* river was allocated, leaving no water at all for instream uses.

Including key nonriparian parties can be useful to reaching agreement; excluding them can be harmful.

Egypt was included in the negotiations because of its preeminence in the Arab world and despite its nonriparian status. Some attribute the accomplishments made during the course in part to President Nasser's support.

In contrast, pressure after the negotiations from other Arab states not directly involved in the water conflict may have had an impact on its eventual demise. Iraq and Saudi Arabia strongly urged Lebanon, Syria, and Jordan not to accept the Plan. Perhaps partially as a result, Lebanon said they would not enter any agreement that split the waters of the Hasbani River or any other river.

All of the water resources in the basin ought to be included in the planning process. Ignoring the relationship between quality and quantity, and between surface water and groundwater, ignores hydrologic reality.

Groundwater was not explicitly dealt with in the Plan and is currently the most pressing issue between Israel and Palestinians. Likewise, tensions have flared over the years between Israel and Jordan over Israel's diverting saline springs into the lower Jordan, increasing the salinity of water on which Jordanian farmers rely.

Even in the absence of an explicit arrangement, some degree of implicit cooperation may be possible, perhaps leading to fairly high stability, if also to suboptimum, water management.

Although the lack of ratified agreement left a legacy of unilateral and generally suboptimum water development in the basin, the implicit arrangement, which resulted, particularly between Israel and Jordan, decreased tensions and added a certain stability between these most active riparians. The "Picnic Table" talks have allowed a venue for some level of technical agreement, and an outlet for minor disputes, for more than 40 years.

C.7.8 Creative outcomes resulting from resolution process

The plan called for water allocations to be determined according to the amount irrigable land each state had within the basin, then it allowed each country to do what it wished with its water, including out-of-basin transfers.

The development plan was created without regard to political borders, guaranteeing a degree of objectivity and engineering efficiency.

The plan incorporated issues of hydrologic variability. For example, Israel was to receive the "residue" after Arab withdrawals, which was sometimes more and sometimes less from the average flow.

C.7.9 Time line

1948 "TVA on the Jordan, Proposals, for Irrigation and Hydroelectric Development in Palestine," by James B. Hays; first Israeli plan for developing Jordan water.

March 1951 First formal plan put forward by Jordan during post-1948 period, presented by Sir M. McDonald and Partners.

1953 United States becomes actively involved in Jordan water management planning. Johnston is appointed by Eisenhower and given the rank of ambassador.

October 1955 Johnston presents "The Unified Development of the Water Resources of the Jordan Valley Region" to Israel, Jordan, Syria, Lebanon, and Egypt – was initially poorly received. Counterproposals put forward: the Cotton Plan for Israel and the Arab Plan for the Arab countries.

1955 Engineering study conducted by Michael Baker, Jr., Inc. and Harza Engineering (American firms) concludes that less water is needed by Jordan than is thought; more water is therefore available for negotiations. An agreement is reached by technical committees.

October 11, 1955 Unified Plan fails to win approval by Arab League and is sent back until plan better protected Arab interests.

October 15, 1954 Letter from Johnston to Assistant Secretary of State Byroade urging that any financial aid in support of the project be in addition to existing aid.

January–February 1955 Johnston returns to the Middle East (Beirut) for talks. On February 19, 1995, Johnston reaches a "preliminary understanding" concerning major elements of the proposed plan with Jordan, Lebanon, Syria, and Egypt. Tentative agreement reached on 300 MCM dam on the Yarmuk and diversion of Yarmuk floodwaters to Sea of Galilee for release to Jordan. Israel would receive approximately 409 MCM/year.

March 10, 1955 Discussion with Israel begins concerning the arrangement; Johnston reassures Israel about its main concern, the nature of the neutral authority which would be established to oversee the allocations of Galilee water.

March 14, 1955 Meeting between Assistant Secretary of State Allen and Ambassador Eban of Israel: Eban says that Allen threatened to withhold aid from Israel if the Israelis did not come to terms with Johnston. In a meeting later that same day w/Secretary of State Dulles, Governor Stassen, Assistant Secretary Allen, and Arthur Gardiner, Johnston brings the issue up for discussion. Allen states that he had "advised Mr. Eban that agreement on the Jordan River problem would furnish a useful basis for aid."

June 1955 Israel agrees to the basic terms of the plan Johnston had set up with the Arabs in Beirut.

1955–1956 Events begin overtaking chances of agreement: Jordanian press reported several times in May 1955 that the project is intended to resettle Palestinian refugees. Public opposition springs up in August 1955; the Jordan National Socialist Party puts out a memo listing several points of opposition.

July 27, 1955 Lebanon expresses its intent not to allow any water from the Hasbani to be distributed.

August 1955 Johnston returns to Middle East for talks with representatives from the Arab states.

August 30, 1955 Jordan states that it would accept Jordan Valley proposals on economic grounds given certain modifications but that a political decision would have to be decided by a subcommittee of Arab states.

September 1955 Meeting with Arab representatives continue, but no decision is reached.

1956 Israel indicates it would be willing to wait and see if Arab states would accept the plan before beginning work on a system to divert water from the upper Jordan.

October 1956 War in Sinai Desert effectively ends any explicit chance of agreement. Implicit agreements managed through ongoing "Picnic Table Talks" between Israel and Jordan.

C.7.10 Negotiations over the Yarmuk

1952 Maqarin Dam first proposed by Miles Bunger, an American attached to the Technical Cooperation Agency in Amman.

1953 Jordan and UNRWA sign an agreement to implement the Bunger Plan, including a dam at Maqarin with a storage capacity of 480 MCM.

Syria and Jordan agree that Syria will receive two-thirds of the hydropower generated in exchange for Jordan receiving seven-eighths of the natural flow of the river.

1953–1955 Johnston Negotiations. Main Plan included a dam 47 meters high with a storage capacity of 47 MCM, to be managed in conjunction with storage in the Sea of Galilee. Arab position argued for the hydropower that a higher dam would produce, and that, "... the water needed for Arab crops should be under direct Arab control." Therefore, a high dam was agreed to, 126 meters high with a storage capacity of 300 MCM. Negotiations were never ratified.

October 28, 1957 Soviet–Syrian Aid Agreement, including provisions for a hydroelectric project in the Yarmuk basin.

1964 Concept of a dam on the Yarmuk reaffirmed at the First (and subsequent) Arab Summit(s) in Cairo, as a component in the All-Arab Diversion Project. Construction begun on lower dam at Mukheiba.

1967 Construction halted as a result of June 1967 war. One side of projected dam site would now abut on Israeli-occupied Golan Heights.

1975 Jordanian Seven Year Plan includes a dam at Maqarin with a storage capacity of 486 MCM, which would generate 20 MW of power.

1977 Jordanian water officials approach their Israeli counterparts through U.S. intermediaries and discuss rebuilding the low dam at Mukheiba. Israelis agree, but elections in that country, and the resulting shift in government, put further negotiations on hold.

1980 President Carter pledges a $9 million USAID loan toward Jordan's plan, in addition to the $10 million that had already been allocated. Congress commits $150 million, on the condition that all riparians agree to resolve their differences over the river. U.S. mediation efforts led by Philip Habib prove fruitless, although some agreement is reached on summer flow allocations, and the plan is indefinitely postponed.

Mid-1980s In absence of an agreement, Syria begins a series of small impoundment dams on the headwaters of the Yarmuk within Syrian territory. By August, 1988, twenty dams were in place with a combined capacity of 156 MCM. That capacity has grown to twenty-seven dams with a combined capacity of approximately 250 MCM and is projected to grow to total storage of 366 MCM by 2010. Israel, meanwhile, increases its Yarmuk withdrawals from the 25 MCM allocated in the Johnston negotiations to 70–100 MCM/year.

1987 Agreement signed by Jordan and Syria, whereby Jordan receives 75 percent of water stored in the proposed dam, while Syria receives 25 percent and all of the 46 MW of hydropower to be generated. World Bank insists that all riparians agree to project before funding is provided – Israel refuses.

1989–1990 Indirect negotiations on the Maqarin Dam are renewed, mediated by Richard Armitage of the U.S. Department of State, with talks focusing on winter flows. Negotiations are put on hold during Gulf War and are not renewed.

C.8 KURA–ARAKS BASIN

C.8.1 Case summary

River basin: Kura and Araks rivers (Figure C.8).
Dates of negotiation: 2000–present.
Relevant parties: Azerbaijan, Armenia, Georgia, Turkey, and Iran (Araks).
Flash point: Collapse of Soviet Union in late 1980s/early 1990s.
Issues: Stated objectives: To eventually form an international management body for the Kura–Araks river basin.
Additional issues: High levels of pollution; Nonwater: Nagorno–Karabakh region.

Criteria for water allocations: None.
Incentives/linkage: Reduction in pollution levels and improvement of regional relations.
Breakthroughs: None.
Status: Still in negotiation phase and moving slowly as Armenia and Azerbaijan relations are cold due to both nations' claims of Nagorno–Karabakh region.

C.8.2 Background

Before the end of the twentieth century and the fall of the Soviet Union, international water resources management in the Kura–Araks river basin was defined by two separate treaties signed by the nations of the region. In 1927, the USSR (Union of Soviet Socialist Republics) signed an agreement with the government of Turkey to share equally all the common water resources along the borders of the two nations. Alongside such an agreement was created the Joint Boundary Water Commission whose charge it was to manage the use of the shared water resources. In 1957, a similar agreement was signed between the USSR and Iran. These two treaties encompass what in 2004 is a goal to be sought after.

With the collapse of the Soviet Union in 1991 and the emergence of Armenia, Azerbaijan, and Georgia as independent states, over forty sections of rivers became transboundary that had not been prior to the breakup. The Kura–Araks basin was no exception to this, and, as the lifeblood of the three nations in terms of agriculture, this was a significant change in the way the region thought about water. After the USSR dissolved, the three countries did not develop a legal framework for the management of the shared water resources of the region, thereby initiating the situation in which the Southern Caucasus finds itself today.

C.8.3 The problem

The reason that Armenia, Azerbaijan, and Georgia are being forced to confront the issue of the Kura–Araks river basin is because of problems of pollution. The rivers are heavily contaminated by chemical, industrial, biological, agricultural, and radioactive pollutants. The failure of wastewater treatment plants plays a major role in this dilemma in that the actual amount of water that is being treated is less than that of a decade ago. The concentrations of contaminants in the Kura–Araks basin reach levels that are much higher than standards in any of the three countries or internationally as well. Azerbaijan, the downstream nation, lacks groundwater resources like Georgia or Armenia and depends on the Kura–Araks basin for the majority of its agricultural, industrial, and household use. As the water flows into Azerbaijan polluted, the Azeris

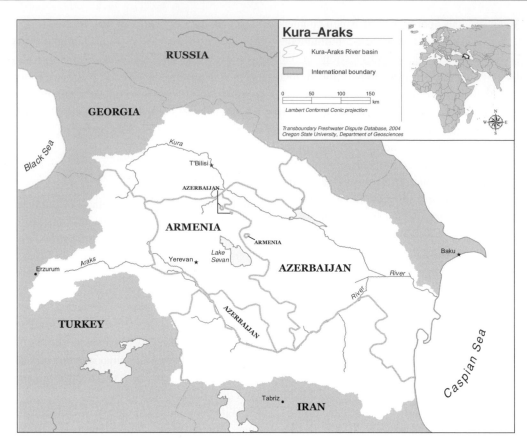

Figure C.8 Map of the Kura–Araks River basin (*Source:* Transboundary Freshwater Dispute Database, 2004).

complain about the contamination that takes place upstream in the other nations.

Compounding the issue is the political unrest between Armenia and Azerbaijan that has existed since 1988. After becoming independent nations in 1991 tensions between the two then-republics of the USSR sprouted into bloody conflict over the Nagorno–Karabakh region, an area embattled with conflict for decades previous. Armenia took over a good portion of the region from Azerbaijan and still controls the region even after a ceasefire took place in 1994. Even 10 years later, the issue has not yet been resolved and this has caused major tensions between the two countries with regard to its diplomatic relations and confronting other issues such as the Kura–Araks dilemma. It has been difficult for the two nations to come to the table to talk about the Kura–Araks rivers when the Nagorno–Karabakh dispute is still underway.

C.8.4 Attempts at conflict management

Due to the relations between Azerbaijan and Armenia, there has not been any advancement toward a regional entity or treaty that would assist in the cooperation of the management of the Kura–Araks river basin. Many bilateral agreements and laws have been signed between Georgia and Armenia and Georgia and Azerbaijan with regard to regulation of water use and management of both quality and quantity of water resources.

There are several international organizations, such as UNDP/GEF, USAID and TACIS, involved in the region to help with water resources management and development. The progress of such programs has been slow as a result of the tensions between Armenia and Azerbaijan, but a foundation is being established for future work between the nations when they are ready.

In 2002, the Regional Environmental Centre for the Caucasus hosted an international conference on "Water Resources Management in the Countries of the South Caucasus" in Tblisi, Georgia, among representatives of environmental agencies within the three governments, NGOs, parliamentary committees, scientists, the EU, international organizations, and donor agencies. The resolution agreed to by the participants took into consideration the following:

- Accelerate the reform of the management of water resources;
- Increase the level of involvement and initiatives by the public and by NGOs;

- Develop an environmental security strategy for water resources especially with regard to the hazardous material industries of oil, mining, and nuclear facilities;
- Develop a regional transboundary water management plan;
- Support a culture of sustainable water use;
- Encourage closer international cooperation in the sustainable use of water resources; and
- Improve the coordination and exchange of information between stakeholders.

These goals, and others, that were agreed to by the participants of the conference are a starting point from which the three nations of the Southern Caucasus can begin to establish good relationships with one another to build trust in order to develop a regional entity and treaty for the improved management of the Kura–Araks rivers.

C.8.5 Outcome

As of yet, there still has been little advancement toward an agreement with regard to the Kura–Araks rivers. It is thought that as long as there is the issue of the Nagorno–Karabakh region at hand, it will be very difficult for the governments to discuss environmental security when national security is still a major issue. Perhaps through building a more secure environment, with better living conditions and potable water, national security threats might prove to be easier to resolve.

C.8.6 Lessons learned

Political tensions between countries do not necessarily prevent governments from coming to the table to talk about issues such as management of their transboundary water resources.

As a result of the Nagorno–Karabakh issue, the relations between Armenia and Azerbaijan have been cold, and neither country has been willing to discuss the Kura–Araks problems to a great degree until the land issue has been resolved. With Georgia acting as a mediator between the two nations, this has slowed down the negotiation process to talks concerning the Kura–Araks, but they have moved forward nevertheless.

C.8.7 Creative outcomes resulting from resolution process

The principle of "parallel unilateralism" was developed here, allowing each collaborating pair of countries to work together, while coordinating the work of the countries that do not. Due to lack of movement from the three primary governments of the Kura–Araks river basin (Armenia, Azerbaijan, and Georgia) toward working together in the management of the river, fifty NGOs came together to form the NGO Coalition of the Kura–Araks in order to start activities between the three countries by cleaning up pollution and educating the public about the current situation.

C.8.8 Time line

1927 Turkey and the Union of Soviet Socialist Republics (USSR) sign the "Treaty on the Beneficial Uses of Boundary Waters" agreeing on a 50 percent/50 percent use of all the shared waters between the two nations. A Joint Boundary Water Commission was formed.

August 1957 Iran and the USSR sign bilateral agreement over the use of the Araks River waters. Similarly to the 1927 accord, each side is to receive half of the amount of water in the river for irrigation and hydropower generation.

1988 War breaks out between the republics of Armenia and Azerbaijan of the USSR over religious divide in the region.

1991 Republics of Armenia, Azerbaijan, and Georgia become independent nations after the collapse of USSR

1994 Ceasefire takes place between countries of Armenia and Azerbaijan. Armenia holds area of Nagorno–Karabakh and many parts of Azerbaijan proper.

1999 Establishment of NGO Coalition of the Kura–Araks, an organization of fifty NGOs from Armenia, Azerbaijan, and Georgia which undertakes the clean-up of contamination and raises awareness among communities in the three nations.

July 2001 First international meeting on the management of the Kura–Araks river basin brings together environmental representatives from the basin governments, NGOs, parliamentary committees, scientists, academics, and international donors.

C.9 LA PLATA BASIN

C.9.1 Case summary

River basin: La Plata (Figure C.9).
Dates of negotiation: La Plata Basin Treaty signed 1969.
Relevant parties: Argentina, Bolivia, Brazil, Paraguay, Uruguay.
Flash point: None.
Issues: Stated objectives: promote and coordinate joint development of the basin; "Hydrovia" proposed in 1989.

Figure C.9 Map of the La Plata River basin (*Source:* Transboundary Freshwater Dispute Database, 2004).

Additional issues: Water-related: Joint management; Non-water: None.

Excluded issues: Treaty does not provide any supralegal authority.

Criteria for water allocations: None.

Incentives/linkage: Possibility of linking water projects with transportation infrastructure.

Breakthroughs: None.

Status: Intergovernmental Coordinating Committee functions; "Hydrovia" technical and environmental studies in February 2004 by Andean Development Corporation.

C.9.2 Background

The La Plata River basin encompasses an area of 3.2 million square kilometers and is among the five largest international rivers basins in the world. It includes territory in Argentina, Bolivia, Brazil, Paraguay, and Uruguay; comprises the Paraná, Paraguay, and Uruguay river systems; and makes up the largest wetland in the world – the Pantanal. The basin is the life sustenance for much of the agricultural and industrial sectors of the

riparian states and has become a source of alternative energy and economic possibility.

The Basin's five riparian states have a history of cooperation and joint management of the watershed and have stressed the river's binding them to each other. Bolivia, Paraguay, and Uruguay's agriculture economies depend on the basin as crucially as the industrial sectors of Argentina and Brazil. Large amounts of grain, beef, wool, timber, and some manufacturing goods are exported from this region to other parts of the world (Elhance, 1999). The 1969 La Plata River Basin Treaty, the umbrella treaty and first to which all of the riparians are signatories, provides a framework for joint management, development and preservation of the basin. Subsequent multilateral and bilateral treaties outline the specifics of economic investment, hydroelectric development, and transportation enhancement.

Following the 1969 multilateral treaty, bilateral hydroelectric development opportunities were explored that gave source to the construction of dams and alternative power plants along the Parana. Today there are 130 dams along the river, two of which are widely known, the Itaipu and the Yacureta.

Itaipu is the largest hydroelectric project in the world and a result of a 1973 bilateral agreement between Paraguay and Brazil. The hydroelectric dam cost the two governments and other international participants US$15 billion and 20 years to construct. The generating capacity is 26,000 MW and supplies 26 percent of all of the electricity for Brazil and 78 percent for Paraguay with zero emissions.

The political and environmental dimensions of the Itaipu make for an interesting case of cooperation over a shared water resource. The land, where the Itaipu dam now sits, was once a source of great controversy between Brazil and Paraguay. Each country declared rights and legal authority over the Guaira Falls, which lies on the border of both countries and to which both claimed ownership and control. In 1957, Brazil, who believed the falls to be within their borders and who wanted to invest in the hydroelectric power of the falls, unilaterally took military control over the region. After 5 years of dispute and disagreement, Brazil and Paraguay finally negotiated the terms of the Itaipu dam. In addition to providing electricity to the two countries, the proposed project would submerge Guaira Falls (Elhance, 1999), thus marking an end to the border dispute.

This conflict negotiation and cooperation between Brazil and Paraguay had ripple effects into areas of conservation and preservation. When the environmental concerns around the construction of the Itaipu basin came to the forefront, the two countries implemented two joint projects, the Gralha Azul and the Mymba Kuera, to minimize the effects of reservoir flooding on the regions ecology and deforestation in the region and moved the wildlife most affected by the dam to biological reserves (American University Trade and Environment Database, 2004).

The Yacyreta Treaty, an agreement between Argentina and Paraguay, to construct a hydroelectric dam downstream from the Itaipu, has not been deemed as successful in its implementation. The treaty was hastily signed in December 1973, very soon after the Itaipu and was similar in content (generated power to be divided evenly between the two nations), except for the Yacyreta allowed for either country to sell power surpluses to a third party (Da Rosa, 1983). This contingency has since caused great confusion and complicated the construction.

The dam, from its inception, has become a "monument to corruption." The project has been unable to fill the reservoir to planned levels and is operating at two-thirds of its capacity because of the environmental repercussions the system would incur if it was at 100 percent capacity. Already US$1.3 billion worth of nongenerated energy has been lost due to delays. In addition, the indigenous populations along the river and beside the dam do not feel they were part of the planning process or were compensated for losses of their own land, and they do not believe they will be allocated power from the hydroelectric plant. At the moment, neither the Paraguayan nor the Argentine governments have the financial resources to allocate for improvements to the construction or to pay remittance to the four thousand families whose lives and environments have been affected by the construction of the dam.

Many bilateral treaties and hydroelectric projects have come out of the 1969 multilateral agreement; the first multilateral economic investment that joins all five riparian states and tests the framework of the La Plata Basin Treaty is termed "Hydrovia." "Hydrovia" is a proposed river transportation project that will dredge and straighten major portions of the Paraná and the Paraguay, including the portions of the river that lie in the Pantanal wetlands. The initial backers of "Hydrovia" ("waterway" in Spanish and Portuguese) were the governments of the La Plata basin states who met in 1988 to discuss the plans for the project (out of which was borne the Intergovernmental Commission on the Paraná–Paraguay Hydrovia). The project would allow year-round barge transportation (current conditions allow only for barges during the 3 dry months) and would open up a major transport thoroughfare for land-locked sections of the riparian states. The proposed waterway would make it possible for barge ships to take the 2,000-mile trip from Argentina and Uruguay ports of the Atlantic to landlocked Bolivia and Paraguay (American University Trade and Environment database, 1999). Environmentalists and those whose livelihoods depend on traditional economies have expressed trepidation at the project.

C.9.3 The problem

A cooperative management body has been in place on the La Plata basin since 1969 and is generally considered a successful and productive organization. At the same time, "Hydrovia" is the largest project for navigational river development proposed to date. Its size and possible impact on the economies and environments of the basin states are beginning to strain the cooperative nature of basin management. The biodiversity of the world's largest wetland, the Pantanal, could be strongly affected by construction of the waterway. Covering over 53,760 square miles in Brazil, Paraguay, and Bolivia, the Pantanal is home to 650 species of birds, 240 varieties of fish, and more than 90,000 types of plants (Bascheck and Hegglin, 2004). Opponents of the project point to loss of biodiversity and significant changes in the hydrology of the Pantanal as reasons why the project should be avoided. The Pantanal currently decreases the occurrence of floods and droughts in the downstream area (Lammers et al., 1994);

maintains the current ecosystem and hydrology there; and is the life sustenance of the people, animals, and wildlife along its banks.

C.9.4 Attempts at conflict management

The La Plata Basin Treaty of 1969 provides an umbrella framework for several bilateral treaties between the riparian states and a direction for joint development of the basin. The treaty requires open transportation and communication along the river and its tributaries, and prescribes cooperation in education, health, and management of "nonwater" resources (e.g., soil, forest, flora, and fauna). The foreign ministers of the riparian states provide the policy direction, and a standing Intergovernmental Coordination Committee is responsible for ongoing administration.

Basin states agree to identify and prioritize cooperative projects and to provide the technical and legal structure to see to their implementation, illustrated best by the 130 dams along the Parana, the construction of the world's largest hydroelectric project, Itaipu, and successive development, infrastructure and transportation projects. The treaty also has some limitations, notably the lack of a supralegal body to manage the treaty's provisions. The necessity to go through each country's legal system for individual projects has resulted in a time lag and lack of implementation.

The 1969 treaty's success has been in the areas of transportation and cooperation, so it is not altogether surprising that the Hydrovia project has been forwarded to the planning stages and that many multilateral and bilateral treaties came out of the 1969 La Plata Basin Treaty. The first meeting of the backers of the project was in April of 1988, from which the Intergovernmental Commission on the Paraná–Paraguay Hydrovia was formed.

C.9.5 Outcome

As positions between supporters and opponents of the project have sharpened, these positions are based on very little information. The Inter-American Development Bank and the United Nations Development Program, in 1997, helped finance a technical and environmental feasibility study by the Intergovernmental Commission on the Paraná–Paraguay Hydrovia. The study included dredging, rock removal, and structural channeling. Through motivation by independent technical critiques and environmental and social action networks the initial studies were discredited. As a result, the future of the Hydrovia is still uncertain. New studies were commissioned by Andean Development Corporation through the Intergovernmental Commis-

sion and were completed in February 2004, but the results have yet to be diffused.

C.9.6 Lessons learned

If riparian states start cooperation from the outset of a conflict, instead of letting it create stronger positions, the economic and joint management prospects are much greater.

Since 1969, the quantity of joint economic ventures in the La Plata basin has allowed for increased cooperation between the riparian nations when many times conflict could have arisen and defeated the benefits the states are receiving today.

If riparian states agree to equal access to transboundary water resources, equal and joint management, investment and distribution of that resource is feasible.

In the water resources sector, neither Brazil nor Argentina has used their economic or military superiority to maintain greater control over water resources or hydroelectric potential.

C.9.7 Creative outcomes resulting from resolution process

The La Plata Basin Treaty has helped bring the five nations together and aided in not only their own disputes but in disputes between sectors. The nations cooperate well, but the treaty nonetheless has been helpful.

Although the Hydrovia project was proposed in 1988, even now in 2004, there is still little movement toward implementing the project due to environmental and social action groups in defending the economic, cultural, and ecological integrity of the basin. In the end, this will allow for a more sustainable project.

C.9.8 Time line

1958 Yacyreta treaty is signed and the first joint Argentine–Paraguayan technical commission is formed to study the possibilities of obtaining hydroelectric energy from the rapids in the Paraná River.

April 1962 Negotiations between Paraguay and Brazil over the development of the rapids on the Paraná River for hydroelectric are interrupted by Brazil, who shows military force, invades, and claims control over the Guaria Falls sight.

1967 Brazilian forces withdraw and a joint Brazilian–Paraguayan commission is formed to examine the development of the region.

April 1969 La Plata Basin Treaty is signed by all five riparian states. The treaty provides a framework for the joint development of the basin; requires open transportation and communication along the river and its tributaries; requires cooperation in education and sanitation; and requires joint management of nonwater resources (soil, forest, flora, and fauna). An Intergovernmental Coordinating Committee is formed and is responsible for ongoing administration. Foreign ministers of the five riparian states are to provide policy initiatives.

April 1973 Itaipu treaty: Brazil and Paraguay announce plans to construct the Itaipu dam; Argentina expresses deep concern for the environmental repercussions of the dam and the effects of the dam on their own planned dam project.

December 1973 Yacyreta treaty: an Argentina–Paraguay organization, Yacyreta, is formed to oversee the construction of the hydroelectric dam and the contributing turbines. 1975 Itaipu dam construction begins.

December 1980 Joint declaration is made by the five riparian Foreign Ministers expressing a need to promote swift development of the resources on the La Plata basin.

April 1988 First meeting of the five riparian states on the proposed Hydrovia, a plan to develop the navigational infrastructure of the Paraná, the Paraguay and the Pantanal, to make an international waterway navigable by large, ocean-going vessels. Intergovernmental Commission on the Paraná-Paraguay Hydrovia formed.

March 1991 Bilateral treaty between Brazil and Uruguay – agree to joint development of the Cuareim River and cooperation in the use of its natural resources.

1991 The Itaipu project, the world's largest hydroelectric plant developed by Brazil and Paraguay on the Paraná River, is in full operation after 20 years of construction and US$15 billion in cost.

September 1994 The Yacyreta turbines begin to produce electrical power for Argentina and Paraguay.

C.10 AUTONOMOUS BINATIONAL AUTHORITY OF LAKE TITICACA

C.10.1 Case summary

River basin: TDPS System: Lake Titicaca, Desaguaduero River, Lake Poopó, Coipasa Salt Lake (Figure C.10).
Dates of negotiation: 1955–1996.
Relevant parties: Bolivia, Peru.
Flash point: None

Figure C.10 Map of Lake Titicaca (*Source:* Transboundary Freshwater Dispute Database, 2004).

Issues: Stated objectives: Management, protection, and control of the basin's water resources.
Additional issues: Water-related: extreme weather conditions, infrastructure projects, pollution; Nonwater: poverty, environmental degradation.
Excluded issues: Public participation.
Criteria for water allocations: None determined, Desaguadero River flow established by the maintenance of lake-level dependent on weather conditions.
Incentives/linkage: Mutual economic development in the region.
Breakthroughs: None.
Status: Autonomous Binational Authority working efficiently moving toward development goals.

C.10.2 Background

Populations have been living around Lake Titicaca for 10,000 years, dating back to the Archaic period. The first communities appeared around Titicaca in 1,200 BC and since then have increased in population and have become more dependent on its water for their livelihood for agriculture and navigation.

A series of natural occurring events took place in the 1980s that pushed the countries of Peru and Bolivia to manage the waters of Lake Titicaca in a more sustainable manner as the vulnerability of the inhabitants of the region was very high in extremely poor conditions that did not need to be exacerbated further. In the rainy seasons of 1982–1983 and 1989–1990, extreme droughts caused hundreds of millions of dollars in damage to the agricultural industry, both crop and animal. The years in between experienced a higher than average rainfall and culminated in the severe floods of 1986–1987, causing, again, over a hundred million dollars of damage to not only the agricultural industry but to infrastructure as well.

C.10.3 The problem

Relations between Peru and Bolivia have always been good, dating back to when they became independent nations in the 1800s. Lake Titicaca has not been a source of contention between the two states but rather a reinforcement of their willing to cooperate with one another when their interests are mutual.

The major problem, therefore, is not about conflict between Bolivia and Peru but how to develop and improve the living conditions of the extremely poor populations who live within the Titicaca basin. Mario Revollo (2001) of the Autonomous Bi-National Authority of Lake Titicaca outlines the four principal problems the lake region suffers.

EXTREME WEATHER EVENTS

As mentioned above, the Lake Titicaca region experiences a high variability in terms of its weather patterns. With such fluctuations in rainfall, the well-being of the inhabitants of the basin is controlled by how much water falls from the sky. And this, from year to year, can change from too much to too little. There is a high level of uncertainty, and risk, living under such conditions.

INSUFFICIENT REGULATORY WORKS

Even though the Lake Titicaca is very large and has significant volume, the hydrological balance of the entire system is very delicate due to the inflow vulnerability as a result of high evaporation. The regulation of the lake's water is deficient in that it does not prioritize sectors of water use and there are insufficient works in place to do so.

ENVIRONMENTAL DEGRADATION

Living beside such a large body of water, people sometimes take for granted the effects pollution can have. Although pollution has never been a regional concern for the two countries, as the volume of the lake is so large, there are several examples of acute cases of pollution near major population centers such as Puno, Peru, and Copacabana, Bolivia. The lack

of sewage treatment plants around the lake causes most waste to be put directly into Titicaca and, as a result, pollution levels have been rising over the decades, thereby contaminating the water.

Other sources of degradation come from the cattle industry that surrounds the lake and the loss of soil due to its impact. With regard to the fishing industry, the introduction of exotic species and the overfishing of both those and indigenous species has left the lake with smaller and smaller fish.

SOCIOECONOMICS

Extreme levels of poverty have existed in the Lake Titicaca basin for several decades now. This has been intensified by the two nations' negative economic growth rate since the late 1990s. Because most of the people who reside around the lake are subsistence farmers, the negative effects of Bolivia and Peru's economic decline have been acute. With everdiminishing and abused natural resources as a result of lack of education in the region, the day-to-day living conditions are not conducive to awareness regarding pollution and sustainability.

C.10.4 Attempts at conflict management

With great vision, Peru and Bolivia have been trying to address the development of the Lake Titicaca region since the 1950s. In 1957, after preliminary declarations by the presidents and foreign ministers, Bolivia and Peru signed the first-ever agreement concerning the waters of Lake Titicaca. It was called the Preliminary Convention for the Study of the Use of the Waters of Lake Titicaca and provided for the "indivisible and exclusive joint ownership of both countries of the waters of the lake," while at the same time creating a joint management entity known as the Joint Sub-commission (Sub-Comisión mixta). The purpose of the Convention was to promote development within the basin of Lake Titicaca in a manner that would not disrupt the flow and volume as to affect the navigational uses of the body of water.

Peru immediately ratified the Convention in 1957, but it took almost 30 years and several severe weather occurrences before, at the end of 1986, Bolivia also ratified the agreement. The economic losses incurred during the drought of 1982–1983 and the floods of 1986–1987, pressured the Bolivian government to ratify in order to improve the management situation of the lake. During the period before ratification, both countries conducted their own research concerning Lake Titicaca but did so in a coordinated way. After ratification occurred, the Joint Subcommission became SUBICOMILAGO, the Joint Subcommission for the Development of the Integrated Region of Lake Titicaca. Entities within each country were formed during

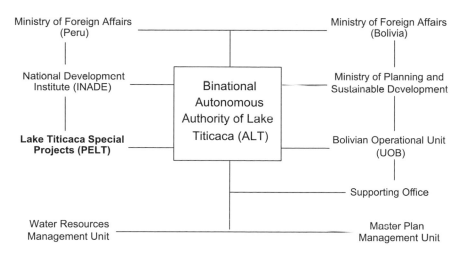

Figure C.11 Lake Titicaca RBO organizational chart.

this same time period, PELT (Lake Titicaca Special Projects) on the Peruvian side and the UOB (Bolivian Operating Unit) on the Bolivian side.

From 1991 to 1993, Peru and Bolivia solicited the cooperation of the European Community in order to help develop a framework for a Binational Master Plan. By 1995, the Binational Master Plan for the Control and Prevention of Floods and for the Use of Resources of the TDPS System (Lake Titicaca, Desaguadero River, Lake Poopo, and Coipasa Salt Lake) had been approved by both countries and, in April of 1996, signed and put into effect by June 1.

During the process of the creation of the Master Plan, diplomatic notes were exchanged between the governments of Peru and Bolivia, which led to the establishment of the Binational Autonomous Authority of Lake Titicaca (ALT).

C.10.5 Outcome

The Autonomous Binational Authority of Lake Titicaca (ALT) was created with the objective to implement and enforce the management, control, and protection of the Lake Titicaca system's water resources as laid out in the Master Plan. Each country has administrative entities that coordinate with the Ministries of Foreign Affairs of both nations and with one another. The technically oriented units of Peru and Bolivia, PELT and UOB, respectively, coordinate the actions of the governments and centralize information (Figure C.11).

Since its inauguration in 1996, ALT has been able to achieve some considerable advancement in the area of regulatory works within the basin. A series of projects was initiated and the first major dam was finished in 2001, near the mouth of the Desaguadero River. These "doors" will attempt to control flood situations when the level of the lake rises above 3,810 meters

above sea level. In creating this dam, irrigation yields have increased on both sides of the border as Peruvians and Bolivians are better able to utilize the lakes water resources.

Although ALT, a concept, has been considered a success story, because of its ability to prevent natural disasters from having large impacts on the local populations around Lake Titicaca and how smoothly the entity operates, there still has been only minimal progress in terms of achieving its goals that it set out to do in the Master Plan. ALT has only been in existence for less than 10 years, so it is a very young entity and, at times, is working in a climate of civil unrest on both sides of the border, which has an influence on its effectiveness.

The major concern, and the central reason and why ALT has not been very effective in the basin, is their lack of programs to include the public in a participatory process in the management of the lake. Without such mechanism in place, there is only so much that ALT can accomplish in an area that is so struck by poverty. A lack of stakeholder participation is hurting the success of the Binational Authority.

ALT has advanced a great degree in a short time and it must be said that the organization has great potential for being one of the model international water basin management institutions in the world.

C.10.6 Lessons learned

Without stakeholder participation in the management of water resources, efficiency and effectiveness are limited.

With little or no stakeholder participation in the management of the Lake Titicaca basin, ALT has been only minimally effective at producing results. It is clear that a more comprehensive system of inclusion of the public is needed to take place in order for the Authority to complete its goals. If three of the four problems identified by the institution deal with the people's actions on the water and land in the basin, then they must be included for optimal functioning of the initiative. Otherwise, gaps and resentment are created by an organization that acts above those who most use the lake.

By viewing the basin as a joint body of water shared equally between countries, much conflict is avoided.

By signing an agreement in 1957, Peru and Bolivia bound themselves into considering Lake Titicaca as a shared body of water, owned by neither country, but both. As a result, there are few, if any, "upstream versus downstream" issues (even though the Desaguadero River does flow into Bolivia from the lake). The countries have worked very well in a cooperative way to manage the lake, both doing their parts. This can be largely attributed to the lake being "owned" by both nations.

C.10.7 Creative outcomes resulting from resolution process

The development of a master plan in conjunction with a joint autonomous management entity that oversees the development of the lake has allowed the two nations to move forward with relative ease once funding was secured for joint ventures.

C.10.8 Time line

1955 Declarations by Presidents and Ministers of the States of Bolivia and Peru to begin diagnostic studies of the Lake Titicaca basin.

1957 Both countries signed the "Preliminary Convention for the Study of the Use of the Waters of Lake Titicaca," an agreement establishing the adoption of a plan to develop the economic uses of Lake Titicaca without altering the navigation and volume of the lake and creating a joint management entity known as SUBCOMILAGO, the Joint Subcommission for the Development of the Integrated Region of Lake Titicaca.

1982–1983 Severe drought causes hundreds of millions of dollars of damage to agricultural industry.

1986–1987 Severe floods cause hundreds of millions of dollars of damage to agricultural industry and infrastructure.

1987 Natural disasters of the previous years promote the ratification of the Preliminary Convention of 1957 and initiates the first meeting of SUBCOMILAGO.

September 1987 With the aid of the European Community, Peru and Bolivia formulate both a plan for the regulation of the waters of Lake Titicaca and a management use plan called "Global Bi-national Master Plan for the Development of the Integrated Region of Lake Titicaca."

1991–1993 Peruvian and Bolivian governments work in cooperation with the European Community to develop a Binational Master Plan for the development of Lake Titicaca.

1995 Binational Master Plan for the Control and Prevention of Floods and for the Use of Resources of the TDPS System (Lake Titicaca, Desaguadero River, Lake Poopo and Coipasa Salt Lake) approved by both nations.

1996 By public international law, the Autonomous Binational Authority of Lake Titicaca (ALT) is created by the governments of Peru and Bolivia.

C.11 LESOTHO HIGHLANDS WATER PROJECT

C.11.1 Case summary

River basin: Senqu River (Figure C.12).

Dates of negotiation: 1978–1986, ongoing negotiation provided for in treaty.

Relevant parties: Lesotho, South Africa

Flash point: Water deficit in South African industrial hub.

Issues: Stated objectives: negotiate technical and financial details of water transfer from Lesotho to South Africa.

Additional issues: Water-related: hydropower for Lesotho internal consumption; Nonwater: general development.

Excluded issues: None.

Criteria for water allocations: Amount for sale negotiated for treaty.

Incentives/linkage: South Africa buys water from Lesotho and finances diversion; Lesotho uses payments and development aid for hydropower generation and general development.

Breakthroughs: Financing arrangement negotiated that allowed for international funding.

Status: Phase I of project completed in 2004, feasibility of Phase II currently being studied

C.11.2 Background

Development in Lesotho has been limited by its lack of natural resources and investment capital. Water is its only abundant resource, which is precisely what regions of neighboring South Africa have been lacking. A project to transfer water from the Senqu River to South Africa had been investigated in the 1950s and again in the 1960s. The project was never implemented due to disagreement over appropriate payment for the water.

C.11.3 The problem

Lesotho, completely surrounded by South Africa, is a state poor in most natural resources, water being the exception. The

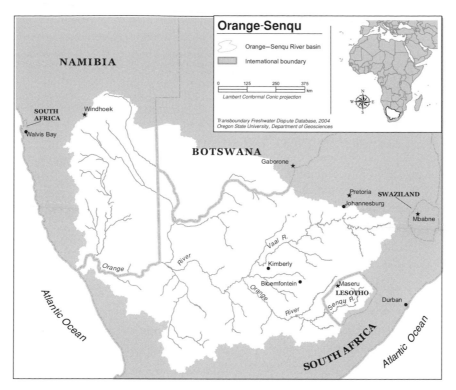

Figure C.12 Map of the Senqu River (Lesotho Highlands Water Project) (*Source:* Transboundary Freshwater Dispute Database, 2004).

industrial hub of South Africa, from Pretoria to Witwatersrand, has been exploiting most of the local water resources for years and the South African government has been in search of alternate sources. The elaborate technical and financial arrangements that led to construction of the Lesotho Highlands Water Project (LHWP) provide a good example of the possible gains of an integrative arrangement, including a diverse "basket" of benefits.

C.11.4 Attempts at conflict management

In 1978, the governments of Lesotho and South Africa appointed a joint technical team to investigate the possibility of a water transfer project. The first feasibility study suggested a project to transfer 35 m³/sec, four dams, 100 kilometers of transfer tunnel, and a hydropower component. Agreement was reached to study the project in more detail, the cost of the study to be borne by both governments.

The second feasibility study, completed in 1986, concluded that the project was feasible and recommended that the amount of water to be transferred be doubled to 70 m³/sec. A treaty between the two states was necessary to negotiate for this international project. Negotiations proceeded through 1986 and the "Treaty on the Lesotho Highlands Water Project between the Government of the Kingdom of Lesotho and the Government of the Republic of South Africa" was signed into law on October 24, 1986.

It is testimony to the resilience of these arrangements that no significant changes were made despite the dramatic political shifts in South Africa at the end of the 1980s until 1990.

C.11.5 Outcome

The Treaty spells out an elaborate arrangement of technical, economic, and political intricacy. A boycott of international aid for apartheid South Africa required that the project be financed, and managed, in sections. The water transfer component was entirely financed by South Africa, which would also make payments for the water that would be delivered. The hydropower and development components were undertaken by Lesotho, which received international aid from a variety of donor agencies, particularly the World Bank. Phase IA of the Lesotho Highlands Water Project was completed in 1998, at a cost of $2.4 billion. Phase IB of the project was completed in early 2004, at a cost of approximately $1.5 billion.

The 1986 treaty provided for the construction of additional phases (II–IV). However, changes in the projection of water demand in South Africa, along with concerns over negative social and environmental impacts of the project, have led to negotiations on the future phases. In 2004 a feasibility study of Phase II began between the nations of South Africa and Lesotho.

Although Environmental Action Plans (EAPs) were carried out for both Phases IA and IB, EAPs for Phase IA were carried out while construction for the phase was already underway. It was in the course of Phase IB EAPs in 1994 that the need for an instream flow requirement became apparent (see http://www.metsi.com/LHWP/ifr.htm#motivating). After studies of the biophysical, social, and economic effects of the project were carried out, an Instream Flow Requirement (IFR) policy was implemented in 2002. In particular, river reaches and communities downstream of the project sites were considered in the assessment, whereas EAPs of Phase I considered only those areas only upstream of the project sites.

C.11.6 Lessons learned

Even with power disparity, there is possibility for agreement over water resources through economic benefits.

South Africa is a much more powerful nation than Lesotho, but Lesotho has abundant water resources, which, through the Highlands Project, will benefit both nations economically and through the provision of water to South Africa. It is possible, even when there is such a wide gap between nations in terms of power, to collaborate for the mutual gain of both countries.

It is more economically sound to begin impact studies before nations start to construct projects.

It was shown through the Lesotho Highlands Water Project that if impact studies are started *after* the initiation of a major hydroproject, the costs for the project go up because necessary components for the project may not have been considered prestudy. For the Phase II of the LHWP, studies are being conducted to judge the feasibility of a project that was designed more than 15 years to ago to investigate in a more comprehensive manner the possible impacts of the project.

Renegotiation clauses in an agreement can prevent issues from arising for the nations involved.

The LHWP treaty also exemplifies the importance of providing for renegotiation of project terms. In the absence of such a provision, the additional phases of the project might have been implemented without adequate consideration of their feasibility

C.11.7 Creative outcomes resulting from resolution process

The Lesotho Highlands Water Project provides lessons in the importance of an integrated approach to negotiating the allocation of a "basket" of resources. South Africa receives cost-effective water for its continued growth, while Lesotho receives revenue and hydropower for its own development.

C.11.8 Time line

1930–1977 Feasibility studies and surveying of water potential in Lesotho.

1978 Joint preliminary feasibility study carried out by consultants from South Africa and Lesotho.

1983–1985 Joint detailed feasibility study.

1986 Lesotho Highlands Water Project Treaty signed by the government of Lesotho and of the Republic of South Africa. Establishment of Joint Permanent Technical Commission to represent two governments.

1990 End of Apartheid Era, South Africa. Construction begins on Phase 1.

1996 Workers protest at the LHWP Site in Butha Buthe; several workers killed and many wounded.

1998 Phase IA completed. First water supply from Lesotho to South Africa.

2004 Phase IB completed.

Present Phase II feasibility study being conducted binationally with 50/50 input and cost-sharing between Lesotho and South Africa.

C.12 MEKONG COMMITTEE

C.12.1 Case summary

River basin: Mekong River (Figure C.13).

Dates of negotiation: Committee formed 1957.

Relevant parties: Cambodia, Laos, Thailand, Vietnam (directly); China, Myanmar (indirectly).

Flash point: None – studies by UN-ECAFE (1952, 1957) and U.S. Bureau of Reclamation provide impetus for creation of Mekong Committee.

Issues: Stated objectives: Promote, coordinate, supervise, and control the planning and investigation of water resources development projects in the Lower Mekong basin.

Additional issues: Nonwater: general political relations between riparians

Excluded issues: China and Myanmar were not included since inception; Cambodia not included from 1978 to 1991.

Criteria for water allocations: Allocations have not been an issue; "reasonable and equitable use" for the basin defined in detail since 1975.

Figure C.13 Map of the Mekong River basin (*Source:* Transboundary Freshwater Dispute Database, 2004).

Incentives/linkage: Financial: extensive funding from international community; Political: facilitated relations between riparians, aid from both east and west despite political tensions.

Breakthroughs: Studies by UN-ECAFE and U.S. Bureau of Reclamation in 1950s.

Status: Mekong Committee established in 1957, became Interim Committee in 1978 with original members except for Cambodia. Early momentum dropped off but has resurfaced with extensive programs and project proposals – extensive data networks and databases established, Committee reratified as Mekong Commission in 1995.

C.12.2 Background

The Mekong is the seventh largest river in the world in terms of discharge (tenth in length), rising in China and then flowing 4,200 kilometers through Myanmar, Laos, Thailand, Cambodia, and finally the extensive delta in Vietnam into the South China Sea. It is also both the first successful application of a comprehensive approach to planning development of an international river and, at the same time, is one of the least developed major rivers in the world, in part because of difficulties inherent in implementing joint management between these the diverse riparians.

In 1947, the United Nations Economic Commission for Asia and the Far East (ECAFE) was created to help with the development of Southeast Asia. A 1952 ECAFE study, undertaken with the cooperation of the four lower riparians – Cambodia, Laos, Thailand, and Vietnam – noted the Mekong's particular potential for hydroelectric and irrigation development. These recommendations could not be acted on until the signing of the Geneva Accords in 1954 ended hostilities in the region.

The U.S. Bureau of Reclamation performed a report on planning and development on the lower basin in 1955–1956, which urged joint management in developing the river, to which the four lower riparians agreed. The study noted the almost total absence of data necessary for river basin planning, emphasized the need to get a program for data collection and analysis under-way immediately, and offered suggestions for the types of programs that should be implemented.

A 1957 ECAFE report concurred with the optimistic potential noted in earlier studies. The report noted that harnessing the main stem of the river would allow hydropower production, expansion of irrigated land, a reduction of the threat of flooding in the delta region, and the extension of navigability of the river as far as northern Laos. As earlier studies had, the ECAFE report emphasized the need for comprehensive development of the river and close cooperation between the riparians in coordinating efforts for projects and management. To facilitate coordination, the report suggested the establishment of an international body for exchanging information and development plans between the riparian states. Ultimately, the report suggested, such a body might become a permanent agency responsible for coordinating joint management of the Mekong Basin. When the report was presented in the tenth-anniversary meeting of ECAFE in Bangkok in March 1957, representatives from the four lower riparian states themselves adopted resolution calling for further study.

C.12.3 The problem

As is common in international river basins, integrated planning for efficient watershed management is hampered by the difficulties of coordinating between riparian states with diverse and often conflicting needs. The Mekong, however, is noted mostly for the exceptions as compared with other basins rather than the similarities. For example, the Mekong is not an exotic stream and consequently does not have the sharp management conflicts between well-watered upstream riparians and their water-poor downstream neighbors as with, for instance, the Euphrates and the Nile. Historically, the two uppermost riparians, China and Myanmar, have not been participants in basin planning, and they have had no development plans that would disrupt the downstream riparians until very recently. Also, because the region is so well watered, allocations *per se* are not a major issue. Finally, negotiations for joint management of the Mekong were not set off by a flash point, as were all of the other examples presented in this work, but rather by creativity and foresight on the part of an authoritative third party – the United Nations – with the willing participation of the lower riparian states

More recently, however, the liberalization of China's economy, population growth, demand for increased agriculture yields, growing household demand of water for consumption and sanitation, and shortages of electricity have incited Chinese officials to look to the potential of the Mekong's upper basin. It is not, therefore, surprising that China would like to fully develop the Upper Mekong basin and has proposed the building of fifteen dams for hydroelectric power (Elhance, p. 197). This unilateral development project alone would have large implications for the downstream riparian states. In the absence of basinwide consensus and cooperation, these unilateral developments have the potential to make the hydropolitics in the Mekong basin much more contentious (Elhance, p. 198). The completion of two major dams on the Chinese part of the Lacang–Mekong mainstream, and the prospect of six or seven more hydropower dams in that area, coupled with the recent in navigability along the Mekong (by blasting the rapids and rocks), underline the urgent need to build and appropriate legal framework and to formulate technical guidelines conducive to turning these potential conflicts into opportunities for sharing benefits (UNESCO-PCCP 2007).

C.12.4 Attempts at conflict management

As we have noted, the 1957 ECAFE study was met with enthusiasm by the lower Mekong riparians. In mid-September 1957, after ECAFE's legal experts had designed a draft charter for a "coordination committee," the lower riparians convened again in Bangkok as a "preparatory commission." The Commission studied, modified, and finally endorsed a statute that legally established the Committee for Coordination of Investigations of the Lower Mekong (Mekong Committee), made up of representatives of the four lower riparians, with input and support from the United Nations. The statute was signed on September 17, 1957.

The Committee was composed of "plenipotentiary" representatives of the four countries, meaning that each representative had the authority to speak for their country. The Committee was authorized to "promote, coordinate, supervise, and control the planning and investigation of water resources development projects in the Lower Mekong Basin." The statute included authority to:

- prepare and submit to participating governments plans for carrying out coordinated research, study, and investigation;
- make requests on behalf of the participating governments for special financial and technical assistance and receive and administer separately such financial and technical assistance as may be offered under the technical assistance program of the United Nations, the specialized agencies, and friendly governments; and
- draw up and recommend to participating governments criteria for the use of the water of the main river for the purpose of water resources development.

It was determined that all meetings must be attended by a representative from each of the four countries and each

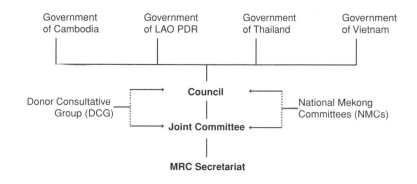

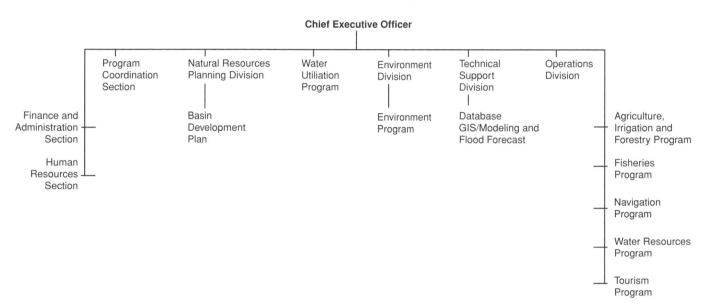

Figure C.14 Organization chart of the Mekong River Committee.

decision must be unanimous. Meetings would be held three to four times a year, and chairmanship would rotate annually in alphabetical order by country (Figure C.14).

The first Committee session was on October 31, 1957, as was the first donation from the international community – 60 million francs (about US$120,000) from France. In late 1957, the Committee, recognizing that data collection was a crucial prerequisite to comprehensive watershed development, asked the UN Technical Assistance Administration to organize a high-level study of the basin. Before the year was out, a mission headed by Lieutenant General Raymond Wheeler (who had been the deputy commander of the Allied bases in the region during World War II and later Chief of the U.S. Army Corps of Engineers) arrived in Bangkok.

The principal recommendation of the Wheeler Mission was that, while reaffirming the great potential of water resources development, suggesting that, properly developed, the river, "could easily rank with Southeast Asia's greatest natural resources," the absence of data required that a series of detailed hydrographic studies precede any construction. The mission recommended a 5-year program of study, to cost approximately US$9 million (Table C.6).

At its second session, February 10–12, 1958, the Mekong Committee adopted Wheeler's program as its own 5-year plan. It also accepted another suggestion of the Wheeler Mission that a permanent advisory board of professional engineers "of worldwide reputation" be established. It likewise noted the desirability of having a full-time director with ancillary staff. ECAFE responded and appointed members to the advisory board, secured Committee approval for the appointment of Dr. C. Hart Schaaf as Executive Agent, who assumed office in mid-1959, and established the Committee Secretariat as an ECAFE adjunct body to which UN staff members could be assigned.

With rapid agreement between the riparians came extensive international support for the work of the Committee – by

Table C.6 *Recommendations of the Wheeler Mission, 1958*

Study or action	Countries/agencies participating	Begun
Preliminary reconnaissance of major tributaries	Japan	1959
Hydrologic and meteorologic observations	United States, France, Great Britain, India	1959
Aerial mapping and leveling	Canada, Philippines	1959
Soil surveys	France	1959
Geological investigations	Australia	1961
Hydrographic survey	UN, Belgium, United States, Great Britain, New Zealand, the Netherlands	1961
Related and special studies[a]	UN, United States, France, Private agencies, Nordic countries	1962
Preliminary planning of projects on main stem	United States, Japan, India, Australia, France	1959
Preparation of basinwide plan	Mekong Committee, aided by ECAFE Secretariat	1959
Appointment of advisory board		1958

[a] Including studies of fisheries, agriculture, forestry, minerals, transportation, and power markets.

1961, the Committee's resources came to $14 million, more than enough to fund field surveys that had been agreed to as priority projects. By the end of 1965, twenty countries, eleven international agencies, and several private organizations had pledged a total of more than $100 million. The Secretariat itself was funded by a special $2.5 million grant made by UNDP. This group of international participants has been dubbed "the Mekong club," which has infused the international community with "the Mekong spirit" (Table C.7).

Along with the collection of physical data and the establishment of hydrographic networks, the Mekong Committee encouraged the undertaking of economic and social studies and the initiation of training programs. In 1961, Professor Gilbert White headed a mission, sponsored by the Ford Foun-dation, which found that, while existing and planned projects would provide water for irrigation and power for industry, these resources could be used to their maximum benefit only with extensive training of the local population. In an important shift from a strictly engineering approach, many of the mission's recommendations have been adopted.

C.12.5 Outcome

The early years were the most productive for the Mekong Committee. Networks of hydrologic and meteorologic stations have been established and have continued to function despite hostilities in the region, as have programs for aerial mapping, surveying, and leveling. Navigation has been improved along

Table C.7 *Studies recommended by the Ford Mission, 1961*

Study or action	Countries/agencies participating	Begun
Addition of skilled personnel to deal with economic and social studies	Mekong Committee, riparians, ECAFE	1962
Development of programs to train personnel for economic and social studies and to use products of river development	Mekong Committee, riparians	1963
Manpower studies	ILO	1966
Fisheries studies	France	1960
Minerals studies	France	1962
Agricultural surveys		n/a
Studies of patterns and levels of living		n/a
Estimates of demands for electric power	France, Resources for the Future, Mekong Committee	1962
Studies of adjustments to floods	UN/TAB, France	1961
Development of agriculture demonstration projects	UN, France, Israel	1962
Establishment of experimental forest		n/a

the main stem of the river, but no major project has yet to be initiated, although dozens have been proposed.

The work of the Committee also helped overcome political suspicion through increased integration. In 1965, Thailand and Laos signed an agreement to develop the power potential of the Nam Ngum River, a Mekong tributary inside Laos. Because most of the power demand was in Thailand, which was willing to buy power at a price based on savings in fuel costs, and because Laos did not have the resources to finance the project, an international effort was mobilized through the Committee to help develop the project. As a sign of the Committee's viability, the mutual flow of electricity for foreign capital between Laos and Thailand was never interrupted, despite hostilities between the two countries.

By the 1970s, the early momentum of the Mekong Committee began to subside for several reasons. First, the political and financial obstacles necessary to move from data gathering and feasibility studies to concrete development projects were often too great to overcome. A 1970 Indicative Basin Plan marked the potential shift between planning and large-scale implementation, including immense power, flood control, irrigation, and navigation projects, and setting out a basin development framework for the next 30 years. In 1975, the riparians set out to refine the Committee's objectives and principles for development in support of the Plan in a "Joint Declaration on Principles," including the first (and so far only) precise definition of "reasonable and equitable use" based on the 1966 Helsinki Rules ever used in an international agreement (International Law Association, 1966). The plan, which included three of the largest hydroelectric power projects in the world as part of a series of seven cascading dams, was received with skepticism by some in the international community (Kirmani 1990, p. 203). At the current time, although many projects have been built along the tributaries of the Mekong within single countries, and despite the update of the Indicative Plan in 1987 and a subsequent "Action Plan" that includes only two low dams, no single structure has been built across the main stem.

Second, although the Committee continued to meet despite political tensions, and even despite outright hostilities, political obstacles did take their toll on their work. Notably, the Committee became a three-member "interim committee" in 1978 with the lack of a representative government in Cambodia. Cambodia rejoined the committee as a full participant in 1991, although the Committee still retains "interim" status. Likewise, funding and involvement from the United States, which had been about 12 percent of the total aid to the Committee, was cut off in June 1975 and has not been restored to significant levels.

Finally, some regional politics between the riparians have been played out through the Mekong Committee. Thailand, with the strongest economy and greatest resource needs, has been pushing in recent years for revisions in the Committee's rules that currently allow an effective veto of Thai projects by downstream riparians. Thailand has found its own funding for four Mekong projects within its own territory and has plans for several more, some of which would probably be opposed by downstream riparians if they were brought before the Mekong Committee. In 1992, Thailand canceled a plenary meeting two days before it was scheduled, and later asked the UNDP to remove the Executive Agent, a request with which the UNDP complied.

Renewed activity came with the signing of the Paris Peace Agreement in 1991, after which Cambodia requested the reactivation of the Mekong Committee. The four lower riparians took up the call and spent the next 4 years determining a future direction for Mekong activities. The results of these meetings culminated finally in a new agreement, signed in April 1995, in which the Mekong Committee became the Mekong Commission. Although it is too early yet to evaluate this renewed body, the fact that the riparians have made a new commitment to jointly manage the lower basin speaks well at least for the resiliency of agreements put into place in advance of hot conflict. It should also be noted that Myanmar and China are still not party to the agreement, effectively precluding integrated basin management.

Although the establishment of the Mekong Committee and its work provide an impressive example of the potential of integrated watershed management on an international scale, its actual accomplishments have not kept pace with its early momentum, likewise providing lessons for the international arena. The 1995 *Agreement Towards Sustainable Development* under the Mekong River Commission lacks the political power and support from China and Myanmar needed to successfully implement all of the goals of the Commission and may mirror past lack of momentum if these two countries are not brought on board.

Since its inception in 1995, the Mekong River Commission has been implementing many programs under its jurisdiction. The following are the programs already underway (Mekong River Commission): Basin Development Plan; Water Utilization Program; Environment Program; Flood Management Program; Capacity-Building Program; Agriculture, Irrigation and Forestry Program; Fisheries Program; Navigation Program; and Water Resources and Hydrology.

Of the few projects that have been implemented within the Mekong River Commission, none have been constructed on the main stem of the river. Two major dams can be found on tributaries of the Mekong: the Pak Moon dam in Thailand, on the Pak Moon River, and the Theun-Hinboun dam, on the Theun River in Laos.

C.12.6 Lessons learned

Establishing an international framework for integrated watershed management well before a flash point makes the task easier and more likely to succeed during later times of stress.

Both the riparians of the Lower Mekong and the international community saw the potential of a well-managed river well before "water stress" led to a crisis. By establishing and utilizing the necessary management infrastructure before respective senses of urgency had the chance to hamper political decision making, the Mekong Committee had already developed a routine of cooperation that proceeded despite later political tensions.

Emphasizing data collection in advance of any construction projects sets the hydrographic stage for more efficient planning and may establish a pattern of cooperation through relatively emotion-free issues.

The insistence of the Wheeler Mission to conduct extensive data-gathering before beginning any construction made both management and political sense.

Solving water-related issues involves both technical and social aspects of development.

The importance of the White Mission was a conceptual shift from a strictly engineering perspective of the challenges of the river to a social view that sought also to address the needs of the riparian population.

The greater the international involvement in conflict resolution, the greater the political and financial incentives to cooperate.

The pace of development and cooperation in the Mekong River watershed over the years has been commensurate with the level of involvement of the international community. Early accomplishments were impressive, impelled in part by strong UN support and a "Mekong spirit" on the part of the "Mekong club" of donors. By the 1970s, the pace of cooperative development began to slacken, partly the result of decreasing involvement by an international community daunted by political obstacles and the size of planned projects.

For an environmentally feasible and sustainably sound management to occur, all riparian states must be present.

The two upper-stream countires – Myanmar and China – need to be involved in international cooperation over the transboundary river basin.

C.12.7 Creative outcomes resulting from resolution process

The early accomplishments of the Mekong Committee, and the particularly ordered approach to the basin – establishment of joint management, data collection, feasibility studies of both technical and social aspects of development, implementation – provide a useful model for any international basin.

The legally intricate question of "reasonable and equitable" use of the basin was defined in detail, the first (and so far only) explicit use of the principles of the 1966 Helsinki Rules in any international agreement.

C.12.8 Time line

1947 United Nations Economic Commission for Asia and the Far East (ECAFE) is created to help with the development of Southeast Asia.

1952 ECAFE study notes Mekong's potential for hydroelectric and irrigation development.

1954 Geneva Accords signed, ending hostilities in the region.

1955–1956 U.S. Bureau of Reclamation report on planning and development in the lower basin urges joint management in developing the river. Four lower riparians – Cambodia, Laos, Thailand, and Vietnam – agree.

1957 ECAFE report concurs with earlier findings. When the report is presented to an ECAFE meeting in March, the riparians themselves call for further study.

September 1957 Riparians negotiate a draft charter for the "Committee for Coordination of Investigations of the Lower Mekong." Statute signed on September 17, 1957, bringing Mekong Committee into legal existence.

Late 1957 Wheeler Mission suggests that first priority be data gathering throughout the basin, in advance of any construction. Wheeler's program adopted as Mekong Committee's first 5-year plan.

1961 White Mission urges that the social aspects of development be investigated commensurate with technical aspects of development. Many of Mission's recommendations for training programs are adopted.

1965 Laos and Thailand sign agreement on power generation project on Nam Ngum River, a Mekong tributary within Laos, by which Thailand agrees to buy surplus power. Exchange of power for foreign capital never discontinued, despite tensions between the two countries.

1970 Indicative Basin Plan describes proposed large-scale development of Mekong basin.

1975 Joint Declaration on Principles signed, including the first precise definition of "reasonable and equitable use," as described in Helsinki Rules, ever used in international agreement.

1978 Mekong Committee becomes a three-member "Interim Mekong Committee," with the lack of a representative government in Cambodia.

1987 Indicative Plan revised and updated.

1991 Cambodia rejoins as full participant, but Committee remains legally "interim."

1991 Greater Mekong Subregion (GMS) Cooperation Program begins for cooperation in development of the region.

1992 Thailand asks UNDP to remove Executive Agent; UNDP complies.

1995 Mekong Committee reratified as Mekong Commission. *Cooperation for the Sustainable Development of the Mekong River Basin* signed by Cambodia, Laos. Thailand, and Vietnam for sustainable development, utilization, conservation, and management of the basin while attempting to bring the two upstream countries, Myanmar and China, into the cooperation.

C.13 MULTILATERAL WORKING GROUP ON WATER RESOURCES (MIDDLE EAST)

C.13.1 Case summary

River basin: All water resources of the Middle East (Figure C.15).

Dates of negotiation: 1992–present.

Relevant parties: United States, European Union, Canada, and France (donor parties) and Russia (sponsoring); bilateral parties (except Syria and Lebanon): Israel, Jordan, Palestine (core parties), Egypt; Periphery: Egypt and Arab states from Gulf and Maghreb.

Flash point: None.

Issues: Stated objectives: help develop capacity for greater efficiency in water supply, demand, and institutions throughout the Middle East, in support of bilateral peace negotiations.

Additional issues: Nonwater: personal ice-breaking and confidence-building.

Excluded issues: Water rights, multiriparian agreements, water quality.

Criteria for water allocations: None.

Incentives/linkage: Financial: donor parties helping to finance feasibility studies and implementation as agreements take place. Political: talks work in conjunction with bilateral negotiations.

Breakthroughs: Question of water rights successfully relegated to bilateral talks; creation of a Palestinian Water Authority accepted by all parties; first Arab proposal for water group and first Israeli proposal for any working group accepted by consensus.

Status: Meetings are ongoing. Due to the outbreak of the second Intifada in 2000 new efforts for cooperation are rare. Concerned countries are focused on keeping the status quo, and protecting the water infrastructures from damage.

C.13.2 Background

By 1991, several events combined to enhance the potential for "hydro-cooperation." The first event was natural but limited to the Jordan basin. Three years of below-average rainfall caused a dramatic tightening in the water management practices of each of the riparians – Israel, Jordan, Lebanon, Syria and the Palestinian Territories – including rationing, cut-backs to agriculture by as much as 30 percent, and restructuring of water pricing and allocations. Although these steps placed short-term hardships on those affected, they also showed that, for years of normal rainfall, there was still some flexibility in the system. Most water decision makers agree that these steps, particularly regarding pricing practices and allocations to agriculture, were long overdue.

The next series of events were geopolitical and region-wide in nature. The Gulf War in 1990 and the collapse of the Soviet Union caused a realignment of political alliances in the Mideast, which finally made possible the first public face-to-face peace talks between Arabs and Israelis, in Madrid on October 30, 1991. This breakthrough was followed by an organizational meeting in Moscow in January 1992, which established a multilateral track that would act alongside the bilateral track. The multilateral track focuses collaboration efforts on five regionally relevant subjects, including the Multilateral Working Group on Water Resources (MWGWR). The "core parties" of this group are Israel, the West Bank/Gaza, and Jordan.

C.13.3 The problem

Until the current Arab–Israeli peace negotiations began in 1991, attempts at Middle East conflict resolution had endeavored to tackle either political or resource problems, always separately. By separating the two realms of "high" and "low" politics, some have argued, each process was doomed to fail. In water resource issues – the Johnston Negotiations of the mid-1950s, attempts at "water-for-peace" through nuclear desalination in the late 1960s, negotiations over the Yarmuk River in the 1970s and 1980s, and the Global Water Summit Initiative of

Figure C.15 Map of all water resources of the Middle East (*Source:* Transboundary Freshwater Dispute Database, 2004).

1991 – all addressed water *qua* water, separate from the political differences between the parties. All failed to one degree or another.

Although political tensions have precluded any comprehensive agreement over the waters of the Middle East, unilateral development in each country has tried to keep pace with the water needs of growing populations and economies. As a result, demand for water resources in most of the countries in the region exceeds at least 90 percent of the renewable supply, the only exceptions being Lebanon and Turkey. All of the countries and territories riparian to the Jordan River – Israel, Syria, Jordan, and the West Bank – are currently using between 95 percent and more than 100 percent of their annual renewable freshwater supply. Gaza exceeds its renewable supplies by 50 percent every year, resulting in serious saltwater intrusion. In recent dry years, water consumption has routinely exceeded annual supply, the difference usually being made up through overdraft of fragile groundwater systems.

In water systems as tightly managed and exploited as those of the Middle East, any future unilateral development is likely to be extremely expensive if based on technology or dangerously

politically volatile if threatening the resources of a neighbor. It has been clear to water managers for years that the most viable options include regional cooperation as a minimum prerequisite.

C.13.4 Attempts at conflict management

Since the opening session of the multilateral talks in Moscow in January 1992, the Working Group on Water Resources, with the United States as "gavel-holder," has been the venue by which problems of water supply, demand and institutions has been raised among the parties to the bilateral talks, with the exception of Lebanon and Syria – Israel, Jordan, and the Palestinian Territories – as well as among the Arab states from the Gulf and the Maghreb. These include Algeria, Bahrain, Egypt, Kuwait, Mauritania, Morocco, Oman, Qatar, Saudi Arabia, Tunisia, United Arab Emirates, and Yemen. Participating in the talks are also "nonregional delegations," including representatives from governments, such as Canada, China, the European Union, Japan, and Turkey, and from donor NGOs, such as the World Bank. The complete list of parties invited to each round includes representatives from Algeria, Australia, Austria, Bahrain, Belgium, Canada, China, Denmark, the European Union, Egypt, Finland, France, Germany, Greece, India, Ireland, Israel, Italy, Japan, Jordan, Kuwait, Luxembourg, Mauritania, Morocco, the Netherlands, Norway, Oman, the Palestinian Territories, Portugal, Qatar, Russia, Saudi Arabia, Spain, Sweden, Switzerland, Tunisia, Turkey, Ukraine, United Arab Emirates, United Kingdom, United Nations, United States, the World Bank, and Yemen.

The two tracks of the current negotiations, the bilateral and the multilateral, are explicitly designed not only to close the gap between issues of politics and issues of regional development but perhaps to use progress in these areas to help catalyze the pace of the other, in a positive feedback loop toward "a just and lasting peace in the Middle East." The idea is that the multilateral working groups would provide forums for relatively free dialogue on the future of the region and, in the process, allow for personal ice-breaking and confidence-building to take place. Given the role of the Working Group on Water Resources in this context, the objectives have been more on the order of fact-finding and workshops rather than tackling the difficult political issues of water rights and allocations or the development of specific projects. Likewise, decisions are made through consensus only.

The Working Group on Water has met five times (Table C.8). The pace of success of each round has vacillated but, in general, has been increasing. The second round, the first of the water group alone, has been characterized as "contentious," with initial posturing and venting on all sides. Palestinians and Jordanians, then part of a joint delegation, first raised the issue of water rights, claiming that no progress can be made on any other issue until past grievances are addressed. In sharp contrast, the Israeli position has been that the question of water rights is a bilateral issue and that the multilateral working group should focus on joint management and development of new resources. Because decisions are made by consensus, little progress was made on either of these issues. Nevertheless, plans were made for continuation of the talks – an achievement in and of itself.

The third round, in Washington, DC, in September 1992, made somewhat more progress. Consensus was reached on a general emphasis for the watersheds that the U.S. State Department had proposed in May, focusing on four subjects: enhancement of water data, water management practices; enhancement of water supply, and concepts for regional cooperation and management.

Progress was also made on the definition of the relationship between the multilateral and bilateral tracks. By this third meeting, it became clear that regional water-sharing agreements, or any political agreements surrounding water resources, would not be dealt with in the multilaterals but that the role of these

Table C.8 *Meetings of the multilateral working group on water resources of the Middle East*

	Dates	Location
Multilateral organizational meeting[1]	January 28–29, 1992	Moscow
Water Talks, Round 2	May 14–15, 1992	Vienna
Water Talks, Round 3	September 16–17, 1992	Washington, DC
Water Talks, Round 4	April 27–29, 1993	Geneva
Water Talks, Round 5	October 26–28, 1993	Beijing
Water Talks, Round 6	April 17–19, 1994	Muscat

[1] After some confusion in numbering, it was eventually officially decided that the multilateral organizational meeting in Moscow represented the first round of the multilateral working groups. Subsequent meetings are therefore numbered correspondingly, beginning with two.

talks was to deal with nonpolitical issues of mutual concern, thereby strengthening the bilateral track. The goal in the Working Group on Water Resources became to plan for a future region at peace and to leave the pace of implementation to the bilaterals. This distinction between "planning" and "implementation" became crucial, with progress being made only as the boundary between the two was continuously pushed and blurred by the mediators.

The fourth round, in Geneva in April 1993, proved particularly contentious, threatening at points to grind the process to a halt. Initially, the meeting was to be somewhat innocuous. Proposals were made for a series of intersessional activities surrounding the four subjects agreed to at the previous meeting. These activities, including study tours and water-related courses, would help capacity-building within while fostering better personal and professional relations.

The issue of water rights was raised again, however, with the Palestinians threatening to boycott the intersessional activities. The Jordanians, who had already agreed to discuss water rights with the Israelis in their bilateral negotiations, helped work out a similar arrangement on behalf of the Palestinians. Agreement was not reached at the time, but both sides agreed later after quiet negotiations in May, before the meeting of the working group on refugees in Oslo. The agreement called for three Israeli–Palestinian working groups within the bilateral negotiations, one of which would deal with water rights. The agreement, in which the Palestinians agreed to participate in the intersessional activities, also called for U.S. representatives of the water working group to visit the region. Although some may have expected the U.S. representatives to take the opportunity of the visit to take a strong proactive position on the issue of water rights, the delegates adhered to the stance that any specific initiatives would have to come from the parties themselves and that agreement would have to be by consensus.

By July 1993, the intersessional activities had begun, including approximately 20 activities as diverse as a study tour of the Colorado River basin and a series of seminars on semiarid lands that focused on capacity building in the region. A series of fourteen courses was designed by the United States and the EU for participants from the region, to range in length from 2 weeks to 12 months and to cover subjects as broad as concepts of integrated water management and as detailed as groundwater flow modeling.

Following a June 1993 agreement in the multilaterals on a joint US/EC proposal to conduct a regional training needs assessment in the Middle East water sector, a team of specialists developed a Priority Regional Training Action Plan. The plan includes a series of fourteen courses to be offered to managers and professionals from the region over 2 years commencing in June 1994. The courses were endorsed at the sixth round of water talks in Oman in April 1994. In the end, 20 courses were given to 275 participants from the Middle East. The courses ranged in duration from 2 weeks to 2 years (Sidebar C.1).

On September 15, 1993, the Declaration of Principles on Interim Self-Government Arrangements was signed by Palestinians and Israelis, which defined Palestinian autonomy and the redeployment of Israeli forces out of Gaza and Jericho. Among other issues, the Declaration of Principles called for the creation of a Palestinian Water Administration Authority. Moreover, the first item in Annex III, on cooperation in economic and development programs, included a focus on cooperation in the field of water, including a Water Development Program prepared by experts from both sides, which will also specify the mode of cooperation in the management of water resources in the West Bank and Gaza Strip and will include proposals for studies and plans on water rights of each party, as well as on the equitable utilization of joint water resources for implementation in and beyond the interim period.

Annex IV describes regional development programs for cooperation, including:

- Development of a joint Israeli–Palestinian–Jordanian Plan for coordinated exploitation of the Dead Sea area;
- The Mediterranean Sea (Gaza) – Dead Sea Canal;
- Regional desalinization and other water development projects;
- Regional plan for agricultural development, including a coordinated regional effort for the prevention of desertification.

Sidebar C.1 Regional Training Action Plan

Water sector level courses

1. Concepts of integrated water resources planning and management
2. Water resources assessment, planning, and management
3. Water quality management
4. Data collection and management systems
5. Alternatives in water resources development
6. Principles and applications of international water law

Water subsector level courses

7. Management of municipal water supply systems
8. Rehabilitation of municipal water supply systems
9. Management of wastewater collection and treatment systems
10. Development of efficient irrigation systems

Specialized courses

11. Environmental impact assessment techniques
12. Groundwater modeling
13. Public awareness campaigns for the water sector
14. Development, management, and delivery of training programs in the water sector

---

The Declaration of Principles also included a description of the mechanisms by which disputes might be resolved. Article XV describes these mechanisms:

1. Disputes arising out of the application or interpretation of this Declaration of Principles, or any subsequent agreements pertaining to the interim period, shall be resolved by negotiations through a Joint Liaison Committee to be established.
2. Disputes that cannot be settled by negotiations may be resolved by a mechanism of conciliation to be agreed on by the parties. The parties may agree to submit to arbitration disputes relating to the interim period, which cannot be settled through conciliation. To this end, upon the agreement of both parties, the parties will establish an Arbitration Committee.

Although the declaration was generally seen as a positive development by most parties, some minor consternation was raised by the Jordanians about the Israeli–Palestinian agreement to investigate a possible Med–Dead Canal. In the working group on regional economic development, the Italians had pledged $2.5 million toward a study of a Red-Dead Canal as a joint Israeli–Jordanian project; building both would be infeasible. The Israelis pointed out in private conversations with the

Jordanians that *all* possible projects should be investigated and that only then could rational decisions on implementation be made.

Although a bilateral agreement, the Declaration of Principles helped streamline a logistically awkward aspect of the multilaterals, as the PLO became openly responsible for the talks and the Palestinian delegations separated from the Jordanians. By the fifth round of water talks in Beijing in October 1993, somewhat of a routine seemed to be setting in, whereby reports were presented on each of the four topics agreed to at the second meeting in Vienna – enhancement of data availability; enhancing water supply; water management and conservation; and concepts of regional cooperation and management – and a new series of intercessional activities was announced.

C.13.5 Outcome

By the fifth round of talks in Beijing in October 1993, the following agreements had been reached in each of the four topics.

ENHANCEMENT OF DATA AVAILABILITY

- Agreement on the need for regional data banks;
- A workshop would be held at USGS facilities in Atlanta as would additional workshops on the subject as part of the United States–EU Priority Training Needs Assessment; and
- A workshop on the standardization of methodologies and formats for data collection would be held.

ENHANCING WATER SUPPLY

- Feasibility studies are being conducted on facilities for the desalination of brackish water, by Japan in Jordan and by the EU in Gaza;
- Canada compiled an exhaustive literature review on water technologies;
- Oman's suggestion was accepted to conduct a survey on the current status of desalination research and technology;
- A Canadian proposal for the installation of a rainwater catchment system in Gaza was accepted, marking the first concrete project to be accepted by the working group.

WATER MANAGEMENT AND CONSERVATION

- Austria ran a seminar on water technologies in arid and semi-arid regions, with special reference to the Middle East;
- The United States organized two seminars jointly sponsored by the water and environment working groups, one on the treatment of wastewater in small communities, and one on drylands agriculture;

- The World Bank is carrying out surveys of water conservation in the West Bank, Gaza, and Jordan.

REGIONAL COOPERATION AND MANAGEMENT

- The UN is organizing a seminar on various models for regional cooperation and management;
- The United States is planning a workshop on weather forecasting;
- Jordan proposed that the working group define a "water charter" for the Middle East to define the principles of regional cooperation and determine mechanisms for water conflict resolution. The proposal was not adopted.

The sixth round of talks was held in Muscat, Oman, in April 1994, the first of the water talks to be held in an Arab country and the first of any working group to be held in the Gulf. Tensions mounted immediately before the talks as it became clear that the Palestinians would use the occasion as a platform to announce the appointment of a Palestinian National Water Authority. Although such an authority was called for in the Declaration of Principles, possible responses to both the unilateral nature and to the appropriateness of the working group as the proper vehicle for the announcement were unclear. Only a flurry of activity prior to the talks guaranteed that the announcement would be welcomed by all parties. This agreement set the stage for a particularly productive meeting. In 2 days, the working group endorsed:

- An Omani proposal to establish a desalination research and technology center in Muscat that would support regional cooperation in desalination research among all interested parties. This marked the first Arab proposal to reach consensus in the working group;
- An Israeli proposal to rehabilitate and make more efficient water systems in small-sized communities in the region. This was the first Israeli proposal to be accepted by any working group;
- A German proposal to study the water supply and demand development among interested core parties in the region;
- A U.S. proposal to develop wastewater treatment and reuse facilities for small communities at several sites in the region. The proposal was jointly sponsored by the water and environmental working groups;
- The Regional Water Data Banks Project, a joint venture with the U.S. Geological Survey to create a data sharing systems in the Middle East. This project would initially focus on bring the Palestinian database up to the speed of those of Jordan and Israel, so that consistent data would be available

Sidebar C.2 Multilateral Working Group A – Water, Energy, and the Environment

I. Surface water basins
   A. Negotiation of mutual recognition of the rightful water allocations of the two sides in Jordan River and Yarmuk River waters with mutually acceptable quality
   B. Restoration of water quality in the Jordan River below Lake Tiberias to reasonably usable standards
   C. Protection of water quality
II. Shared groundwater aquifers
   A. Renewable freshwater aquifers – southern area between the Dead Sea and the Red Sea
   B. Fossil aquifers – area between the Dead Sea and the Red Sea
   C. Protection of the water quality of both
III. Alleviation of water shortage
   A. Development of water resources
   B. Municipal water shortages
   C. Irrigation water shortages
IV. Potentials of future bilateral cooperation, within a regional context where appropriate
   [Includes Red Sea–Dead Sea Canal; management of water basins; and interdisciplinary activities in water, environment, and energy.]

to inform and recommend local and regional decision making.

• Implementation of the United States–EU regional training program, as described in Sidebar C.2.

As we have mentioned in this section, the working group officially welcomed the announcement of the creation of the Palestinian Water Authority and pledged to work with the Authority on multilateral water issues.

In 1995, the core parties formed the Executive Action Team (EXACT), a thus far extremely successful initiative to manage, coordinate, and promote project implementation. With the United States through the U.S. Geological Survey as gavelholder and executive secretary, it is composed of two representatives from each core party and each donor party. Since 1995, EXACT has met biannually to plan, coordinate, and direct project implementation. Since its inception, EXACT has met twice a year and focused on implementing thirty-nine recommendations involving the following activities:

• Trainings for water managers and field technicians: database development, interpretation of water quality network data, interpretation of surface-water quality network data, inter-

pretation of surface water network data, and installation and operation of hydrometeorological and stream gauging stations, statistical analysis and laboratory quality assurance plans (Executive Action Team Multilateral Working Group on Water Resources).

• The establishment of mobile laboratories staffed by trained technicians in the field; twenty-five regional labs now participate in a semiannual standard reference sample.

• Joint database for rainfall data.

• Inventory of wastewater-related concerns. Water data collection, storage, and retrieval systems have been established within the Palestinian Water Authority, and those of the Israeli Hydrological Service and the Jordan Ministry of Water and Irrigation have been improved and enhanced.

The greatest success has been the ongoing communication despite fluctuations in bilateral negotiations.

Progress has been made in bilateral negotiations between Jordan and Israel as well. In September of 1993, the two states agreed to work toward an agenda for peace talks. The sub-agenda for these talks, established on June 7, 1994, included several water-related items, notably in the first heading listed (in advance of security issues, and border and territorial matters), Group A – Water, Energy, and the Environment (see Sidebar C.2).

Following these bilateral talks, the two sides signed the Treaty of Peace between the State of Israel and the Hashemite Kingdom of Jordan in 1994. The parties agreed to recognize the rightful allocations to both of them from the Jordan River, Yarmuk River, and the Araba–Arava aquifer. A Joint Water Committee was established composed of three members from each country, to monitor water use, enforce regulations, and develop new cooperation activities.

Talks in 1996 succeeded in creating a number of structures in the areas of data availablility, water management and conservation, and regional cooperation and management. Norway agreed to sponsor the establishment of a "Declaration of Principles for Cooperation among Core Parties on Water-Related Matters and New and Additional Water Resources." This declaration made advances in the area of water management by establishing The Waternet Project, a project to develop computerized water information systems. A common information system, Waternet Information System (WIS) was inaugurated to assist the core parties in linking local information networks to a regional computer information network and to establish a Regional Waternet and Research Center in Amman, Jordan that will maintain this project, stimulate cooperation, and initiate new and joint activities.

The United States agreed to assist the MWGWR in the creation of a Public Awareness and Water Conservation Project, which produced a video and student resource book for youth that highlights the importance of water issues in the region. This group project is done in collaboration with EXACT (Public Awareness and Water Conservation).

Luxembourg collaborated with the MWGWR to establish a project on Optimization of Intensive Agriculture under Varying Water Quality Conditions in order to demonstrate how brackish and saline water can be used for sustainable farming in Beit-Hanoun, Gaza.

Middle East Desalination Research Center (MEDRC) was established in Muscat, Oman, in December of 1996 to conduct, facilitate, promote, coordinate, and support basic and applied research in water desalination to reduce the cost of desalination and improve the quality (see www.medrc.org).

C.13.6 Conclusion

Given the length of time that the region has been enmeshed in bitter conflict, the pace of accomplishment of the peace process has been impressive, no less so in the area of water resources. This may be due in part to the structure of the peace talks, with the two complementary and mutually reinforcing tracks – the bilateral and the multilateral. As noted earlier, past attempts at resolving water issues separate from their political framework, dating from the early 1950s through 1991, have all failed to one degree or another. Once the taboo of Israelis and Arabs meeting openly in face-to-face talks was broken in Madrid in October 1991, the floodgates were open, as it were, and a flurry of long-repressed activity on water resources began to take place outside of the official peace process. This included several academic conferences on Middle Eastern water resources in, among other places, Canada, Turkey, Illinois, Washington, DC (3), and, notably, the first Israeli–Palestinian conference on water resources in Geneva; unofficial "Track II" dialogues in Nevada, Cairo, and Idaho; the establishment by the IWRA of the "Middle East Water Commission" to help facilitate research on the subject; and organization of the Middle East Water Information Network (MEWIN) to coordinate regional data collection. Although this flurry of water-related activity may have been moderately helpful in generating ideas outside of the constraints of the official process, and more so in fostering better personal relations between the water professionals of the region, many negotiators involved with the official process suggest limited influence, usually because no mechanism exists to encourage dialogue between the tracks. (The term "Track II" refers to those activities outside of the official negotiations. There may be some confusion, because in the case of the Middle East peace talks, the official process is likewise divided in two – the bilateral negotiations and the multilateral working groups.)

Despite the relative success of the multilateral working group on water, and given its stated objective to deal with non-political issues of mutual concern, one might wonder where the process might go from here. The working group has performed admirably in the crucial early stages of negotiations as a vehicle for venting past grievances, presenting various views of the future, and, perhaps most important, allowing for personal "de-demonization" and confidence-building on which the future region at peace will be built. Currently, however, there is some frustration on the part of many of the participants that it is not, by design, a vehicle for actually resolving any of the issues of conflict. The contentious topics of water rights and allocations, which some argue must be solved before proceeding with any cooperative projects, are relegated to the bilateral negotiations, where they take a relatively lower priority. Likewise, the principles of integrated watershed management are difficult to encourage: water quantity, quality, and rights all fall within the purview of different negotiating frameworks – the working group on water, the working group on the environment, and the various bilateral negotiations, respectively. There is slightly more overlap than the institutional setting might indicate. Several of the regional delegates sit on both bilateral and multilateral groups, and each of the states have some sort of steering committee, which fosters communication. Furthermore, the U.S. team includes members who participate in both the water and the environment working groups, which helps ensure that issues of water quantity and quality are not entirely separated. Finally, and perhaps somewhat related, are the limitations imposed by Syrian and Lebanese refusal to participate in any of the multilateral working groups. The result of this omission means that a comprehensive settlement of the conflicts related to the Jordan or Yarmuk Rivers are precluded from discussions (Sidebar C.2).

C.13.7 Lessons learned

In attempts at resolving particularly contentious disputes, solving problems of politics and resource use is best accomplished in two mutually reinforcing tracks.

The most useful lesson of the multilateral working group on water resources is the handling of water and political tensions simultaneously in the bilateral and multilateral working groups, respectively, each track helping to reinforce the other. This lesson has been learned after a long history of failure to solve water problems outside of their political context.

The first task of water negotiations between particularly hostile riparians may be simply to get individuals together talking about relatively neutral issues.

The working group has performed admirably in the crucial early stages of negotiations as a vehicle for venting past grievances, presenting various views of the future, and, perhaps most important, fostering personal relations and confidence-building. Where traditional negotiations might have tried to tackle issues of water rights and allocations initially, those directing the working group negotiations recognized the greater initial value of seminars, field trips, and workshops on relatively neutral issues. These activities also provided practice in reaching consensus as a group.

This process has an alternative side, though, in that if carried on too long, it may leave a gap when a vehicle for resolving the difficult issues is called for.

Inclusion of donor and observer parties can generally be helpful, although coordination is necessary.

Both donor and observer parties have helped the process by funding and/or performing feasibility studies, holding workshops, and organizing field trips. The World Bank has also helped to prioritize the needs of the core basins through a series of questionnaires and country reports. Some frustration has been expressed, though, that countries have occasionally embarked on projects without coordinating with the sponsors of the talks.

Successful negotiations might include an eventual simultaneous narrowing and broadening of focus, to move from the neutral topics necessary in early stages of negotiation, to dealing with the contentious issues at the heart of a water conflict. Concepts of integrated water management may also be included.

Although relatively neutral topics were vital in the early stages of the negotiations, some shift may be in order to be able to handle watershed-wide problems such as water rights and allocations. This narrowing of focus might be accompanied by a simultaneous broadening, to include all issues of water rights, quantity, and quality relevant to a basin within one framework.

Track II dialogues lose much of their utility if there is no mechanism for feeding ideas generated into the main negotiating track.

Despite a flurry of water-related studies, conferences, and alternative track dialogue, and despite some creative ideas and thinking that resulted outside of the pressures of official negotiations, sponsors of the multilaterals report little influence of this activity on the official talks, probably because few meetings have a mechanism for feeding the ideas generated directly to the parties concerned.

C.13.8 Creative outcomes resulting from resolution process

The most creative outcome of the current negotiations is probably the structure of the two tracks of the negotiations: the bilateral negotiations, which deal with explicitly political issues from the past, and the multilateral working groups, which help define a common vision of the future. Each track helps reinforce the other, catalyzing the pace toward a comprehensive peace settlement.

Early emphasis of the working group on water resources was on comparatively neutral topics and workshops, not on contentious political aspects of the water conflict. The talks foster a relatively open exchange of ideas by, for example, having no official minutes and relying on consensus for all decision making. The consensus approach gives ensures a level of egalitarianism in the working group by giving each party an effective veto over each issue. This encourages dividing issues into small, manageable portions on which all parties will agree but also discourages attempts at solving larger, more difficult issues.

C.13.9 Time line

October 30, 1991 First public, face-to-face peace talks between Arabs and Israelis are held in Madrid. Talks begin as bilateral between Israel and each of its neighbors.

January 28–29, 1992 Multilateral organizational meeting in Moscow. Peace process is designed along two tracks – the bilateral negotiations, involving separate direct negotiations between Israel and each of its neighbors, and the multilateral negotiations revolving around five regional subjects, including water resources. Goal is to allow framework for defining future of the region, as well as to include peripheral Arab states, other countries, and donor NGOs.

May 14–15, 1992 First meeting of Multilateral Working Group on Water Resources in Vienna (dubbed the "second" round of multilaterals). Little practical progress made due to venting and posturing on all sides. Palestinians and Jordanians first raise issue of water rights; Israel's position is that water rights are a bilateral issue. World Bank asks each party to compile a program for regional water resources development, following three possible scenarios: no outside investment, current government plans, and unlimited resources. These scenarios would be examined in the United States for any commonalities, which could be culled to induce cooperation. Only decision reached is to plan for next round of talks.

September 16–17, 1992 Third round of water talks in Washington, DC. Agreement on four general subjects for multilateral talks on water: enhancement of water data, water management practices, enhancement of water supply, and concepts for regional cooperation and management. Role of multilaterals clarified to plan for future region at peace, not to implement specific agreements.

April 27–29, 1993 The fourth working group on water meeting in Geneva proves tense, following a disagreement over a Palestinian request that water rights be included in multilateral talks, otherwise the Palestinians would boycott intersessional activities.

May 1993 Israelis and Palestinians agree to discuss water rights in the Occupied Territories within the framework of the Bilateral Negotiations and Palestinians agree to participate in intersessional activities. This agreement, which came about in discussions at the working group on refugees meeting in Oslo, also called for American representatives of the water working group to visit the region.

September 15, 1993 Declaration of Principles signed between Israelis and Palestinians, which includes several water-related items, including the creation of a Palestinian Water Administration Authority and a Water Development Program. The Program would include investigations of development of regional agricultural and desalination projects, and a Med-Dead Canal.

October 26–28, 1993 Fifth round of Working Group on Water Resources meets in Beijing. Presentations are made in each of four topics and several projects are agreed to; priority needs assessment is presented and courses are approved.

April 17–19, 1994 Sixth round of Working Group meets in Muscat, Oman. The meeting is productive after all parties agree to welcome a Palestinian announcement of the creation of a Palestinian Water Authority in the autonomous territories of Gaza and Jericho (Israel agrees *provided* it will not be seen as a precedent in other territories). Other endorsements include:

- an Omani proposal to establish a desalination research and technology center; an Israeli proposal to lead an effort of water conservation and rehabilitation of municipal water systems;
- a German offer to study regional supply and demand;
- a U.S. proposal to perform a study of wastewater treatment and reuse; and
- United States and EU implementation of a regional water training program tthat began in June 1994.

June 7–9, 1994 Bilateral talks take place between Israel and Jordan in Washington, DC. Subagenda items are determined for talks leading to a Treaty of Peace, including several water-related topics.

November 1994 At the meeting in Athens, Greece, the parties approved the Implementation Plan of the Regional Water Data Banks Project.

January 1995 Regional Water Data Bank Project initiated.

June 1995 Meeting of Multilateral Working Group in Amman, Jordan.

May 1996 Meeting of Multilateral Working Group in Hammamet, Tunisia.

December 1996 Established the Middle East Desalination Research Center.

1996 Established the Public Awareness and Water Conservation Project, the Optimization of Intensive Agriculture under Varying Water Quality Conditions Project managed by Luxembourg and the Waternet Project with aid from the Norwegian Government.

C.14 THE NILE WATERS AGREEMENT

C.14.1 Case summary

River basin: Nile River (Figure C.16).

Dates of negotiation: 1920–1959 – Treaties signed in 1929 and 1959.

Relevant parties: Egypt, Sudan (directly); other Nile riparians (indirectly)

Flash point: Plans for a storage facility on the Nile.

Issues: Stated objectives: negotiate an equitable allocation of the flow of the Nile River between Egypt and Sudan; develop a rational plan for integrated watershed development.

Additional issues: Water-related: upstream versus downstream storage; Nonwater: general Egypt–Sudan relations.

Excluded issues: Water quality; other Nile riparians.

Criteria for water allocations: Acquired rights plus even division of any additional water resulting from development projects.

Incentives/linkage: Financial: Funding for Aswan High Dam; Political: Fostered warm relations between Egypt and new government of Sudan.

Breakthroughs: 1958 coup in Sudan by pro-Egypt leaders made agreement possible.

Status: Ratified in 1959. Allocations between Egypt and Sudan upheld until today. Other riparians, particularly Ethiopia, are planning development projects that may necessitate renegotiating a more inclusive treaty. Nile Basin Initiative, established in 1999, includes all basin nations.

C.14.2 Background

In the early 1900s, a relative shortage of cotton on the world market put pressure on Egypt and the Sudan, then under a British–Egyptian condominium, to turn to this summer crop, requiring perennial irrigation over the traditional flood-fed

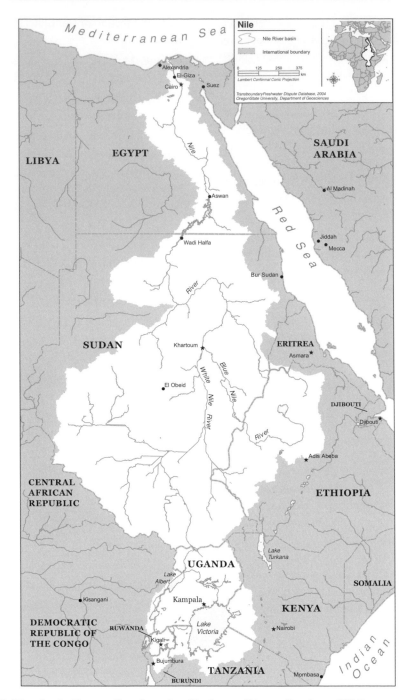

Figure C.16 Map of the Nile River basin (*Source:* Transboundary Freshwater Dispute Database, 2004).

methods. The need for summer water and flood control drove an intensive period of water development along the Nile, with proponents of Egyptian and Sudanese interests occasionally clashing within the British foreign office over whether the emphasis for development ought to be further upstream or down.

With the end of World War I, it became clear that any regional development plans for the Nile basin would have to be preceded by some sort of formal agreement on water allocations.

In 1920, the Nile Projects Commission was formed, with representatives from India, the United Kingdom, and the United States. The Commission estimated that, of the river's average flow of 84 BCM/year, Egyptian needs were estimated at 58 BCM/year. Sudan, it was thought, would be able to meet irrigation needs from the Blue Nile alone. The Nile flow fluctuates greatly, with a standard deviation of about 25 percent. In recognition of this fact, an appendix was added that suggested that any gain or shortfall from the average be divided evenly between Egypt and Sudan. The Commission's findings were not acted on.

The same year saw publication of the most extensive scheme for comprehensive water development along the Nile, now known as the Century Storage Scheme. The plan, put forth by the British, included a storage facility on the Uganda-Sudan border, a dam at Sennar to irrigate the Gezira region south of Khartoum, and a dam on the White Nile to hold summer flood water for Egypt.

The plan worried some Egyptians, and was criticized by nationalists, because all the major control structures would have been beyond Egyptian territory and authority. Some Egyptians saw the plan as a British means of controlling Egypt in the event of Egyptian independence.

C.14.3 The problem

As the Nile riparians gained independence from colonial powers, riparian disputes became international and consequently more contentious, particularly between Egypt and Sudan. The core question of historic versus sovereign water rights is complicated by the technical question of where the river would be controlled best – upstream or downstream.

C.14.4 Attempts at conflict management

In 1925, a new water commission made recommendations based on the 1920 estimates, which would lead finally to the Nile Waters Agreement between Egypt and Sudan on May 7, 1929. Four BCM/year was allocated to Sudan but the entire timely flow (from January 20 to July 15) and a total annual amount of 48 BCM/year was reserved for Egypt. Egypt, as the downstream state, had its interests guaranteed by:

• Having a claim to the entire timely flow. This meant that any cotton cultivated in Sudan would have to be grown during the winter months.
• Having rights to onsite inspectors at the Sennar Dam, outside of Egyptian territory.
• Being guaranteed that no works would be developed along the river or on any of its territory, which would threaten Egyptian interests.
• In accord with this agreement, one dam was built and one reservoir raised with Egyptian acquiescence.

The Aswan High Dam, with a projected storage capacity of 156 BCM/year, was proposed in 1952 by the new Egyptian government, but debate over whether it was to be built as a unilateral Egyptian project or as a cooperative project with Sudan kept Sudan out of negotiations until 1954. The negotiations that ensued carried out with Sudan's struggle for independence as a

backdrop, focused not only on what each country's legitimate allocation would be but also on whether the dam was even the most efficient method of harnessing the waters of the Nile.

The first round of negotiations between Egypt and Sudan took place between September and December 1954, even as Sudan was preparing for its independence, scheduled for 1956. The positions of the two sides are summarized below.

EGYPTIAN POSITION
Existing needs should take priority. These were described as being 51 BCM/year for Egypt and 4 BCM/year for Sudan, out of an average flow of 80 BCM/year as measured at Aswan.

Any remainder from development projects should be divided as a percentage of each country's population after subtracting 10 BCM/year for evaporation losses. The respective population and growth rates led to an Egyptian formula for 22/30 of the remainder, or 11 BCM/year for Egypt, and 8/30, or 10 BCM/year for Sudan.

There should be one large storage facility, the High Dam at Aswan.

Total allocations would therefore be 62 BCM/year for Egypt and 8 BCM/year for Sudan.

SUDANESE POSITION
Sudan insisted on using the standard value of 84 BCM/year for average Nile discharge and that Egypt's acquired rights were for 48 BCM/year, not the 51 BCM/year that Egypt claimed.

Sudan also suggested that their population was actually 50 percent larger than Egypt had estimated, and that resulting population-based allocations should be adjusted accordingly, giving Sudan at least one-third of any additional water.

Storage facilities should be smaller and upstream, as envisioned in the Century Storage Scheme. Consequently, if Egypt insisted on one large project, with comparatively high evaporation losses, these losses should be deducted from Egypt's share.

Total allocations, therefore, should be approximately 59 BCM/year (69 BCM/year less evaporation) for Egypt and 15 BCM/year for Sudan.

Negotiations were broken off inconclusively, then briefly, and equally inconclusively, resumed in April 1955. Relations then threatened to degrade into military confrontation in 1958 when Egypt sent an unsuccessful expedition into territory in dispute between the two countries. In the summer of 1959, Sudan unilaterally raised the Sennar dam, effectively repudiating the 1929 agreement.

Sudan attained independence on January 1, 1956, but it was with the military regime that gained power in 1958 that Egypt adopted a more conciliatory tone in the negotiations, which resumed in early 1959. Progress was speeded in part by the

Table C.9 *Water allocations from Nile negotiations*

Position	Egypt (BCM/year)	Sudan (BCM/year)
Egyptian[1]	62.0	8.0
Sudanese[2]	59.0	15.0
Nile Waters Treaty (1959)[3]	55.5	18.5

[1] The Egyptian position assumed an average flow of 80 BCM/year and divided approximately 10 BCM/year in evaporation losses equally.
[2] The Sudanese position assumed an average flow of 84 BCM/year and deducted evaporation from the Egyptian allocations.
[3] The Treaty allowed for an average flow of 84 BCM/year and divided evaporation losses equally.

fact that any funding that would be forthcoming for the High Dam would depend on a riparian agreement. On November 8, 1959, the Agreement for the Full Utilization of the Nile Waters (Nile Waters Treaty) was signed (Table C.9).

C.14.5 Outcome

The Nile Waters Treaty had the following provisions:

- The average flow of the river is considered to be 84 BCM/year. Evaporation and seepage were considered to be 10 BCM/year, leaving 74 BCM/year to be divided.
- Of this total, acquired rights have precedence and are described as 48 BCM/year for Egypt and 4 BCM/year for Sudan. The remaining benefits of approximately 22 BCM/year are divided by a ratio of 7 1/2 for Egypt (approx. 7.5 BCM/year) and 14 1/2 for Sudan (approx. 14.5 BCM/year). These allocations total 55.5 BCM/year for Egypt and 18.5 BCM/year for Sudan.
- If the average yield increases from these average figures, the increase would be divided equally. Significant decreases would be taken up by a technical committee, described below.
- Because Sudan could not absorb that much water at the time, the treaty also provided for a Sudanese water "loan" to Egypt of up to 1,500 MCM/year through 1977.
- Funding for any project that increases Nile flow (after the High Dam) would be provided evenly, and the resulting additional water would be split evenly.
- A Permanent Joint Technical Committee to resolve disputes and jointly review claims by any other riparian would be established. The Committee would also determine allocations in the event of exceptional low flows.
- Egypt agreed to pay Sudan £E 15 million in compensation for flooding and relocations.

Egypt and Sudan agreed that the combined needs of other riparians would not exceed 1,000–2,000 MCM/year, and that any claims would be met with one unified Egyptian–Sudanese position. The allocations of the Treaty have been held to the present.

Ethiopia, which had not been a major player in Nile hydropolitics, served notice in 1957 that it would pursue unilateral development of the Nile water resources within its territory, estimated at 75 percent to 85 percent of the annual flow, and suggestions were made recently that Ethiopia may eventually claim up to 40,000 MCM/year for its irrigation needs both within and outside of the Nile watershed. No other state riparian to the Nile has ever exercised a legal claim to the waters allocated in the 1959 treaty.

Ever since the signing of the Nile Basin Treaty of 1959, there have been various cooperative activities that have taken place between nations within the Nile River Basin. From 1967 to 1992, the United Nations Development Program (UNDP) supported HYDROMET, a project designed to collect hydrometeorologic information within the basin. In 1993, the Technical Cooperation Committee for the Promotion of the Development and Environmental Protection of the Nile Basin (TECCONILE) was formed at the same time as the first of ten Nile 2002 conferences were launched with the idea to create informal dialogue between riparian nations.

Nile-COM, the Council of Ministers of Water Affairs of the Nile Basin States, in 1997 was allowed by the World Bank to direct and coordinate donor activities within the basin, which led the Council to work in cooperation with organizations such as the UNDP, the World Bank, and the Canadian International Development Agency (CIDA). In May of 1999, the Nile Basin Initiative (NBI) was launched with the understanding that a cooperative effort in the development and management of Nile waters will bring the greatest level of mutual benefit on the region. All nations of the basin, Burundi, Democratic Republic of the Congo, Egypt, Eritrea, Ethiopia, Kenya, Rwanda, Sudan, Tanzania, and Uganda, joined the organization. The objectives for the NBI (see http://www.nilebasin.org/Documents/TACPolicy.html) include the following:

- To develop the water resources of the Nile in a sustainable and equitable way to ensure prosperity, security, and peace for all of its peoples;
- To ensure efficient water management and the optimal use of resources;
- To ensure cooperation and joint action between the riparian countries, seeking win–win gains;
- To target poverty eradication and promote economic integration; and

- To ensure that the program results in a move from planning to action.

In May 2004, the "Nile Transboundary Environmental Action Project," the first of eight basinwide projects under the NBI, was launched in Sudan. Sudanese president, General Omar El-Bashir, declared, "Since environmental hazards are not restricted within geographical boundaries, local and international efforts are required to overcome the dangers and threats in the environmental arena. This project is providing solutions to these problems" (http://www.nilebasin.org/pressreleases.htm#launch).

C.14.6 Lessons learned

Shifting political boundaries can turn intranational disputes into international conflicts, exacerbating tensions over existing issues.

Similar to the Indus, the disappearance of British colonialism turned national issues international, making agreement more difficult.

Downstream riparians are not necessarily at a political disadvantage to their upstream neighbors.

Although in many cases relative riparian positions result in comparable power relationships, with upper riparians having greater hydropolitical maneuverability, Egypt's geopolitical strength was able to forestall upstream attempts to sway its position.

The individuals or governments involved can make a difference in the pace of the negotiations.

Negotiations made little progress between 1954 and 1958, even given Sudan's independence in 1956. It was only after pro-Egyptian General Ibrahim Abboud took power in a coup in 1958 that negotiations moved toward resolution, finally gaining for Sudan water allocations greater than those of their initial bargaining point.

C.14.7 Creative outcomes resulting from resolution process

The measure for water allocations is rather elegant, incorporating existing uses as well as providing a measure (population) for allocating additional sources.

Some financing arrangements were creative, with Egypt agreeing to finance water enhancement projects in Sudanese territory, in exchange for the water that would be made available. Provisions were made for Sudan to pick up responsibility for up to 50 percent of costs in exchange for up to 50 percent of the water, when their water needs required.

C.14.8 Time line

1920 Nile Projects Commission formed, offers allocation scheme for Nile riparians. Findings were not acted on.

1920 Century Storage Scheme put forward, emphasizing upstream, relatively small-scale projects. Plan is criticized by Egypt.

1925 New water commission is named.

May 7, 1929 Commission study leads to Nile Waters Agreement between Egypt and Sudan.

1952 Aswan High Dam proposed by Egypt. Promise of additional water necessitates new agreement.

September–December 1954 First round of negotiations between Egypt and Sudan. Negotiations end inconclusively.

1956 Sudan gains independence. Egypt is more conciliatory with government after 1958 coup.

November 8, 1959 Agreement for the Full Utilization of the Nile Waters (Nile Waters Treaty) signed between Egypt and Sudan.

1967–1992 Launch of Hydromet regional project for collection and sharing of hydrometeorologic data, supported by UNDP.

1993 Formation of TECCONILE (Technical Cooperation Committee for the Promotion of the Development and Environmental Protection of the Nile Basin) to address development agenda for the Nile basin.

1993 First of ten Nile 2002 Conferences for dialogue and discussions between riparians and the international community, supported by CIDA (Canadian International Development Agency),

1995 Nile River Basin action plan created within TECCONILE framework, supported by CIDA.

1997–2000 Nile riparians create official forum for legal and institutional dialogue with UNDP support. Three representatives from each country (legal and water resource experts) and a panel of experts draft a "Cooperative Framework" in 2000.

1997 Formation of Nile-COM, a council of the Ministers of Water from each of the riparian nations of the Nile basin.

1998 First meeting of the Nile Technical Advisory Committee (Nile-TAC).

May 1999 Nile Basin Initiative established as a cooperative framework between *all* riparians (excluding Eritrea) for the sustainable development and management of the Nile.

May 2004 First basinwide project under NBI, the "Nile Transboundary Environmental Action Project," launched in Sudan.

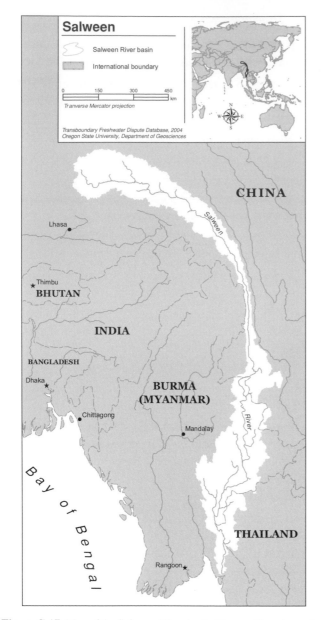

Figure C.17 Map of the Salween River basin (*Source:* Transboundary Freshwater Dispute Database, 2004).

C.15 SALWEEN RIVER

C.15.1 Case summary

River basin: Salween River (Figure C.17).

Dates of negotiation: Joint working group established in 1989.

Relevant parties: Myanmar, Thailand (directly); China (indirectly).

Flash point: None.

Issues: Stated objectives: promote and coordinate joint development of hydropower projects within the Salween basin.

Additional issues: Water-related: possibility of out-of-basin transfers to Thailand; Nonwater: river flows through regions of ethnic unrest and drug trade; collaboration and support of a government in Myanmar that violates human rights; dam project could detrimentally impact the environment and disrupt the livelihoods of local peoples.

Excluded issues: China has not been included in any planning.

Criteria for water allocations: None.

Incentives/linkage: Possibility of linking water projects with transportation infrastructure.

Breakthroughs: None.

Status: Talks are in most-preliminary stage; meetings continue although no plan for the basin, nor any main-stem project, has yet been established.

C.15.2 Background

The Salween River (known as the Nu in Chinese) originates in the Tibetan plateau and drains an area of 320,000 km^2 in China, Myanmar, and Thailand before it flows into the Gulf of Martaban. Totaling 2,413 kilometers, it is the longest undammed river in mainland Southeast Asia. More than 10 million people from at least thirteen different ethnic groups depend on the Salween watershed for their livelihoods: fisheries are a major source of dietary protein, and the river's nutrients nourish vegetable gardens in the dry season and fertilize farmland. The Nujiang, the section of the Salween that flows through China, is found in the Three Parallel Rivers area, a rich center of biodiversity recognized by UNESCO as a World Heritage Site. Despite the fact that studies since the 1950s have identified tremendous hydropower potential, the Salween is a relatively undeveloped basin – with only one major hydroelectric project at Baluchaung. However, it is likely that with economic development and more political integration in the region, development pressure in the river basin will increase, and there will be more demands to use the waters for irrigation, urban and industrial uses, and navigation. The power companies of Thailand and Myanmar, as well as private Japanese concerns, have pursued individual feasibility studies but it is only since the 1970s that the potential of the basin as a whole has been investigated.

Since he took power in 2001, Thailand's Prime Minister, Thaksin Shinawatra, has reversed past policy of distancing Thailand from Myanmar and is pursuing a policy of conciliation, cooperation, and public support. Thai businesses are encouraged to invest in Myanmar, Thailand has agreed to construct a bridge across the border to boost trade and tourism, and is proceeding with a hydroelectric dam project on the Salween River. Thailand's northeast region has always had inadequate water resources and in the past decade the rest of the country has also been suffering from water scarcity due in large part to massive deforestation. However, Thailand already has 28 large dams, 800 small dams, and 1,000 low-capacity reservoirs, and is unlikely to extract more water from its own sources. As a result, the government has decided to channel water from Myanmar to solve its needs for irrigation and drinking water and as a source of electrical power. In 1992, eight major hydroelectric dam projects were selected, some of which are entirely in Myanmar and others are on shared sections within the Salween River basin.

C.15.3 The problem

China, Myanmar, and Thailand do not yet have an agreement on the use of the Salween, thus allowing each of them free use of the river. Each of these countries has unilateral plans to construct dams and development projects along the Salween, but these sets of plans are not compatible.

Since December 2002, the Myanmar Military and the Electricity Generation Authority of Thailand (EGAT) have been discussing the possibility of constructing large dam projects on the Salween. Between October 1998 and the end of March 1999 several teams of experts – Thais from the MDX Power Co. and Burmese from a firm called Aye Chan Aye, assisted by about twenty Japanese specialists from the Electronic Power Development Corporation (EPDC) – inspected three sites in the Salween gorges about 120 kilometers from the Thai border carrying out geological studies, test bores, and feasibility studies. Depending on the site, the size and design of the structure, and the output of the hydraulic turbines, the cost of the dams would range from $3.0 to $3.4 billion, and it is not clear where Thailand and Myanmar would get funding. Total energy production is estimated at 3,400 MW, a quarter of which would go to Myanmar, and Thailand would purchase the rest.

Some of the sites that have been studied are Tasang (in southern Shan State; estimated capacity at 3,300–3,600 MW; feasibility studies have been completed), Weigyi (on the Thai–Burmese border, west of Mae Sariang; estimated capacity at 4,540 MW; preliminary studies have been completed), and Dagwin (just below Weigyi; estimated capacity at 792 MW; preliminary studies have been completed). Of these, the planned dam at Tasang is proposed to be more than 180 meters high, making it one of the largest dams in Southeast Asia.

As much as 10 percent of the Salween water could be diverted via channels and existing rivers, across a distance of 300 kilometers, to join the Kok and Ping rivers in Thailand. There have not been estimates made as to the cost of transporting this water but would probably be high.

Thai and Myanmar officials have been working together discretely in an insurgent area where the Myanmar army has persecuted the Shan civilian population. This part of the Shan state is the operational base of the armed Shan nationalist resistance movement, which is opposed to the junta in Rangoon. With an already large number of Shan people being forced from the region, environmental groups and local populations are worried that the dam project will only exacerbate the problem. But, as Myanmar has a serious need for energy and has also felt the effects of drought, the junta has been willing to cooperate with Thailand (Le Monde Diplomatique, February 2000).

Environmental groups expressed concerns about the ecological effects of the projects, and human rights advocates warned against coinvesting with a military junta that is oppressive, unpredictable, and might not respect benefit-sharing agreements. Nonetheless, in August 2004, Thailand and Myanmar agreed to set up a joint venture to construct five hydropowered dams in the Salween river basin, beginning with Tasang dam.

Meanwhile, in 2003 China announced plans to build thirteen hydropower projects on the Nujiang River in China. More than eighty environmental and human rights groups in Thailand and Myanmar petitioned China to consult downstream countries before proceeding with the project. In April 2004, Prime Minister Wen Jiabao purportedly suspended plans for the massive dam system and ordered officials to conduct a review of the hydropower project and an environment impact assessment. However, Li Yunfei, the director of the Nu River Power Bureau, said he had not heard of any changes and was still working on the project. According to Chinese media, the thirteen dams would have a total generating capacity of 21.32 million kW. Because electricity shortages forced some factories to close this past summer, the promise of a new power facility capable of generating this much electricity is very tempting. China is relying heavily on hydropower to meet its soaring demand for electricity, and officials plan to triple installed hydroelectric capacity to 270,000 MW by 2020 (http://www.irn.org/programs/nujiang/).

C.15.4 Attempts at conflict management

In June 1989, following a visit of a Thai government delegation to Rangoon, a joint technical committee was established between Thailand and Myanmar, made up primarily of representatives from the power companies of the two countries. Since that time, the committee has continued to meet and to pursue feasibility studies, but no project or management body has been implemented nor a basinwide plan created. China has not to date been included in discussions, nor has it included Thailand and Myanmar in its plans for projects on the Nu River.

Although there have been meetings and negotiations at the state level, local populations have not been included in the decision-making process. Thus, while efforts are being made in terms of river planning to avoid interstate conflicts, large-scale water projects may create or exacerbate intrastate conflicts.

C.15.5 Outcome

As mentioned, the Salween is a basin in its earliest stages of development. What is noteworthy is that technical and management discussions have been proceeding in advance of major development projects, which allows for integrated management almost from the beginning.

Discussions have included issues outside of hydropower, and studies have suggested linkages among power, irrigation and drinking water diversions, barge transportation, and related surface infrastructure. Complicating management issues is the fact that sections of the watershed include regions of ethnic unrest and tensions brought about by the international drug trade. Nevertheless, the basin offers the opportunity for integrated management to be implemented in advance of any flash point brought about by unilateral development.

C.15.6 Lessons learned

Tensions are created when a country within a basin acts unilaterally without consulting other nations.

Thailand and Myanmar have been working together for some time on the development of the Salween River Basin, but China has been acting unilaterally, potentially constructing up to thirteen dams on the upper stem of the river. Without working with the two downstream nations, China risks creating conflict with Thailand and Myanmar.

Upstream nations with superior strength can hinder joint management of river basins.

China, with far more military might and economic power than both Thailand and Myanmar combined, has little incentive to work jointly with them in the management of the Salween River. Thailand and Myanmar's water resources from the Salween may be at great risk depending on what China decides to do on the upper part of the river.

The importance of water cooperation/economic development can supercede working with an oppressive regime.

Even though Myanmar is controlled by a junta that is blamed for human rights violations, Thailand is still willing to cooperate with their government in order to promote regional management of the Salween River. For Thailand, the development of the Salween River and the benefits received from such development takes precedence over working with an oppressive regime. National sovereignty to protect water resources goes beyond international pressure.

Lack of inclusion of populations of a shared river basin in the decision-making processes may cause conflicts.

The local populations in both Thailand and Myanmar have not been included in decision-making processes with regards to major hydroelectric projects. Whereas Thailand and Myanmar may work cooperatively to avoid conflict, large-scale projects may create or exacerbate intrastate conflicts.

C.15.7 Creative outcomes resulting from resolution process

Even before a joint management entity has been created between the basin nations, cooperation exists between the countries far ahead of major projects, thereby avoiding conflicts between the Thailand and Myanmar, even though China is not a party to talks.

C.15.8 Time line

1979 Electricity Generating Authority of Thailand (EGAT) initiates fourteen projects to divert water from the tributaries of the Kong and Salween international rivers.

1985 Japan International Cooperation Agency (JICA) presents their study of the Khun Yuam Development Project to the National Committee on Energy. Included in the study are ten hydropowered dam projects on the Yuam, Mae Rid, and Ngao Rivers.

January 1989 Thai Cabinet appoints a committee responsible for the hydropower dam projects on the Thai–Burmese border.

April 1989 Representatives from the Thai Committee on the hydropower dam projects on the Thai–Burmese border discuss the projects with the Myanmar Electric Power Enterprise, and together set up a joint committee.

July 1989 Top officials from the National Committee on Energy visit Rangoon. The two countries enter into an agreement of cooperation in water development projects, and establish a coordinating team with the National Myanmar Electric Power Enterprise and the National Committee on Energy playing a key role.

November 1989 Coordinating team calls for the first meeting in Bangkok. Seven hydropowered dam projects are proposed. Thailand responsible for the study of the Khlong Kra Project, and Burma for the Mae Sai Project.

December 1989 EGAT lists the Lama Luang and Nam Ngao hydropower dam projects in its 17-year power development plan (1990–2006). EGAT Executive Board approves the two projects that are based on the Khun Yuam Project of Jica. Under the plan, the two dams are to be completed by 2000.

August 1990 Coordinating team meets for a second meeting in Bangkok where it agrees to speed up preliminary study of the remaining five dam projects.

1991 Coordinating team meets in Rangoon and decides that National Committee on Energy will ask the EPDC of Japan to conduct the feasibility study of the dam projects.

May 1991 EPDC agrees to join the project and sends a survey team to Thailand.

March 1992 EPDC completes study and proposes eight dam projects along the Thai–Burmese border.

August 1992 Thai Cabinet gives approval to a plan to solve the water crisis in the Chao Phraya River Basin, which encompasses the Salween Diversion Scheme.

Janary 1993 United Nation's People Organization (UNPO) holds a human rights conference at The Hague. The Shan State calls for international cooperation in condemning Slorc for violence against the Shan people and for collaborating with the Thai Government on the Salween Dam projects.

October 1993 Gen. Saw Bo Mya, leader of Karen National Union (KNU), declares at Manerplaw that the KNU is against the Salween Dam projects. He states they are willing to use armed force if peaceful protests prove useless.

March 1994 The House Committee, led by northern MP Songsuk Pakkasem, announce they will organize a seminar to improve comprehension about the Salween water diversion scheme.

1997 Signing of Thai–Myanmar Memorandum of Understanding, which justifies the construction of large hydroelectric dams for electricity generation "for the mutual benefits of the peoples of the Kingdom of Thailand and the Union of Myanmar" (WRM Bulletin, 2000).

October 2002 Thai Cabinet endorses the draft Inter-Governmental Agreement on Regional Power Trade in the Greater Mekong Sub-Region. The cooperation strives to enhance economic relations and improve environmental protection in Myanmar, Thailand, Cambodia, China, Laos, and Vietnam (http://www.irn.org/programs/mekong/030605.ratification.html).

June 2003 Thai Cabinet gives its approval for the ratification of a power supply pact between the six Greater Mekong subregion countries

December 2003 Groups in Thailand and Myanmar protest China's plans for thirteen large dams on the Nu/Salween River.

March 2004 Approximately eighty environmental and human rights organizations protested China's proposed dam projects on the Nu River (Yardley, 2004).

April 2004 China Premier Wen Jiabao suspends dam plan for the Nu River.

August 2004 Thailand and Burma agree to set up a joint venture for the construction of five hydropowered dams in the Salween River Basin, beginning with Tasang dam.

C.16 ORGANIZATION FOR THE DEVELOPMENT OF THE SENEGAL RIVER (OMVS)

C.16.1 Case summary

River basin: Senegal River (Figure C.18).

Dates of negotiation: Organization formed 1972.

Relevant parties: Mali, Mauritania, Senegal (directly); Guinea (indirectly).

Flash point: None – Independence of countries provided opportunity for multilateral development.

Issues: Stated objectives: develop the basin by facilitating closer coordination beyond the water and agricultural sectors.

Additional issues: Water-related: hydropower, artificial flooding; Nonwater: poverty, Guinea's relationship with OMVS states.

Excluded issues: None.

Criteria for water allocations: "Principle of benefit sharing" looks at benefits instead of water allocation.

Incentives/linkage: Financial: Cost-sharing plan based on each state's exploitation benefits; Political: Joint management has built trust between basin states.

Breakthroughs: Basin states ignored unilateral approach in favor of a joint management system.

Status: Continual progress to strengthen agreements between basin states, the most recent the signing of the Water Charter in 2002; attempts to increase the participation of Guinea in basinwide decisions through the OMVS.

C.16.2 Background

The Senegal River, the second-largest river in Western Africa, originates in the Fouta Djallon Mountains of Guinea, where its three main tributaries, the Bafing, Bakoye, and Faleme contribute 80 percent of the river's flow. After originating in Guinea, the Senegal River then travels 1,800 kilometers crossing Mali, Mauritania, and Senegal on its way to the Atlantic Ocean.

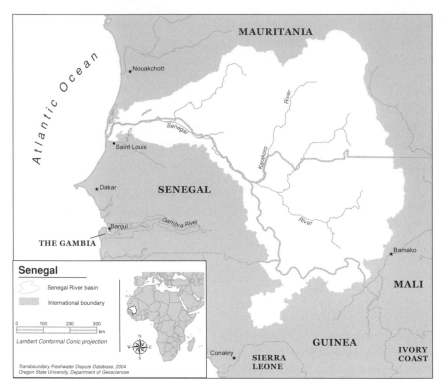

Figure C.18 Map of the Senegal River basin (*Source:* Transboundary Freshwater Dispute Database, 2004).

Following the independence of the basin countries, tension remained in the region due to the instability of the political powers and the influence of neocolonial states such as the United States and the Soviet Union. Throughout the turmoil following World War II into the 1970s, the Senegal River continued to be a common link between the basin countries. There was a desire between them to cooperate in the management of the basin so that all countries would benefit from its development. This aspiration has been carried into the twenty-first century as Guinea, Mali, Mauritania, and Senegal work cooperatively toward more effective basin management.

The river is a key resource for all three countries. Large herds of cattle, camels, goats, and sheep migrate season to season across these borders and herders rely on this water source to sustain their herds. The basin region receives only an average of 660 millimeters of rainfall per year and the Senegal River waters represent the key to agriculture in the region. On the left bank, the surface area of community-based irrigated fields grew from 20 hectares in 1974 to 7,335 hectares in 1983 to 12,978 hectares in 1986. After agriculture, fishing is the largest economic activity in the region. Other river-based economic activities include sugar cane production, rice farming, and, to a lesser extent, mining,

Two dams, Manantali and Diama, were built in 1986 and 1988, respectively, in order to provide fresh water for agri-culture and municipal uses and, in the case of the Manantali Dam, eventually to produce hydroelectric power for the region. These two dams were part of an economic growth strategy for the region that would reduce the investment risk and reduce poverty by increasing income-generating activities. The Manantali Dam was put on line in 2002 and is now supplying the three basin countries with 547 GWH/year.

C.16.3 The problem

There are a few areas of concern in regards to the Senegal River Basin. The first is that of the climate. Beginning in the 1960s, the region suffered a continuous drop in rainfall until the mid-1980s. Because of the local populations' dependence on rainfall for crops, the droughts caused severe disruption in the economies of the basin states. The impacts to the economy were a result of the effects of the drought on the environment. Erosion, saltwater intrusion, drop in groundwater, vegetation loss among other impacts were felt in the entire region, resulting in the exodus of large numbers of inhabitants from the rural areas toward the cities. The extreme poverty in the region makes these populations very vulnerable to changes in the climate.

In the 1960s and 1970s this problem led the countries of the Senegal River basin to look at ways to work together to mitigate the disastrous affects of severe droughts. Unlike other international water bodies, cooperation over this basin did not grow out of a conflict over use of the Senegal River resources. Instead the catalyst for cooperation was the vulnerability of the populations of the basin states. These four countries believe that collaboration on the development of this resource would improve the standard of living of all involved.

Problems in the basin today focus on the detrimental health and environmental and agricultural impacts of the two dams. Seasonal flooding and water movement decreased dramatically after the dams were built. This has caused an increase in the incidence of numerous waterborne diseases: diarrhea, schistosomiasis, and malaria. The reduction of flooding also prevents pollution from industrial agricultural from flushing out of the basin. The dams have caused a reduction in pastureland, degradation of river fisheries, increased soil salinity, and riverbank erosion. Traditional agricultural and pastoral productions systems have been superceded by irrigated, in some cases industrial size, agriculture. This emphasis on irrigation has created problems for social cohesion and access to land in some areas of the river basin, and in some cases artificial flooding has been so poorly planned that it has wiped out crops.

The basin countries expected to see a decrease in the rural–urban drift once irrigation was more feasible, but this has not happened. Politics also play an important role in decisions regarding the yearly artificial flood levels; they often vary according to the policies of current basin country governments (see Organization for the Development of the Senegal River 1972, pp. 448–461 for more information on the positive and negative effects of the dams).

A separate, but equally worrisome issue is the pressure on this resource from a rapidly increasing river basin population: 16 percent of the population of the three river basin countries – Mali, Mauritania, and Senegal – live in this basin, and this population is growing at a rate of 3 percent per year.

C.16.4 Attempts at conflict management

During the 1960s, as the newly independent African states established their national identities and put in place their national infrastructures, there was a large movement among the Senegal River basin states to act jointly in the development of the basin.

The first step to this mutual development of the Senegal River basin came in 1963 when, after several meetings between the basin states, the four countries signed the Bamako Convention, which recognized the Senegal River as an international waterway and created the Interstate Committee (CIE, Comité Inter-Etats pour l'Amenagement du fleuve Sénégal) to oversee

its development. The CIE's main goal was to use a multilateral approach when developing technical capacity and financial support for the Senegal River. The CIE and the basin states were encouraged to present individual proposals to international aid agencies for funding, but not without approval of the CIE. This would guarantee coordinated management of the basin and prevent any unilaterally beneficial projects. The Bamako Convention, and a year later, the Dakar Convention, introduced a framework for future development of the basin for mutual benefit of all countries.

In meetings between the heads of state of the four countries in 1965 and 1966, proposals were made to improve the already existing infrastructure of the CIE and move beyond the Senegal River basin to look at linkages between other West African rivers. The goal was to reinforce the idea of cooperation through the development and integration of the region's economies. However, due to tensions that arose between Senegal and Guinea in January of 1967, Guinea suspends its participation in the CIE. Mali and Mauritania, still interested in the regional integration of the Senegal River basin, managed to bring all four heads of states back to the negotiating table in November of 1967 in Bamako, Mali, to revive collaboration.

In late 1967, several ministerial meetings took place to revive the idea of cooperation. These meetings resulted in the Labé Convention, signed on March 24, 1968, which created the Organization of Boundary States of the Senegal River (OERS, Organisation des Etats Riverains du Sénégal). The goals of OERS were more comprehensive than those of the CIE. Because its objectives were not limited to the valorization of the basin, the member states attempted to politically and economically integrate the basin through the standardization of legislation, the improvement of education, and the further breaking down of borders to allow increased trade and labor movement. This initiative demonstrated the interest these four countries had in treating the river basin as an international resource.

The economic cooperation of the member states of OERS advanced with various ministerial-level meetings in the transportation and economic sectors, but when political instability of the basin occurred in 1970, difficulties arouse within the organization. The nature of OERS was such that decisions were made unanimously, and when Guinea was absent for two meetings in 1971 (due to regional political instability), negotiations came to a halt. Consequently, Guinea withdrew from OERS in 1972 and the organization became defunct.

The desire for regional integration remained with Mali, Mauritania, and Senegal, however, and the three countries established the Organization for the Development of the Senegal River (OMVS, Organisation pour la Mise en Valeur du fleuve Sénégal). One of the most important aspects of this convention is that Guinea did not participate, but it did not object either, which has allowed the process to move forward with less difficulty. The organization created two types of shared infrastructure, physical and institutional, which were designed to accomplish the goals of OMVS:

- The development of food security for the populations of the basin.
- The reduction in the economic vulnerability of OMVS states to external factors such as climate changes.
- The acceleration of the economic development of member states the preservation of ecosystem balance in the subregion and particularly in the basin.
- Secure and improved revenue of the valley populations.

By the early 1990s when Senegal began to realize that the OMVS objectives were not going to be obtained, they drew up The Master Plan for the Integrated Development of the Left Bank. This plan shifted development focus from the establishment of irrigated agriculture to a more integrated form of development. They would continue to promote irrigated agriculture without jeopardizing other uses of the water such as flood-recessional farming, while at the same time promoting an artificial yearly flood. However, the OMVS felt that this plan was an affront to the authority of the OMVS and did not allow its implementation.

C.16.5 Outcome

The history of cooperation over this river basin has led to numerous multilateral agreements, projects, and organizations over the past 25 years. The Manantali Dam has been working at full capacity since May 2003, providing each of the basin countries with electricity based on the amount the invested in the dam project. Mali is receiving 52 percent of the benefits, while Mauritania receives 15 percent and Senegal 33 percent. An additional benefit of this dam has been the fiber optic cables used for the transmission lines, which telecommunications companies can also use (see Organization for the Development of the Senegal River 1972, pp. 448–461, for a more complete list of positive and negative effects of the dams).

The Senegal River Charter, signed in 2002, sets the principles and procedures for allocating water between the various use sectors, defines procedures for the examination and acceptance of new water use projects, determines regulations for environmental preservation and protection, and defines the framework and procedures for water user participation in resource management decision-making bodies.

Following this charter, the GEF funded a 4-year Water and Environmental Management Project to provide a framework for sustainable development and transboundary land–water management in the Senegal river basin. The project has five goals: (1) to build capacity, (2) to effectively manage data and knowledge, (3) to complete the Transboundary Diagnostic Analysis and Strategic Action Program, (4) to act on local priorities, and (5) to initiate public participation and awareness.

C.16.6 Lessons learned

Stakeholder participation should be included at all levels of decision-making processes for optimal mutual gain.

When local populations were not included on the decision-making processes within the Senegal River Basin, there tended to be frustration, confusion, and economic losses directly as a result of not participating. Participation by all stakeholders can only benefit all groups involved in making agreements more sustainable, mutually beneficial, and efficient.

Lack of participation of all basin nations weakens the overall negotiations and creates opportunity losses for those not participating.

Guinea, not party to the OMVS organization, has not experienced the development benefits of the other three countries in the basin. As a result, they are lacking water resource management infrastructure, a reliable energy source, and water supplies.

Mutually beneficial projects and integrated investments create good neighbors.

As a result of the OMVS and the design and implementation of joint projects, the relations between the countries has improved and economic development has increased thereby making cooperation rather than conflict a meeting point with regards to the Senegal River.

C.16.7 Creative outcomes from resolution process

The mutually beneficial design of the OMVS and how it redistributes the economic benefits based on how much each country puts into the project creates incentives and equality in the development process.

Even though Guinea dropped out of the cooperation process officially in 1967, Mali, Mauritania, and Senegal have allowed them to be an observer thereby reducing the potential for conflict within the basin. Full participation would be ideal, but under the circumstances, it is better to have Guinea present and make the process transparent rather than exclude them altogether.

PASIE, Plan d'Attenuation et de Suivi des Impact sur l'Environnement, was formed in 1998, an entity whose sole purpose to investigate the environmental impacts of the development and distribution of power from the Manantali hydroelectric power station.

C.16.8 Time line

July 25, 1963 Guinea, Mali, Mauritania, and Senegal signed the Bamako Convention for the Development of the Senegal River Basin thereby creating the Interstate Committee.

February 7, 1964 Signing of the Dakar Convention.

November 1965 and 1966 The heads of state of Guinea, Mali, Mauritania, and Senegal meet to discuss the promotion of regional integration

January 1967 Guinea suspends participation with the Interstate Committee due to tensions with Senegal.

November 1967 Through the efforts of Mali and Mauritania, the four heads of state met again in Bamako.

May 26, 1968 Through the Labé Convention, the Interstate Committee is replaced with the Organization of the Boundary States of the Senegal River (OERS).

1971 Due to political instability, Guinea does not attend two OERS meetings and later withdraws from the organization.

June 1971 The Council of Ministers acknowledge the problems arising within OERS.

March 11, 1972 Mali, Mauritania, and Senegal declare OERS void and sign the Nouakchott Convention creating Organization for the Development of the Senegal River (OMVS) reconfirming it's international status and the pursuit of OERS same objectives.

December 1975 Restructuring of OMVS into three entities: Heads of State Summit, Council of Ministers, and the High Commission (technical).

December 12, 1978 Convention recognizing the Legal Status of Jointly-Owned Structures.

March 12, 1982 Convention recognizing the Financing of Jointly-Owned Structures.

1992 OMVS–Guinea protocol signed outlining framework for cooperation in projects of mutual interest, including permitting Guinea to attend OMVS meetings as an observer.

1995 Agriculture Sector Adjustment Program – aims to provide food security, rural incomes, natural resource management through deregulation, reduced state intervention, and land-tenure reform.

1998 Creation of the Environment Impact Mitigation and Monitoring Program (PASIE) by OMVS.

2000 Establishment of the Environmental Observatory by OMVS to monitor environmental change in the Basin as part of PASIE.

2002 Water Charter signed and ratified by Mali, Mauritania, and Senegal

2002 Water and Environmental Management Project funded by GEF

2003 Heads of States Summit in Nouakchott, Mauritania. Guinea participates.

2004 First Inter-Ministerial meeting between Guinea and OMVS member states in Dakar, Senegal.

C.17 TIGRIS–EUPHRATES BASIN

C.17.1 Case summary

River basin: Tigris–Euphrates (Figure C.19).

Dates of negotiation: Meetings from the mid-1960s to the present.

Relevant parties: Iraq, Syria, Turkey.

Flash point: Filling of two dams during low-flow period results in reduced flow to Iraq in 1975.

Issues: Stated objectives: negotiate an equitable allocation of the flow of the Euphrates River and its tributaries among the riparian states.

Additional issues: Water-related: water quality considerations, Orontes River; Nonwater: Syrian support for PKK Kurdish rebels.

Excluded issues: Relationship between Tigris and Euphrates.

Criteria for water allocations: None determined.

Incentives/linkage: Financial: none; Political: none.

Breakthroughs: Adana Agreement (2001), in which Syria agrees to ban PKK rebels from country; Protocol of coordination between Turkey and Syria (2001) on respective development projects (GAP and GOLD).

Status: Bilateral and tripartite negotiations continue with mixed success – no final agreement to date.

C.17.2 Background

Bilateral and tripartite meetings, occasionally with Soviet involvement, had been carried out between the three riparians since the mid-1960s, although no formal agreements had been reached by the time the Keban and Tabqa dams began to fill late in 1973, resulting in decreased flow downstream. In mid-1974, Syria agreed to an Iraqi request that Syria allow an additional flow of 200 MCM/year from Tabqa. The following year, however, the Iraqis claimed that the flow had been dropped from the normal 920 m^3/sec to an "intolerable" 197 m^3/sec and asked that the Arab League intervene. The Syrians claimed that less than half the river's normal flow had reached its borders that year and, after a barrage of mutually hostile

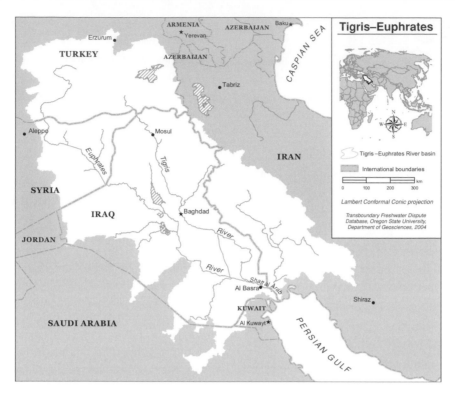

Figure C.19 Map of the Tigris–Euphrates basin (*Source:* Transboundary Freshwater Dispute Database, 2004).

statements, pulled out of an Arab League technical committee formed to mediate the conflict. In May 1975, Syria closed its airspace to Iraqi flights and both Syria and Iraq reportedly transferred troops to their mutual border. Only mediation on the part of Saudi Arabia was able to break the increasing tension, and on June 3, 1975, the parties arrived at an agreement that averted the impending violence. Although the terms of the agreement were not made public, Iraqi sources are cited as privately stating that the agreement called for Syria to keep 42 percent of the flow of the Euphrates within it borders and to allow the remaining 58 percent through to Iraq.

C.17.3 The problem

In 1975, unilateral water developments came very close to leading to warfare along the Euphrates River. The three riparians to the river – Turkey, Syria, and Iraq – had been coexisting with varying degrees of hydropolitical tension through the 1960s. At that time, population pressures drove unilateral developments, particularly in southern Anatolia (Turkey), with the Keban Dam (1965–1973), and in Syria, with the Tabqa Dam (1968–1973) (see Lowi, 1991, p. 108). Additional tensions between Turkey and Syria involving Syrian support for Kurdish separatists (Kurdish Worker's Party, or PKK) and Turkey's military support for Israel have exacerbated the water dispute (*Mideast Mirror*, 1998). Military tensions flared between Turkey and

Iraq in 1997, as Turkey invaded northern Iraq to attack Kurdish rebels in the area (*Mideast Mirror*, 1997). In August of 1998, Turkey threatened military action against Syria if it continued to support the PKK (*Middle East Newsfile*, 1998).

C.17.4 Attempts at conflict management

The Southeast Anatolia Development Project (GAP is the Turkish acronym) has given a sense of urgency to resolving allocation issues on the Euphrates. GAP is a massive undertaking for energy and agricultural development that, when completed, will include the construction of twenty-one dams and nineteen hydroelectric plants on both the Tigris and the Euphrates. 1.65 million ha of land are to be irrigated and 26 billion kWh will be generated annually with an installed capacity of 7,500 MW. If completed as planned, GAP could significantly reduce downstream water quantity and quality.

A Protocol of the Joint Economic Committee was established between Turkey and Iraq in 1980, which allowed for Joint Technical Committee meetings relating to water resources. Syria began participating in 1983, although meetings have been intermittent at best.

A 1987 visit to Damascus, Syria, by Turkish Prime Minister Turgut Ozal reportedly resulted in a signed agreement for the Turks to guarantee a minimum flow of 500 m^3/sec across the border with Syria. According to Kolars and Mitchell, this total of 16 BCM/year is in accordance with prior Syrian requests. However, according to Naff and Matson, this is also the amount that Iraq insisted on in 1967, leaving a potential shortfall. A tripartite meeting among Turkish, Syrian, and Iraqi ministers was held in November 1986 but yielded few results.

Talks among the three countries were held again in January 1990, when Turkey closed the gates to the reservoir on the Ataturk Dam, the largest of the GAP dams, essentially shutting off the flow of the Euphrates for 30 days. At this meeting, Iraq again insisted that a flow of 500 m³/sec cross the Syrian–Iraqi border. The Turkish representatives responded that this was a technical issue rather than one of politics and the meetings stalled. The Gulf War that broke out later that month precluded additional negotiations.

In their first meeting after the war, Turkish, Syrian, and Iraqi water officials convened in Damascus in September 1992 but broke up after Turkey rejected an Iraqi request that flows crossing the Turkish border be increased from 500 m³/sec to 700 m³/sec. In bilateral talks in January 1993, however, Turkish Prime Minister Demirel and Syrian President Assad discussed a range of issues intended to improve relations between the two countries. Regarding the water conflict, the two agreed to resolve the issue of allocations by the end of 1993. Prime Minister Demirel declared at a press conference closing the summit that "there is no need for Syria to be anxious about the water issue. The waters of the Euphrates will flow to that country whether there is an agreement or not" (cited in Gruen, 1993). Despite this pledge, no agreement was reached in the allocated timeframe.

In February 1996, a joint Syria–Iraq water coordination committee convened in Damascus, where the two sides discussed what would be a fair and reasonable distribution of the Euphrates and Tigris among Turkey, Syria, and Iraq. In this meeting, Syria and Iraq decided to coordinate their positions on the water dispute. In May of the same year, Turkey called on Syria to engage in talks over water. Turkey wanted to resolve the dispute by dividing water by cultivated land, whereas Syria wanted to divide the water equally (Gruen, 1993).

Tension between Syria and Turkey escalated in late 1998 over Kurdish rebels. To avert invasion by Turkey, Syria agreed to ban the PKK from Syria (Ilter, 2000) with the signing of the Adana Agreement on October 20, 1998 (*Mideast Mirror*, 2000).

C.17.5 Outcome

In 2001, Syria and Iraq held talks about the water of the Euphrates and restated their commitment to take a united stand on the issue in any negotiations with Turkey (*Technical Review Middle East*, 2001). In August of 2001, Syria and Turkey

agreed on a protocol of cooperation for Turkey's GAP and Syria's corresponding GOLD (General Organization for Land Development) projects.

Despite these strides, the situation remains unresolved. As of 2003, Turkey would not sign a final accord regarding the sharing of waters with Syria and Iraq (United Press International, 2003). Since the ousting of Saddam Hussein in Iraq by U.S.-led forces, the newly appointed Minister of Water Resources Abdul, Latif Rasheed, has stated that previous problems in trying to come to agreement on allocation of Tigris and Euphrates waters were due to the bad relations developed by the previous leadership. The new Iraqi government hopes to reach an agreement with Turkey and Syria over the waters (Hafidh, 2003). These new developments in the region may play a large role in the future of the sharing of the Euphrates.

C.17.6 Lessons learned

When one riparian holds the most geographic and military power, equitable agreements are difficult to reach.

With the large majority of water originating in Turkey and Turkey having the most advanced military power, it has less incentive to work cooperatively with Syria and Iraq and to approach negotiations with a "basket of benefits" outlook.

When mostly bilateral talks are used to attempt to resolve issues, the most powerful country typically maintains its power.

In bilateral talks, Turkey has succeeded in maintaining its power in the water dispute. Syria lost one of its "playing cards" in overall negotiations when it signed the Adana Agreement.

Unilateral development of water resources leads to increasing tension over water.

Developments in the basin have been made unilaterally without the cooperation of other riparian countries. This has increased resentment of downstream riparians who had no say in developments that occurred upstream.

C.17.7 Creative outcomes resulting from resolution process

Although no final resolution has been reached, the protocol for cooperation between Syria and Turkey is a step away from unilateral development. In addition, the signing of the Adana Agreement banning PKK rebels from Syria helped break the deadlock between countries.

C.17.8 Time line

Late 1973 Keban and Tabqa dams fill.

Mid-1974 Syria agrees to Iraqi requests to allow additional flow of 200 MCM/year to Iraq.

1975 Iraq claims that flow in Euphrates has dropped from normal 920 m³/sec to 197 m³/sec and requests that Arab League intervene; Syria claims less than average flow and drops out of Arab League.

June 3, 1975 Mediation by Saudi Arabia leads to agreement that averts war (though agreement is not made public); Syria uses 42 percent of water and allows 58 percent to flow to Iraq.

1980 Protocol of the Joint Economic Committee is established between Turkey and Iraq, which allows for Joint Technical Committee meetings relating to water resources.

1983 Syria begins participating in the Joint Economic Council.

November 1986 Tripartite meeting among Turkish, Syrian, and Iraqi ministers with few results.

1987 Turkish Prime Minister visits Damascus and signs agreement for Turks to guarantee 500 m³/sec across the border to Syria.

January 1990 Talks among three countries held when Turkey begins filling the Ataturk Dam, shutting off flow to the Euphrates for 30 days; Iraq insists that 500 m³/sec reaches its border; Gulf War breaks out.

September 1992 Turkish, Syrian, and Iraqi water officials convene in Damascus but break up after Turkey rejects Iraqi request that flows crossing the Turkish border increase from 500 m³/sec to 700 m³/sec.

January 1993 Bilateral talks between Turkish Prime Minister and Syrian President where a range of issues are discussed to improve country relations; the two countries agree to resolve the issue of Euphrates water allocation by the end of 1993.

February 1996 Joint Syria–Iraq water coordination committee convenes in Damascus; here the two sides discuss what would be a fair and reasonable distribution of the Euphrates and Tigris among Turkey, Syria, and Iraq; Syria and Iraq decide to coordinate their positions on the water dispute.

August 1998 Turkey threatens military action against Syria if it continues to harbor PKK rebels.

October 1998 Adana Agreement signed by Turkey and Syria, in which Syria agrees to ban PKK rebels from the country.

January 2001 Syria and Iraq hold talks to establish water sharing; restate commitment to coordinate efforts in negotiations with Turkey.

August 2001 Syria and Turkey agree on a protocol of cooperation for Turkey's GAP and Syria's corresponding GOLD (General Organization for Land Development) projects.

2003 Iraqi President Saddam Hussein ousted by U.S.-led forces; later, new leadership states intentions to reach agreement with Turkey and Syria regarding allocation of the Tigris and Euphrates waters.

C.18 UNITED STATES–MEXICO SHARED AQUIFERS

C.18.1 Case summary

River basin: Aquifers that straddle the United States /Mexico boundary (Figure C.20).

Dates of negotiation: United States–Mexico Water Treaty signed 1944; groundwater negotiations since 1973.

Relevant parties: Mexico, United States.

Flash point: Salinity crisis of 1961–1973 raised groundwater as important issue not detailed in 1944 treaty.

Issues: Stated objectives: develop an equitable apportionment of shared aquifers.

Additional issues: Water-related: pollution; Nonwater: none.

Excluded issues: None.

Criteria for water allocations: None.

Incentives/linkage: None.

Breakthroughs: None.

Status: Talks have been ongoing since 1973.

C.18.2 Background

The border region between the United States and Mexico has fostered its share of surface-water conflict, from the Colorado River to the Rio Grande–Rio Bravo River Basin. It has also been a model for peaceful conflict resolution, notably the work of the International Boundary and Water Commission (IBWC), the supralegal body established to manage shared water resources as a consequence of the 1944 United States–Mexico Water Treaty. Yet the difficulties encountered in managing shared surface water pale in comparison to trying to allocate groundwater resources – each aquifer system is generally not understood as gathering information on aquifers is very costly and the science of groundwater is still inexact. This makes negotiations over a shared aquifer system very difficult.

Mumme (1988) has identified twenty-three sites in contention in six different hydrogeologic regions along the 3,300 kilometers of shared boundary. Although the 1944 Treaty mentions the importance of resolving the allocations of groundwater between the two states, it does not do so. In fact, shared surface-water resources were the focus of the IBWC until the early 1960s, when a U.S. irrigation district began draining saline groundwater into the Colorado River and deducting the quantity of saline water from Mexico's share of fresh water.

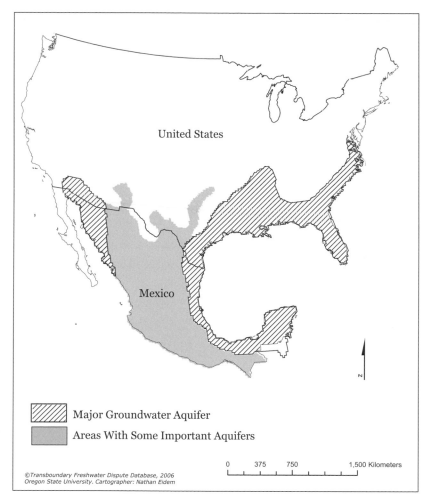

United States

Mexico

Major Groundwater Aquifer

Areas With Some Important Aquifers

0   375   750   1,500 Kilometers

©Transboundary Freshwater Dispute Database, 2006
Oregon State University. Cartographer: Nathan Eidem

Figure C.20 Map of shared aquifers between Mexico and United States (*Source:* Transboundary Freshwater Dispute Database, 2006).

In response, Mexico began a "crash program" of groundwater development in the border region to make up the losses.

C.18.3 The problem

The complications of groundwater are exemplified in the border region between the United States and Mexico where, despite the presence of an active supralegal authority since 1944, groundwater issues have yet to be resolved. Mentioned as vital in the 1944 Treaty, and again in 1973, the difficulties in quantifying the ambiguities inherent in groundwater regimes has eluded legal and management experts ever since.

C.18.4 Attempts at conflict management

Ten years of negotiations resulted in a 1973 addendum to the 1944 Treaty – Minutes 242 of the IBWC, which limited groundwater withdrawals on both sides of the border, and committed each nation to consult the other regarding any future

groundwater development. In all of the Minutes added to the 1944 Treaty since its inception, Minutes 242 is still the only agreement between the two nations with regard to groundwater pumping.

Stephen Mumme (2004) states that there are three main reasons why Minutes 242 has had trouble being advanced as its agreement intended. First, and maybe most importantly, was that there was not the political support to carry out Minutes 242. A rift between state and federal government over whose authority it was to control water rights played a key role and when there are ninety-six seats in the House of Representatives from the border region, this makes it difficult to pass any legislation going against those states.

Second, it is possible that Minutes 242 did not refer to groundwater quality in general but more pointedly at salinity. This may have averted governments from pursuing appropriate studies.

Third, the terms of reference of both Minutes 242 and the 1944 Treaty are not very clear. The wording of the agreements does not have enough definition to promote decisive acts and leaves much to be questioned.

C.18.5 Outcome

Even after three decades of having problems with Minutes 242 and groundwater issues, there does not appear to be a movement toward a new agreement referring to the United States–Mexico shared aquifers anytime soon, although Mumme (2004) states that it is likely that "some form of systematic cooperation will emerge" between stakeholders in more local areas along the border.

C.18.6 Lessons learned

Even if conditions for agreement are good, this does not guarantee that issues will be resolved.

It is testimony to the complexity of international groundwater regimes that despite the presence of an active authority for cooperative management, and despite relatively warm political relations and few riparians, negotiations have continued since 1973 without resolution.

Difference of opinion of federal and state governments can impede cooperation.

After Minutes 242 was agreed on by both Mexico and the United States, the differences between the United States federal government and the government of the States bordering Mexico most likely played a role in the lack of cooperation between the two nations with regards to groundwater resources.

C.18.7 Creative outcomes resulting from resolution process

Treaty includes mechanism to modify terms and even topics covered, essentially allowing for adaptive management, without renegotiating entire treaty.

C.18.8 Time line

1944 United States–Mexico Water Treaty signed. IBWC expanded to include water allocation within its responsibilities.

1973 Minute 242 on groundwater signed between Mexico and the United States.

1983 La Paz Agreement signed creating technical working groups that addressed water quality among other environmental concerns.

1993 North American Free Trade Agreement (NAFTA) signed.

1994 NAFTA implemented.

D International water pricing: An overview and historic and modern case studies

Kristin M. Anderson and Lisa J. Gaines

D.1 THE VALUE OF WATER: AN OVERVIEW OF MAJOR ISSUES

Managing water conflicts ultimately concerns values. Values and perceptions toward how to use and prioritize water in various locations change over time as social demographics and needs change. Today's enhanced discussion of water as an economic good, among other characteristics, has brought more attention to how water services are priced and how markets might be a means to prioritize and/or reallocate water uses. Because such prioritization and reallocation decisions involve trade-offs among values, they are often at the heart of water conflicts. This appendix briefly describes various values associated with water uses and conflicts. It then provides studies on international water pricing in treaties.

D.2 THE DUBLIN STATEMENT AND UNITED NATIONS AGENDA 21

The Dublin Statement, issued from the International Conference of Water and the Environment (ICWE) held in Dublin, Ireland, in January 1992, was a primary catalyst of the debate over treatment of water as an economic good (ICWE, 1992). Resulting from the call from 500 participants from 100 nations for fundamental new approaches to the management of freshwater resources, the Dublin Statement included within it the principle that "water has an economic value in all its competing uses and should be recognized as an economic good" (ICWE, 1992, Guiding Principle No. 4). This was the first explicit recognition of water as an economic good, and this principle is often found quoted in literature that has ensued since its establishment.

Shortly thereafter, this same idea was adopted by the Plenary in Rio de Janiero at the United Nations Conference on Environment and Development in June 1992, with some additions to the statement. Agenda 21, emanating from that meeting states, "Integrated water resources management is based on the perception of water as an integral part of the ecosystem, a natural resource and a social and economic good, whose quantity and quality determine the nature of its utilization" (UNCED, 1993).

D.3 THE MANY VALUES OF WATER

Water is not strictly limited to the status of an economic good. It is also a social good, and it has cultural and religious value as well. In *The World's Water*, Peter Gleick illustrates the characteristics of water outside of being an economic good (Gleick, 2003). Gleick outlines some of water's major values as follows:

Water is a social good. Access to clean water is vital to people. Water quality affects public health in the short and the long term. Water supply management for populations involves the building of large infrastructure. Such works are best handled with public oversight.

Water is an economic good. Water is a scarce resource with value in competing uses. Allocation of water resources could be optimized to maximize benefits to society.

Water has ecological value. Water is essential not only for humans but also for all life. Changing the hydrology of ecosystems threatens populations of many species.

Water has religious, moral, and cultural value. Water figures into cultural and religious identities as part of rituals and symbolism. Moral values may come into play with property rights issues when people feel they morally have a right to water (Gleick, 2003).

D.4 GLOBALIZATION, PRIVATIZATION, AND COMMODIFICATION OF WATER

Globalization, privatization, and commodification of water are all major topics of water policy debates today. Gleick, in *The New Economy of Water*, reviews how these have changed the way water is being treated in these debates (Gleick et al., 2002).

Commodification is the transformation of a formerly non-market good to a market good. Although water has, on a smaller scale, had a market value in the past, with the issue of the Dublin Statement (ICWE, 1992) on water and changes in global markets, the commodification of water has increased (Gleick et al., 2002). Water delivery and management of its various uses has always entailed costs. However, these costs have not always been made explicit. The Dublin Statement has increased awareness and made more explicit the notions of cost to deliver water services.

Globalization refers to the process of integrating markets internationally. The uneven distribution of water across the globe, coupled with newly opened global markets, has raised awareness that water is an item that has been and is being traded on the global scale. Water can be traded as a bulk good or as a value-added product as bottled water. Bottled water sales have been increasing noticeably in the last decade (Gleick et al., 2002). As the case studies in this appendix show, water trade as a bulk quantity is also occurring. In addition, water is embedded in goods that are traded among countries and thus also traded virtually.

Privatization of water involves transferring control of all or parts of water systems from public into private hands. Privatization of water has become a volatile and sometimes misleading term. Many see the term as meaning private ownership and control of water. However, privatization has really referred to new forms of management of water services by private and public sectors in varying degrees of partnerships. These are referred to as private public partnerships. At best, these new partnerships have brought more explicit performance standards for water service operations. In addition, some believe that private public partnerships can be more efficient than government run services and they can secure more capital more quickly. On the other hand, as Gleick and others note, there are potential downsides. Among many risks of privatization that Gleick outlines are: that privatization may result in social inequities, public ownership of the water itself may be at risk, ecosystem impacts could be ignored, and water use efficiency and water quality may not be as valued (Gleick et al., 2002). Throughout the world, water remains primarily under public control. The question is mostly over how that public control is realized – through regulation or direct government management.

D.5 COMPLEXITIES IN THE ECONOMIC BEHAVIOR OF WATER

The question of whether water can actually be treated as a true economic good is debated. Looking at water resources from a big-picture perspective, it appears that by treating water as an

Sidebar D.1 Unique Characteristics of Water

- Water is essential.
- Water is scarce.
- Water is fugitive, meaning that its availability varies with time; it needs to be stored for certain uses.
- Water is a system; the hydrologic cycle is interconnected, and interruption to one part will affect the rest.
- Water is bulky and not easily transportable; movement of water is in most cases too expensive to transport and still meet buyer's willingness to pay.
- Water has no substitute; no other economic good can replace it.
- Water is not freely tradable; because of the combination of it being essential but too bulky to easily move, trade is difficult.
- Water is complex.
  - It is a common good and public good that cannot be owned privately. The societal dependency is high.
  - Water is also bound by its location of origin and its natural conveyance system.
  - Water has high production and transaction costs.
  - The market for water is not homogeneous. Willingness to pay is different for different users.
  - There are macroeconomic interdependencies among water-using activities.
  - There is a threat of market failure in water supply. Because of its bulk, economies of scale lead to monopolies in water services.
- Water has a high merit value relating to our perceptions of beauty, well-being, and health (Savenije, 2001).

economic good, pricing will improve overall allocations and encourage sustainable use. Dinar and Subramanian state that on both individual and social levels, if price reflects the value of the resource, water use efficiency will improve (Dinar and Subramanian, 1997).

Some argue that water cannot be treated like other economic goods because of its unique characteristics. Savenije outlines several characteristics (Sidebar D.1) of water as a resource that, together, illuminate how it is not an ordinary economic good (Savenije, 2001). These characteristics of water as a resource lead it to behave differently from ordinary economic goods. To be effective, water pricing schemes need to be able to handle these complexities.

D.6 WATER AS A HUMAN RIGHT

As a response to the Dublin Statement identifying water as an economic good, there has been much outcry about the need

Table D.1 *Water price ranges for various sectors and countries in the analysis (in 1996 US$)*

		Minimum		Maximum
Agriculture				
Fixed (per hectare per year or season)	0.164	Bottom of range for India; range based upon state and crop	213.64	Top of range for Taiwan
Variable (per cubic meter)	0.0001	Bottom of range for Spain	0.398	Top of range for Tanzania
Domestic				
Fixed (per household per year or month)	0.075	Bottom of range for Madagascar	1937	Top of range for Portugal
Variable (per cubic meter)	0.0004	Bottom of range for Spain	2.58	Top of range for France
Industry				
Fixed (per plant per year or month)	1.67	Bottom of range for Sudan	2,705	Top of range for Portugal
Variable (per cubic meter)	0.0004	Bottom of range for Spain	7.82	Australia

Note: Values from Dinar and Subramanian, 1997, pp. 7–8.

to treat water as a human right (Baillat, 2005). Because water is essential to life, and there are no substitutes for it, there is concern that treating it as an economic good will leave certain people without access to needed freshwater resources. But others note that if the costs of water are not made more explicit, the poor of the world and others will not realize increased access to water. Scanlon, Cassar, and Nemes provide a review of this topic that covers many of the arguments found in the literature (Scanlon et al., 2004). In their review of international laws, conventions, and judicial decisions, they find that the idea of water as a human right has not been clearly defined by international instruments. The idea of water as a human right is implicit in existing fundamental human rights laws, and it is explicitly included only in nonbinding instruments.

D.7 ACTUAL PRICING OF WATER

Dinar and Subramanian present a picture of international water prices in 1997 (Dinar and Subramanian, 1997, pp. 6–8). Although these figures are now somewhat dated, no similar more recent such compilation was found, and the information presented is still quite helpful in understanding international and regional water price variations and price variations between water use sectors. Water price ranges in agricultural, domestic, and industry sectors for twenty-two countries are listed. Table D.1 provides a summary of the minimum and maximum values reported in their analysis.

Dinar and Subramanian state that the fixed prices of water have variable denominators in different countries (year, area, crop, water velocity, etc.), which make those figures particularly hard to compare, whereas the variable prices are more easily compared. They also state that variable prices in agricultural and domestic sectors are fairly similar in all countries, while industrial prices vary more based on the value of industry to different cultures and their inclusion of pollution costs in the price of industrial water.

D.8 FULL COST RECOVERY

The principle of full cost recovery (FCR) is a valuable principle to review in a general discussion of the economics of water, as it illuminates the various costs associated with water use. The principle is considered to be an option for handling increasing water scarcity and all of its effects, including environmental and human effects. This principle is described in a review of water pricing in the European Union (EU) carried out by the European Environmental Bureau (EEB) (Roth, 2001). FCR would include consideration of all of the following:

Operational and maintenance costs
Capital costs
Opportunity costs
Resource costs
Social costs
Environmental damage costs
Long-run marginal costs period (Roth, 2001)

Integrating social and environmental costs in an FCR framework would involve making a certain quantity available to every person and employing the Polluter Pays Principle (PPP). The PPP, in which those creating pollution have to pay the entire environmental cost, internalizes those environmental damages, rather than leaving environmental costs as externalities where the public ends up paying the cost (with health care bills or otherwise). This is often quite difficult to quantify (Roth, 2001).

The basic premise of FCR is that the representation of the true cost of water in all sectors will cause users to value water at its real cost and will help to allocate water where it is most valued. Since water infrastructures are also capital intensive, private markets are unlikely to provide all the necessary capital. Thus subsidies are usually involved in some way with long term water infrastructures. While subsidies have been part of all wealthy countries' water infrastructures, once established, subsidies are hard to change even after the water infrastructure has helped achieve socio-economic goals. FCR has brought attention to the need to give transparency on subsidies as they will continue to be needed and to be a part of conflict management strategies for water.

D.9 CONCLUSION

The essential role of water to life on earth makes its treatment as an economic good extremely complex. Discussion of the value of water necessarily involves the consideration of a great breadth of factors, several of which have been touched upon above.

D.10 CASE STUDIES

A set of case studies was compiled to highlight characteristics of water pricing. Case studies of international water transfers and trades were sought, but certain domestic water pricing case studies where pricing is well defined were also included. Both historic and recent case studies are presented. As water pricing is a deeply complex issue and influenced by a wide array of factors ranging from local to global in scale, a clear and concise comparison from one case study to another is difficult. Rather, this summary of many real-world scenarios, where costs/prices are at least partially stated, is provided to give the reader a perspective of the range of ways in which the price of water has been considered in water provisions to date.

D.10.1 Historical cases

The Transboundary Freshwater Dispute Database (TFDD) is a project of Oregon State University's Department of Geosciences that includes information on international freshwater treaties and water events as well as spatial information on international basins. The International Freshwater Treaties Database, a component of the TFDD, contains information on more than 400 international freshwater-related agreements, covering the years 1820–2002 (TFDD, 2006). The database was searched to find historical cases that defined the value of water in water transfers. Queries were conducted for agreements with the criteria of water quantity as the principal issue area and capital as a linkage within the agreement. All information presented in this section is derived from the TFDD unless otherwise noted in the text.

Some agreements with stated price considerations define the value of raw water explicitly. Others tie the value of the water to irrigation or hydropower. In some cases, the value of the raw water is implied to be zero. There does not appear to be a consistent way of valuing water in these historic cases. Conversions to 2005 U.S. dollars were made for easier understanding of the price terms of agreements. Calculations do not represent a rigorous analysis and are intended to provide a general idea of modern day equivalent value.

AGREEMENT: AMENDED TERMS OF AGREEMENT BETWEEN THE BRITISH GOVERNMENT AND THE STATE OF JIND, FOR REGULATING THE SUPPLY OF WATER FOR IRRIGATION FROM THE WESTERN JUMNA CANAL
Basin: Indus
Date: September 16, 1892
Parties: Great Britain, State of Jind (a state in India)
Summary: The British Government agreed to supply the State of Jind with water from the Hansi Branch of the Western Jumna Canal through ten main distributaries. Gauges would be placed at each distributary, and the British were to be in charge of monitoring these.
Price considerations: The construction of the distributaries would be done at the cost of the British government, but when completed, it would be handed to the Jind State with the exception of some parts. The distributaries would be kept in repair by the Jind State, and a deduction from the annual charge would be made accordingly. In return for the irrigation water, the State of Jind agreed to pay the British Government an amount annually based on an area of 50,000 acres, and a rate per acre calculated as the average of similar rates in other British territories. The price for the water was set at 2.4 rupees (approximately US$17 in the 2005 equivalent[1]) per acre, with deductions for maintenance, establishment savings, and fees to local agricultural supervisors. Prices are listed in the treaty as shown in Table D.2.

AGREEMENT: AGREEMENT BETWEEN THE BRITISH GOVERNMENT AND THE PATIALA STATE REGARDING THE SIRSA BRANCH OF THE WESTERN JUMNA CANAL
Basin: Ganges
Date: August 29, 1893
Parties: Great Britain, State of Patiala (a Punjab state in India)

---

[1] Rupee to pound 1892 exchange rate obtained from Catão and Solomou, Effective exchange rates and the classical gold standard adjustment, *American Economic Review*, **95**(4),1259–1275, September 2005. Conversion of 1892 pounds to 2005 U.S. dollars by Officer and Williamson, Computing 'real value' over time with a conversion between UK Pounds and U.S. dollars, 1830–2005, *MeasuringWorth.com*, August 2006.

Table D.2 *Water prices in the 1892 amended terms of agreement between the British government and the State of Jind, for regulating the supply of water for irrigation from the Western Jumna canal*

Gross sum payable		Rs.
50,000 acres and Rs. 2.4 per acre		120,000
Deductions	Rs.	
(1) Maintenance and repairs	5,000	
(2) Establishment savings	3,500	
(3) Fees to Lambardars and Patwares	6,000	
		14,500
Net amount payable per annum		105,500 (2005 US$760,000)

Note: The cost of the raw water is tied to its irrigation value and is not stated explicitly.

Summary: The British Government planned to build infrastructure to supply water from the Western Jumna Canal to British and Patiala territory. The British government was to have exclusive control of the project as it was built; 5 years after completion, control of Patiala distributaries would be transferred to the Patiala State.

Price considerations: The British and Patiala governments agreed to share the cost of infrastructure building and maintenance, with the British billing the Patiala for their portion of the infrastructure annually. After completion of the waterworks, the Patiala state would then also pay the British government an annual sum for the Jumna water based on amount of land irrigated. The price of water per year agreed on is shown in Table D.3.

The estimated acreage was based on the amount of land irrigated during the year from the British Distributaries and the relative proportions of the supplies actually passed in the British and Patiala Distributaries during the same period. The Patiala State agreed to furnish the British government with half-yearly statements for each harvest with information regarding area irrigated, income derived, and working expenses related to the Patiala distributaries.

Each state was entitled to all revenue that would be assessed on account of irrigation or sales of water for other purposes in their own villages, regardless of whether the water was supplied from British or Patiala distributaries.

AGREEMENT: FINAL WORKING AGREEMENT RELATIVE TO THE SIRHIND CANAL BETWEEN GREAT BRITAIN AND PATIALA, JIND, AND NABHA

Basin: Sirhind Canal

Date: February 24, 1904

Parties: Great Britain, State of Patiala, State of Jind, State of Nabha

Summary: Flow allocations from the Sirhind Canal were agreed to in the following proportions: Patiala, 83.6 percent; Nabba, 8.8 percent; Jind, 7.6 percent.

Price considerations: Costs for the establishment of infrastructure and management were to be borne in the same proportion as the flow allocations. In addition, charges for water

Table D.3 *Water prices in the 1893 agreement between the British government and the Patiala state regarding the Sirsa Branch of the Western Jumna canal*

Estimated area in acres	Rate of seigniorage per acre	Value in 2005 US$[a]
<42,000	Nil	Nil
42,000–43,999	One anna[b]	1 anna = 1/16 rupee = 1893 £0.0042 = 2005 US$0.45
44,000–45,999	Two annas	US$0.90
46,000–47,999	Three annas	US$1.35
48,000	Four annas	US$1.80

[a] Rupee to pound 1893 exchange rate obtained from Catão and Solomou, Effective exchange rates and the classical gold standard adjustment, *American Economic Review*, **95**(4), 1259–1275, September 2005. Conversion of 1893 pounds to 2005 U.S. dollars by Officer and Williamson, Computing "real value" over time with a conversion between UK pounds and U.S. dollars, 1830–2005, *MeasuringWorth.com*, August 2006.

[b] An anna is a former currency. One anna (1/16 rupee).

supplied to British villages from the Patiala Branches were "not to exceed the charges which are livable under the schedule of rates in force on the British and Signatory States." Further language in the text of the agreement states that measurements of land irrigated would be made to determine rates. Although these rates were not laid out explicitly in the agreement, they may be similar to rates found in earlier agreements between the British government and states within India.

AGREEMENT: CONVENTION REGARDING THE WATER SUPPLY OF ADEN BETWEEN GREAT BRITAIN AND THE SULTAN OF ABDALI

Basin: Aden (groundwater)

Date: April 11, 1910

Parties: Great Britain, Aden (Yemen)

Summary: The Sultan of Abdali, on behalf of himself and his heirs, agreed to grant the British Government sole use of a piece of land east of Wadi-As-Saghir in perpetuity for the purpose of developing a groundwater resource. The land allocated to developing the groundwater resource would be approximately 110 acres in area. The Sultan of Abdali agreed to guarantee no contamination of the wells, to provide facilities to construct and maintain the works, and to safeguard the work and those working there. In return, the British Government agreed to pay a monthly rate.

Price considerations: Great Britain was to pay 3,000 rupees (approximately US$17,400 in the 2005 equivalent[2]) per month. No quantity of water was specified for this fee. If the water was diminished due to damage by Subehis or Abdalis, a maximum amount of 15 rupees (2005 US$111) per 100,000 gallons would be paid. In addition, 1,000 rupees (2005 US$5,800) would be paid to land owners for land on which the wells were dug.

AGREEMENT: EXCHANGE OF NOTES BETWEEN THE UNITED KINGDOM AND ITALY RESPECTING THE REGULATION OF THE UTILIZATION OF THE WATERS OF THE RIVER GASH

Basin: Gash

Date: June 15, 1925

Parties: Great Britain, Italy

Summary: The governments of Great Britain and Italy approved of the agreement between the Governor of the

Colony of Eritrea and the Acting Governor General of the Sudan regarding water from the Gash flowing from Eritrea into the Kassala province of the Sudan. Eritrea was allocated 65 million m^3 of water (or a mean discharge of 15 m^3 per second for 50 days) to irrigate the Tessenei plain, an area of approximately 20,000–25,000 hectares. In order to safeguard the interests of both Eritrea and Kassala in times of water shortage, the water was to be divided as follows:

Since it would not be for the practical advantage of either territory to divide the very small supplies, we would leave the first 5 cubic metres per second at the complete disposal of Tessenei. The division of the supply from 5 up to 20 cubic metres per second should be made in such proportionately progressive manner that, when 20 cubic metres per second is reached, the partition will be 10 cubic metres per second to each.

It was agreed that water in excess of 65 million m^3 would flow to the Kassala province. Later on April 18, 1951, Eritrea and Sudan signed an agreement to reaffirm established water quantities, with Eritrea receiving a maximum of 65 million m^3. The new agreement was signed by Eritrea and Sudan as independent nations.

Price considerations: The Sudan Government agreed to pay the Government of Eritrea annually based upon profits from irrigated land. Sudan would pay 20 percent of the sum of profits made due to cultivation by irrigation in excess of £50,000 (approximately US$2.65 million in the 2005 equivalent[3]). The yearly amount due would be calculated using the yearly statements of the Kassala Cotton Company.

AGREEMENT: THE INDUS WATERS TREATY 1960 BETWEEN THE GOVERNMENT OF INDIA, THE GOVERNMENT OF PAKISTAN AND THE INTERNATIONAL BANK FOR RECONSTRUCTION AND DEVELOPMENT

Basin: Indus

Date: September 19, 1960

Parties: India, Pakistan, International Bank for Reconstruction and Development

Summary: India and Pakistan signed this agreement regarding the use of different parts of the Indus River system. The water system is divided primarily into the "Eastern Rivers," including the Sutlej, the Beas, and the Ravi; and the "Western Rivers," including the Indus, the Jhelum, and the Chenab. All the waters of the Eastern Rivers were to be available for the unrestricted use of India, and Pakistan was to receive

[2] Rupee-to-pound 1910 exchange rate obtained from Catão and Solomou, Effective exchange rates and the classical gold standard adjustment, *American Economic Review*, **95**(4), 1259–1275, September 2005. Conversion of 1910 pounds to 2005 U.S. dollars by Officer and Williamson, Computing "real value" over time with a conversion between UK Pounds and U.S. dollars, 1830–2005, *MeasuringWorth.com*, August 2006.

[3] Conversion of 1925 UK pounds to 2005 U.S. dollars by Officer and Williamson, Computing "real value" over time with a conversion between UK Pounds and U.S. dollars, 1830–2005, *MeasuringWorth.com*, August 2006.

unrestricted use of the Western Rivers. India and Pakistan agreed to let all the waters allocated to the other party to flow freely.

Price considerations: Because the water from Western Rivers and other sources were designed to replace water that would have previously been provided by the Eastern Rivers to Pakistan, India agreed to make a fixed payment of £62,060,000 (approximately US$1.33 billion in 2005 equivalent[4]) toward the cost of infrastructure necessary to deliver Western rivers water to Pakistan. This was to be paid in ten equal annual installments. A transition period of ten years in duration was outlined in which the replacement of Eastern River water with Western River water in Pakistan would be completed. If the transition period were extended, India would be repaid a portion of its payment. Volumes of water are not discussed, and costs agreed to in the treaty involve infrastructure building; the value of the water is implied to be zero.

AGREEMENT: EXCHANGE OF NOTES CONSTITUTING AN AGREEMENT BETWEEN THE UNITED STATES OF AMERICA AND MEXICO CONCERNING THE LOAN OF WATER OF THE COLORADO RIVER FOR IRRIGATION OF LANDS IN THE MEXICALI VALLEY

Basin: Colorado

Date: August 24, 1966

Parties: United States, Mexico

Summary: In order to relieve a critical water shortage in the Mexicali Valley, the United States agreed to release an additional 40,535 acre-feet of water from the Colorado River beyond the annual allocation to Mexico (annual allocation determined in 1994 water treaty). The International Boundary and Water Commission was supposed to agree to a schedule for water deliveries in the 1967 calendar year. In the case that runoff in the Colorado River Waters in the United States from April to July 1967 was expected to exceed 8.5 million acre-feet, the 40,535 acre-feet reserved for Mexico would be held for 3 years.

Price considerations: Mexico agreed to reimburse the United States at market value for any decrease in power generation at Hoover and/or Glen Canyon Power Plant that would be caused by the loss of power resulting from the release of the agreed 40,535 acre-feet. The value of the water is tied solely to its hydropower generation capabilities; the value of the raw water in this agreement is implied to be zero.

AGREEMENT: AGREEMENT BETWEEN THE GOVERNMENT OF THE REPUBLIC OF SOUTH AFRICA AND THE GOVERNMENT OF PORTUGAL IN REGARD TO THE FIRST PHASE OF DEVELOPMENT OF THE WATER RESOURCES OF THE CUNENE RIVER BASIN

Basin: Cunene

Date: January 1, 1969

Parties: South Africa, Portugal

Summary: This agreement was designed to optimize utilization of the water resources of the Cunene River basin, and aimed to achieve several benefits as follows:

Regulation of the flow of the Cunene
Improvement of hydroelectric power generation at Matala
Irrigation and water supply for human and animal needs in the middle Cunene
Water supply for human and animal needs in South West Africa and irrigation in Ovanboland
Hydroelectric power at Ruacana

The parties agreed to pursue four major works as follows:

A dam at Gove for Cunene flow regulation
A dam at Caluque for Cunene flow regulation
A pumping scheme at Caluque from the Cunene for human and animal needs in South West Africa and irrigation in Ovanboland
A hydroelectric power station at Ruacana for power in South West Africa

Price considerations: The costs for each work were allocated to the party who would benefit from them. The building of the Gove dam was the responsibility of the Portugese government. South Africa was to participate in the financing of the Gove dam in all parts related to flow regulation and not hydroelectric power. Its financial obligations to Portugal were limited to R 8,125,000 (approximately US$39 million in the 2005 equivalent[5]). The cost of construction of the works at Caluque was to be entirely the responsibility of South Africa, as they would be benefiting from the water. South Africa agreed to pay the Portugese government R 220,000 (2005 US$1.1 million) as compensation from the ground occupied by the Caluque dam. South Africa also agreed to be entirely responsible for the costs of construction and operation of the Ruacana power station, though they were granted use of Portugese territory occupied by the

---

[4] Conversion of 1960 pounds to 2005 U.S. dollars by Officer and Williamson, Computing "real value" over time with a conversion between UK pounds and U.S. dollars, 1830–2005, *MeasuringWorth.com*, August 2006.

[5] South African Rand to U.S. dollar 1969 exchange rate obtained from Williamson, S. H. (2006). Exchange rate between the United States dollar and forty other countries, 1913–2005. EH.Net (supported by Economic History Association). Conversion of 1969 Rand to 2005 U.S. dollars by Officer and Williamson, Computing 'real value' over time with a conversion between UK pounds and U.S. dollars, 1830–2005, *MeasuringWorth.com*, August 2006.

Ruacana works for free by the Portuguese government. South Africa was, however, to pay the Portuguese government a royalty based on the forecast of power generation at Ruacana. This was estimated to be the same cost as the payments for construction of the Gove dam.

No specific charge for raw water is stated in this agreement.

D.10.2 Modern cases

Information on modern case studies was sought from a much broader array of resources. The following studies are summaries of various agreements or provisions of water transfers or allocations. The price considerations vary significantly from one study to the next but represent an array of real-world situations that illuminate how the cost of water is handled in different circumstances.

AGREEMENT: AGREEMENT OF WATER TRANSPORTATION TO THE TURKISH REPUBLIC OF NORTHERN CYPRUS FROM TURKEY

Parties: Turkey, North Cyprus

Date: Agreement signed in 2003, but water transport had been taking place since 1998

Issue: Transport of water from Manavgat River in Turkey to North Cyprus

Summary: North Cyprus has experienced decreased precipitation in the past few decades. Overdrafting of groundwater resources has led to saltwater intrusion in aquifers. The use of saline water for irrigation has resulted in the killing of citrus crops and made some water unsafe for drinking. One of the main solutions considered to the water shortage is the importation of water from Turkey.

In 1999, a Harvard Institute for International Development paper outlined the financial feasibility of importing 40,000 m^3 of water by tankers per year from the Manavgat River in Turkey to North Cyprus (Biçak and Jenkins, 1999). Annual demand for water in North Cyprus is 106.6 million m^3, 82 percent of which is for agriculture. Safe yield from aquifers is about 74 million m^3, and rivers and dams can provide approximately 13 and 7 million m^3, respectively. Thus the water deficit of North Cyprus is approximately 12.5 million m^3 per year, and this deficit had been accommodated by over-pumping of the aquifers, leading to salt water intrusion (Biçak and Jenkins, 1999).

Price considerations: Biçak and Jenkins (1999) lay out in detail the total estimated costs for all parts of water import by water tanker. The cost of transportation of the water is US$0.4 per m^3. With infrastructure investment added to this figure, the cost is estimated as US$0.79 per m^3. Leakage of 30 per-

cent in the distribution system would increase total cost to US$1.13 per m^3. None of these costs account for any charge for the raw water, but the authors estimate that Turkey would charge US$0.15 per m^3.

In a paper presented at the Water for Life in the Middle East Conference in 2004, Mithat Rende, the head of the Department of Regional and Transboundary Waters in the Ministry of Foreign Affairs of Turkey, reviewed the history and outlook for water transfer from Turkey to North Cyprus (Rende, 2004). He stated that initially Turkey had signed an agreement in 1997 for transport via water bags, a technology out of Norway, with Nordic Water Supply, with a price of 55 cents per m^3. However, that technology failed (Morgan, 2002).

The current agreement for water transport is with an Israeli company to transport water via a "new technology" to Cyprus, the agreed cost of which would be 60 cents per m^3. In addition, the project of water transfer via pipeline to North Cyprus was approved by the Turkish government in 1998, and movement toward that goal is anticipated. Following this 2004 conference, the *Turkish Daily News* reported that a $9.5 billion deal had been signed to construct the 78-km pipeline (*Turkish Daily News*, 2005).

AGREEMENT: INTERGOVERNMENTAL AGREEMENT BETWEEN TURKEY AND ISRAEL

Parties: Turkey, Israel

Date: March 2004

Issue: Transport of water from Manavgat River in Turkey to Israel

Summary: For several years, Israel and Turkey have discussed the option of water transport from Turkey's Manavgat River to Israel. Increasingly dry conditions in the late 1990s and early 2000s prompted Israel to more seriously pursue an agreement with Turkey. In August of 2002, Israel agreed to buy 50 million m^3 of water annually for 20 years. However, a price was not determined at that time. After a few years of negotiations, an agreement was signed in March 2004. Israel and Turkey agreed to a "water for arms" deal, in which Turkey would supply water to Israel, and Israel would provide certain high-tech weapons to Turkey (*U.S. Water News Online*, 2002).

In April of 2006, movement forward on this water transport project halted. Both governments agreed it was not feasible, but hoped to return to it in the future (*U.S. Water News Online*, 2006). The reasons cited for such a decision were the rising price of oil and the privatization of the water treatment facilities on the Manavgat River, both of which have contributed to raising the price of the water transport project (*Turkish Daily News*, 2006).

Price considerations: Although Israel and Turkey appeared to keep the price component of the negotiation mostly private, a few figures have been published. Citing Blanche (2001), Feehan (2001) states that Turkey was asking for US$0.23 per m^3 for the water, making overall cost to Israel US$0.55–0.60 per m^3. However, Israel was hoping to get a price of US$0.15 per m^3 for the water, with the overall price with tanker transportation to be US$0.50–0.55 per m^3. The Washington Institute for Near East Studies states the estimated cost of water imported from Turkey to be around US$0.80 per m^3 (Washington Institute for Near East Studies, 2003). In all prices mentioned, it does appear that Turkey was planning to charge for the raw water.

A special note on Turkey: In the last decade, Turkey has actively pursued becoming a leader in exporting water. It is the only Middle Eastern country with a substantial natural supply of water. It has massive water projects within its borders, such as the GAP, or Southeast Anatolia Project. It also has the Manavgat Water Supply Project, which it has hoped would become a hub for water exports to other countries in the region.

Turkey has sought to promote its water export vision as an instrument for peace in the Middle East (*Turkish Daily News*, 1999). It has held a plan of ultimately being able to harness much of the total supply of the Manavgat water, which is nearly 5 billion m^3 per year (Morris, 2000), and is reported to have discussed water exports with many countries, including Cyprus, Israel, Libya, Malta, Greece, and Jordan. Its vision has not yet been realized, some of which has been attributed to poor governance of water policy (*Turkish Daily News*, 2001).

AGREEMENT: TREATY ON THE LESOTHO HIGHLANDS WATER PROJECT BETWEEN THE GOVERNMENT OF THE REPUBLIC OF SOUTH AFRICA AND THE GOVERNMENT OF LESOTHO

Parties: Lesotho, South Africa

Date: October 24, 1986

Issue: Creation of massive works in Lesotho to transfer water from the Senqu/Orange River to South Africa

Summary: Lesotho, a small country bordered on all sides by South Africa, has a relative abundance of water compared to its population. Water of the Senqu/Orange basin is of high water quality. South Africa suffers from water shortages. The objectives of the Lesotho Highlands Water Project (LWHP) are to provide a high-quality water source for South Africa and to create hydropower and revenue for Lesotho from the transfer of the water (Lesotho Highland Development Authority, 2006). The LHWP is a massive infrastructural undertaking that has many phases of development and has involved many contractors to complete the work.

When it is completed, it is estimated that 40 percent of the Senqu River flow will be transferred to South Africa, and 70 m^3/sec will be available (Baillat, 2005). Planning for the LHWP was completed in 1986. The start of Phase I happened in 1987, and contracts were first awarded in 1988. In the span of 1997–99, the Katse Dam was completed, water was delivered to South Africa, and hydropower was inaugurated (Trans-Caledon Tunnel Authority, 2006b). Although serious problems with bribery by contractors have come to light in recent years (*Africa News*, 2006; *Comtex News Network*, 2004; *Global News Wire*, 2003), the project is still seen as a successful needs-based water transfer.

Price considerations: In the Treaty on the Lesotho Highlands Water Project (Lesotho and South Africa, 1986), South Africa agrees to be financially responsible for implementation, operation, and maintenance of that part of the project relating to water delivery to South Africa [Lesotho and South Africa, 1986, Article 10(1)], while Lesotho agrees to be responsible for implementation, operation, and maintenance of the part of the project relating to the hydropower generation in the Kingdom of Lesotho [Lesotho and South Africa, 1986, Article 10(2)]. Although the water is not explicitly priced, royalty payments are made by the government of South Africa to the government of Lesotho. The determination of royalty payments from South Africa to Lesotho is based mostly on a comparison to an alternative project called the Orange-Vaal Transfer Scheme (OVTS), in which water from the Orange River in South Africa would have been transferred to the Vaal Dam. Cost analyses of the OVTS and the LHWP showed the LHWP to be a more cost-effective option. The payment of royalties is, at least in large part, recognition of this cost difference, with the estimated difference between the two projects being shared 44 percent to 56 percent between South Africa and Lesotho, respectively. The Trans-Caledon Tunnel Authority (TCTA), the South African agency managing the project, highlights on their Web site that "Africa does not pay for the water. Lesotho does receive a financial benefit but for different reasons" (TCTA, 2006a). Baillat, in a review of the LHWP, states that "For South African officials, water is not an international commodity" (Baillat, 2005, p. 14). However, *Africa News* reports that water is called "white gold" in Lesotho, and says it is the largest single source of foreign exchange (*Africa News*, 2004).

The royalty payments consist of both a fixed and variable component. The fixed component is based on the calculated difference between the estimated benefit of the OVTS project and the LHWP and is paid monthly to Lesotho through the year 2045. The variable component will be paid in perpetuity as long as South Africa receives water. The calculation of this

Table D.4 *Water deliveries and royalty payments, 1999–2005*

Year	Water Delivered	Royalty Payments
1999–2000	540	M147
2000–2001	574	M158
2001–2002	584	M183
2002–2003	585	M206
2003–2004	687	M208
2004–2005	314	M102

Note: Deliveries are in millions of cubic meters, and royalty payments are in millions of Maluti. Figures from LHDA (2004)

component relates to the difference in electricity, operation, and maintenance costs [Lesotho and South Africa, 1986, Article 12(10b–c)].

Water deliveries and royalty payments made from 1999–2005 are shown in Table D.4. As of 2004, royalties made up about 6 percent of government revenue. In addition to royalty payments, hydropower revenues from 1998 to 2004 had contributed M297 million and substituted for electricity imports from South Africa (Healing, 2005).

AGREEMENTS: TEBRAU AND SCUDAI RIVERS
WATER AGREEMENT AND JOHORE RIVER WATER
AGREEMENT

Parties: Malaysia, Singapore

Date: 1961 and 1962 for each agreement, respectively

Issue: Transfer of water from Johore State of Malaysia to Singapore

Summary: Singapore is an island state that, though it receives a significant amount of rainfall, is water stressed due to its low per-capita availability of water. To meet these supply shortfalls, Singapore imports water from neighboring Malaysia. Singapore receives approximately 40 percent of its raw water supply from the state of Johore in Malaysia through a pipeline (Onn, 2005). An original agreement between the Sultan of Johore and the Town of Singapore was signed in 1927 (Johore and Singapore City Council, 1927), when both Singapore and Malaysia were colonies of Britain. This initial agreement was succeeded by agreements in 1961 and 1962. Under the 1961 agreement, known as the "Tebrau and Scudai Water Agreement," Singapore could draw 86 million gallons of water per day (mgd) from the Tebrau and Scudai Rivers and the Pontian and Gunung Pulai Reservoirs. This agreement expires in 2011 (Johore and Singapore, 1961). Under the 1962 agreement, the "Johore River Water Agreement," Singapore can draw up to 250 mgd of water from the Johore River. This agreement expires in 2061 (Johore

and Singapore, 1962). Both agreements were upheld in the 1965 Separation Agreement, in which both Singapore and Malaysia became separate independent countries (Malaysia and Singapore, 1965). In 1990, an agreement to draw additional water in excess of 250 mgd from the Johore River was signed and expires in 2061 (Johore and Singapore, 1990). In addition to receiving raw water from Malaysia, Singapore also returns treated water to Malaysia. The terms of both the 1961 and 1962 agreements have provisions for review of water prices after 25 years (Johore and Singapore, 1961, section 17; Johore and Singapore, 1962, section 14), and water price has been hotly debated in the past decade.

Price considerations: Under the 1927 agreement, Singapore paid nothing for raw water from Johore, but it was responsible for the cost of the infrastructure to transport, store, and treat the water. In both the 1961 and 1962 agreements, Singapore agreed to pay Malaysia 3 sen (RM 0.03) for every 1,000 gallons (4,546 m^3) drawn from the Johore state [Johore and Singapore, 1961, section 16(i); Johore and Singapore, 1962, section 13(1)]. In return, Malaysia pays Singapore 50 sen (RM 0.50) for every 1,000 gallons of treated water [Johore and Singapore, 1961, section 16(ii); Johore and Singapore, 1962, section 13(2)]. Singapore is responsible for the cost of infrastructure. Provisions of the agreements allow for prices to be modified according to the purchasing power of money and labor, power, and material costs for supplying water (Johore and Singapore, 1961, section 17; Johore and Singapore, 1962, section 14). Prices were not revised upon the first such opportunities in 1986 and 1987 for the 1961 and 1962 agreements, respectively (Onn, 2005).

In recent years, much argument about the price of the water has ensued between Singapore and Malaysia governments. Segal, in a master's thesis reviewing the water situation between Malaysia and Singapore, states that political tensions between the two states have existed since Singapore's independence, and the issue of water has been used as a bargaining tool. Ethnic tensions exist between the states and lead to conflict. In addition, Malaysia's experience of water shortages and its uncertainty about its own future water needs make water negotiations difficult (Segal, 2004).

Malaysia wanted to raise the price of water to Singapore to 60 sen per 1,000 gallons (Ng, 2001, and Yian, 2001, as cited in Segal, 2004). Singapore states that in September, 2001, it proposed to revise the current price to 45 sen per 1,000 gallons and agreed to Malaysia's price of 60 sen per 1,000 gallons for additional water to be supplied after current contracts expire (Singapore Ministry of Foreign Affairs, 2006). Despite these negotiations, a new agreement has not yet been reached at the time of this writing.

Singapore has also been developing plans for alternative freshwater resources including desalination and recycled water ("NEWater") (Onn, 2005).

ARRANGEMENT: TRANSPORT OF WATER TO VARIOUS SMALL ISLAND STATES

Parties: Various mainland states and small island nations
Date: Early 1980s to present
Issue: Provision of emergency supplementary freshwater supplies to small island states
Summary: The Pacific Islands Applied Geoscience Commission (SOPAC) categorizes island freshwater resources into conventional and nonconventional groups. Conventional resources include rainwater, surface water, and groundwater. Nonconventional resources require a greater level of technology to supply. These include desalination, importation and the reuse of wastewater or the use of saline water for nonpotable uses (SOPAC, 2006). Several Pacific Island states have experienced water shortages using their own conventional, or naturally available, sources and have handled these shortages by the importation of water via tanker. Nauru, an island fully exploited for its phosphate deposits, received around 30 percent of its water as return cargo in ships returning from delivering phosphate exports (Jacobson and Hill, 1988; *This American Life*, 2003). Many other states are known to have received water in tankers in recent decades (UNESCO, 1991).

Price considerations: Island states have few cost-effective options for supplementing water supply beyond natural island resources. Unless located very near a mainland, imported bulk water can only practically be done by tanker. Although the price of all these types of water transfers is not known precisely, both the United Nations Educational, Scientific, and Cultural Organization (UNESCO) and the United Nations Environment Programme (UNEP) report prices of a few water imports by barge to island nations. Citing Meyer (1987), UNESCO states that transportation costs for water tankers are between US$1.50 to US$3.50 (1985 value) per m³, depending on distance traveled and the size of the tanker. In addition, loading costs are between US$0.20 and US$0.75 per m³ and oil removal costs are between US$0.05 and US$0.20 per m³. For small island states, large tankers are often not practical because of port needs. In some cases, barges towed behind small ships are best. In the mid-1980s, the cost for transporting water from Dominica over distances of 100 to 1,000 km ranged from US$1.40 to US$5.70 per m³ for barges and US$1.60 to US$3.30 per m³ for ships between 20,000 and 80,000 dead weight tonnage (dwt). UNEP specifically gives the cost of transporting water from Dominica to

Antigua as US$20 per 1,000 gallons (UNEP, 1998). Citing Brewster and Buros (1985), UNESCO also gives figures for transport from Puerto Rico to St. Thomas in the early 1980s. The cost of water transport via tanker and barges with capacities ranging from 3,800 to 11,500 dwt over the distance of 100 km was US$4.65 per m³.

UNEP (1998) reports costs of transporting water in the Bahamas between Andros Island and New Providence as US$3.41 per 1,000 gallons including fuel costs. When shore costs are included, the total cost is approximately US$5.41 per 1,000 gallons. UNEP also states that economies of scale, when water is transported using larger tankers continuously over the long term, reduce the cost.

Though the development of water bags, known as "Medusa bags," as described in the case study of water transfer from Turkey to North Cyprus, were a very hopeful development for water importation to islands; that technology has not yet succeeded, as bags were shown to burst when in use (Morgan, 2002). As such, transportation by water tanker, as was available at the time these figures were reported, is still the most viable option. The technology of this type of water transfer has not likely changed much, and the prices stated in the 1987 study may not be dissimilar to those that would be expected for a similar transport in the current day.

AGREEMENT: DONGJIANG WATER SUPPLY AGREEMENT

Parties: China, Hong Kong
Date: April 2006
Issue: Transfer of water from Guangdong Province in China to Hong Kong
Summary: Hong Kong has received a significant portion of its freshwater supplies from the Guangdong Province of China since the 1960s. Agreements over the provisions of this water transfer have gone through many iterations. In 1989, an agreement between Hong Kong and Guangdong secured 690 million to 1.1 billion m³ per year, increasing from 1995 to 2008. Dongjiang, or the East River, is the source of the water from the Guangdong Province that is supplied to Hong Kong. Transferred through a series of dams and open channels and a pipeline, it supplies over 70 percent of the total freshwater demand of Hong Kong (Hong Kong Water Supplies Department, 2006b).

A new water supply deal was signed in April of 2006. The new deal specifies an annual supply of 1.1 billion m³ per year to Hong Kong and allows for seasonal fluctuations, with Hong Kong alerting Guangdong authorities of demand. This should aid in minimizing overflow of reservoirs, which amounts to a large overall loss of water to Hong Kong (Ng, 2006).

Price: The price of water, which was specified as HK$3.085 per m³ in the 1989 agreement, remained at that price in the 2006 agreement. Because of rising prices on the mainland, it is considered a savings to Hong Kong to maintain the earlier rate. This is the price of the delivered raw water to Hong Kong and includes costs of infrastructural resources and investment for projects done to protect water supplies and improve water quality (Ng, 2006). Hong Kong water users pay for water based on a tiered fee system, which is designed to subsidize lower volume users (Hong Kong Water Supplies Department, 2006a).

LEGISLATION: WATER RESOURCES PROTECTION ACT OF 1999 IN NEWFOUNDLAND AND LABRADOR, SIMILAR LEGISLATION IN OTHER CANADIAN PROVINCES

Parties: Canadian provinces

Date: Legislation in 1999 for Newfoundland and Labrador, late 1990s and early 2000s for other provinces

Issue: Canadian exports of bulk water

Summary: Bulk water exports from Canada have been considered, but public outcry about this issue has led provinces to pass legislation banning the bulk export of water. One such example is *The Water Resources Protection Act of 1999* (Newfoundland, 1999, chapter W-4.1), in which the Government of Newfoundland and Labrador prohibited bulk water removal. Debate on this topic did not end with the legislation, and the government commissioned a review of the current legal, trade, environmental, and economic aspects of bulk water exports (Government of Newfoundland and Labrador, 2001).

In this particular situation, the definition of water as either an economic good or a noneconomic good is pinnacle to the debate. At the time of this writing, it is as yet unclear how the prohibition of bulk water exports such as this fit into international trade agreement such as NAFTA and GATT trade provisions (refer to Baillat, 2005, and Gleick, 2003, for a detailed discussion on this topic).

Price considerations: As part of the government report on bulk water exports, an economic feasibility study was commissioned (Feehan, 2001). The main conclusions of the study as stated in the government report are as follows:

Most bulk water export operations are capital intensive.

Tanker transport costs, at moderate or high rates, make bulk water export uneconomic.

At "low" tanker costs, a few bulk export operations might be commercially viable, if aimed at displacing desalinated water in the U.S. southeast. Profit margins, however, would likely be very thin.

Rationalized U.S. water policies (such as eliminating subsidies for agriculture) or further improvements in desalinization techniques would eliminate any chance of a U.S. market.

There might be some opportunities for the supply of bulk water to bottling plants located outside North America. Competition there would be stiff.

The potential employment and royalty benefits of a bulk water export project are relatively small.

Only a few sites, if any, would be commercially viable.

Alaskan bulk water ventures have been proposed over the past six or seven years, and there are still no exports. (Government of Newfoundland and Labrador, 2001)

Feehan's economic feasibility report includes figures for the direct use and marginal value of water as well as projected costs for transport by tanker. These figures are quite extensive and can be obtained from the original report. He summarizes his synthesis of all this cost data into what he calls a "back-of-the-envelope" overall cost estimate:

Assuming that relatively large tankers, 250,000 to 325,000 dwt, are used; that tanker day-rates tend to the middle and low ranges given in Table 6 [day-rates in the table range from US$0.08 per m³ to US$0.42 per m³]; and that the markets are about a 15-day return trip away; then the tanker costs would be US$1.25 to US$2.50 per m³. Adding an allowance of US$0.10 to US$0.50 for non-tanker costs, gives US$1.35 to US$3.00 a cubic metre. Longer distances, delays due to weather or ice or technical conditions, or tight tanker markets could add substantially to those figures. On the other hand, a return to a slack tanker market, strategic location of a facility or the possibility of some partial back-haul cargo could perhaps result in a somewhat lower cost.

For the remainder of this report, the estimates of US$1.35 to US$3.00 per m³ meter is a reasonable point of reference for the cost of harvesting and shipping water from the province to Florida, Texas and the Caribbean. (Feehan, 2001)

In addition, he later states that there would also be environmental and public costs for exporting water, which he does not attempt to quantify.

Related issue: The legality of government prohibitions to bulk water exports are being challenged by Sun Belt Water, Inc., a company based out of California who in 1991 signed a contract for bulk water delivery from Canada to the American Southwest. Shortly thereafter, the government of British Columbia killed this contract with similar actions as outlined above by the Government of Newfoundland and Labrador. In 1999, Sun Belt Water, Inc. filed a claim under Chapter of the North American Free Trade Agreement to challenge the actions of the government of British Columbia (Sun Belt Water, Inc., 2006). This dispute is yet unresolved at the time of this writing.

DIRECTIVE: DIRECTIVE 2000/60/EC OF THE EUROPEAN
PARLIAMENT AND OF THE COUNCIL (EU WATER
FRAMEWORK DIRECTIVE)

Parties: EU member countries

Date: October 23, 2000

Issue: Water resources management in EU member countries

Summary: In recognition of scarce water resources in European countries, the European Union put forth a Framework Directive in 2000 to outline provisions for water resource management in member countries. Article 9 of the Directive addresses recovery of costs for water. The full text of Article 9 is as follows:

Recovery of costs for water services

1. Member States shall take account of the principle of recovery of the costs of water services, including environmental and resource costs, having regard to the economic analysis conducted according to Annex III, and in accordance in particular with the polluter pays principle.

    Member States shall ensure by 2010

    – that water-pricing policies provide adequate incentives for users to use water resources efficiently, and thereby contribute to the environmental objectives of this Directive,

    – an adequate contribution of the different water uses, disaggregated into at least industry, households and agriculture, to the recovery of the costs of water services, based on the economic analysis conducted according to Annex III and taking account of the polluter pays principle.

    Member States may in so doing have regard to the social, environmental and economic effects of the recovery as well as the geographic and climatic conditions of the region or regions affected.

2. Member States shall report in the river basin management plans on the planned steps towards implementing paragraph 1 which will contribute to achieving the environmental objectives of this Directive and on the contribution made by the various water uses to the recovery of the costs of water services.

3. Nothing in this Article shall prevent the funding of particular preventive or remedial measures in order to achieve the objectives of this Directive.

4. Member States shall not be in breach of this Directive if they decide in accordance with established practices not to apply the provisions of paragraph 1, second sentence, and for that purpose the relevant provisions of paragraph 2, for a given water-use activity, where this does not compromise the purposes and the achievement of the objectives of this Directive. Member States shall report the reasons for not fully applying paragraph 1, second sentence, in the river basin management plans. (European Parliament and the Council of the European Union, 2000)

Price considerations: Although the EU Water Framework Directive does not specifically state full cost recovery as a requirement, it does state that cost recovery for "water services, including environmental and resource costs" be taken into account by member states (discussion of full cost recovery is included in the overview of the value of water). The European Environmental Bureau (EEB) wrote a report outlining how a pricing policy would need to be devised to meet the requirements of the Directive (Roth, 2001). In the report, costs that need to be recovered for full cost recovery to occur include the following (as listed earlier in section D8):

Operation and maintenance costs
Capital costs
Opportunity costs
Resource costs
Social costs
Environmental damage costs
Long run marginal costs (Roth, 2001)

Roth states that, "Without making the full costs of water use clear to the users by integrating them into the water price, any water pricing policy is thus in breach with the main principles supposed to underlie EU environmental policy."

The study showed that the level of the water price in EU countries is generally lower than the cost recovery level. However, the pricing in most countries does play a role in achieving environmental goals. Pricing water appropriately will be different in all different sectors including household, industry, and agricultural sectors, each having a different set of influencing factors. In addition, influencing factors will vary on different scales and in different locations, all of which will need to be considered in pricing. As outlined in the first section of the appendix, EEB identifies several factors necessary to consider in an EU water pricing policy, which include the following:

Public awareness and participation
Full cost recovery that includes the costs for environmental
    damage
Metering and volumetric pricing schemes
Increasing block schedules with blocks adjusted to social needs
Seasonal variation where appropriate
Earmarking of water charges
Only a minimum of fixed and minimum charges
Information for water users
An understandable water bill
Transparency
A gradual transition to the new pricing scheme (Roth, 2001)

The basis for full cost recovery is an economic analysis that was to be completed for each river basin in 2004 (Lanz and Scheuer, 2001). Though the EU Framework Directive tasks member countries with cost recovery and promotes appropriate pricing, water itself is not defined as a commodity. In a leaflet explaining the Water Framework Directive, the European Commission (EC) defines water as a "heritage" when explaining the "fair price" of water. The EC states that though water is not a commercial product, the pricing of it should be done in such a way to encourage sustainable use. This includes using the Polluter Pays Principle. The Directive also provides an affordable price for people in need (European Commission, 2002).

Water prices currently vary throughout the EU. A summary of average water prices in many countries in 2005 was calculated by the NUS Consulting Group (2006). Denmark and Germany had the highest reported price, at an average of US$225 per m^3. The Netherlands, France, Belgium, and the United Kingdom had prices from US$149 per m^3 to US$190 per m^3. Finland and Italy were in the US$103 per m^3 to 115 per m^3 range. Sweden and Spain were the European countries shown with the lowest price of water, from US$86 to 93 per m^3. Most of the EU countries represented in the table have high water prices relative to other large countries including Canada, the United States, and Australia. The United States and Canada had the lowest reported prices, at US$66 per m^3 and US$79 per m^3, respectively. Prices for all countries had gone up noticeably from the NUS report done in 2005 (NUS Consulting Group, 2005).

LEGISLATION: LAW 9,433, THE ESTABLISHMENT OF THE
NATIONAL WATER RESOURCE POLICY

Party: Brazil

Date: 1997

Issue: Domestic water pricing strategy

Summary: The World Bank has promoted the French model of privatization of water systems, where there is public ownership and there is mixed public and private management (Ouyahia, 2006). The Brazil Country Management Unit of the World Bank completed a study on bulk water pricing in Brazil (Asad et al., 1999). They recommended that Brazil use water pricing to promote sustainability and efficient use and allocation of resources. Though the report states that economic efficiency and full cost recovery are objectives, full cost recovery may not be feasible, and they propose full cost recovery for operations and maintenance and partial cost recovery for investments. To accomplish this, they advise establishing bulk water tariffs for each of the major water use sectors. Brazil has pursued bulk water pricing, adapting the French example to the legislative structures of Brazil (Lanna, 2003). Both state and union level laws institute water systems. Law No. 9,433 of 1997 established the National Water Resource Policy, which adopted the following principles:

Water is public property;

Water is a limited natural resource, which has economic value;

When there is a shortage, priority in the use of water resources is given to human consumption and the watering of animals;

The management of water resources should always allow for multiple uses of water;

The river basin is the territorial unit for the implementation of the National Water Resources Policy and the actions of National Water Resources Management System;

The management of water resources should be decentralized and should involve participation by the Government, the users, and the communities. (Brazil, 1997, Chapter I, Article 1)

Price considerations: The National Water Resource Policy also outlines fees for water use in Section IV of Chapter IV of the policy as follows:

Art. 19. The objectives of charging for the use of water resources are the following:

I – To recognize water as an economic good and to provide users an indication of its real value;

II – To provide incentives for the rational use of water;

III – To obtain financial resources for financing the programs and activities included in the Water Resource Plan.

Art. 20. Water resource uses subject to an award shall be charged for pursuant to the terms of article 12 of this law.

Sole Paragraph. (Vetoed)

Art. 21. In establishing the sums to be charged for the use of water resources, the following must be taken into consideration, among other items:

I – Derivation, capture and extraction of water, volumes removed and the variation system;

II – Emissions of drainage and other liquids or gaseous waste, volumes emitted and the variation system, and the physical-chemical, biological and toxicity characteristics of the effluent.

Art. 22. Sums collected by charging for the use of water resources shall be applied on a priority basis in the watershed in which they were generated and shall be used:

I – for financing studies, programs, projects and works included in the Water Resources Plan;

II – for paying startup expenses and for the administrative financing of the bodies and entities forming part of the National Water Resources Management System.

1. Application of the expenditures provided for in Section II of this Article shall be limited to seven and one-half percent of the total amount collected.

2. The sums provided for in the main body of this Article may be applied to sunk costs in projects and works that change, taking

into consideration the benefits to the community, and the quality, quantity of and the discharge system from a body of water.

Source: Accessed at http://faolex.fao.org/docs/texts/12967ENG.doc (Food and Agriculture Organization of the United Nations legal database)

Although implementation of water pricing policies will take many years, there were already pricing strategies in place for certain water sectors at the time the policy was written. Lanna (2003) identifies four different types of uses to price as follows:

The use of water available in the environment – bulk water, as a factor of production or final consumer good,

The use of water available in the environment as waste receptor,

The use of water diversion, regulation, transport, treatment and distribution (supply service to domestic, agricultural, industrial users, etc.) and

The use of collection services, transport, treatment and final disposal of sewage. (Lanna, 2003)

Prices for the latter two uses are fairly well established, according to Lanna, as set in agricultural and sanitary sectors, while prices for the first two uses are not established through the country, but local cases exist as examples.

Pricing in agricultural and water supply and sanitation sectors

In the agricultural sector, Law 89.496 of 1984 specifies that water tariffs for public irrigation projects be set at the sum of two coefficients, K1 and K2. K1 is supposed to reflect the capital costs of the project with a 50-year repayment period and is an annual set value for all of Brazil. K2 is designed to include the operation and maintenance costs and is estimated based on the volume of water used. Although this system theoretically would work, administrative shortfalls can lead to strange actual charges. In 1995, tariffs for irrigation ranged from US\$3 to US\$40 for 1,000 m^3 (Azevedo, 1997).

In the water supply and sanitation sector, users have paid a monthly fee for water, which rarely covered costs (Azevedo, 1997). New pricing is anticipated with the submission of a new bill regarding a national environmental sanitation policy that includes provisions for transparent rate calculations and user involvement (*BNamericas.com*, 2005).

Bulk water pricing and accounting for impacts to the environment

The State of Ceará was the first to implement water pricing policies regarding the first two types of water prices (bulk water and impacts on environment) that are in line with the 1997 National Water Resource Policy. As of December, 1999, charges for water were R\$0.012 per m^3 consumed by the concessionaires that have the delegation of public supply service for clean water and R\$0.67 per m^3 for water consumed for industrial uses and users. In August, 2000, in order to include the cost of electric energy consumption at pumping stations the value was established as R\$0.028 per m^3 to be charged for the use of bulk water by the public service concessionaires supplying clean water (Lanna, 2003).

LEGISLATION: COUNCIL OF AUSTRALIAN GOVERNMENTS (COAG) FRAMEWORK FOR WATER REFORM, 1994; INTERGOVERNMENTAL AGREEMENT ON A NATIONAL WATER INITIATIVE (NWI), 2004

Parties: Various parties within Australia

Date: Original agreement 1994, updated 2004

Issue: Water pricing and trading within Australia

Summary: Due to Australia's arid environment, the government found it necessary to implement a framework for water reform in 1994 to manage scarce water supplies. Along with provisions for education, environmental requirement, and institutional reform, the 1994 framework addressed water trading and pricing. Water pricing was to be based on consumption-based pricing, full cost recovery, and transparency of subsidies. In addition, there were to be formal determinations of water entitlements and allocations. Trading of these entitlements and allocations were allowed within physical and ecological constraints of watersheds. The framework promoted the development of water markets to achieve goals of sustainable use and efficiency (Environment Australia, 2002). A system of water trading markets emerged.

The 2004 National Water Initiative (NWI) builds on the 1994 framework and covers a full range of objectives, some of which are specific to water trading and pricing. The NWI works toward the removal of institutional barriers to water trade. Under the NWI, water trading will not be restricted to within watersheds. Water pricing will be used to achieve economic efficiency and sustainable use of water resources, infrastructure, and government resources. Pricing will also facilitate functioning of water markets and provide mechanisms for the release of unallocated water (Australian National Water Commission, 2006).

Price considerations: The Australian Bureau of Statistics analyzed results of the country's water trading program in 2004–2005 (Australian Bureau of Statistics, 2006). The Bureau reports that in that year, a total of 1,802 permanent water trades with a total of 248 Gigaliters (GL) of water and 13,456 temporary water trades with a total of 1,053 GL of water were carried out in the country. Victoria had by far the highest number of water trades, both permanent and temporary. It also had the highest volume of water traded temporarily, though Western Australia had the highest volume of water traded permanently.

Though the average prices for many permanent and temporary water trades were listed as "not available" in the report, some figures were cited. The average price per megaliter for permanent trades in Queensland was AU$1,750 and in Western Australia was AU$680. For temporary trades, the average price per megaliter in New South Wales was AU$96 and for Western Australia was AU$80 (Australian Bureau of Statistics, 2006).

ARRANGEMENT: BULK WATER SUPPLY EXPORTS TO BUYERS FROM PRIVATE ENTITIES

Parties: Corporate bulk water suppliers and buyers

Date: Ongoing offers

Issue: Selling of bulk water as a commodity by private entities

Summary: As the door has opened to treat water as a commodity, naturally private commercial interests have made developments in this market. Water Bank appears to be a hub of water selling and trading (Water Bank, 2007). Water Bank acts as a water rights broker and merger and acquisition specialist, and they state on their Web site that they are dedicated to the buying, selling, and trading of the following:

Water rights
Water investments
Water utilities
Spring water
Bottled water
Bottling companies
Property and water
Geothermal water
Bulk water
Irrigation district water
Water from state trust lands
Abstraction licenses

Both Feehan (2001) and Baillat (2005) mention the development of Water Bank as part of a review of recent developments. Feehan states that as of August 2001, there had been no bulk water sales through Water Bank, though there had been "lots of talk." Baillat, reporting in 2005, states that few international trade operations have occurred through Water Bank. On its site, Water Bank reports being called on by the states of Texas and Florida, as well as FEMA for bottled water supplies needed in the 2005 hurricane season. It states that it has more than 375 sources of water worldwide for which it can arrange deals (Water Bank, 2007).

Other companies have developed to provide bulk water procurement and transportation. Water Exports NZ Limited offers water from Mount Aspiring National Park on the west coast of the South Island of New Zealand (Water Exports NZ Limited, 2007). The company's Web site states that they will provide both bulk and bottled water, after necessary infrastructure, including an 11 km pipeline from the water source to Jackson Bay on the west coast, is completed. Persons or companies interested in becoming venture partners can fill out a form on their Web site. As of January 2007, Water Exports NZ Limited was seeking partners to establish bottling facilities, distribution networks, and vendor services.

Flow, Inc., based out of South Carolina, offers on its Web site long-term bulk water supply from the Charleston area (Flow, Inc., 2007; EC Europe, 2007). The company offers up to 77 million gallons per day of high-quality untreated water and 42 million gallons per day potable treated water from the excess municipal supply of Charleston. Flow, Inc., offers up to 20-year contracts. In a 1994 opinion piece in *National Geographic* magazine, the president of the company, Eugene P. Corrigan, Jr., stated the significance of the source water's location at Bushy Park Reservoir near Charleston. Bushy Park is located at the 33° North latitude line, directly across from Gibraltar and the Suez Canal return route to Arabian Gulf oil ports. Surplus water can be delivered as backhaul loads in returning crude carrier tankers (Corrigan, 1994).

Although other similar commercial or corporate entities exist, these serve as examples of private industry involvement in water commodities.

Price considerations: In the private arena, price figures for water are not readily accessible. One could surmise that each case is subject to considerable negotiation, and it is expected that commercial entities would choose to keep their price negotiations undisclosed. However, some general information was obtained about price considerations from various commercial entities.

Water Bank, serving as a broker for water deals, does list its service prices on its Web site, and they include sellers costs (e.g., water audit = $600 per tract, with additional contiguous tracts up to $100 each; title search and report = $200; closing costs = $800 each side; attorney review = $150; declaration = $200; brokerage commission = 10 percent + gross receipts tax), buyer's costs (e.g., application filing and regulatory agency research = $120 per hour; closing costs = $800; attorney review = $150), reimbursable costs (copies, photography, maps, mileage, and the like, mostly at cost), and indirect costs (15 percent of the reimbursable costs) (Water Bank, 2007).

Within its newsletter archives, Water Bank has a newsletter that covers issues associated with bulk water exports

(Davidge, 1994). With regard to pricing of such exports, Davidge, lists the following factors as critical:

Length of contract (most important factor due to depreciation and amortization)
Volume of water delivered
Distance of delivery
Security of source
Cost of transportation device
Cost of facilities at source and delivery points
Permitting and compliance costs
Operating cost of transport system. (Davidge, 1994)

Water Exports NZ Limited does not state prices for their water procurement and transportation services, though it lists its approximate annual sales with the Export Bureau as US$10,000,000 (Export Bureau, 2007). Eugene P. Corrigan, Jr., of Flow, Inc., states that the dominant cost is for carriage (ship charter) of water, which the company overcomes by backhaul shipments in the return leg of tanker voyages. In addition, there is the cost of payroll for workers and investors (Corrigan, 2007, personal communication). In the 1994 *National Geographic* opinion piece, Corrigan stated that 1 tankerful of oil (1/month) could be exchanged for 30 tankerfuls of water (1/day). Regarding the price of water, which Corrigan calls, "Possibly the world's most guarded proprietary figure in the Arabian Gulf," he states that it is quoted at $1.25 to $12 per m^3 (or $4.73 to $45.42 per thousand gallons) (Corrigan, 1994).

Future more detailed studies of water pricing will necessitate thorough investigation of pricing structures in the private sector.

E Treaties with groundwater provisions

Kyoko Matsumoto

E.1 LEVEL 3 TREATIES*

Treaty name	Category	Date	No. of Parties	Countries	GW Reference
Treaty of peace between the state of Israel and the Hashemite Kingdom of Jordan, done at Arava/Araba crossing point	Quantity (quality)	10/26/1994	Bilateral	Israel, Jordan	Article IV
Johnston Negotiations	Quality	12/31/1955	Multilateral	Israel, Jordan, Syria, Lebanon	3. Division of Water
Convention regarding the Water Supply of Aden between Great Britain and the Sultan of Abdali	Quantity	4/11/1910	Bilateral	Great Britain, Aden (Yemen)	Entire agreement
Treaty concerning the state frontier and neighborly relations between Iran and Iraq and protocol	Quantity	6/13/1975	Bilateral	Iran, Iraq	Article 4
Convention on cooperation for the protection and sustainable use of the River Danube	Quality	6/29/1994	Multilateral	Austria, Bulgaria, Croatia, Germany, Hungary, Republic of Moldova, Romania, Slovakia, Ukraine, European Economic Community	Article 2 (1)
Mexico–United States agreement on the permanent and definitive solution to the salinity of the Colorado River basin (International Boundary and Water Commission Minutes No. 242)	Quantity	8/30/1973	Bilateral	Mexico, United States	Article 5
The Israeli–Palestinian Interim Agreement on the West Bank and the Gaza Strip: Protocol Concerning Civil Affairs	Quantity	9/28/1995	Bilateral	Israel, Palestine Autonomy	Annex III Article 40. Schedule 8,10

* See page 273 for definitions of levels.

Treaty name	Category	Date	No. of Parties	Countries	GW Reference
Convention on environmental impact assessment in a transboundary context, Espoo	Quantity	9/10/1997	Multilateral	Albania, Austria, Byelarus, Belgium, Bulgaria, Canada, Croatia, Czechoslovakia, Denmark, Finland, France, Germany, Greece, Hungary, Iceland, Ireland, Italy, Luxembourg, Moldova (Republic of), the Netherlands, Norway, Poland, Portugal, Romania, Russian Federation, Slovakia, Spain, Sweden, Switzerland, Ukraine, United Kingdom, United States	Appendix I
Convention on the protection, utilization and recharging of the Geneva Aquifer between Canton of Geneva in Switzerland and the department of Haute-Savoie in France	Quantity (quality)	9/6/1977	Bilateral	Swiss, France	Chapter 1-Article 1, Chapter 4, Article 9

E.2 LEVEL 2 TREATIES

Treaty name	Category	Date	No. of Parties	Countries	GW Reference
Agreement between Persia and Turkey concerning the fixing of the frontier line	Territory/ boundary	1/23/1932	Bilateral	Persia, Turkey	Exchange of Notes
Joint declaration of principles for utilization of the waters of the lower Mekong basin, signed by the representatives of the Governments of Cambodia, Laos, Thailand, and Vietnam to the committee for coordination of investigations of the lower Mekong basin	Quantity	1/31/2975	Multilateral	Cambodia, Laos, Thailand, Vietnam	Article XXIII

(*continued*)

(continued)

Treaty name	Category	Date	No. of Parties	Countries	GW Reference
Statute of the Committee for Co-Ordination of Investigations of the Lower Mekong Basin Established by the Governments of Cambodia, Laos, Thailand, and the Republic of Viet-Nam in Response to the Decision Taken by the United Nations Economic Commission for Asia and the Far East	Quantity	10/31/1957	Multilateral	Kampuchea, Laos, Thailand, Vietnam	Article XXIII
Draft agreement on water quality management of Zapadnaya Dvina/Daugava River basin	Physical relationships	11/12/1997	Multilateral	Byelarus, Latvia, Russian Federation	Introduction
Treaty between the United States of America and Mexico Relating to the Waters of the Colorado and Tijuana Rivers, and of the Rio Grande	Others	11/14/1944	Bilateral	United States, Mexico	Article 4
Protocol Amending the 1978 Agreement between the United States of American and Canada on Great Lakes Water Quality, as Amended on October 16, 1983.	Physical relationships	11/18/1987	Bilateral	Canada, United States	Annex 16
Convention creating the Niger Basin Authority and Protocol	Others	11/21/1980	Multilateral	Benin, Cameroon, Chad, Côte D'Ivoire, Guinea, Mali, Niger, Nigeria, Upper Volta	Article 4 (d)
Agreement between The Federal Republic of Germany and the EEC, on the one hand, and, the Republic of Austria, on the other, on cooperation and management of water resources in the Danube basin	Quality	12/1/1987	Multilateral	Germany (GFR), Austria, EEC	Article 2
Provisions relating to the Belgian–German frontier established by a six-nation delimitation commission in execution of the Versailles Treaty	Physical relationships	11/6/1922	Bilateral	Belgium, German	Subsections 1 and 3
Arrangement between Germany and Belgium concerning the common frontier	Territory/ boundary	11/7/1929	Bilateral	Belgium, German	Article 65
Agreement between Finland and Sweden Concerning Frontier Waters	Physical relationships	12/15/1971	Bilateral	Finland, Sweden	Article 2
Convention on the Protection of the Rhine against chemical pollution	Physical relationships	12/3/1976	Multilateral	Germany (GFR), France, Luxembourg, Netherlands, Switzerland, European Economic Community	Article 7–2

Treaty name	Category	Date	No. of Parties	Countries	GW Reference
France–Federal Republic of Germany–Luxembourg–Netherlands– Switzerland: Convention on the protection of the Rhine against pollution by chlorides	Physical relationships	12/3/1976	Multilateral	Germany, Luxembourg, Netherlands, Switzerland	Article 7
Treaty of Peace with Italy, Signed at Paris, on February 10, 1947	Territory/ boundary	2/10/1947	Multilateral	Italy, France (primarily), and the Allied Powers	Annex 5
Agreement between the governments of Great Britain and France with regard to the Somali Coast	Territory/ boundary	2/2/1888	Bilateral	UK, France	Article 1
Exchanges of notes between the United Kingdom and France constituting an agreement relating to the Boundary between the Gold Coast and the French Sudan	Territory/ boundary	3/18/1904	Bilateral	UK, France	Aticle III
Agreement on joint activities in addressing the Aral Sea and the zone around the Sea crisis, improving the environment, and enduring the social and economic development of the Aral Sea region	Quality	3/26/1993	Multilateral	Kazakhstan, Kyrgyzstan, Tajikistan, Turkmenistan, Uzbekistan	Article 1
Convention between Switzerland and Italy concerning the protection of Italo-Swiss Waters against pollution	Quality	4/20/1972	Bilateral	Switzerland, Italy	Article 1
State Treaty between the Grand Duchy of Luxembourg and the land Rhineland-Palatinate in the Federal Republic of Germany concerning the construction of a hydroelectric powerplant on Sauer at Rosport/Ralingen	Physical relationships	4/25/1950	Bilateral	Luxembourg, Germany (GFR)	Article 6,10
Agreement concerning water-economy questions between the government of the Federal People's Republic of Yugoslavia and the Government of the People's Republic of Bulgaria	Others	4/4/1958	Bilateral	Yugoslavia, Bulgaria	Article1
Convention and Statutes relating to the development of the Chad basin	Others	5/22/1964	Multilateral	Cameroon, Chad, Niger, Nigeria	Article 4
Agreement between the Republic of Syria and the Hashemite Kingdom of Jordan concerning the utilization of the Yarmuk waters.	Water right	6/4/1953	Bilateral	Jordan, Syria	Article 8

(continued)

(*continued*)

Treaty name	Category	Date	No. of Parties	Countries	GW Reference
State treaty between the Grand Duchy of Luxembourg and the Land Rhineland-Palatinate in the Federal Republic of Germany concerning the construction of hydroelectric power installations on the Our (with annexes)	Physical relationships	7/10/1958	Bilateral	Luxembourg, Germany (FRG)	Annex II
Agreement between the Government of the Polish People's Republic and the Government of the Union of Soviet Socialist Republics concerning the use of water resources in frontier waters	Territory/ boundary	7/17/1964	Bilateral	USSR, Poland	Article 2(3)
Exchange of notes between France and Great Britain relative to the boundary between the Gold Coast and French Sudan	Territory/ boundary	7/19/1906	Bilateral	Great Britain, France	Article 41 (3)
Process-Verbal from the meeting of Yugoslav and Greek delegations at Stari Dojran, to determine the manner and plan of collaboration concerning hydroeconomic studies of the drainage basin of Lake Dojran	Physical relationships	9/1/1957	Bilateral	Yugoslav, Greek	Section A ii(d), Section B (d)
Agreement between the Government of the Fededal People's Republic of Yugoslavia and the Government of the Hungarian People's Republic Together with the Statute of the Yugoslav-Hungarian Water Economy Commission	Others	8/8/1955	Bilateral	Hungary, Yugoslavia	Article 1
African Convention on the conservation of nature and natural resources	Others	9/15/1968	Multilateral	Algeria, Cameroon, Central African Republic, Congo, Cote D'Ivoire, Djibouti, Egypt, Ghana, Kenya Liberia, Madagascar, Malawi, Mali, Morocco, Mozambique, Niger, Nigeria, Rwanda, Senegal, Seychelles, Sudan, Swaziland, Togo, Tunisia, Uganda, Tanzania, Zaire	Article V water
Franco–Italian convention concerning the supply of water to the Commune of Menton	Physical relationships	9/28/1967	Bilateral	France, Italy	Article I
Convention between the French Republic and the Federal Republic of	Physical relationships	7/4/1969	Bilateral	France, Germany	Article 2

Treaty name	Category	Date	No. of Parties	Countries	GW Reference
of Germany concerning development of the Rhine between Strasbourg/Kehl and Lauterbourg/Neuburgweier					
Exchange of Notes between the United Kingdom and Italy respecting the regulation of the utilisation of the waters of the river Gash	Physical relationships	6/15/1925	Bilateral	United Kingdom, Italy	Question 4.5
UN/ECE protocol on water and health to the 1992 convention on the protection and use of transboundary watercourses and international lakes	Others	6/17/1999	Multilateral	Albania, Armenia, Belgium, Bulgaria, Croatia, Cyprus, Czech Republic, Denmark, Estonia, Finland, France, Georgia, Germany, Greece, Hungary, Iceland, Italy, Latvia, Lithuania, Luxembourg, Malta, Monaco, the Netherlands, Norway, Poland, Portugal, Republic of Moldova, Romania, Russian Federation, Slovakia, Slovenia, Spain, Sweden, Switzerland, Ukraine, United Kingdom	Articles 2, 3, 5, 6–5(b)
Statute of the River Uruguay	Physical relationships	2/26/1975	Bilateral	Argentina, Uruguay	Chapter IX

E.3 LEVEL 1 TREATIES

Treaty name	Category	Date	No. of Parties	Countries	GW Reference
Treaty between Germany and Poland for the Settlement of Frontier Questions	Territory/ boundary	1/27/1926	Bilateral	Germany, Poland	9.23.6.1922
Agreement between the USSR and Afghanistan	Territory/ boundary	1/18/1958	Bilateral	USSR, Afghanistan	Article 1
Agreement between France and Great Britain relative to the Frontier between French and British possessions from the Gulf of Guinea to the Niger	Water right	10/19/1906	Bilateral	Great Britain, France	Annex III
Convention between the French Republic and the Federal Republic of Germany on the development of the upper course of the Rhine between Basel and Strasbourg	Physical relationships	10/27/1956	Bilateral	France, Germany	Article 4
Convention between the government of the French Republic and the Swiss Federal Council Concerning protection	Physical relationships	11/16/1962	Bilateral	France, Switzerland	Article 1

(continued)

(*continued*)

Treaty name	Category	Date	No. of Parties	Countries	GW Reference
of the waters of Lake Geneva against pollution					
Agreement between the USSR and Czechoslovakia	Territory/ boundary	11/30/1956	Bilateral	USSR, Czechoslovakia	Article 1, Paragraph 2
Austria–Czechoslovakia treaty regarding the settlement of frontier legal questions	Territory/ boundary	12/13/1928	Bilateral	Austria, Czechoslovakia	Article 4
Agreement between the Government of the Federal People's Republic of Yugoslavia and the Government of the People's Republic of Albania concerning water economy questions, together with the statue of the Yugoslav-Albanian Water economic commission and with the protocol concerning fishing in frontier lakes and rivers	Others	12/5/1956	Bilateral	Albania, Yugoslavia	Article1
Agreement between Egypt and Italy concerning the establishment of frontiers between Cyrenaica and Egypt	Territory/ boundary	12/6/1925	Bilateral	Egypt, Italy	Article 5, 6
Exchange of notes constituting an agreement between the United Kingdom of Great Britain and Northern Ireland and Egypt regarding the utilization of profits from the 1940 British government cotton buying commission and the 1941 joint Anglo-Egyptian cotton buying commission to finance schemes for village water supplies	Physical relationships	12/7/1946	Bilateral	Great Britain, Egypt	Enclosure
Convention on the protection and use of transboundary watercourses and international lakes, Helsinki	Quality	3/18/1992	Multilateral	Albania, Austria, Belgium, Bulgaria, Croatia, Denmark, Estonia, Finland, France, Germany, Greece, Hungary, Italy, Latvia, Lithuania, Luxembourg, Moldova, the Netherlands, Norway, Poland, Portugal, Romania, Russia, Spain, Sweden, Switzerland, United Kingdom	Article 1, Annex III (d)
Agreement between the Government of the Czechoslovak Republic and the Government of the Polish People's Republic concerning the use of water resources in frontier waters	Territory/ boundary	3/21/1958	Bilateral	Czechoslovakia, Poland	Article 2

Treaty name	Category	Date	No. of Parties	Countries	GW Reference
Treaty between France and Switzerland, regulating fishing in Lake Geneva	Others	3/9/1904	Bilateral	France, Switzerland	Article 6
Treaty between the government of the Union of Soviet Socialist Republics and the imperial government of Iran concerning the Regime of the Soviet–Iranian Frontier and the procedure for the settlement of frontier disputes and incidents	Territory/ boundary	5/14/1957	Bilateral	USSR, Iran	Article 1
Agreement between the Federal Republic of Nigeria and the Republic of Niger Concerning the Equitable Sharing in the Development, Conservation and Use of Their Common Water Resources	Physical relationships	7/17/1986	Bilateral	Niger, Nigeria	Article 9
Agreement between Poland and the German Democratic Republic	Territory/ boundary	7/6/1950	Bilateral	Poland, German Democratic Republic	Article 2
Protocol on Shared Watercourse systems in the Southern African Development community (SADC) region	Others	8/28/1995	Multilateral	Angola, Botswana, Lesotho, Malawi, Mozambique, Namibia, South Africa, Swaziland, Tanzania, Zambia, Zimbabwe	Article 1
Revised Protocol on Shared Water Courses in the Southern African Development Community	Others	8/7/2000	Multilateral	Angola, Botswana, Republic of the Congo, Lesotho, Malawi, Mauritius, Mozambique, Namibia, Seychelles, South Africa, Swaziland, Tanzania, Zambia, Zimbabwe	Article 1
Treaty of limits between Portugal and Spain	Territory/ boundary	9/29/1864	Bilateral	Spain, Portugal	Article XXVIII.
Agreement of cooperation between the United States of America and the United Mexican States regarding pollution of the environment along the inland international boundary by discharges of hazardous substances	Others	7/18/1985	Bilateral	Mexico, United States	Article 1

* *Notes:*

Level 3 Treaties deal specifically with groundwater regulations, including allocation, quality provisions, and/or protection of land.

Level 2 Treaties briefly mention groundwater provisions of management; water rights of groundwater are assigned to a state although specificity of allocation is absent.

Level 1 Treaties indirectly mention groundwater; no specific provisions for management.

Source: Matsumoto, 2002.

F   Treaties with water quality provisions

Meredith Giordano

F.1 INTERNATIONAL WATER AGREEMENTS

	Category*	Agreement title	Date	Parties	Basin(s)	Water quality reference	Reference excerpt/summary
1	One	Convention on cooperation for the sustainable use of the Danube River	06/29/94	Albania, Austria, Bulgaria, Croatia, Czech Republic, Germany, Hungary, Italy, Moldova, Poland, Romania, Slovakia, Slovenia, Switzerland, Ukraine, Yugoslavia	Danube	Entire Document	SUMMARY: One of the primary objectives of the Draft Agreement is to coordinate on fundamental water management issues in order to maintain and improve the water quality of the Danube River. The Agreement details numerous multilateral coordination efforts to ensure "efficient water quality protection and sustainable water use and thereby prevent, control and reduce transboundary impact" (Article 5). Areas of cooperation include specific water resources protection measures, emission limitations and inventories, action programs, monitoring, reporting, information exchange, warning and emergency plans. The International Commission for the Protection of the Danube River is established to implement the objectives and provisions of the Agreement. General guidance is provided on water quality objectives and criteria and hazardous substances or listed, but the signatory nations only agree that they will define specific water quality standards.

* See page 307 for definitions of categories.

Category	Agreement title	Date	Parties	Basin(s)	Water quality reference	Reference excerpt/summary
2 One	1978 Agreement between the [United States] and [Canada] on Great Lakes water quality (as amended through October 16, 1983)	11/22/78	United States, Canada	St. Lawrence	Entire Document	SUMMARY: The purpose of the Agreement is "to restore and maintain the chemical, physical, and biological integrity of the waters of the Great Lakes Basin Ecosystem" (Article II). To carry out this objective, the Parties agree to several measures. Minimum concentration levels are established for specific chemical, physical, microbiological, and radiological substances (Article IV and Annex 1). Collaborative efforts concerning research, standards, programs and other measures are outlined (Article V and VI). Required programs include abatement, control and prevention of municipal and industrial discharge, eutrophication (Annex 3), and pollution from other land and offshore activities (Annex 4–8); joint contingency planning (Annex 9); coordinated surveillance and monitoring (Annex 11); maintenance of hazardous polluting substances lists (Annex 10/Appendix 1 and Appendix 2); and measures to control the input of persistent toxic substances (Annex 12). The existing International Commission, together with two newly established boards, is tasked with implementing the terms of the Agreement.
2a Amendment to above agreement	Protocol Amending the 1978 Agreement between the [United States] and [Canada] on Great Lakes water quality as Amended on October 16, 1983	11/18/87	United States, Canada	St. Lawrence	Entire Document	SUMMARY: Several revisions are made to the original Agreement. Additional water quality programs include the establishment of remedial actions and lakewide management plans (revised Annex 2), abatement and control of pollution from all contaminated sediments, and assessment and control of pollution from contaminated groundwater and

(*continued*)

(*continued*)

Category	Agreement title	Date	Parties	Basin(s)	Water quality reference	Reference excerpt/summary
						subsurface sources. A supplement to Annex 1 is added to the agreement that concerns Interim Objectives for Persistent Toxic Substances. Several revisions are made to existing Annexes and new Annexes are added to the Agreement including Annex 13 Pollution from Non-Point Sources, Annex 14 Contaminated Sediment, Annex 15 Airborne Toxic Substances, Annex 16 Pollution from Contaminated Groundwater, and Annex 17 Research and Development.
3 One	[Mexico]– [United States] Agreement on the permanent and definitive solution to the salinity of the Colorado River Basin (International Boundary and Water Commission Minutes No. 242)	08/30/73	United States, Mexico	Colorado, Rio Grande, Tijuana, Rio Bravo	Entire Document	SUMMARY: Specific salinity levels are outlined for water delivered from the United States to Mexico as well as measures for mitigating future salinity problems.
4 One	Colorado River salinity agreement effected by Minutes No. 241 of the International Boundary and Water Commission, [United States] and [Mexico]	07/14/72	United States, Mexico	Colorado River	Entire Document	SUMMARY: The resolution outlines measures to be taken by the United States to reduce the salinity of Colorado River Waters entering Mexico, including annual discharge of drainage waters from designated locations in the United States and minimum rates of flow for water entering Mexico.

	Category	Agreement title	Date	Parties	Basin(s)	Water quality reference	Reference excerpt/summary
4a	Amendment to above agreement	Agreement extending Minutes No. 241 of the International Boundary and Water Commission, [United States] and [Mexico]	04/30/73	United States, Mexico	Colorado River	Entire Document	SUMMARY: Extends the terms of Minutes No. 241 and revises certain discharge rates.
5	Two	Revised protocol on shared watercourses in the Southern African Development Community	08/07/00	Angola, Botswana, Lesotho, Malawi, Mauritius, Mozambique, Namibia, Seychelles, South Africa, Swaziland, Tanzania, Zambia, Zimbabwe	Buzi, Chiloango, Congo, Etosha-Cuvelai, Incomati, Kunene, Limpopo, Maputo, Okavango, Orange, Pungwe, Ruvuma, Save, Umbeluzi, Zambezi	Articles 1, 3, 4	SUMMARY: The overall objective of the agreement is to promote closer cooperation in terms of the management, protection, and utilization of shared watercourses. Water quality is noted in two definitions in Article 1("management of shared watercourses" and "pollution of shared watercourses"). In the General Provisions, the parties agree to exchange information concerning water quality. The Specific Provisions includes several clauses addressing water quality. These clauses address joint/individual steps to reduce and control pollution; the harmonization of related policies and legislation; and mutual steps for the prevention, reduction, and control of pollution (e.g., joint establishment of water quality objectives; techniques to address point and nonpoint pollution; and lists of substances to be banned, controlled, or investigated). The parties also agree to prevent the introduction of detrimental alien species, to protect and preserve the aquatic environment, and to prevent /mitigate harmful conditions. Like the original protocol, waste discharge permitting is required.

(*continued*)

(continued)

	Category	Agreement title	Date	Parties	Basin(s)	Water quality reference	Reference excerpt/summary
6	Two	The [Israeli]-[Palestinian] interim agreement on the West Bank and Gaza Strip	09/28/95	Israel, Palestinian Authority	Jordan, West Bank Aquifers	Annex III (Protocol concerning Civil Affairs) Article 40 and Schedules 8–11	SUMMARY: Article 40 states that the two parties agree to coordinate their management efforts of water and sewage in order to prevent the deterioration of water quality, to utilize the water resources in a sustainable manner and to take measures to prevent harm to the water and sewage systems in their respective areas (3b,c,h, and 21–23). A Joint Water Commission is established to implement Article 40 and is tasked, among other things, with the protection of water resources and water and sewage systems (12c) and with the development of a Protocol concerning the quantity and quality of water supplied from one Party to another (19). Both sides also agree to reimburse the other for any unauthorized use of or sabotage to water and sewage systems that affects the other party (24). Schedule 9 establishes joint teams whose duties include the rectification of problems related to water quality in the West Bank (4d,i). Schedule 11, concerning the Gaza Strip, discusses the establishment of a subcommittee to handle, among other issues, the "mutual prevention of harm to water resources" (8).
7	Two	Protocol on shared watercourse systems	08/28/95	Angola, Botswana, Lesotho, Malawi, Mozambique, Namibia, South Africa, Swaziland, Tanzania, Zambia, Zimbabwe	Buzi, Chiloango, Congo, Etosha-Cuvelai, Incomati, Kunene, Limpopo, Maputo, Okavango, Orange, Pungwe, Ruvuma, Save, Umbeluzi, Zambezi	Articles 2, 3	SUMMARY: The parties agree to exchange information and data concerning water quality; to require individuals to obtain State permits in order to discharge waste into shared watercourses, which will be granted only after the State has determine the discharge will not adversely affect the watercourse regime; to take necessary steps to prevent the introduction of aquatic species that may have detrimental effects on the ecosystem; and to maintain and protect shared watercourse systems and related facilities in

Category	Agreement title	Date	Parties	Basin(s)	Water quality reference	Reference excerpt/summary
						order to prevent pollution or environmental degradation. The parties also agree to establish appropriate institutions that will promote measures for the protection of the environment and prevention of environmental degradation, assist in developing a list of substances that should be banned or controlled, promote environmental impact assessments of water projects, and monitor the navigational impacts on water quality and the environment.
8 Two	Agreement on the cooperation for the sustainable development of the Mekong River Basin	04/05/95	Cambodia, Laos, Thailand, Vietnam	Mekong	Chapter 3: Articles 1, 3, 7, 8, and Chapter 4: Articles 18, 24	SUMMARY: The four signatories agree to cooperate in terms of water management to minimize the harmful effects from natural occurrences and human-made activities. When one State is notified that its activities are damaging the waters, the activities shall stop pending further investigation. Where harmful activities cause substantial damage to one or more riparians from the use of and/or discharge, the parties involved shall investigate all relevant factors, the cause, damage, and responsibility in compliance with the principles of international law. All related disputes should be resolved in accordance with the Agreement and in conformity with the United Nations Charter. The Council of the Mekong River Commission is responsible for making policies and decisions related to the protection of the environment and aquatic conditions of the Mekong River basin. The Joint Committee of the Mekong River Commission is responsible for conducting appropriate studies and assessments for the protection of the Mekong River basin environment.

(*continued*)

(*continued*)

	Category	Agreement title	Date	Parties	Basin(s)	Water quality reference	Reference excerpt/summary
9	Two	Treaty of between [Israel] and [Jordan]	10/26/94	Israel, Jordan	Jordan, Yarmuk, Araba/Arava groundwater	Article 6 and Annex III	ARTICLE 6 (4): "... the Parties agree to search for ways to alleviate water shortages and to co-operate in the following fields: b. prevention of contamination of water resources; ..." Article III WATER QUALITY AND PROTECTION (Summarized): The two parties agree to utilize existing national laws to protect the shared waters of the Jordan and Yarmuk Rivers and the Arava/Araba groundwater; to jointly monitor water quality along the frontier; to prohibit untreated wastewater disposal; to supply to the other country water of equal quality as that used in the same location by the supplying country; to begin desalinization of certain saline springs within four years; and to protect water systems in each country's own territory that will be supplied to the other country.
10	Two	Agreement on joint activities in addressing the Aral Sea and the zone around the Sea crisis, improving the environment, and enduring the social and economic development of the Aral Sea region	03/26/93	Kazakhstan, Kyrgyzstan, Tajikstan, Turkmenistan, Uzbekistan	Aral Sea, Syr Darya, Amu Darya	Articles 1, 3	ARTICLE 1: "States-participants recognize as common objectives: ....maintaining the required water quality in the rivers, reservoirs, and springs, due to an, in future, preventing the release into these bodies of industrial and urban waste waters, and polluted and mineralized collector and drainage waters; ... improving the sanitary and medico-biological living conditions, especially for the sea zone residents, and addressing the urgent problem of a clean drinking water supply for the region; .... ARTICLE 3: "The Russian Federation participates in the Interstate Council work as an observer in addressing the Aral Sea crisis and the rehabilitation of the disaster zone. It also provides the required financial and technical assistance in water

Category	Agreement title	Date	Parties	Basin(s)	Water quality reference	Reference excerpt/summary
						treatment, creating the domestic- and drinking-water supply system in the region and fighting desertification. The Russian Federation also cooperates in the scientific and technical spheres, in designing projects of regional significance, in creating the environment monitoring system, . . ."
11 Two	Convention between [Germany] and the [Czech and Slovak Republic] and the [European Economic Community] on the International Commission for the Protection of the Elbe	10/08/90	Federal Republic of Germany, Czech and Slovak Federative Republic, European Economic Commission	Elbe	Entire Document	SUMMARY: Under the purview of the International Commission for the Protection of the Elbe, the contracting parties agree to cooperate to prevent pollution of the Elbe River and its drainage area. The Agreement lays out a general guidelines for the Commission which include the proposal of water quality objectives, standards, and measures; proposal and implementation of investigative, conservation, and disaster preparedness projects; promotion of information exchange; and preparation of environmental protection regulations.
12 Two	Agreement on cooperation on management of water resources in the Danube basin	12/01/87	Austria, Germany (FRG)	Danube	Articles 1–7	SUMMARY: With regard to water quality, the Agreement covers projects related to the "protection of the aquatic environment including the groundwater, in particular the prevention of pollution, and the discharge of waste water and heat" (Article 1). The Parties agree to ensure the projects on frontier waters do not adversely affect the condition of water resources of the other state. If a project may adversely impact the condition of the other state's water, then the other state shall be notified in advance and provided ample time to respond to the proposed project (Articles 3, 4).

(continued)

(*continued*)

Category	Agreement title	Date	Parties	Basin(s)	Water quality reference	Reference excerpt/summary
						The Parties agree to coordinate water quality measurements on frontier waters and alarm, intervention and notification plans (Articles 6, 7). A Standing Committee on Management of Water Resources is established under the Agreement. The Committee may address such issues as minimum discharge requirements, measures to improve or protect the aquatic environment, and methods to establish type and extent of water pollution (Article 7).
13 Two	Agreement on the action plan for the environmentally sound management of the common Zambezi River system.	05/28/87	Botswana, Mozambique, Tanzania, Zambia, Zimbabwe	Zambezi	References made throughout the document	SUMMARY: Water quality is an integral part of the Zambezi Action plan and is referenced throughout the document. Section I of Annex I describes current water quality problems. Section II outlines general actions related to water quality and other water issues including information collection, assessments, monitoring, and legislative actions. Short-term water quality objectives and programs are identified in Section III of Annex I and described in the project listing found in Appendix I. Projects that include references to water quality are: ZACPRO 3, ZACPRO 6, ZACPRO 13, ZACPRO 14 and ZACPRO 19.
14 Two	Convention creating the Niger Basin Authority	01/21/80	Benin, Cameroon, Chad, Côte d'Ivoire, Guinea, Mali, Niger, Nigeria, Upper Volta	Niger	Article 4	ARTICLE 4 (2): For the purpose set out in the above paragraph (1) the "AUTHORITY" shall notably undertakes [sic] in harmony with the development plans of States relating to the Niger Basin and in accordance with the general objectives of integrated development of the Basin, the following activities: . . . (d) Environment control and

Category	Agreement title	Date	Parties	Basin(s)	Water quality reference	Reference excerpt/summary
						preservation: (i) Protection of the environment comprising the establishment of norms and measures applicable to the States in the alternative uses of waters in the Basin; (ii) Flood control; (iii) Construction and maintenance of dikes; (iv) Prevention and control of drought and desertification; (v) Prevention and control of soil erosion and sedimentation; (vi) Setting up of structures and works for land development including salt water and drainage control.
15 Two	Joint declaration of principles for utilization of the waters of the lower Mekong basin, signed by [Cambodia], [Laos], [Thailand], and [Vietnam] to the Committee for Coordination of Investigations of the Lower Mekong basin	01/31/75	Cambodia, Laos, Thailand, Vietnam	Mekong	Chapter III: Articles IV, VIII, XIX, XXV	ARTICLE IV: The Basin States shall ensure the conservation of the Basin water resources by taking every reasonable necessary measure to: 1. Maintain their flow and quality; 2. Prevent their misuse, waste, and pollution; . . ." ARTICLE VIII "Every reasonable measure shall be taken by the Basin States to ensure the coordinated control of the Basin water resources, including . . . reduction of salt water intrusion. . . ." ARTICLE XIX "Every reasonably necessary measure shall be taken by the riparian State diverting mainstream waters . . . to restrict the pollution of the return flow." ARTICLE XXV "When developing its Basin water resources, each Basin State shall take such measures as are practicable and reasonably necessary to avoid or minimize detrimental effects upon the ecological balance of the Basin, or any part thereof."

(*continued*)

(*continued*)

	Category	Agreement title	Date	Parties	Basin(s)	Water quality reference	Reference excerpt/summary
16	Two	Agreement between [Australia] (acting on its own behalf and on behalf of [Papua New Guinea]) and [Indonesia] concerning administrative border arrangements as to the border between Papua New Guinea and Indonesia	11/13/73	Papua New Guinea, Indonesia	Sepik, Fly	Articles 5, 12	ARTICLE 5 SETTLEMENT: "It shall be an agreed objective to discourage the construction of villages or other permanent housing within a two kilometer zone on each side of the border." ARTICLE 12 POLLUTION: "The Governments agree that when mining, industrial, forestry, agricultural or other projects are being carried out in the respective border areas the necessary precautionary measures shall be taken to prevent serious pollution of rivers flowing across the border. There shall be consultations, if so requested, on measures to prevent pollution, arising from such activities, of rivers on the other side of the border."
17	Two	Agreement between [Romania] and the [USSR] on the joint construction of the Stinca-Costesti Hydraulic Engineering Scheme on the River Prut and the establishment of the conditions for its operation (with Protocol)	12/16/71	USSR, Romania	Prut	Main Agreement: Article 16, Protocol: Articles 5, 8	ARTICLE 16: "Each Party shall ensure the measures are taken in its territory to prevent and combat pollution of the waters of the river Prut." PROTOCOL, ARTICLE 5 (3): "Each Party shall be obliged to ensure, in accordance with the health requirements, a permanent minimum discharge of 2.5 cubic metres per second below the hydraulic engineering scheme." ARTICLE 8: "(1) The Parties shall not carry out works or take measures which would cause any deterioration in the water quality of the river Prut existing on the date of conclusion of this Protocol. (2) The direct discharge into the storage lake of waste water and of matter or substances that could pollute the water shall be prohibited. In special cases if may be effected, solely on the approval of the Mixed Commission."

	Category	Agreement title	Date	Parties	Basin(s)	Water quality reference	Reference excerpt/summary
18	Two	Treaty between [Austria] and [Czechoslovakia] concerning the regulation of water management questions relating to frontier waters	12/07/67	Czechoslovakia, Austria	Danube	Articles 3, 4, 5 and Annex 1, Article 2	ARTICLE 3: "(1) Each Contracting State undertakes to refrain from carrying out, without the consent of the other Contracting State, any measures relating to frontier waters within the meaning of article 1(a) which would adversely affect water conditions in the territory of the other Contracting State.... (4) Where it is necessary to prevent the pollution of frontier waters, the Contracting States shall endeavor to introduce improvements and shall arrange for the purification of waste water arising from new sources." ARTICLES 4 (2) and 5 (3)/(5) describe general objectives and measures concerning the cleaning of the bed and banks of frontier waters. The Austrian–Czechoslovak Frontier Water Commission, established by the Agreement, is tasked with implementing the terms of the Agreement, including matters related to water quality and cleaning of the frontier waters (ANNEX 1, ARTICLE 2).
19	Two	Agreement concerning the River Niger commission and the navigation and transport on the River Niger	11/25/64	Benin, Cameroon, Chad, Côte d'Ivoir, Guinea, Mali, Niger, Nigeria, Upper Volta	Niger	Article 12	ARTICLE 12: "In order to achieve maximum co-operation in connection with the matters mentioned in Article 4 of the Act of Niamey, the riparian States undertake to inform the Commission as provided for in Chapter I of the present Agreement, at the earliest stage, of all studies and works upon which they propose to embark- They undertake further to abstain from carrying out on the portion of the River, its tributaries and sub-tributaries subject to their jurisdiction any works likely to pollute the waters, or any

(continued)

(*continued*)

	Category	Agreement title	Date	Parties	Basin(s)	Water quality reference	Reference excerpt/summary
							modification likely to affect biological characteristics of its fauna and flora, without adequate notice to, and prior consultation with, the Commission."
20	Two	Agreement between [Poland] and the [USSR] concerning the use of water resources in frontier waters	07/17/64	Poland, USSR	Vistula	Articles 3, 4, 9, 10, 11	SUMMARY: The purpose of the Agreement is to promote cooperative water resources management in such areas as water quality investigation and protecting surface and groundwaters against "the introduction into the waters, directly or indirectly, of solid, liquid or gaseous substances and heat in such quantities as may cause physical, chemical and biological changes which limit or prevent the normal utilization of the said waters for communal, industrial, agricultural, fishery or other purposes" (Articles 3, 4). The parties agree not to undertake projects that could harm the use of water resources by the other party nor to discharge of sewage and other water into the frontier waters without mutual consent. The parties also agree to jointly measure water quality; to develop common quality standards and pollution control procedures, if necessary; to endeavor to keep frontier waters clean; to employ appropriate water purification procedures, and to refrain from discharging any sewage which may cause harmful pollution to the frontier waters.
21	Two	Convention and Statutes Relating to the Development of the Lake Chad Basin	05/22/64	Cameroon, Chad, Niger, Nigeria	Lake Chad	Chapter II Article 5 (second paragraph)	ARTICLE 5: ". . . . In particular, the Member States agree not to undertake in that part of the Basin falling within their jurisdiction any work in connection with the development of water resources or the soil likely to have a marked

Category	Agreement title	Date	Parties	Basin(s)	Water quality reference	Reference excerpt/summary
						influence upon the system of the water courses and levels of the Basin without adequate notice and prior consultations with the Commission, provided always that the Member States shall retain the liberty of completing any plans and schemes in the course of execution or such plans and schemes as may be initiated over a period of 3 years to run from the signature of the present Convention."
22 Two	Act regarding navigation and economic cooperation between the states of the Niger basin	10/26/63	Benin, Cameroon, Chad, Côte d'Ivoir, Guinea, Mali, Niger, Nigeria, Upper Volta	Niger	Article 4	ARTICLE 4: "The riparian States undertake to establish close co-operation with regard to the study and the execution of any project likely to have an appreciable effect on certain features of the regime of the River, its tributaries and sub-tributaries, their conditions of navigability, agricultural and industrial exploitation, the sanitary conditions of their waters, and the biological characteristics of their fauna and flora."
23 Two	Indus Waters Treaty	09/19/60	India, Pakistan	Indus	Article IV	ARTICLE IV (10): "Each Party declares its intention to prevent, as far as practicable undue pollution of the waters of the Rivers/which might affect adversely uses similar in nature to those to which the waters were put on the Effective Date, and agrees to take all reasonable measures to ensure that, before any sewage or industrial waste is allowed to flow into the Rivers, it will be treated, where necessary, in such manner as not materially to affect those uses: Provided that the criterion of reasonableness shall be the customary practice in similar situations on the Rivers."

(continued)

(*continued*)

	Category	Agreement title	Date	Parties	Basin(s)	Water quality reference	Reference excerpt/summary
24	Two	State treaty between [Luxembourg] and [West Germany] concerning the construction of hydroelectric power-installations on the Our	07/10/58	Germany (FRG), Luxembourg	Our	Article 2	ARTICLE 2 "... Nothing shall be done to interfere with the water resources of the Our in such a way as to impair the operation, in accordance with article 1, of the power-plants covered by the Concession. Thus, water may not be taken from watercourses in the catchment area of the Our above the installations in such a way as to cause such impairment, nor may the water above the installations be polluted or chemically contaminated in a manner detrimental to the operation of the plants. No claims arising out of offences committed by third parties may be made against the Contracting Countries."
25	Two	Agreement between the [Czechoslovakia] and [Poland] concerning the use of water resources in frontier waters	03/21/58	Czechoslovakia, Poland	Oder	Articles 2, 3, 8, 9	ARTICLE 2 (2): "For the purposes of this Agreement, the term "questions relating to the use of water resources" refers, in particular to ... (b) Discharge of flood waters, drifting of ice, pollution abatement, and conservation of natural resources in relation to the water economy. ARTICLE 3: "(4) The Contracting Parties have agreed to abate the pollution of frontier waters and to keep them clean to such extent as is specifically determined in each particular case in accordance with the economic and technical possibilities and requirements of the Contracting Parties. (5) When installations discharging polluted water into frontier waters are constructed or reconstructed, treatment of the waste water shall be required." ARTICLE 8 (1): "The Contracting Parties shall: (a) Exercise control over work carried out under this Agreement, over the

Category	Agreement title	Date	Parties	Basin(s)	Water quality reference	Reference excerpt/summary
						diversion of water and over the extraction of material from stream beds and shall inspect the quality of the water;" ARTICLE 9 concerns the appointment of plenipotentiaries to carry out the agreement.
26 Two	Treaty between the [USSR] and [Iran] concerning the regime of the Soviet—Iranian frontier and the procedure for the settlement of frontier disputes	05/14/57	USSR, Iran	Tedzen, Atrak, Araks, Hariud	Article 10	ARTICLE 10: "1. The Contracting Parties shall ensure that frontier waters are maintained in the proper state of cleanliness and are kept free of any artificial pollution and fouling. 2. Frontier watercourses shall be cleaned out on the sectors where such work is jointly considered essential by the competent authorities of the two Contracting Parties. The cost of cleaning in such cases shall be equally divided between the two Contracting Parties. 3. The cleaning of those sectors of frontier water which are situated wholly in the territory of one of the Contracting Parties shall be carried out by that Party as necessary, at its own expense. 4. In cleaning out frontier waters, the earth, stones, trees and other objects removed shall be thrown out to such a distance from the bank or levelled down in such a way as to avoid any danger that the banks might fall in, or the river bed be polluted, and so as to prevent the flow of water from being obstructed in time of flood."
27 Two	Treaty between [Czechoslovakia] and [Hungary] concerning the regime of state frontiers	10/13/56	Czechoslovakia, Hungary	Danube	Article 15	ARTICLE I5: "(1) The beds of frontier watercourses shall be cleaned out on sectors to be determined jointly by the competent authorities of the Parties. (2) In cleaning out the beds of frontier watercourses, the substances removed shall be

(*continued*)

(*continued*)

	Category	Agreement title	Date	Parties	Basin(s)	Water quality reference	Reference excerpt/summary
							placed at such a distance as to prevent any subsidence of the banks, any obstruction of the beds or any reduction in the flow of water."
28	Two	Treaty between the [Hungary] and [Austria] concerning the regulation of water economy questions in the frontier region	04/09/56	Hungary, Austria	Boundary Waters between Austria and Hungary	Article 2	ARTICLE 2 (7): "In order to prevent the pollution of frontier waters, the Contracting Parties shall endeavour to ensure that factories, mines, industrial plants and similar installations, as well as residential communities, drain waste water into the said waters only after suitable purification. When new installations of that nature are built, they shall be required to take appropriate measures to purify waste water."
29	Two	State Department document concerning the Johnston Negotiations (negotiations did not result in an agreement)	12/31/55	Israel, Jordan, Syria, Lebanon	Jordan	Section 3 Division of Water	Section 3 addresses the issue of salinity in Lake Tiberias and discusses the possibility of preventing flow from certain saline springs into the Lake.
30	Two	Agreement between [Yugoslavia] and [Romania] concerning questions of water control on water control systems and watercourses on or intersected by the state frontier, together with the statue of the Yugoslav–Romanian water control commission	04/07/55	Romania, Yugoslavia	Danube, Tisza	Articles 1, 2. (Article 2 of attached Statute of the Water Control Commission reiterates objectives noted in Article 1 of the overall Agreement)	ARTICLE 1: "Water control questions, measures and works on water control systems and watercourses and in valleys and depressions on or intersected by the State frontier which may affect the regime and quality of the waters and which are of interest to both Contracting States shall be examined and regulated by the two Contracting States in accordance with the provisions of this Agreement. The provisions of this Agreement relate to the following questions: . . . (e) Protection of waters against pollution; . . . (i) protection against erosion. . . ." ARTICLE 2 (2): The erection of any new installations and the

Category	Agreement title	Date	Parties	Basin(s)	Water quality reference	Reference excerpt/summary
						execution of any new works, in the territory of either Contracting State, which may change the existing regime of the waters, interfere with the free discharge of the waters where it now exists, change the quality of the waters, or cause flooding on water control systems or watercourses or in valleys or depressions on or intersected by the State frontier shall be referred to the Mixed Commission for examination."
31 Two	Agreement between [Syria] and [Jordan] concerning the utilization of the Yarmuk waters	06/04/53	Jordan, Syria	Yarmuk	Article 10	ARTICLE 10: "The Commission shall have the following duties, the enumeration of which is not meant to be restrictive: . . . (h) To study methods of preventing silting in the reservoir and the contamination of its waters, as well as of combating malaria, and to make appropriate recommendations to the two Governments."
32 Two	Treaty between the [USSR] and [Hungary] concerning the regime of the Soviet-Hungarian state frontier and final protocol	02/24/50	USSR, Hungary	Danube	Articles 16, 17	ARTICLE 16: "1. Frontier watercourses shall be cleaned out in sectors where such work is jointly considered essential by the competent authorities of the Contracting Parties. The cost of cleaning in such cases shall be divided equally between the two Contracting Parties. 2. The cleaning of frontier waters in sectors situated wholly in the territory of one of the Contracting Parties shall be carried out by that Party at its own expense as need arises. 3. In cleaning out frontier waters, earth and stones removed shall be thrown out to such a distance from the bank, and levelled down in such a way, as to avoid any danger of subsistence of the banks or choking up of the

(*continued*)

(*continued*)

Category	Agreement title	Date	Parties	Basin(s)	Water quality reference	Reference excerpt/summary
						river bed and to prevent the flow of water form being obstructed in time of flood." ARTICLE 17: "The competent authorities of the Contracting parties shall take steps to maintain the frontier waters in such due state of cleanliness as to prevent the waters from being poisoned or polluted by acids or refuse from factories or industrial establishments, or from being fouled by any other means."
33 Two	Treaty between the [USSR] and [Romania] concerning the regime of the Soviet–Romanian state frontier and final protocol	11/25/49	USSR, Romania	Danube	Articles 16, 17	ARTICLE 16: "1. Frontier watercourses shall be cleaned out in sectors where such work is jointly considered essential by the competent authorities of the Contracting Parties. The cost of cleaning in such cases shall be divided equally between the two Contracting Parties. 2. The cleaning of frontier waters in sectors situated wholly in the territory of one of the Contracting Parties shall be carried out by that Party at its own expense as need arises. 3. In cleaning out frontier watercourses, the earth removed shall be dumped on the banks or at dumps on the river in such a way as to avoid any subsidence of the banks, choking up of the river-bed, or obstruction of the flow of water in time of flood." ARTICLE 17: "The competent authorities of the Contracting Parties shall take steps to maintain the frontier waters in such due state of cleanliness as to prevent the waters from being poisoned or polluted by acids or refuse from factories or industrial establishments, or from being fouled by any other means."

	Category	Agreement title	Date	Parties	Basin(s)	Water quality reference	Reference excerpt/summary
34	Two	Convention between [Germany and [Lithuania] regarding the maintenance and administration of the frontier waterways	01/29/28	Germany, Lithuania	Memel, Kurische Haff	Articles 15, 17, 19, 21, 22, and 24	ARTICLE 15: "The construction of weirs, water-mills or any other installations liable to change the direction of a frontier waterway or influence the water level may only be undertaken with the approval of both States; this shall also be required for the utilisation of frontiers waterways for the discharge of industrial waste waters." ARTICLE 17: "Solid and viscous substances and dead cattle may not be thrown into the frontier waterways, nor left so close to the latter that they are liable to fall in or be washed in." ARTICLE 19: "The waterways shall be cleared and kept in good order by each State within its territorial sectors." ARTICLE 21: "(1)The material cleared from the river shall be deposited at such a distance from the bed that there is no risk of its falling back." (Specific guidance is then provided for certain tributaries). ARTICLE 22 address costs associated with cleaning and maintaining frontier waterways. ARTICLE 24 addresses the inspection of cleaning and maintenance work.
35	Two	Convention regarding the water supply of Aden between [Great Britain] and the Sultan of Abdali	04/11/10	Aden (Yemen), Great Britain	Groundwater resources	Agreement concerns groundwater. Water quality is discussed in Section II.	"Sir Ahmed Fadthl Mohsin on behalf of himself, his heirs and successors hereby agrees:- . . . (II) not to do or allow to be done anything that will reduce or contaminate the supply of water yielded by the wells sunk on the above site, i.e., the working of wells by machinery and the throwing of dirt within a distance of 400 feet of the above site."

(*continued*)

(continued)

	Category	Agreement title	Date	Parties	Basin(s)	Water quality reference	Reference excerpt/summary
36	Three	Agreement Between [Kazakhstan], [Kyrgyz], [Uzbekistan] on use of water and energy resources of Syr Darya basin	03/17/98	Kazakhstan, Kyrgyz, Uzbekistan	Aral Sea	Article X	ARTICLE X: "To provide further improvement of the management and use of the water and energy resources and the enhancement of economic relations aimed at guaranteed water supply in the basin, the Parties agree to consider jointly the following issues: . . . Reduction and discontinuation of polluted water discharges in the water sources of the Syr Darya basin."
37	Three	Agreement Between [Kazakhstan], [Kyrgyz], [Uzbekistan] on cooperation in the area of environment and rational nature use	03/17/98	Kazakhstan, Kyrgyz, Uzbekistan	Aral Sea	Article 2	ARTICLE 2: "The Parties shall cooperate: . . . i) in protection, rational use and prevention against pollution of the transboundary water resources. . . ."
38	Three	Joint Water Commission terms of reference	01/01/96	South Africa, Mozambique	Incomati, Maputo, Umbeluzi, Limpopo	Article 3	ARTICLE 3 (1): "The functions and powers of the Commission shall be to advise the Parties on all technical matters relating to - . . . (h) the control of the quality of water resources of common interest and the prevention of pollution and soil erosion affecting such water resources. . . ."
39	Three	Agreement between [Angola], [Botswana] and [Namibia] on the establishment of a Permanent Okavango River Basin Water Commission (OKACOM)	09/16/94	Angola, Botswana, Namibia	Okavango	Article 4	ARTICLE 4: "The functions of the Commission shall be to advise the Contracting Parties on: . . . (4.5) The prevention of the pollution of water resources and the control over aquatic weeds in the Okavango River Basin. . . ."

	Category	Agreement title	Date	Parties	Basin(s)	Water quality reference	Reference excerpt/summary
40	Three	Agreement between [Namibia] and [South Africa] on the establishment of a Permanent Water Commission	09/14/92	Namibia, South Africa	Orange	Article 3	ARTICLE 3 (1): "The functions and powers of the Commission shall be to advise the Parties on: . . . (e) the prevention of and control over the pollution of common water resources and soil erosion affecting such resources. . . ."
41	Three	Treaty on the establishment and functioning of the Joint Water Commission between [South Africa] and [Swaziland]	03/13/92	South Africa, Swaziland	Incomati, Maputo	Article 3	ARTICLE 3 (1): "In addition to any other functions or powers conferred on the Commission by the Parties, the functions and powers of the Commission shall be to advise the Parties on all technical matters relating to- . . . (g) the prevention and exercise of control over the pollution of water resources of common interest and soil erosion affecting such resources. . . ."
42	Three	Treaty on the development and utilization of the water resources of the Komati River Basin between [South Africa] and [Swaziland]	03/13/92	South Africa, Swaziland	Komati	Article 13, 14	Article 13: "The Parties agree to take all reasonable measures to ensure that the design, construction, operation and maintenance of the Project are compatible with the protection of the existing quality of the environment and, in particular, shall pay due regard to the maintenance of the welfare of persons and communities immediately affected by the Project." Article 14 (5): The Parties shall use their best endeavours to − (a) minimize waste and non-beneficial use of water from the Komati River Basin within their respective territories; and ensure that the necessary steps are taken within their respective territories to prevent water pollution and to minimise soil erosion within the said basin."

(continued)

(*continued*)

	Category	Agreement title	Date	Parties	Basin(s)	Water quality reference	Reference excerpt/summary
43	Three	Treaty on the Lesotho Highlands Water Project between [South Africa] and [Lesotho]	10/24/86	South Africa, Lesotho	Orange	Article 6, 7, 8, 15	ARTICLE 6 (15): "Lesotho shall take the necessary measures to prevent or abate any significant pollution of the water to be delivered to South Africa. The Parties shall consult through the Joint Permanent Technical Commission with a view to reaching agreement with regard to the defrayment of the reasonable costs for prevention or abatement of pollution caused by adverse effects of the Project." ARTICLE 7(22) and ARTICLE 8 (10) state that both the Lesotho Highlands Development Authority and the Trans-Caledon Tunnel Authority "shall effect all necessary catchment conservation measures as well as all measures necessary to prevent pollution of the water to be delivered to South Africa and pollution caused by the adverse effects of the implementation of the Project." ARTICLE 15: "The Parties agree to take all reasonable measures to ensure that the implementation, operation and maintenance of the Project are compatible with the protection of the existing quality of the environment. . . ."
44	Three	Agreement on Paraná river projects	10/19/79	Argentina, Brazil, Paraguay	Paraná	Section 5	SECTION 5 (j): "In accordance with the commitments undertaken in the system of the Treaty on the River Plate Basin, and in view of the existing respective legislation in this regard, the three Governments, insofar as it is pertinent to each, shall undertake efforts, in the context of the application of this Note, to preserve the environment, the fauna and flora, as well as the quality of the waters of the Paraná

Category	Agreement title	Date	Parties	Basin(s)	Water quality reference	Reference excerpt/summary
						River, avoiding its contamination and assuring, at the least, the present conditions of health in the areas of influence of both projects. In this respect, they shall likewise promote the creation of new national parks and the improvement of existing parks."
45 Three	Protocol concerning the delimitation of the river frontier between Iran and Iraq	06/13/75	Iran, Iraq	Bnava Suta, Qurahtu, Gangir, Alvend, Kanjan	Article 8	ARTICLE 8: "1. Rules governing navigation in the Shatt al'Arab shall be drawn up by a mixed Iranian-Iraqi Commission, in accordance with the principle of equal rights of navigation for both States. 2. The two Contracting Parties shall establish a commission to draw up rules governing the prevention and control of pollution in the Shatt al'Arab. 3. The two Contracting Parties undertake to conclude subsequent agreements on the questions referred to in paragraphs 1 and 2 of this article."
46 Three	Treaty between the [Netherlands] and [Germany] concerning the course of the common frontier, the boundary waters, real property situated near the frontier, traffic crossing the frontier on land and via inland waters, and other frontier questions (Frontier Treaty).	04/08/60	Netherlands, The Federal Republic of Germany	Meuse, Rhine	Article 58	ARTICLE 58 (2): "In performing the obligations undertaken in paragraph 1, the Contracting Parties shall in particular take or support, within an appropriate period of time, all measures required: . . . (e) To prevent such excessive pollution of the boundary waters as may substantially impair the customary use of the waters by the neighbouring State."

(*continued*)

(continued)

	Category	Agreement title	Date	Parties	Basin(s)	Water quality reference	Reference excerpt/summary
47	Three	Agreement concerning water economy questions between the governments of [Yugoslavia] and [Bulgaria].	04/04/58	Bulgaria, Yugoslavia	Danube	Articles 1, 2	ARTICLE 1: "(1) The Contracting Parties undertake, pursuant to the provisions of the Agreement, to examine and resolve all questions of water economy, including measures and works which may affect the quantity and quality of the waters and which are of interest to both or either of the Contracting Parties. (2) The provisions of this Agreement shall, in so far as the Contracting Parties are interested in accordance with paragraph 1 of this article, apply water-economy questions, measures and works on rivers, tributaries and river basins followed or intersected by the State frontier, and in particular to: (e) Protection of the waters against pollution. . . . ." ARTICLE 2 "The Contracting Parties undertake: (1) Each in its own territory and jointly in the case of rivers and tributaries followed or intersected by the State frontier, to maintain in good condition the beds of rivers and of tributaries and all installations; . . . ."
48	Three	Agreement between [Yugoslavia] and [Albania] concerning water economy questions, together with the statute of the Yugoslav– Albanian water economy questions, together with the statute of the	12/05/56	Yugoslavia, Albania	Crni Drim, Beli Drim, Bojana, Lake Skadar	Article 1 and Annex II (b)	ARTICLE 1: 1. "The Contracting Parties undertake, pursuant to the provisions of this Agreement, to examine and to resolve by agreement all questions of water economy, including measures and works which may affect the quantity and quality of the water and which are of interest to both or either of the Contracting Parties, having due regard to the maintenance of a common policy in water economy relations and recognizing the rights and obligations arising out of such

Category	Agreement title	Date	Parties	Basin(s)	Water quality reference	Reference excerpt/summary
	Yugoslav-Albanian Water Economic Commission and with the protocol concerning fishing in Frontier lakes and rivers.					policy. 2 (c) The provisions of this Agreement shall apply to all water economy questions, measures and works on watercourses which form the State frontier and watercourses, lakes and water systems which are intersected by the State frontier (especially Lake Ohrid, the Crni Drim, the Beli Drim, Lake Skadar and the Bojana), and which are of interest to both Contracting Parties, and in particular to:(c) The discharge of water, drainage and similar measures; . . . " ANNEX II (b) "Biological measures relating to fishing . . . The provisions shall also cover measures to settle questions of protecting lake and river water from pollution."
49 Three	Agreement between [Yugoslavia] and [Hungary] together with the statute of the Yugoslav-Hungarian water economy commission	08/08/55	Hungary, Yugoslavia	Mura, Drava, Maros, Tisza, Danube	Articles 1, 2	ARTICLE 1: "(1) The Contracting Parties undertake, pursuant to the provisions of this Agreement, to examine and resolve by agreement all questions of water economy, including measures and works which may affect the quantity and quality of the water and which are of interest to both or either of the Contracting Parties, having due regard to the maintenance of a common policy of water economy relations and recognizing the rights and obligations arising out of such policy. (2) The Provisions of this Agreement shall, . . . , apply to all water economy questions measures and works on watercourses which form the State frontier and watercourses and water systems intersected by the State frontier, and in particular to: . . . (f) Protection of the waters against pollution; . . . (i) Protection against soil erosion." ARTICLE 2:

(*continued*)

(*continued*)

	Category	Agreement title	Date	Parties	Basin(s)	Water quality reference	Reference excerpt/summary
							"The Contracting Parties undertake: (1) Each in its own territory and jointly . . . , to maintain in good condition the beds of watercourses and all installations."
50	Three	Treaty between [Germany] and [Poland] for the settlement of frontier questions	01/27/26	Germany, Poland	Boundary Waters between Germany and Poland	Article 30	ARTICLE 30: "Subject to reciprocity, the two contracting States shall each on its own side take all the measures provided for by the laws of the country with a view to maintaining the frontier waterways and frontier waters in a clean condition. . . ."
51	Three	Treaty between [Great Britain] and the [United States] relating to boundary waters and boundary questions	01/11/09	United States, Great Britain	Boundary Waters between Canada and the United States	Article IV	ARTICLE IV: " . . . It is further agreed that the waters herein define as boundary waters and waters flowing across the boundary shall not be polluted on either side to the injury of health or property on the other."

F.2 U.S. INTERSTATE COMPACTS

	Category	Agreement title	Date	Parties	Basin(s)	Water quality reference	Reference excerpt/summary
52	One	Ohio River Valley Water Sanitation Compact	06/30/48	Ohio, Indiana, Illinois, Kentucky, New York, Tennessee, Pennsylvania, West Virginia	Ohio	Entire Document	SUMMARY: The parties agree to cooperate in the control of future pollution and abatement of existing pollution in the Ohio basin and to enact necessary legislation to maintain the waters in satisfactory, sanitary condition. The parties recognize that no single standard for treatment can be universally applicable. Therefore the guiding principles of the compact is that "pollution by sewage or industrial wastes, originating within a signatory state shall not injuriously affect the various uses of the interstate

Category	Agreement title	Date	Parties	Basin(s)	Water quality reference	Reference excerpt/summary
						waters...." (Article IV). Effluent standards are established for municipal and industrial wastes into the Ohio basin, including its tributaries. A Commission is established to administer and enforce the compact. The Commission will conduct water quality surveys and report on pollution problems, investigate water quality matters and issue orders as necessary to correct problems identified.
53 One	Interstate Sanitation Commission	06/24/41	New York, New Jersey, Connecticut	Interstate coastal, estuarial, and tidal waters	Entire Document	SUMMARY: The signatory States pledge to cooperate in the control of future and abatement of existing pollution; to enact appropriate legislation to maintain the waters in a satisfactory sanitary condition and to render safe waters used for bathing or recreational purposes. A Commission is established to carry out the Agreement / provisions and is tasked with grouping the designated waters into two general classifications described in the Compact (although supplemental classes and effluents standards may be developed by the Commission). Effluent standards are described for each of the two classes. Effluent standards for stream tributaries flowing into the tidal waters are also described. The parties agree to prohibit the pollution described in the compact and to enact suitable legislation to carry out the objectives of the compact. The Commission has the authority to investigate matters of compact compliance and enforce the provisions of the agreement.

(continued)

(continued)

	Category	Agreement title	Date	Parties	Basin(s)	Water quality reference	Reference excerpt/summary
54	One	Rio Grande Compact	03/18/38	Texas, Colorado, New Mexico	Rio Grande	Articles III, XI	SUMMARY: Article III details Colorado's water delivery obligations. Within this article, water quality standards (in terms of sodium content) are set for contributions from a particular closed basin. Article XI states that the two states agree that previous water quality disputes are resolved upon ratification of the Compact, but that signature of the Compact does not prevent future litigation if the water quality is to change nor does it imply that the two states admit that irrigation causes increased salinity for which the user is responsible.
55	Two	Bear River Compact	12/22/78	Utah, Idaho, Wyoming	Bear	Articles III, IX	SUMMARY: Article III states that Bear River Commission, established by the Compact, is responsible, among other things, for cooperating with state and federal agencies concerning interstate water pollution matters. Article IX of the compact allows for water exchanges provided that water quality is not compromised.
56	Two	Red River Compact	05/12/78	Texas, Arkansas, Oklahoma, Louisiana	Red	Article I (c) and (d); Article II, Sec. 2.10 (a); Article III, Sec 3.01 (i) and (j); Article X, Sec 10.02 (b); Article XI	SUMMARY: The compact includes objectives to promote projects for and enforce laws related to the control and abatement of natural deterioration and pollution of the basin's waters. Definitions of pollution and natural deterioration are included in the compact (Article III). In addition to relying on State regulations and laws to abate and control pollution, the signatory States agree to cooperate with one another and with federal agencies to alleviate the natural deterioration of the basin's waters. The States also agree to maintain records concerning the types of amounts of discharge. A

Category	Agreement title	Date	Parties	Basin(s)	Water quality reference	Reference excerpt/summary
						Commission with responsibilities to coordinate with federal, state and other agencies to abate and control natural deterioration and pollution; to recommend "reasonable" water quality objectives; to investigate and resolve disputes concerning interstate pollution utilizing applicable Federal statutes to resolve interstate pollution problems.
57 Two	Great Lakes Basin Compact	07/24/68	Ohio, Quebec, Indiana, Ontario, Illinois, Michigan, New York, Minnesota, Wisconsin, Pennsylvania	Great Lakes	Articles I, VI, VII	ARTICLE I: "The purposes of this compact are through means of joint or cooperative action: 1. To promote the orderly, integrated, and comprehensive development, use, and conservation of thew water resources of the Great Lakes Basin; . . ." ARTICLE VI: "The Commission (the Great Lakes Commission established by the compact) shall have power to: . . . B. Recommend methods for the orderly, efficient, and balanced development, use, and conservation of the water resources of the Basin or any portion thereof to the party states and to any other governments or agencies having interests in or jurisdiction over the Basin or any portion thereof . . . G. Recommend uniform or other laws, ordinances, or regulations relating to the development, use and conservation of the Basin's water resources to the party states or any of them and to other governments, political subdivisions, agencies or intergovernmental bodies. . . ." ARTICLE VII: "Each party state agrees to consider the action the Commission recommends in respect to: . . . B. Measures for combating pollution."

(*continued*)

(*continued*)

	Category	Agreement title	Date	Parties	Basin(s)	Water quality reference	Reference excerpt/summary
58	Two	Susquehanna River Basin Compact	07/17/68	Maryland, New York, Pennsylvania	Susquehanna	Articles 1 (Section 1.3),3 (Sections 3.1,3.3, 3.4, 3.5), 4 (Section 4.2a), 5, 6 (Section 6.8),11 (Section 11.5), 14 (Section 14.2), Article 15 (Section 15.2)	SUMMARY: One of the main objectives of the compact is to promote coordinated management of the Delaware Basin, which includes water quality management. The compact establishes a Commission with the broad powers concerning the development and coordination of plans, policies, and projects for the management, control, conservation of the interstate waters; establishment of standards; conducting/sponsoring research; compilation of data and publishing reports concerning water quality. Article 5 specifically addresses the Commission's water quality management and control responsibilities and powers including those related to investigating water quality, constructing of appropriate facilities, and establishing and enforcing standards. The parties agree to prohibit and control future pollution, to abate existing pollution, and to maintain the waters in satisfactory condition in accordance with compact terms, enacting appropriate legislation as needed.
59	Two	Kansas–Oklahoma Arkansas River Basin Compact, 1965	03/31/65	Kansas, Oklahoma	Arkansas	Article I(D), Article II(H), Article IX, Article XI (B2)	SUMMARY: One of the major goals of the Compact is to "encourage" pollution abatement programs in the two states. Both man-made and natural pollutants are to be addressed. The compact defines pollution as certain properties within or discharges into the water that adversely impact public health/safety or the beneficial uses of the water. In support of the principle of reducing pollution the States agree to cooperatively investigate and control interstate pollution

Category	Agreement title	Date	Parties	Basin(s)	Water quality reference	Reference excerpt/summary
						problems. Moreover, the two parties agree not to rely on the provision of water as a substitute for waste water treatment for the purposes of water quality control. The provisions of the Federal Water Pollution Control Act are to be used in the event that pollution problems cannot be resolved within the purview of the Compact. Finally, the Kansas–Oklahoma Arkansas River Commission, created by the Compact, is responsible for collecting, analyzing and reporting on water quality data.
60 Two	Delaware River Basin Compact	07/07/61	Delaware, New York, New Jersey	Delaware	Articles 1 (Section 1.3), 3 (Sections 3.1, 3.2, 3.6), 4 (Section 4.2a), 5, 10 (Section 10.5), 13 (Section 13.2), 14 (Section 14.2)	SUMMARY: One of the main objectives of the compact is to promote coordinated management of the Delaware Basin, which includes water quality management. The compact establishes a Commission with the broad powers to develop and coordinate plans, policies, and projects for the management, control, and conservation of the interstate waters; establish standards; conduct/sponsor research; compile data and publish reports concerning water quality. Article 5 specifically addresses the Commission's water pollution responsibilities and powers including those related to investigating water quality and establishing and enforcing standards. The parties agree to prohibit and control future pollution, to abate existing pollution, and to maintain the waters in satisfactory condition in accordance with compact terms, enacting appropriate legislation as needed.

(*continued*)

(*continued*)

	Category	Agreement title	Date	Parties	Basin(s)	Water quality reference	Reference excerpt/summary
61	Two	Klamath River Basin Compact	09/11/57	Oregon, California	Klamath	Article VII	SUMMARY: As part of the Compact, the two States agree to cooperate in terms of pollution abatement and control programs. Pollution is defined as a reduction in the quality of waters in one State that "materially and adversely affects beneficial uses" of the waters to the other State. Each State is required to abate and control interstate pollution problems in accordance with its own state laws. In addition, the Klamath Basin Compact Commission, created by the Compact, is tasked with the responsibilities of coordinating with state, local and federal agencies to promote effective laws and regulations for reducing and controlling pollution in the basin; recommending reasonable water quality standards; disseminating information to the public concerning water quality; and investigating and resolving conflicts concerning interstate pollution.
62	Two	Pecos River Compact	12/03/48	G746	Pecos	Article IV	Article IV (b):"New Mexico and Texas shall cooperate with agencies of the United States to devise and effectuate means of alleviating the salinity conditions of the Pecos River."
63	Two	New England Interstate Water Pollution Control Act	06/16/47	Connecticut, Maine Massachusetts, New Hampshire, Rhode Island, Vermont	Interstate inland and tidal waters	Entire Document	SUMMARY: The signatory States pledge to provide for the abatement of existing pollution, the control of future pollution and maintenance in a satisfactory condition of interstate inland and tidal waters. A Commission is established control water quality in shared waters. The duties of the Commission include the establishment of water quality

Category	Agreement title	Date	Parties	Basin(s)	Water quality reference	Reference excerpt/summary
						standards for various classes of use and the development and maintenance of a water quality sampling and testing network. The Commission is also tasked with investigating water quality compliance problems and enforcing its established water quality standards.

[*] *Notes:*

Category One: Agreements with the most detailed water quality provisions specifying standards, action plans, and/or comprehensive management frameworks.

Category Two: Agreements that defined water quality related actions but lacked specific standards or a comprehensive management framework.

Category Three: Agreements that simply outlined an indefinite commitment to some aspect of water quality management.

Source: "Managing the Quality of Transboundary Rivers: International Principles and Basin-level Practice." *Natural Resources Journal* 43 (1): 111–136.

G Treaties that delineate water allocations

Aaron T. Wolf

G.1 BOUNDARY WATERS AGREEMENTS

Main/subbasin(s)	Parties/date of treaty	Title of treaty	Method for water allocations[1]	Comments[1]
Boundary waters between Canada and United States	Great Britain (for Canada), United States 1/11/1910	Treaty between Great Britain and the United States relating to boundary waters and boundary questions	Existing uses protected; equal shares of benefits (not necessarily of water). Order of precedence for uses: domestic and sanitary; navigation; power and irrigation.	Niagara: No diversion above Falls; 20,000 cfs to United States and 36,000 cfs to Canada for hydropower. St. Mary and Milk: Both rivers treated as single unit, with overall equal apportionment to each party; Canada retains prior rights to minimum 500 cfs on St. Mary during irrigation season, United States does likewise on Milk.
Boundary waters between Mexico and the United States/Colorado, Tijuana, Rio Grande (Rio Bravo)	Mexico, United States 5/21/1906 2/3/1944	Utilization of waters of Colorado and Tijuana Rivers and of the Rio Grande (Rio Bravo)	Full rights to some tributaries, partial rights (by thirds) to others, half rights to main stem of boundary rivers. Minimum flows guaranteed to cross-boundary streams. Uses prioritized by: domestic, agriculture, electric power, other industry, navigation, fishing, other beneficial uses.	Rio Grande: 1906 treaty assures Mexico 60,000 acre-feet/year, mostly in summer, according to set schedule. 1944 treaty allocates full rights to some tributaries, partial rights (by thirds) to others, half rights to main stem. Any shortages due to drought can be made up in following cycle. Colorado: Mexico guaranteed minimum flow of 1,500,000 acre-feet/year. Tijuana: Commission agrees to study "equitable distribution." Allocations "are not to be construed as a recognition of any claims to said waters."
Colorado	Mexico, United States 8/24/1966	Exchange of notes constituting an agreement concerning the loan of waters of the Colorado River for irrigation of lands in the Mexicali Valley	United States "loans" water for irrigation to Mexico during one dry year in exchange for value of lost power generation.	United States provides 40,535 acre-feet above 1944 Treaty allocations during September and December 1966 (after an especially dry year), but retains an equal amount the following year (or over 3 years if low flow). Mexico pays market value for lost power generation at Hoover and Glen Canyon dams. Treaty explicitly mentions that no precedent is being set.

Colorado	Mexico, United States 9/30/1973	Mexico–United States-Agreement on the permanent and definitive solution to the salinity of the Colorado River (Minute #242)	Reaffirms 1944 agreement for 1,500,000 acre-feet/year to flow to Mexico, but describes salinity and quality of flow. Also restricts some groundwater pumping of shared aquifers.	
Boundary waters between Austria and Bavaria/Blaserbach, Dollmannbach, Durrach, Kesselbach (Danube)	Austria, Bavaria 10/16/1950	Agreement between the Austrian Federal Government and the Bavarian State Government concerning the diversion of water in the Rissbach, Durrach and Walchen Districts	Five tributaries to Isar divided: one allowed to flow freely to Bavaria, two can be freely developed by Austria, and two can be developed by Austria, provided minimum flows to Bavaria between August and March.	Austria is upstream on all these tributaries to Isar, but becomes a downstream riparian when Isar flows to Danube and back into Austria. Upstream/downstream relationships seem not so valid – each tributary divided uniquely, but all follow basin plan. Allocations can be modified if dams are built. "Notwithstanding this agreement," each maintains its "respective position regarding the legal principles of international waters."
Boundary waters between Austria and Czechoslovakia/Danube	Austria, Czechoslovakia 12/7/1967	Treaty between the Republic of Austria and the Czechoslovak Socialist Republic concerning the regulation of water management questions relating to frontier waters	"Existing water rights in respect of frontier waters and the obligations connected therewith shall remain unaffected;" all others to be worked out within States or through Commission.	
Boundary waters between Austria and Hungary	Austria, Hungary 4/9/1956	Treaty between the Hungarian People's Republic and the Republic of Austria concerning the regulation of water economy questions in the frontier region	Rights to use of one half of the natural (not enhanced by artificial means) flow to each party from rivers which flow along the boundary, "without prejudice to acquired rights;" upstream state of watercourses, which intersect boundary may not decrease flow by more than one third; no development without joint approval.	

(continued)

(*continued*)

Main/subbasin(s)	Parties/date of treaty	Title of treaty	Method for water allocations[1]	Comments[1]
Boundary waters between Czechoslovakia and Hungary/Danube, Tisza	Czechoslovakia, Hungary 4/16/1954	Agreement between the Czechoslovak Republic and the Hungarian People's Republic concerning the settlement of technical and economic questions relating to frontier watercourses	Each State has rights to half the natural (excluding artificially increased) discharge, "without prejudice to acquired rights," of frontier watercourses; no development which might affect discharge or the bed.	
Boundary waters between Iran and Iraq/Tigris	Iran, Iraq 12/26/1975	Agreement between Iran and Iraq concerning the use of frontier watercourses	Equal parts.	Flows of the Bnava Suta, Qurahtu, and Gangir rivers are divided equally. Flows of the Alvend, Kanjan Cham, Tib, and Duverij are divided based on a 1914 commission report on the Ottoman/Iranian border "and in accordance with custom."
Euphrates	Iraq, Kuwait 2/11/1964	Agreement between Iraq and Kuwait concerning the supply of Kuwait with fresh water	Iraq agrees to supply Kuwait with 120 million imperial gallons per day without compensation, and to discuss additional needs if necessary.	Water source is unspecified in the agreement.
Ganges	Bangladesh, India 11/5/1977 12/12/1996	Treaty between the Government of the Republic of India and the Government of the People's Republic of Bangladesh on sharing of the Ganga/Ganges waters at Farakka	Schedule is established for dry months – January 1 – May 31, which allocates the flow at Farakka: flow of 70,000 cusecs or less – 50% to India, 50% to Bangladesh; 70,000–75,000 cusecs – 35,000 cusecs to Bangladesh, rest to India; 75,000 cusecs or more – 40,000 cusecs to India, rest to Bangladesh.	1977 agreement was only to last for 5 years. Short-term agreements reached in 1982 and 1985; the latter lapsed in 1988. A final agreement was reached December 1996.

Basin	Parties/Date	Agreement	Allocation	Notes
Gash	Italy (Eritrea) and United Kingdom (Sudan) 6/12/1925 4/8/1951	Notes exchanged between the United Kingdom and Italy respecting the regulation of the utilisation of the waters of the River Gash; and 1951 amending letters	Eritrea can divert all water from a flow up to 5 m³/sec, about half the flow above 5 m³/sec, and a maximum of 17 m³/sec, or a total of 65 MCM/year. The rest flows to Sudan.	Sudan paid Eritrea a share of what was received for cultivation in the Gash Delta – 20% of any sales over £50,000 (payments discontinued with British control of Eritrea). One of few agreements which explicitly favors upstream riparian.
Ili/Horgos	China, Russia 6/12/1915	Protocol between China and Russia for the delimitation of the frontier along the River Horgos	Upper reaches: Prior rights for Chinese outpost; lower reaches: prior rights for existing canals, rest to be shared equally.	China "binds itself" to withdraw only the water necessary for one outpost in upper reaches (within Chinese territory), otherwise, water will go to existing canals with remainder to be shared equally.
Pasvik (Patsjoki)/ Pasvik (Patsjoki), Jakobselv (Vuoremajoki)	Finland, Norway 2/14/1925	Convention between the kingdom of Norway and the republic of Finland concerning the waters of the Pasvik (Patsjoki) and the Jakobselv (Vuoremajoki)	Equal shares of shared boundary waters, absolute sovereignty over tributaries, where both banks are within single territory.	Jakobselv (Vuoremajoki) and parts of Pasvik (Patsjoki) form boundary – the waters from these are divided equally. Absolute rights for tributaries of the Pasvik (Patsjoki) which have both banks in one state are retained by that state.
Rhine/Lake Constance	Austria, Germany, Switzerland 4/30/1966	Agreement regulating the withdrawal of water from Lake Constance	Requires notification and agreement for withdrawals over 750 l/sec within the catchment area, or 1,500 l/sec outside.	Must notify of withdrawals and "afford one another good time to express their views," and to submit to arbitration if disagreement. "Withdrawals ... shall not be deemed to justify any claim to the provision of water in a specific volume or of a specific quality."
Roya	Italy, France 10/14/1972	Franco–Italian convention concerning the supply of water to the Commune of Menton	Italy allows 400 l/sec withdrawal from alluvial aquifer for French town; Italian town can tap into delivery pipeline for 100 l/sec.	Italian government grants 70-year concession to Menton to be governed by Italian law on water-related issues. Menton deposits 10 million lire for security against concession.
West Bank and Gaza Aquifers	Israel, Palestine 9/28/1995	Israeli–Palestinian Interim Agreement	Population and consumption patterns – Israel recognizes Palestinian water rights, and agrees to provide 28.6 MCM/year additional water towards future Palestinian needs of 70–80 MCM/year.	Final allocations and rights to be determined in final status negotiations. Interim accord marks first time prior rights relinquished in an agreement, first joint management of aquifer systems, and first treaty which allows for future market mechanism, provided water is not subsidized.

(continued)

(continued)

Main/subbasin(s)	Parties/date of treaty	Title of treaty	Method for water allocations[1]	Comments[1]
Zarumilla	Ecuador, Peru 5/22/1944	Declaration and exchange of notes concerning the termination of the process of demarcation of the Peruvian-Ecuadorian frontier	Prior rights for Ecuadorian villages.	"Peru undertakes . . . to guarantee the supply of water necessary for the life of the Ecuadorian villages on the right bank of the so-called old bed of the river Zarumilla . . ." in conjunction with boundary delineation.

G.2 RIVER DEVELOPMENT AGREEMENTS

Main/subbasin(s)	Parties/date of treaty	Title of treaty	Method for water allocations[1]	Comments[1]
Araks, Atrak	Iran, USSR 8/11/1957	Agreement between Iran and the Soviet Union for the joint utilization of the frontier parts of the rivers Aras and Atrak for irrigation and power . . .	50% of all potential water and power resources on the shared portions of the two rivers.	Provides for "separate and independent division and transmission of water and power in each party's territory," along with joint data-gathering. Also, each party has rights to potential even ". . . if the activities of one of the parties . . . are slower than those of the other."
Boundary waters between Canada and United States/Columbia, Kootenai	Canada, United States 9/16/1964	Treaty relating to cooperative development of the water resources of the Columbia River Basin (with annexes)	Equal share of benefits – cooperative management for flood control and hydropower. Water may not be diverted out-of-basin (except for some specified in treaty), but power may (for compensation).	Equal share of benefits from power generation. United States pays Canada for benefits of flood control (payment can be in cash or in electric power) and, in 1964 Exchange of Notes, agrees to pay US$254,000,000 for entitlement. Canada granted diversions from Kootenai to Columbia and from Columbia to Kootenai, provided minimum flows are maintained.
Cunene	Portugal (Angola), South Africa (Southwest Africa) 7/2/1926	Agreement between the Government of the Union of South Africa and the Government of the Republic of Portugal regulating the use of the water of the Cunene River	Up to half of flood water may be diverted to Southwest Africa from above dam.	Dam to be constructed in Portuguese territory with shared cost. No charge for diversion if for subsistence, but payment would be made to Portuguese government if water used for "purposes of gain."

312

Basin	Agreement	Provisions	Notes
Cunene	Portugal (Angola), South Africa (Southwest Africa) 1/21/1969	Agreement between the Government of South Africa and the Government of Portugal in regard to the first phase of development of the water resources of the Cunene River Basin	Diversion solely for water for human and animal requirements in South West Africa and initial irrigation in Ovamboland, limited to one half of the flow or 6 m³/s.	"Humanitarian" part of larger project for hydropower. South Africa pays for water diversion and compensation to Portugal for land flooded as a result of dam (also royalties for hydropower generated).
Douro	Portugal, Spain 8/11/1927	Convention between Spain and Portugal to regulate the hydroelectric development of the international section of the River Douro	Roughly equal sections of the international stretch of the Douro are allocated to each for development. No diversions permitted, except "for reasons of public health," and only with joint agreement.	Separate, but equal and coordinated development.
Ganges/Bagmati, Gandak	India, Nepal 12/4/1959	Agreement between his majesty's government of Nepal and the government of India on the Gandak irrigation and power project	Diversions for project — irrigation and power generation — are laid out in a monthly schedule of water requirements, with about 60% to Nepal (5,760–16,060 cusecs) and 40% to India (3,690–14,600 cusecs). Nepal retains rights to irrigate with any water above these project requirements.	Broad "basket" of benefits to each side: land acquisition, power generation, capital resources (primarily from India), irrigation water, and transportation facilities.
Ganges/Kosi	India, Nepal 12/19/1966	Amended agreement between his Majesty's Government of Nepal and the Government of India concerning the Kosi project	Nepal retains right to divert upstream water, "as may be required from time to time." India has right to regulate balance.	Broad "basket" of benefits, including irrigation/hydropower project, navigation, fishing, and aforestation (India plants trees in Nepal to contain sedimentation).
Indus	India, Pakistan 5/4/1948	Interdominion agreement between the Government of India and the Government of Pakistan, on the canal water dispute between East and West Punjab	Rights are not determined, but India agrees, "without prejudice to its legal rights," to reduce flows of tributaries at a rate which would allow Pakistan to develop alternative sources.	India was to reduce flow from upper Indus basin rivers progressively, to allow Pakistan to "develop areas where water is scarce and which were under-developed in relation to Parts of West Punjab." Pakistan agreed to pay for some water sources.

(continued)

(continued)

Main/subbasin(s)	Parties/date of treaty	Title of treaty	Method for water allocations[1]	Comments[1]
Indus	India, Pakistan, World Bank 9/19/1960	The Indus waters treaty	River divided geographically: three eastern tributaries to India, three western tributaries to Pakistan.	Considerations were made for some withdrawals in other state's tributaries, in order of priority: domestic, nonconsumptive, agriculture, hydropower. Agreement was phased in and India paid for some Pakistani works deemed "replacement."
Jordan/Yarmuk	Jordan, Syria 6/4/1953	Agreement between the Republic of Syria and the Hashemite Kingdom of Jordan concerning the utilization of the Yarmuk waters	Dam would be built to guarantee 10 m^3/sec. minimum flow to Jordan, about seven eighths of natural flow of river. Syria relinquishes rights to tributaries between dam and 250m contour, receives 75% of hydropower.	Jordan was to cover 95% of costs, and provide 80% of workforce; Syria the remainder. Dam was never built, although plans were said to have been revived in August 1996.
Jordan	Israel, Jordan, Lebanon, Syria Finalized 1/1/1956, never ratified	Johnston Accord	Allocations of Jordan based on survey of irrigable land within basin: Israel, 31%; Jordan, 56%; Lebanon, 3%; Syria, 10%.	Allocations were based on irrigable land within basin; then each could do what it wished with water. Each tributary had one state without designated flow, to accommodate fluctuating supply. Accord was never ratified for political reasons.
Jordan/Yarmuk, shared aquifers	Israel, Jordan 10/26/1994	Treaty of peace between the State of Israel and the Hashemite Kingdom of Jordan	Allocations of Yarmuk and Jordan based on Johnston accord; agreed in conjunction with joint development projects. Water from shared aquifers allocated on basis of prior use.	"Rightful allocations" divide waters on the basis of historic rights plus future projects. Creative management: land and water historically used by Israel leased from Jordan; in absence of storage facility, Yarmuk water "loaned" to Israel in summer, returned to Jordan from Jordan River during winter.

314

Basin	Parties/Date	Agreement	Provisions
Mekong/Lower Mekong	Cambodia, Laos, Thailand, Vietnam 1/31/1975	Joint declaration of principles for utilization of the waters of the lower Mekong basin	"Equality of right" does not mean equal shares of water, but equal right to use water on basis of economic and social needs. Domestic and urban uses should have a preference; existing uses are protected. All parties must agree to any out-of-basin transfers. Groundwater with hydrologic connection to main stream is covered by agreement. Agreement based on 1957 establishment of Mekong Committee – renewed in 1995. Allocations are based, verbatim, on eleven parameters of 1966 Helsinki Rules definition of "reasonable and equitable shares" plus addition of benefit – cost ratio of each project.
Nile/Atbara Nile/Semliki, Isango	Great Britain, Italy – 1891, 1925 Great Britain, Ethiopia – 1902 Great Britain, Congo – 1906	Series of protocols, agreements, and exchanges of notes	Agreements required any upstream development be "in consultation" with Great Britain. 1925 exchange of notes offers British support for Italian concession for railway in Eritrea, Ethiopia, and Somaliland, and recognition of "exclusive character of Italian economic influence" in area to be covered by railway, in exchange for Great Britain gaining concession to build barrage at Lake Tana and, recognizing the "prior hydraulic rights of Egypt and the Sudan," an agreement by Italy not to modify the flow. "Prior hydraulic rights" – Great Britain made agreements with upstream riparians to allow Nile tributaries to flow uninterrupted to Sudan and Egypt. Water for "subsistence" of local populations may be used, and existing uses are protected.
Nile	Egypt, United Kingdom 5/7/1929	Exchange of notes between his majesty's Government in the United Kingdom and the Egyptian Government in regard to the use of the waters of the River Nile for irrigation purposes (Nile Waters Agreement)	Entirely protects existing, downstream uses – no irrigation or power works are to be built on the river which would reduce the quantity of water arriving in Egypt, modify the date of its arrival, or lower its level. If Egypt were to develop projects in Sudan to enhance flow, agreement would have to be reached beforehand with local authorities, although Egypt would retain direct control of such works. Prior rights – restricts amount Sudan may use in order to guarantee to Egypt the water needed for existing agriculture.

315

(continued)

(continued)

Main/subbasin(s)	Parties/date of treaty	Title of treaty	Method for water allocations[1]	Comments[1]
Nile	Egypt, Sudan 11/8/1959	Agreement between the Government of the United Arab Republic and the government of Sudan	Prior rights ("present acquired rights") for natural flow, plus benefits of Aswan Dam divided, based on population, on a ratio of 14.5 to Egypt, 7.5 to Sudan. Water from future projects, and the costs borne, would be divided equally.	If benefits of projects are greater than expected, they are to be divided equally. Egypt paid 15 million Egyptian pounds to Sudan for compensation for flooding and relocation from Aswan Dam; Sudan was to loan 1.5 BCM/year to Egypt until 1977. Both states agreed to develop joint position before negotiating with any other riparian.
Orange/Senqu	Lesotho, South Africa 11/7/1986	Treaty on the Lesotho Highlands Water Project between the Government of the kingdom of Lesotho and the Government of the republic of South Africa	Lesotho agrees to provide increasing water delivery to South Africa, from 57 MCM/ year in 1995 until 2,208 MCM/ year after 2020. Lesotho receives hydropower and capital payment from project.	A boycott of international aid for apartheid South Africa required that the project be financed, and managed, in sections. The water transfer component was entirely financed by South Africa, which would also make payments for the water that would be delivered. The hydropower and development components were undertaken by Lesotho, which received international aid from a variety of donor agencies, particularly the World Bank.

G.3 SINGLE PROJECT AGREEMENTS

Main/subbasin(s)	Parties/date of treaty	Title of treaty	Method for water allocations[1]	Comments[1]
Aden groundwater	Great Britain, Sultan of Abdali (Aden) 4/11/1910	Terms of a convention regarding the water supply of Aden between Great Britain and the Sultan of the Abdali	Great Britain buys groundwater from Sultan of the Abdali.	Sultan gives Great Britain land in perpetuity and guarantees safety of headworks. Great Britain agrees to pay 3,000 rupees/month if works unmolested; otherwise 15 rupees/100,000 gallons. Early groundwater agreement.

Ebro/Lake Lanoux, Font-Vive, Carol	France, Spain 7/12/1958 (revised 1/27/1970)	Agreement between the Government of the French Republic and the Spanish Government relating to Lake Lanoux	France diverts water out-of-basin, then tunnels same volume back before Carol reaches boundary; guarantees minimum 20 MCM flow timed for Spanish irrigation.	French hydropower project which moves water out-of-basin, then returns through tunnel before boundary. Arbitration for this project led to an important international precedent when a Tribunal ruled in 1957 that "territorial sovereignty . . . must bend before all international obligations," effectively negating the water rights doctrine of "absolute sovereignty," while admonishing downstream state from the right to veto "reasonable" upstream development, negating the "natural flow" principle.
Indus/Sirhind Canal	Great Britain, Patiala, Jind, Nabha 8/12/1903	Final working agreement relative to the Sirhind canal between Great Britain and Patiala, Jind and Nabha	Available supply, and development costs, divided by percentage: Patiala, 83.6; Nabha, 8.8; Jind, 7.6. British villages receive water sufficient to irrigate the same proportion of its lands as of other villages nearby.	If the flow allocations cannot be met, the engineer may reduce flows proportionally, or may deliver full proportion to one, then shut off entirely while the others receive their full allotments.
Näätämo/Näätämo, Gandvik	Finland, Norway 4/25/1951	Agreement between the Governments of Finland and Norway on the transfer from the course of the Näätämo (Neiden) River to the course of the Gandvik River . . .	Water diverted between basins for power generation in Norway, which agrees to compensate Finland for lost water power.	Fish habitat and timber transport are also described.
Niagara	Canada, United States 5/20/1941; 10/27/1941	Exchange of notes between the Government of the United States and the Government of Canada constituting an arrangement concerning temporary diversion for power purposes of additional waters of the Niagara river above the Falls	5,000 cfs additional diversion to the United States and 3,000 cfs to Canada agreed to for hydropower generation during war effort; raised an additional 7,500 cfs to the United States and 6,000 cfs to Canada in addendum.	Despite war effort, protecting the "scenic beauty of this great heritage of the two countries" is described as the primary obligation of the two countries.
Niagara	Canada, United States 2/27/1950	Treaty between the United States of America and Canada relating to the uses of the waters of the Niagara River	Equal amount of water for power generation, and equal share of cost, to each country. Minimum flow of river delineated	Benefits of tourism versus hydropower: 100,000 cfs minimum during "show times" at Falls – summer daylight hours; otherwise 50,000 cfs. "Primary obligation to preserve and enhance scenic beauty . . ."

317

---

[1] All units are reported as in original documents. One gallon = 3.61 liters; one acre-foot = 1,233 cubic meters; one cfs (cusecs) = 0.0283 cubic meters/second (cumecs).

Bibliography

Abu-Nimer, M. (1996). Conflict resolution in an Islamic context. *Peace and Change*, **21**(1), 22–40.

Adams, A. (2000). River Senegal: Flood management and the future of the valley. Drylands Issue Paper (E93), International Institute for Environment and Development.

Adams, R. M. (1974). The evolution of urban society, In *Man and water: Social sciences and management of water resources*, edited by L. D. James. Lexington: Lexington University Press, p. 43.

Africa News. (2004). Lesotho: Phase I of Highlands Water Project now fully operational. March 16.

Africa News. (2006). South Africa: Top Lesotho water boss charged. February 12.

Agrarian Research and Training Institute (ARTI). (1986). *Proceedings of a Workshop on Water Management in Sri Lanka*, 8th ed., edited by S. Abeyratne, P. Ganewatte, and D. J. Merrey, ARTI documentation series No. 10. Colombo, Sri Lanka: ARTI.

Agrawal, A. (2002). Common resources and institutional sustainability. In *The drama of the commons*, edited by E. Ostrom, T. Dietz, N. Dolšak, P. C. Stern, S. Stonich, and E. U. Weber. Washington, DC: National Academy Press, pp. 41–86.

Agrawal, A., and Gibson, C. C. (1999). Communities and natural resources: Beyond enchantment and disenchantment. *World Development*, **27**(4), 629–641.

Alam, U. (2002). Questioning the water wars rational: A case study of the Indus Waters Treaty. *Geographical Journal*, **168**(4), 354–364.

Alearts, G. J. (1999). The role of external support agencies. *River Basin Management: Proceedings of the UNESCO International Workshop*, The Hague, October 27–29, 1999, edited by E. Mostert, E. van Beek, N. W. M. Bouman, E. Hey, H. H. G. Savenije, and W. A. H. Thissen. IHP-V Technical Document in Hydrology No. 31, Paris: UNESCO.

Alearts, G. J. (2001). *Institutions for river basin management: The role of external support agencies in developing countries*. Washington, DC: World Bank.

Alearts, G., and LeMoigne, G. (editors). (2002). *Integrated water management at river basin level: An institutional development perspective*. Washington, DC: RFF for the World Bank.

Allan, J. A., (editor). (1996). *Water, peace and the Middle East: Negotiating resources in the Jordan basin*, 19th ed. London and New York: Tauris Academic Studies; St. Martin's [distributor].

Allan, J. A. (1998a). "Virtual water": An essential element in stabilizing the political economies of the Middle East. In *Transformations of Middle Eastern natural environments: Legacies and lessons*, edited by J. Albert, M. Bernhardsson, and R. Kenna. Bulletin Series No. 103, November (9). New Haven: Yale School of Forestry and Environmental Studies, pp. 141–149.

Allan, J. A. (1998b). *Water resources, prevention of violent conflict and the coherence of EU policies in the Horn of Africa*. Discussion paper, London: SOAS and Saferworld.

Allan, J. A. (2001). *The Middle East water question: hydro-politics and the global economy*. London: I B Tauris.

Allan, J. A. (2002). Hydro-peace in the Middle East: Why no water wars? A case study of the Jordan River basin. *SAIS Review*, **22**(2, Summer/Fall), 255–272.

Allee, D. J., and Abdalla, C. W. (1989). *Policy education to build local capacity to manage the risk of groundwater contamination*. A. E. Staff Report 89-29 (August), Development of Agricultural Economics, Cornell University, Ithaca, New York, pp. 2B.31–2B.40.

Allouche, J. (2005). Water nationalism: An explanation of the past and present conflicts in Central Asia, the Middle East, and the Indian Subcontinent? Unpublished Ph.D. dissertation, Universite de Geneve.

Ambroise-Rendu, Marc (1992). *Le Monde*, February 16–17.

American Academy of Arts and Sciences. (1994). Introduction. *Bulletin*, **48**(2, November). pp. 1–4.

American University Trade and Environment Database. (1999). *Hydrovia Canal plan and environment*. Available online at http://www.american.edu/TED/hidrovia.htm.

American University Trade and Environment Database (2004). *Itaipu Dam*. Available online at http://www.american.edu/TED/itaipu.htm.

Amery, H. A. (2001). Islamic water management. *Water International*, **26**(4, December), 481–489.

Amery, H. A. (2002). Water wars in the Middle East: A looming threat. *Geographical Journal*, **168**(4), 313–323.

Amery, H., and Wolf, A. (editors). (2000). *Water in the Middle East: A geography of peace*. Austin: University of Texas Press.

Amy, D. (1987). *The politics of environmental mediation*. New York: Columbia University Press.

Anand, P. B. (2004). *Water and identity: An analysis of the Cauvery River water dispute*. BCID Research Paper Number 3, Bradford, UK: Bradford Centre for International Development Series.

Anderson, E. W. (1994). *Hydropolitics, conflict analysis and management*. Paper for the International Water Resources Association's VIII Congress, Cairo, November 21–25, 1994.

Anderson, T. L., and Snyder, P. (1997). *Water markets: Priming the invisible pump*. Washington, DC: Cato Institute.

Arcadis Euroconsult. (2000). *Transboundary water management as an international public good*. Stockholm: Ministry for Foreign Affairs, Sweden. Available at http://www.fritzes.se.

Arnold, J. L. (1988). *The evolution of the 1936 Flood Control Act*. Fort Belvoir, VA: USACE Office of History.

Asad, M., Azevedo, L. G., Kemper, K. E., and Simpson, L. D. (1999). *Management of water resources: Bulk water pricing in Brazil*. World Bank Technical Paper No. 432.

Ashton, P. J. (2000). Southern African water conflicts: Are they inevitable or preventable? In *Water wars: Enduring myth or impending reality*, edited by H. Solomon and A. R. Turton. Africa Dialogue Monograph Series No. 2. Durban, South Africa: The African Centre for the Constructive Resolution of Disputes (ACCORD), pp. 62–105.

Ashton, P. J., and Turton, A. R. (2006). Water and security in Sub-Saharan Africa: Emerging concepts and their implications for effective water resource management in the Southern African region. In *Globalisation and environmental challenges*, edited by H. G. Brauch, J. Grin, C. Mesjasz, N. C. Behera, B. Chourou, U. O. Spring, P. H. Liotta, and P. Kameira-Mbote. Berlin: Springer Verlag.

Ashworth, A. E. (1968). The sociology of trench warfare 1914–18. *British Journal of Sociology*, **19**(4, December), 407–423.

319

Ashworth, T. (1980). *Trench warfare, 1914–1918: The live and let live system.* New York: Holmes & Meier.

Aslov, S. M. (2003). *IFAS Initiatives in the Aral Sea Basin.* 3rd World Water Forum, Kyoto; March 16–23, 2003.

Attia, H. (1985). Water-sharing rights in the Jerid Oases of Tunisia. In *Property, social structure and law in the modern Middle East,* edited by A. E. Mayer. Albany, NY: State University of New York Press, pp. 85–106.

Australian Bureau of Statistics. (2006). *Water Access Entitlements, Allocations and Trading, 2004–05.* Open Document 4610.0.55.003. Accessed at http://www.abs.gov.au/AUSSTATS/abs@.nsf/ProductsbyTopic/E2D1678343AFE4BDCA257205002428DD?OpenDocument.

Australian National Water Commission. (2006). *National water initiative.* Accessed at http://www.nwc.gov.au/nwi/index.cfm#pricing.

Avis, C. (2003). *Danube River basin strategy for public participation in river basin management planning 2003–2009.* May 12, 2003, WWF, funded by UNDP/GEF under the Danube Regional Project.

Avruch, K. (1998). *Culture and conflict resolution.* Washington, DC: U.S. Institute of Peace.

Axelrod, R. (1984). *The evolution of cooperation.* New York: Basic Books.

Azar, E. E. (1980). The conflict and peace data bank (COPDAB) project. *Journal of Conflict Resolution,* **24**(1), 143–152.

Azevedo, L. G. (1997). Brazil. In *Water pricing experiences: An international perspective.* World Bank Technical Paper No. 386. Washington DC.

Bahr, J. (1988). Personal communication.

Baillat, A. (2004). *Power asymmetries along international watercourses: Hydropolitics in the Himalayan kingdoms.* Fifth Pan-European Conference Standing Group on International Relations, The Hague; September 9–11, 2004.

Baillat, A. (2005). *Hydropolitics of international water transfers: The challenges of water resources commodification.* A paper prepared for the ISA Conference, March 1–5.

Bandyopadhayay, J. (2002). Water management in the Ganges–Brahmaputra Basin: Emerging challenges for the 21st century. In *Conflict Management of Water Resources,* edited by M. Chatterji, S. Arlosoroff, and G. Guha. Burlington, VT: Ashgate, pp. 179–218.

Barber, B. (1985). *Strong democracy: Participatory politics for a new age.* Berkeley, CA: University of California Press.

Barraque, B. (2000). Participatory Processes in Water Management (PPWM). *Proceedings of the Satellite Conference to the World Conference on Science: International Conference on Participatory Processes in Water Management,* June 23–30, 1999, Budapest, Hungary, edited by József Gayer. UNESCO-IHP/IA2P/IWRA. Paris: UNESCO. IHP-V.

Bascheck, B., and Hegglin, M. (2004). *Plata/Parana River basin, a case study.* Swiss Federal Insititute for Environmental Science and Technology.

Baviskar, A. (1995). *In the belly of the river: Tribal conflicts over development in the Narmada Valley.* Oxford: Oxford University Press.

Biçak, H. A., and Jenkins, G. P. (1999). *Costs and pricing policies related to transporting water by tanker from Turkey to North Cyprus.* Development Discussion Paper No. 689. Cambridge, MA: Harvard Institute for International Development.

Beaumont, P. (1991). Transboundary water disputes in the Middle East. Paper presented at a conference on transboundary waters in the Middle East, Ankara.

Beaumont, P. (2000). Conflict, coexistence, and cooperation: A study of water use in the Jordan basin. In *Water in the Middle East: A geography of peace,* edited by H. Amery, and A. Wolf. Austin: University of Texas Press.

Beecher, J. A. (2000). Privatization, monopoly, and structured competition in the water industry: Is there a role for regulation? *Water Resources Update,* **117**, 13–20.

Bennett, J. W. (1974). Anthropological contribution to the cultural ecology and management of water resources. In *Man and water: Social sciences and management of water resources,* edited by L. D. James. Lexington: Lexington University Press, p. 67.

Bennett, L. L., and Howe, C. W. (1998). The interstate river compact: Incentives for noncompliance. *Water Resources Research,* **34**(3), 485–495.

Bennett, L., Ragland, S., and Yolles, P. (1998). Facilitating international agreements through an interconnected game approach: The case of river basins.

In *Conflict and cooperation on trans-boundary water resources,* edited by R. Just and S. Netanyahu. The Netherlands: Kluwer Academic Publishers.

Ben-Shachar H. (1989). Introduction In *Economic Cooperation in the Middle East,* edited by G. Fishelson. Boulder: Westview Press.

Bercovitch, J. (1986). International mediation: A study of incidence, strategies, and conditions of successful outcomes. *Cooperation and Conflict,* **21**, 155–168.

Betlem, I. (1995). *River basin planning and management.* Erowater Horizontal Report, Research Report No. 5. RBA Delft University of Technology.

Bhatnagar, B. (1992). Participatory development in the World Bank: Opportunities and concerns. World Bank Workshop on Participatory Development, Washington, DC.

Bingham, G. (1986). *Resolving environmental disputes: A decade of experience.* Washington, DC: Conservation Foundation.

Bingham, G., and Orenstein, S. G. (1989). The role of negotiation in managing water conflicts. In *Managing water related conflicts: The engineer's role,* edited by W. Viessman, Jr. and E. T. Smerdon. New York: American Society of Civil Engineers, pp 38–53.

Bingham, G., Wolf, A., and Wohlgenant, T. (1994). *Resolving water disputes: Conflict and cooperation in the U.S., Asia, and the Near East.* Washington, DC: U.S. Agency for International Development.

Biswas, A. K. (1970). *History of hydrology.* Amsterdam: North-Holland Publishing.

Biswas, A. (1992). Indus Water Treaty: The negotiating process. *Water International,* **17**(44), 201–209.

Biswas, A. K. (1993). Management of international waters: Problems and perspective. *Water Resources Development,* **9**(2), 161–188.

Biswas, A. K. (1995). World Water Council highlight editorial. *Water Resources Development,* **11**(2), 101–102.

Biswas, A. K. (1999). Management of international waters. *Water Resources Development,* **15**, 429–441.

Biswas, A. K., and Hashimoto, T. (editors). (1996). *Asian international waters: From Ganges–Brahmaputra to Mekong.* Oxford: Oxford University Press.

Biswas, A. K., and Tortajada, C. (editors). (2005). *Water pricing and public-private partnership.* New York: Routledge.

Biswas, A. K., and Tortajada, C. (editors). (2006). *Impacts of mega-conferences on the water sector.* New York: Springer.

Blackmore, D. (2002). Reforming the water sector, what Australia has done. In *Integrated water management at river basin level: An Institutional Development Perspective,* edited by G. Alearts and G. LeMoigne. Washington, DC: RFF for the World Bank.

Blaikie, P., Cannon, T., Davies, I., and Wisner, B. (1994). *At risk: natural hazards, peoples' vulnerability and disasters.* New York: Routledge.

Blanche, E. (2001). Turkey's water sales to Israel signal new tension over precious resource. *Daily Star.* October 2001, Daily Star archives.

Blatter, J., and Ingram, H. (editors). (2001). *Reflections on water: New approaches to transboundary conflicts and cooperation.* Cambridge, MA: MIT Press.

Bleed, A. S. (1990). Platte River conflict resolution. In *Managing water-related conflicts: The engineer's role,* edited by W. Viessman, Jr. and E. T. Smerdon. New York: American Society of Civil Engineers, pp. 131–140.

Blomquist, W., Heikkila, T., and Schlager, E. (2004). Building the agenda for instituional research in water resource management. *Journal of the American Water Resources Association,* **40**(4), 925–936.

BNamericas.com. (2005). Brazil: Bill to establish a national environmental sanitation policy submitted. May 31. Accessed at http://www.irc.nl/page/23948

Bolte, J. P., Hulse, D. W., Gregory, S. V., and Smith, C. (2004). Modeling biocomplexity – Actors, landscapes and alternative futures. Keynote speech. *Proceedings from the International Environmental Modelling and Software Society, Conference on Complexity and Integrated Resources Management,* University of Osnabrück, Germany, 14–17 June 2004.

Booth, William. (1991). Did Maya Tap Water for Power? *Washington Post.* February 18, 1991.

Brewster M. R., and Buros O. K. (1985). Non-conventional water resources: Economics and experiences in developing countries (I). *Natural Resources Forum,* **9**(1), 65–75.

Brittain, R. (1958). *Rivers, man and myths: From fish spears to water mills.* Garden City, NY: Doubleday & Company, Inc.

British Broadcasting Corporation (BBC). (2005). Thousands flee Kenyan water clash. BBC News Africa, January 24, 2005. Available at http://news.bbc.co.uk/2/hi/africa/4201483.stm.

Brochmann, M. (2006). *Cooperation in international river basins. An empirical analysis of the impact of sharing a river basin on cooperation.* Paper presented at the 14th Norwegian National Conference in Political Science, Bergen, 4–6 January 2006.

Bruch, C., Jansky, L., Nakayama, M., and Salewicz, K. A. (editors). (2005). *Public participation in the governance of international freshwater resources.* Tokyo: United Nations University Press.

Buck, S. J., Gleason, G. W., and Jofuku, M. S. (1993). The institutional imperative: Resolving transboundary water conflict in arid agricultural regions of the United States and the commonwealth of independent states. *Natural Resources Journal*, **33**, 595–628.

Bulkley, J. W. (1995). Integrated watershed management: Past, present, and future. *Water Resources Update*, **100**, 7–18.

Bulloch, J., and Darwish, A. (1993). *Water wars: Coming conflicts in the Middle East.* London: St. Edmundsbury Press.

Burchi, S., and Mechlem, K. (2005). *Groundwater in international law: Compilation of treaties and other legal instruments.* Rome: Food and Agriculture Organization of the United Nations (FAO), FAO Legislative Study No. 86.

Burchi, S., and Spreij, M. (2003). *Institutions for International Freshwater Management.* Report for the FAO Development Law Service. SC-2003/WS/41 Paris: FAO Legal Office UNESCO/IHP/WWAP. Available at http://webworld.unesco.org/water/wwap/pccp/cd/pdf/legal_tools/institutions_for_int_freshwater_management_2.pdf.

Butts, K. H. (1997). The strategic importance of water. *Parameters*, **27**, 65–83.

Cady, F., and Soden, D. L. (2001). The legal-institutional analysis model and water policymaking in a bi-national setting. *Journal of the American Water Resources Association*, **37**(1), 47–56.

Cai, X., and Rosegrant, M. W. (2002). Global water demand and supply projections, Part I. A modeling approach. *Water International*, **27**(2), 159–169.

Cano, G. (1985). The "Del Plata" basin: Summary chronicle of its development process and related conflicts. In *Management of international river basin conflicts*, edited by E. Vlachos, A. Webb, and I. Murphy. Proceedings of a Workshop held at the Institute for Applied Systems Analysis, Laxenberg, Austria, September 22–25, 1986.

Cano, G. (1989). The development of the law in international water resources and the work of the International Law Commission. *Water International*, **14**, 167–171.

Caponera, D. (1985). Patterns of cooperation in international water law: Principles and institutions. *Natural Resources Journal*, **25**(3), 563–587.

Caponera, D. (1987). International Water Resources Law in the Indus Basin. In *Water resources policy for Asia*, edited by M. Ali. Boston: Balkema, 509–515.

Caponera, D. A., (1995). Shared waters and international law. In *The peaceful management of transboundary resources*, edited by Gerald H. Blake et al., 121–126.

Carius, A., Dabelko, G. D., and Wolf, A. T. (2004). *Water, conflict, and cooperation.* Policy briefing paper for the United Nations and Global Security Initiative of the United Nations Foundation. Available at http://www.un-globalsecurity.org/pdf/Carius_Dabelko_Wolf.pdf.

Carius, A., Feil, M., and Taenzler, D. (2003). *Addressing environmental risks in Central Asia: Risks, policies, capacities.* Report for the Organization for Security and Cooperation in Europe, the United Nations Development Programme, and the United Nations Environment Programme, Bratislava: OSCE, UNDP, and UNEP.

Carmo Vaz, A. (1999). Problems in the management of international river basins – the case of the Incomati. *River basin management: Proceedings of the UNESCO international workshop*, The Hague, 27–29 October 1999, edited by E. Mostert. IHP-V Technical Document in Hydrology No. 31.

Carmo Vaz, A., and Lopez P. (2000), The Incompati and Limpopo, international river basins: A view from downstream. In H. H. G. Savenije, P. van der Zaag, and A. T. Wolf (editors), Management of shared river basins. *Water Policy*, **2**(2), 99–112.

Catão, L. A. V., and Solomou, S. N. (2005). Effective exchange rates and the classical gold standard adjustment, *American Economic Review*, **4**(95), 1259–1275.

Cernea, M. M. (1992). *Putting people first: Sociological variables in rural development.* Washington, DC: World Bank.

Chakraborty, R. (2004). Sharing of river waters among India and its neighbors in the 21st century: War or peace? *Water International*, **29**(2), 201–208.

Chatterji, M., Arlosoroff, S., and Guha, G. (editors). (2002). *Conflict management of water resources.* Burlington, VT: Ashgate.

Chenoweth, J. L., and Feitelson, E. (2001). Analysis of factors influencing data and information exchange in international river basins: Can such exchanges be used to build confidence in cooperative management? *Water International*, **26**(4), 499–512.

Chinatown (1974). Paramount Pictures. Directed by Roman Polanski. Written by Robert Towne. Performed by Jack Nicholson, Faye Dunaway, and John Huston.

CIDA. (1999). Public Participation in Environmental Impact Assessments in Developing Countries: Index of Useful Resources. Available online at http://www.acdi-cida.gc.ca.

Clark, E. H., Bingham, G., and Orenstein, S. G. (1991). Resolving water disputes: Obstacles and opportunities. *Resolve*, **23**(1), 1–10.

Clark, G. (1952). Ecological zones and economic stages. In *Prehistoric Europe: The economic basis*. London: Metheun. In *Man and water: Social sciences and management of water resources*, edited by L. D. James (1974). Lexington: Lexington University Press, p. 47.

Clean Air Act (CAA). (1970). 42 U.S.C. s/s 7401 et seq. Available at http://www.epa.gov/airprogm/oar/caa/index.html.

Clean Water Act (CWA). (1977). 33 U.S.C. ss/1251 et seq. Available at http://www.access.gpo.gov/uscode/title33/chapter26_.html.

Cohen, R. (1993). An advocate's view. In *Culture and negotiation: The resolution of water disputes*, edited by G. O. Faure, and J. Z. Rubin. London: Sage Publications.

Comtex News Network. (2004). Lesotho "Will continue to fight graft." September 10.

Conca, K. (2006). *Governing water: Contentious transnational politics and global institution building.* Cambridge, MA: MIT Press.

Conca, K., and Dabelko, G. D. (editors). (2002). *Environmental peacemaking.* Washington, DC: Woodrow Wilson Press.

Conca, K., Wu, F., and Mei, C. (2006). Global regime formation or complex institution building? The principled content of international river agreements. *International Studies Quarterly*, **50**, 263.

Conley, A. H., and van Niekerk, P. H. (2000). Sustainable management of international waters: The Orange River case. *Water Policy*, **2**, 131–149.

Cooley, J. K. (1984). The war over water. *Foreign Policy*, **54**, 3–26.

Correlates of War (COW) project. (2006). University Park, PA: Pennsylvania State University. http://cow2.la.psu.edu.

Corrigan, E. P., Jr. (1994). Oil for water. *National Geographic*, **10**(1), 9.

Coser, L. (1959). *The functions of social conflict.* New York: Free Press.

Crawford, S. (1988). *Mayordomo: Chronicle of an acequia in northern New Mexico.* Albuquerque: University of New Mexico Press.

Creighton, J., Delli Priscoli, J., and Dunning, M. (1983). *Public involvement techniques: A Reader of ten years experiences at the Institute for Water Resources.* U.S. Army Corps of Engineers IWR Research Report 87-R-1. Ft. Belvoir, VA: Institute for Water Resources.

Creighton, J. (1998a). Use of values in public participation in the planning process. In *Public involvement: A reader of ten years experience at the Institute for Water Resources*, edited by J. Creighton, J. Delli Priscoli, and C. M. Dunning. U.S. Army Corps of Engineers IWR Research Report 82-R-1. Alexandria, VA: Institute for Water Resources.

Creighton, J. (1998b). A thought process for designing public involvement. In *Public involvement: A reader of ten years experience at the Institute for Water Resources*, edited by J. Creighton, J. Delli Priscoli, and C. M. Dunning. U.S. Army Corps of Engineers IWR Research Report 82-R-1. Alexandria, VA: Institute for Water Resources, p. 131.

Creighton, J. (1999). Tools and techniques for effective public participation in water resources decisions. *Participatory Processes in Water Management (PPWM): Proceedings of the Satellite Conference to the World Conference on Science: International Conference on Participatory Processes in Water Management*, June 28–30,1999, Budapest, Hungary, edited by Gayer. UNESCO-IHP/IA2P/IWRA. Paris: UNESCO. IHP-V.

Creighton, J., and Delli Priscoli, J. (2004). *Public involvement and teaming in planning: A training manual*. Ft. Belvoir, VA: U.S. Army Corps of Engineers Institute for Water Resources.

Da Rosa, J. E. (1983). Economics, politics, and hydroelectric power: The Parana River Basin. *Latin American Research Review*, **68**(3), 77–107.

Daniels, S. E., and Walker, G. B. (2001). *The Collaborative Learning Approach*. Westport, CT: Praeger.

DANUBE International Commission for the protection of the Danube River. Available online at www.icpdr.org.

Davidge, Ric. (1994). *Water exports*. Newsletter of WaterBank.com. Accessed at http://www.waterbank.com/Newsletters/nws12.html.

Davis, U., Maks, A., and Richardson, J. (1980). Israel's water policies. *Journal of Palestine Studies*, **9**(2), 3–31.

Deason, J. P., Schad, T. M., and Sherk, G. W. (2001). Water policy in the United States: A perspective. *Water Policy*, **3**, 175–192.

Delaware River Basin Commission (2006). (Updated 19 January 2007). Available at http://www.state.nj.us/drbc/.

Dellapenna, J. (1994). Treaties as instruments for managing internationally-shared water resources: Restricted sovereignty vs. community of property. *Case-Western Reserve Journal of International Law*, **26**, 27–56.

Dellapenna, J. (1995). Building international water management institutions: The role of treaties and other legal arrangements. In *Water in the Middle East: Legal, political, and commercial implications*, edited by J. A. Allan and C. Mallat. London and New York: Tauris Academic Studies, pp. 55–89.

Dellapena, J. (1997). Personal communication.

Dellapenna, J. W. (2001). The customary international law of transboundary fresh waters. *International Journal of Global Environmental Issues*, **1**(3/4), 264–305.

Delli Priscoli, J. (1975). Citizen Advisory Groups and Conflict Resolution in Regional Water Resources Planning.

Delli Priscoli, J. (1976). Public participation in regional intergovernmental water resources planning: Conceptual frameworks and comparative case studies, Unpublished Ph.D. dissertation, Georgetown University, Washington, DC.

Delli Priscoli, J. (1983). Retraining the modern civil engineer. *The Environmentalist*, **3**, 137–146.

Delli Priscoli, J. (1988). Conflict resolution in water resources: Two 404 general permits. *Journal of Water Resources Planning and Management*, **114**(1), 66–77.

Delli Priscoli, J. (1989). Public involvement, conflict management: Means to EQ and social objectives. *Journal of Water Resource Planning and Management*, **115**(1), 31–42.

Delli Priscoli, J. (1990). *Public involvement, conflict management and dispute resolution in water resources and environmental decisions making*. Institute for Water Resources, USACE, IWR Working Paper 90 ADR WP 2.

Delli Priscoli, J. (1994). *Conflict resolution, collaboration and management in the international and regional water resources issues*. 7th Congress of the International Water Resources Association (IWRA), Cairo, Egypt, November 1994.

Delli Priscoli, J. (1996). *Conflict resolution, collaboration and management in international water resources*. Institute of Water Resources (IWR) and U.S. Army Corps of Engineers (USACE), Working Paper 96-ADR-WP-6. Alexandria, VA: IWR/USACE.

Delli Priscoli, J. (1998a). Chapters 2 and 3. In *Strengthening the capacity of the United Nations in prevention and resolution of international environmental conflicts*, edited by J. M. Trolldalen and T. Wennesland. Oslo: CESAR.

Delli Priscoli, J. (1998b). Water and civilization: Using history to reframe water policy debates and to build a new ecological realism, *Water Policy*, **1**(6), 623–636.

Delli Priscoli, J. (1999/2000). Foreword, in Management of Shared River Basins, ed. Savenije, van der Zaag, and Wolf. *Water Policy*, **2**, 1–2.

Delli Priscoli, J. (2000a). *Water security, interdependency, dependence and vulnerability*. Stockholm 2000 Symposium, Institute for Water Resources, U.S. Army Corps of Engineers, August 2000, Stockholm.

Delli Priscoli, J. (2000b). What is public participation? *Proceedings of the Satellite Conference to the World Conference on Science: International Conference on Participatory Processes in Water Management*. June 28–30, 1999. Budapest, Hungary, edited by J. Gayer. UNESCO-IHP/IA2P/IWRA Technical Documents in Hydrology No. 30. Paris: UNESCO-IHP-V.

Delli Priscoli, J. (2001a). *Participation, river basin organizations and flood management*. ESCAP, Bangkok, October 2001.

Delli Priscoli, J. (2001b). *River basins organizations*. ESCAP.

Delli Priscoli, J. (2004). *River basin organizations, RBO: U.S. experiences*. World Bank Training on building RBOs, Abuja, Nigeria, June 2004.

Delli Priscoli, J. (2005a). *Overview: Global water policy and institutional trends*. Keynote lecture at water conference for CEOs of water utilities and political leaders in Australia, August 2005. Sydney: BOG World Water Council, Institute for Water Resources USACE.

Delli Priscoli, J. (2005b). *The Columbia-Snake River – Nch'i-Wana; Water Infrastructure and Development*, Presentation during Stockholm Water Week, Stockholm, Sweden, August 25, 2005.

Delli Priscoli, J., and Creighton, J. (2004). *Second Ten Years*. Alexandria, VA: IWR.

Delli Priscoli, J., and Hassan, F. (1997). *Water and Civilization*. UNESCO IHP. Paris.

Delli Priscoli, J., and Moore, C. (1985). *Executive training course in conflict management*. Institute for Water Resources, USACE, Alexandria, VA.

Delli Priscoli, J., and Moore, C. (1988). *ADR training*. Alexandria, VA: USACE.

Delli Priscoli, J., and Montville, J. (1994). *Report on preventing Slovak–Hungarian ethnic violence*. Washington, DC: Center for Strategic and International Studies.

Dillman, J. (1989). Water rights in the occupied territories. *Journal of Palestine Studies*, **19**(1), 46–71.

Dinar, A., and Wolf, A. T. (1994a). Economic potential and political considerations of regional water trade: The western Middle East example. *Resources and Energy Economics*, **16**(4, Winter), 335–356.

Dinar, A., and Wolf, A. T. (1994b). International markets for water and the potential for regional cooperation: Economic and political perspectives in the western Middle East. *Economic Development and Cultural Change*, **43**(1, October), 43–66.

Dinar, A. (1998). Water policy reforms, information needs and implementation obstacles. *Water Policy*, **1**(4), 367–382.

Dinar, A. (editor). (2000). *The political economy of water pricing reforms*. Oxford: Oxford University Press.

Dinar, A., and Subramanian, A. (editors). (1997). *Water pricing experiences: An international perspective*. World Bank Technical Paper No. 386. Washington, DC: World Bank.

Dinar, S. (2004). Treaty principles and patterns: Negotiations over international rivers. Unpublished Ph.D. dissertation, Johns Hopkins, Baltimore, MD.

Dinar, S., and Dinar, A. (2003). Recent developments in the literature on conflict, negotiation and cooperation over shared international freshwater. *Natural Resources Journal*, **43**(4, Fall), 1217–1286.

Diplas, P. (2002). Integrated decision making for watershed management: Introduction. *Journal of the American Water Resources Association*, **38**(2), 337–340.

Donahue, J. M., and Johnston, B. R. (editors). (1998). *Water, culture, and power: Local struggles in a global context*. Washington, DC: Island Press.

Dryzek, J., and Hunter, S. (1987). Environmental mediation for international problems. *International Studies Quarterly*, **31**, 87–102.

Dukhovny, V., and Ruziev, U. (1999). Aral Sea. *River Basin Management: Proceedings of the UNESCO International Workshop*, The Hague, October 27–29, 1999, edited by E. Mostert. IHP-V Technical Document in Hydrology No. 31.

Dukhovny, V., and Sokolov, V. (2002). International arrangement to manage the Aral Sea Basin. In *Integrated water management at river basin level: An institutional development perspective*, edited by G. Alearts and G. LeMoigne. Washington, DC: RFF for the World Bank.

Dukhovny, V., and Sokolov, V. (2003). *Lessons on cooperation building to manage water conflicts in the Aral Sea basin*. Paris: UNESCO IHP Technical Documents in Hydrology, PCCP Series No. 11.

Dworsky, L., and Allee, D. (1997). *A critique of the Great Lakes Water Quality Agreement on its 25th anniversary and a discussion of the Great Lakes basin ecosystem as a management tool.* Report from a Seminar; Cornell University. Available at http://www.on.ec.gc.ca/glwqa/critique-e.html.

Eaton, J., and Eaton, D. (1994). Water utilisation in the Yarmouk–Jordan, 1192–1992. In *Water and peace in the Middle East,* edited J. Isaac and H. Shuval. Amsterdam: Elsevier Publishers, pp. 93–106.

EC Europe. (2007). Bulk water (SR11378). Accessed at http://www.eceurope.com/showrooms/details.htm?itemID = 11378&session. Accessed January 13, 2007.

Eckstein, G. E. (2004) Protecting a hidden treasure: The UN International Law Commission and the international law of transboundary ground water resources. *Sustainable Development Law and Policy,* **1,** 6–12.

Eckstein, G., and Eckstein, Y. (2003). A hydrogeological approach to transboundary ground water resources and international law. *American University International Law Review,* **19,** 201–258.

Eckstein, O. (1958). *Water resources development: The economics of project evaluation.* Cambridge, UK: Cambridge University Press.

Eckstein, O., and Krutilla, J. (1958). *Multipurpose river development.* Baltimore: Johns Hopkins Press.

Eckstein, Y., and Eckstein, G. E. (2005). Transboundary aquifers: Conceptual models for development of international law. *Journal of Groundwater,* **43**(5), 679–690.

ECLAC. (1997). *Network for cooperation in integrated water resource management for sustainable development in Lain America and the Caribbean.* ECLAC, Circular No. 6, December 1997.

Economic Commission for Europe. (1999a). Draft guidelines on public participation in water management. MP. WAT/2000/6, add.1, December 20, 1999.

Economic Commission for Europe. (1999b). The need for a strategy and framework for compliance with agreements on transboundary waters and guidelines on public participation in water management. MP. WAT/2000/4, December 21, 1999.

Economic Commission for Europe. (2000). Geneva Strategy and Framework for Monitoring Compliance with Agreements on Transboundary Waters, MP.WAT/2000/5/add. 1, February 8, 2000.

Economic Commission for Latin America and the Caribbean. (1997). *Network for Cooperation in Integrated Water Resources Management for Sustainable Development in Latin America and the Caribbean,* ECLAC, Circular (6), December.

Elhance, A. P. (1999). *Hydropolitics in the 3rd world: Conflict and cooperation in international river basins.* Washington, DC: United States Institute of Peace.

Elhance, A. P. (2000). Hydropolitics in the third world: Conflict and cooperation in international river basins. *International Negotiation,* **5** (2, February).

Endangered Species Act (ESA). (1973). 7 U.S.C. 136;16 U.S.C. 460 et seq. Available at http://www.access.gpo.gov/uscode/title16/chapter35_.html

Environment Australia. (2002). *Australia's Water Reforms.* Available at http://www.deh.gov.au/commitments/wssd/publications/water.html#download.

Environmental Programme for the Danube Basin. (1992–1994). Reports dated August 1992–June 1994.

ESCAP. (1999a). ESCAP-UNDP guidelines for participatory planning of rural infrastructure. New York: United Nations.

ESCAP. (1999b). Regional cooperation in the twenty first century on flood control and management in Asia and the Pacific. New York: United Nations.

European Commission. (2002). *Water is life: Water framework directive.* Informational summary leaflet. Available at http://ec.europa.eu/environment/water/water-framework/pdf/leaflet_en.pdf.

European Parliament and the Council of the European Union. (2000). Directive 2000/60/EC of 23 October 2000 establishing a framework for Community action in the field of water policy.

Executive Action Team (EXACT) Multilateral Working Group on Water Resources. (2006). ExactFactsheet.pdf. Available at http://www.exact-me.org

Export Bureau. (2007). Water Exports NZ Limited. Accessed at http://www.exportbureau.com/company_report.html?code=28672&name=water_exports_nz_limited. January 13.

Faigman, D. L. (1999). *Legal alchemy – The use and misuse of science in the law.* New York: W. H. Freeman and Co.

Falkenmark, M. (1986). Fresh waters as a factor in strategic policy and action. In *Global resources and international conflict: Environmental factors in strategic policy and action,* edited by A. H. Westing. New York: Oxford University Press, pp. 85–113.

Falkenmark, M. (1989). Middle East hydropolitics: Water scarcity and conflicts in the Middle East. *Ambio,* **18**(6), 350–352.

Falkenmark, M., Lundquist, J., and Widstrand, C. (1989). Macro-scale water scarcity requires micro-scale approaches: Aspects of vulnerability in semi-arid development. *Natural Resources Forum,* **14,** 258–267.

Faure, G. O., and Rubin, J. Z. (1993). *Culture and negotiation: The resolution of water disputes.* London: Sage Publications.

Feehan, J. (2001). *Export of bulk water from Newfoundland and Labrador: A preliminary assessment of economic feasibility.* A report prepared for the Government of Newfoundland and Labrador.

Feitelson, E. (2006). Impediments to the management of shared aquifers: A political economy perspective. *Hydrogeology Journal,* **14,** 319–329.

Feitelson, E., and Chenoweth, J. (2002). Water poverty: Towards a meaningful indicator. *Water Policy,* **4,** 263–281.

Feitelson, E., and Haddad, M. (editors). (1995). *Joint management of shared aquifers.* Harry S. Truman Research Institute, Hebrew University and the Palestinian Consultancy Group, Jerusalem.

Feitelson, E., and Haddad, M. (1998). *Identification of joint management structures for shared aquifers.* Technical Paper No. 415. Washington, DC: World Bank.

Feitelson, E., and Haddad, M. (editors). (2000). *Management of shared groundwater resources: The Israeli–Palestinian case with an international perspective.* Boston: Kluwer Academic Publishers and Ottawa: International Development Research Centre.

Finger, M., and Allouche, J. (2002). *Water privatisation: Trans-national corporations and the re-regulation of the water industry.* London and New York: Spon Press.

Fischhendler, I. (2004). Legal and institutional adaptation to climate uncertainty: a study of international rivers. *Water Policy,* **6,** 281–302.

Fischhendler, I. (2008). Ambiguity in transboundary environmental dispute resolution: The Israeli–Jordanian water agreement. *Journal of Peace Research,* **45**(1), 91–110.

Fischhendler, I., and Feitelson, E. (2005). The formation and viability of non-basin transboundary water management: The case of the U.S.–Canada boundary water. *Geoforum,* **36,** 792–804.

Fisher, F. M., Arlosoroff, S., Eckstein, Z., Haddadin, M., Hamati, S. G., Huber-Lee, A., Jarrar, A., Jayyousi, A. Shamir, U., and Wesseling, H. (2002). Optimal water management and conflict resolution: The Middle East Water Project. *Water Resources Research,* **38**(11), 25-1–25-17.

Fisher, R., and Ury, W. L. (1981). *Getting to yes.* London: Hutchinson Business.

Fisher R., and Ury W. (1991). *Getting to yes: Negotiating agreement without giving in,* 2nd ed. New York: Penguin Books.

Flow, Inc. (2007). Water to the world. Accessed at http://www.flowinc.com. January 13.

Frederiksen, H. D. (1992). *Water resources institutions: Some principles and practices,* Papers 191. World Bank – Technical Papers.

Freeman, D. M. (2000). Wicked water problems: Sociology and local water organizations in addressing water resources policy. *Journal of the American Water Resources Association,* **36**(3), 483–491.

Frey, F. W. (1993). The political context of conflict and cooperation over international river basins. *Water International,* **18**(1), 54–68.

Friends of the Earth, Middle East. (2005). *Good water neighbors: A model for community development programs in regions of conflict.* Amman, Bethlehem, and Tel Aviv: EcoPeace.

Frohlich, N., and Oppenheimer, J. (1994). Alienable privatization policies: The choice between inefficiency and injustice. In *Water quantity/quality management and conflict resolution,* edited by A. Dinar and E. Loehman. Westport, CT: Praeger Publishers.

Furlong, K., and Gleditsch, N. P. (2003). The boundary dataset. *Conflict Management and Peace Science,* **20**(1), 93–117. Data available at http://www.prio.no/cscw/datasets.

Furlong, K., Gleditsch, N. P., and Hegre, H. (2006). Geographic opportunity and Neomalthusian willingness: Boundaries, shared rivers, and conflict. *International Interactions,* **32**(1), 79–108.

García-Acevedo, M. R. (2001). The confluence of water, patterns of settlement, and constructions of the border in the Imperial and the Mexicali valleys (1900–1999). In *Reflections on water: New approaches to transboundary conflicts and cooperation*, edited by J. Blatter and H. Ingram. Cambridge, MA: MIT Press.

Giannias, D. A., and Lekakis, J. N. (1996). Fresh surface water resource allocation between Bulgaria and Greece. *Environmental and Resource Economics*, **8**(4, December), 473–483.

Giordano, M. (2002). Water quality management in international river basins. Unpublished Ph.D. dissertation, Oregon State University, Corvallis, OR.

Giordano, M., Giordano, M., and Wolf, A. T. (2002). The geography of water conflict and cooperation: Internal pressures and international manifestations. *Geographical Journal*, **168**(4), 293–312.

Giordano, M. A., and Wolf, A. T. (2003). Sharing waters: Post-Rio international water management. *Natural Resources Forum*, 27, 163–171.

Glantz, M. H. (1998). Creeping environmental problems in the Aral Sea basin. In Kobori, I. and M. H. Glantz (editors). *Central Eurasian Water Crisis: Caspian, Aral and Dead Seas*. Tokyo: United Nations University Press, pp. 25–28.

Gleditsch, N. P., and Hamner, J. (2001). *Shared rivers, conflict, and cooperation*. Paper presented at the 42nd Annual Convention of the International Studies Association, Chicago, IL, 21–24 February 2001, and The Fourth European International Relations Conference, University of Kent, Canterbury, September 8–10.

Gleditsch, N. P., Furlong, K., Hegre, H., Lacina, B., and Owen, T. (2006). Conflicts over shared rivers: Resource scarcity or fuzzy boundaries? *Political Geography*, **25**, 361–382.

Gleick, P. H. (1993). Water and conflict: Fresh water resources and international security. *International Security*, **18**(1), 79–112.

Gleick, P. (1994). Water, war, and peace in the Middle East. *Environment*, **36**(3), 6–15, 35–42.

Gleick, P. (1996). Basic water requirements for human activities: Neeting basic needs. *Water International*, **21**(2), 83–92.

Gleick, P. H. (1998). *The world's water 1998–1999: The biennial report on freshwater resources*. Washington, DC: Island Press.

Gleick, P. H. (2003). *The world's water: Biennial report on freshwater resources 2002–2003*. Washington, DC: Island Press.

Gleick, P., Wolff, G., Chalecki, E. L., and Reyes, R. (2002). *The new economy of water: The risks and benefits of globalization and privatization of fresh water*. Oakland: Pacific Institute.

Glick, T. F. (1970). *Irrigation and society in medieval valencia*. Cambridge, MA: Belknap Press, pp. 31, 34–37, 52–53, 176–77, 198–206.

Global Environment Facility. (2000). *Environmental protection and sustainable integrated management of the Guaraní Aquifer*. Proposal for Project Development Funds, March, pp. 4–7.

Global Environment Facility. (2001). Project Report Document, Report No. PID10124, 20 November.

Global News Wire. (2003). Ex-Lesotho water chief must serve 18 years. *South African Press Association*. April 9.

Golden, T. (1992). *New York Times*. Mexico moves ahead with embattled dam project in Mayan area. March 15.

Goldfarb, W. (1997). Teaching water resources policy to university science and engineering students: Opportunities and challenges. *Journal of the American Water Resources Association*, **33**(2), 255–259.

Gonzalez, A., and Rubio, S. (1992). *Optimal interbasin water transfers in Spain*. Presented at a workshop, Sharing Scarce Fresh Water Resources in the Mediterranean Basin: An Economic Perspective, Padova, Italy, April 23–24.

Gopalakrishnan, G., Tortajada, C., and Biswas, A. K. (editors). (2005). *Water institutions: Policies, performance, and prospects*. New York: Springer.

Gooch, G. D., and Stålnacke, P. (2006). *Integrated transboundary water management in theory and practice: Experiences from the new EU eastern borders*. London: IWA Publishing.

Goslin, I. V. (1977). International river compacts: Impact on Colorado. In *Water needs for the future – Political, economic, legal, and technological issues in a national and international framework*, edited by V. Nanda. Boulder, CO: Westview Press.

Government of the People's Republic of Bangladesh (1976). *White paper on the Ganges water dispute*. September.

Government of Newfoundland and Labrador. (2001). *Export of Bulk Water from Newfoundland and Labrador: A Report of the Ministerial Committee Examining the Bulk Export of Water*. October. Accessed at http://www.gov.nf.ca/publicat/ReportoftheMinisterailCommittee ExaminingtheExportofBulkWater.PDF.

Green Cross International. (2000). *National Sovereignty and International Watercourses*. Green Cross International, Geneva, March 2000.

Grey, D. (2008). *Infrastructure for achieving water security*. Keynote presentation for World Bank, African Water Week, Tunis, March 26, 2008.

Gruen, G. (1993). Recent negotiations over the waters of the Euphrates and Tigris. *Proceedings of the International Symposium on Water Resources in the Middle East: Policy and Institutional Aspects*, Urbana, IL, October 24–27.

Gunderson, L. H, and Pritchard, L. (2002). *Resilience and the behavior of large-scale systems*. Washington, DC: Island Press.

Gurr, T. (1969). *Why men rebel*. Princeton: Princeton University Press.

Gurr, T. R. (1985). On the political consequences of scarcity and economic decline. *International Studies Quarterly*, **29**(1), 51–75.

Gyuk, I. (1977). Resources and the dynamics of cultures. *Water International*, **2**(1, March), 8–10.

Haddadin, M. J. (2001). Water scarcity impacts and potential conflicts in the MENA region. *Water International*, **26**(4, December), 460–470.

Haddadin, M. J., and Shamir, U. (2003). *Jordan case study*. Paris: UNESCO IHP Technical Documents in Hydrology, PCCP Series No. 15.

Hafidh, H. (2003). Iraq wants to clinch water deal with Syria, Turkey. *Environmental News Network*. Updated 16 September. Available at http://www.enn.com/news/2003-09-16/s_8435.asp.

Hamerlynk, O. (editor). (2000). An alternative to the water management of the Senegal River. In *The World Commission on Dams*. London: Earthscan Publishers. ATW: Can't find this; if nec., cut from ref's.

Hamner, J., and Wolf, A. (1998). Patterns in international water resource treaties: The transboundary freshwater dispute database. *Colorado Journal of International Environmental Law and Policy*, 1997 Yearbook. pp. 157–177.

Haught, J. F. (1996). Christianity and ecology. In *This sacred Earth: Religion, nature, environment*, edited by R. S. Gottlieb. New York: Routledge.

Haynes, K. E., and Whittington, D. (1981). International management of the Nile–Stage three? *Geographical Review*, **71**(1), 17–32.

Hayton, R., and Utton, A. (1989). Transboundary groundwaters: The Bellagio draft treaty. *Natural Resources Journal*, **29**(Summer), 663–720. Available at http://uttoncenter.unm.edu/pdfs/Bellagio_Draft_Treaty_E.pdf.

Healing, J. (2005). Development flows from vast water project. *Sunday Times (South Africa)*, October 16.

Henry, N. (1991). *Washington Post*. Arid Botswana keeps its democracy afloat. March 21.

Hera, G., Sanchez-Fresneda, C., Esanola, J. M., and Pozo de Castro, M. (2002). The public water administration and the role of the Confederaciones Hidorgaphics in Spain, In *Integrated water management at river basin level: An institutional development perspective*, edited by G. Alearts and G. LeMoigne. Washington, DC: RFF for the World Bank.

Hewage, A. (1999). Potential in the Mahaweli river basin. *River basin management: Proceedings of the UNESCO International Workshop*, The Hague, October 27–29, 1999, edited by E. Mostert. IHP-V Technical Document in Hydrology No. 31.

Heyns, P. (2005). *Governance of a shared and contested resource: A case study of the Okavanga River basin*. International Ecosystem Symposium on Ecosystem Governance, Kwa Maritane Bush Lodge, Pilansberg South Africa, October 10–13, CSIRO, Pretoria, RSA.

Hickey, E. (1992). *Washington Times*. Conflict is his peace of act. March 10, Section E.

Hiryi, R. (2002). River basin based water management in Tanzania. In *Integrated water management at river basin level: An institutional development perspective*, edited by G. Alearts and G. LeMoigne. Washington, DC: RFF for the World Bank.

Hoffman, J. (2001). Interstate Commission on the Potomac River Basin (ICPRB) briefing, October 9. Alexandria, VA: USACE, Institute for Water Resources.

Homer-Dixon, T. (1991). On the threshold: Environmental changes as causes of acute conflict. *International Security*, **16**(2), 76–116.

Homer-Dixon, T. (1994). Environmental scarcities and violent conflict: Evidence from cases. *International Security*, **19**(1), 5–40.

Homer-Dixon, T. (1996). Strategies for studying causation in complex ecological–political systems. *Journal of Environment and Development,* **5**(2), 132–148.

Homer-Dixon, T. (1999). *Environment, scarcity, and violence.* Princeton, NJ: Princeton University Press.

Hong Kong Water Supplies Department. (2006a). Revenues and pricing. Accessed at http://www.wsd.gov.hk/en/text/accessinfo/rpt9899/chapter12.htm. December 15, 2006.

Hong Kong Water Supplies Department. (2006b). Water from Guangdong. Accessed at http://www.wsd.gov.hk/en/html/water/hkwchn.htm on December 15.

Hori, H. (1993). Development of the Mekong River Basin, its problems and future prospects. *Water International,* **18**(2), 110–115.

Housen-Couriel, D. (1994). *Some examples of cooperation in the management and use of international water resources.* Jerusalem: Harry S. Truman Research Institute for the Advancement of Peace.

Howe, C. (1996). Water resources planning in a federation of states: Equity versus efficiency. *Natural Resources Journal,* **36**(1), 29–36.

Howe, C. W., Schurmeier, D. R., and Shaw, W. D., Jr. (1986). Innovative approaches to water allocation: The potential for water markets. *Water Resources Research,* **22**, 439–445.

Hubbard, P. J. (1961). *Origins of the TVA.* Nashville: Vanderbilt University Press.

Huddle, F. (1972). *The Mekong Project: Opportunities and problems of regionalism.* Washington, DC: U.S. Government Printing Office.

Huisman, P., de Jung, J., and Wieriks, K. (2000). Transboundary cooperation in shared river basin: Experiences from the Rhine, Meuse, and North Sea. In *Management of shared river basins,* edited by Savenije, van der Zaag, and Wolf. *Water Policy,* **2**, 83–97.

ICWE. (1992). *Development issues for the 21st century.* The Dublin Statement and Report of the Conference, Dublin, Ireland, January 26–31, 1992. Geneva: World Meteorological Organization.

ICPDR. (2002). List of major decisions of the ICPDR (from October 1998 onwards), ICPDR Homepage. Updated November 26, 2002. Available at http://www.icpdr/doccener/basicdocuments/listofmajordecisions.

IFAS. (2006). The executing agency of the international fund for the saving of the Aral Sea for implementation of the GEF and ASBP projects. Available at http://www.aral.uz.

Ilter, K. (2000). Analysts expect no drastic change in Turco–Syrian relations. *Turkish Daily News,* June 12.

Ingram, H. (1971). Patterns of politics in water resources development. *Natural Resources Journal,* **11** (January): 102–118.

Ingram, H. (1990). *Water politics: Continuity and change.* Albuquerque: University of New Mexico Press.

Ingram, H. M., Laney N. K., and Gillilan D. M. (1995). *Divided waters: Bridging the U.S.–Mexico border.* Tucson, AZ: University of Arizona Press.

Ingram, H. M., and White, D. (1993). International Boundary and Water Commission: An institutional mismatch for resolving transboundary water problems. *Natural Resources Journal,* **33**.

International Boundary and Water Commission. (2006). Available at http://www.ibwc.state.gov. Updated January 19, 2007.

International Joint Committee. (2006). Available at http://www.ijc.org. Updated January 16, 2007.

International Law Association. (1966). *Helsinki rules on the uses of the waters of international rivers.* Report of the Fifty-Second Conference, Helsinki, August 14–20, 1966, (London, 1967), pp. 484–532.

Interstate Commission on the Potomac River Basin. (2005). Available at http://www.potomacriver.org.

IRN, Nam Teum 2 Dam, Theun River, Laos: Beyond big Dams, an NGO guide to the WCD. Available at www.irn.org.

Isaac J., and Selby, J. (1996). The Palestinian water crisis: Status projections and potential for resolution. *Natural Resources Forum,* **20**, 17–26.

Islam, M. R. (1987). The Ganges water dispute: An appraisal of a third party settlement. *Asian Survey,* **27**(8, August), 918–934.

IUCN. (2003). *The Senegal River: Release of an artificial flood to maintain traditional floodplain production systems.* Water and Nature Initiative of the World Conservation Union-case studies. Gland: Switzerland.

IWMI. (2001). International water resources management in a river basin context: Institutional strategies for improving agricultural water management, Workshop Malang, East Java, Indonesia, January 15–19, 2001. Available at www.cgiar.org.

Iyer, R. (1999). Conflict Resolution: Three River Treaties. *Economic and Political Week, Bombay,* June 12, pp. 1510–1512.

Jacobson, G., and Hill, P. J. (1988). *Hydrogeology and groundwater resources of Nauru Island, Central Pacific Ocean.* Bureau of Mineral Resources Record NQ 1988/12, Australian Government.

Jägerskog, A. (2003). *Why states cooperate over shared water: The water negotiations in the Jordan River basin.* Sweden: Linköping University.

James, L. D. (1974). *Man and water: Social sciences and management of water resources.* Lexington: Lexington University Press, p. 47.

Jarvis, T. (2006). Transboundary groundwater: Geopolitical consequence, common sense, and the law of the hidden sea. Unpublished Ph.D. dissertation, Oregon State University, Corvallis, OR.

Jarvis, T., Giordano, M., Puri, S., Matsumoto, K., and Wolf, A. (2006). International borders, ground water flow, and hydroschizophrenia. *Ground Water,* **43**(5), 764–770.

Johnson, N., Revenga, D., and Echeverria, J. (2001). Managing water for people and nature. *Science,* **292**(11 May), 1071–1072.

Johnson, R. (1999). The Colorado. *River Basin Management: Proceedings of the UNESCO International Workshop,* The Hague, October 27–29, 1999, edited by E. Mostert. IIHP-V Technical Document in Hydrology No. 31.

Johnson, R. (2001). *Redefining Brazil's water management system: The cases of the Paraiba do Sul and Curu River basins,* COPPE Hydrology and Environmental Studies Laboratory, Federal University of Rio de Janeiro, Brazil, World Bank Workshop Washington, DC, August 28.

Johnson, R. (2002). The river basin approaches in Brazil: The Pariba do Sul and the Ceara Cases. In *Integrated water management at river basin level: An institutional development perspective,* edited by G. Alearts and G. LeMoigne. Washington, DC: RFF for the World Bank.

Johore, Sultan of and Singapore City Council. (1927). The Agreement as to Certain Water Rights in Johore. December 5. Extracted from Administration Report of the Singapore Municipality for the Year 1927.

Johore (Government of the State of) and Singapore (the Public Utilities Board of). (1990). November 24.

Johore (Government of the State of) and Singapore (City Council of). (1961). The Tebrau and Scudai River water agreement. September 1.

Governments of Malaysia and Singapore (1962). The Johore River water agreement. September 29.

Jovanovic, D. (1985). Ethiopian interests in the division of the Nile river waters. *Water International,* **10**(2, June), 82–85.

Jovanovic, D. (1986a). Response to discussion by M. M. A. Shahin of the paper "Ethiopian interests in the division of the Nile River waters." *Water International,* **11**(1, March), 20–22.

Jovanovic, D. (1986b). Response to discussion of the paper "Ethiopian interests in the division of the Nile River waters." *Water International,* **11**(2, June), 89.

Jung, C. G. (1968). *Analytical psychology: Its theory and practice.* New York: Random House.

Just, R., and Netanyahu, S. (editors). (1998). *Conflict and cooperation on trans-boundary water resources.* Amsterdam, The Netherlands: Kluwer Academic Publishers.

Kaczmark, B. (2002). The Organization of Water Management with the Agencies de l'eau. In *Integrated water management at river basin level: An institutional development perspective,* edited by G. Alearts and G. LeMoigne. Washington, DC: RFF for the World Bank.

Kaijser, A. (2002). System building from below: Institutional change in Dutch water control systems, *Technology and Culture,* **43**(3, July), 521–548.

Kally, E. (1989). The potential for cooperation in water projects in the Middle East at peace. In *Economic cooperation in the Middle East,* edited by G. Fishelson. Boulder: Westview Press, pp. 303–326.

Kalter, R. (1971). *Criteria for federal evaluation of resources investments.* Washington, DC: U.S. Water Resources Council.

Kardoss, L. (1999). The Danube. *River Basin Management: Proceedings of the UNESCO International Workshop,* The Hague, October 27–29, 1999, edited by E. Mostert. IHP-V Technical Document in Hydrology No. 31.

Kattelmann, R. (1990). Conflicts and cooperation over floods in the Himalaya–Ganges region. *Water International,* **15**(4), 189–194.

Kaufman, E. (2002). Innovative problem-solving workshops. In *Second track/citizens' diplomacy: Concepts and techniques for conflict transformation*, edited by J. Davies and E. Kaufman. Lanham, MD: Rowman & Littlefield, pp. 171–247.

Kenney, D. S. (1995). Institutional options for the Colorado River. *Journal of the American Water Resources Association*, **31**(5), 837–850.

Kenney, D. (editor). (2005). *In search of sustainable water management*. Northampton, MA: Edward Elgar Publishing.

Kenney, D. S., and Lord, W. B. (1994). *Coordination mechanisms for the control of interstate water resources: A synthesis and review of the literature*. Task 2 Report for the ACT–ACF Coordination Mechanism Study, Institute for Water Resources, USACE, Alexandria, VA, July 1994.

Kenworthy, T. (1996). *Washington Post*. High Tide in the Grand Canyon. March 27, A17.

Keohane, R. O. (1989). *International institutions and state power*. Boulder: Westview Press.

Khan, M. Y. (1990). Boundary water conflict between India and Pakistan. *Water International*, **15**(4), 1995–1999.

Khan, T. (1994). Challenges facing the management and sharing of the Ganges. *Transboundary Resource Report*. Spring 1994.

Khassawneh, A. (1995). The International Law Commission and Middle East waters. In *Water in the Middle East: Legal, political, and commercial implications*, edited by J. A. Allan and C. Mallat. London and New York: Tauris Academic Studies, pp. 21–28.

Khlobystov, V. (1999). North Caucasus Region. *River Basin Management: Proceedings of the UNESCO International Workshop*, The Hague, October 27–29, 1999, edited by E. Mostert. IHP-V Technical Document in Hydrology No. 31.

Kibaroğlu, A., and Olcay Unver, I. H. (2000). An institutional framework for facilitating cooperation in the Euphrates–Tigris River basin. *International Negotiation*, **5**(2), 311–330.

Kibaroğlu. A. (2002a). *Building a regime for the waters of the Euphrates–Tigris River basin*. The Netherlands: Kluwer Law International.

Kibaroğlu, A. (2002b). *Management and allocation of the waters of the Euphrates–Tigris basin: Lessons drawn from global experiences*. Ankara: Bilkent University.

Kibaroğlu, A. (2002c). *Transboundary water issues in the Euphrates–Tigris river basin: Prospects for cooperation.*

Kirmani, S. S. (1990). Water, peace and conflict management: The experience of the Indus and Mekong river basins. *Water International*, **15**(4, December), 200–205.

Kliot, N. (1995). Building a legal regime for the Jordan–Yarmouk river system: Lessons from other international rivers. In *The peaceful management of transboundary resources*, edited by G. Blake, W. Hildesley, M. Pratt, R. Ridley, and C. Schofield. London and Dordrecht: Graham and Trotman/Martinus Nijhoff, pp. 187–202.

Kliot, N., Shmueli, D., and Shamir, U. (1999). *Institutional frameworks for the management of transboundary water resources*, Vols. 1–2, Technion, Israel: Water Research Institute.

Kolars, J. (1992). Trickle of hope: Negotiating water rights is critical to peace in the Middle East. *Sciences*, **32** (6), 16–21.

Kolars, J., and Mitchell, W. (1991). *The Euphrates River and the Southeast Anatolia Development Project*. Carbondale and Edwardsville: Southern Illinois University Press.

Kriesberg, L. (1988). Strategies of negotiating agreements: Arab–Israeli and American–Soviet cases. *Negotiation Journal*, January, 19–29.

Krishna, R. (1995). International watercourses: World Bank experience and policy. In *Water in the Middle East: Legal, political, and commercial implications*, edited by J. A. Allan and C. Mallat. London and New York: Tauris Academic Studies, pp. 29–54.

Krutilla, J. V. (1969). *The Columbia River Treaty – The economics of an international river basin development*. Baltimore, MD: Published for the Resources for the Future by John Hopkins Press.

Kulshreshtha, S. N. (1993). *World water resources and regional vulnerability: Impact of future changes*. Laxenburg, Austria: IIASA, RR-93-10.

Kultida, S. (2003). Cabinet gives green light for ratification of power accord. *International Rivers Network*, Available at http://www.irn.org/programs/mekong/030605.ratification.html. Updated June 5, 2003.

Kwaku Kyem, P. A. (2004). Of intractable conflicts and participatory GIS applications: The search for consensus amidst competing claims and institutional demands. *Annals of the Association of American Geographers*, **94**(1), 37–57.

Lammers, O., Moore, D., and Preakle, K. (1994). Considering the Hidrovia: A preliminary report on the status of the proposed Paraguay/Parana waterway project. *Working Paper 3*. Berkley CA: International Rivers Network, July.

Lancaster, T. (1990). *The econometric analysis of transition data*. Cambridge, UK: Cambridge University Press.

Langton, S. (1999). *Participatory Processes in Water Management (PPWM): Proceedings of the Satellite Conference to the World Conference on Science International conference on participatory processes in water management*. June 28–30, 1999. Budapest, Hungary, edited by J. Gayer. UNESCO-IHP/IA2P/IWRA. Paris: UNESCO-IHP-V.

Lanna, A. E. (2003). Water charges in Brazil: Implementation and perspectives. In *Water pricing and public-private partnership in the Americas*. Washington, D.C: Sustainable Development Department of the Inter-American Development Bank.

Lanz, K., and Scheuer, S. (2000). *EEB handbook on EU water policy under the Water Framework Directive*, Brussels: European Environmental Bureau.

Lasswell, H. (1971). *A preview of policy sciences*. New York: American Elsevier.

Laylin, J., and Bianchi, R. (1959). The role of adjudication in international river disputes: The Lake Lanoux case. *American Journal of International Law*, **53**, 30–49.

Lederach, J. P. (1995). *Preparing for peace: Conflict transformation across cultures*. Syracuse, NY: Syracuse University Press.

Lee, K. N. (1995). *Compass and gyroscope: Integrating science and politics for the environment*. Washington, DC: Island Press.

Lee, D. J., and Dinar, A. (1995). *Review of integrated approaches to river basin planning, development and management*. World Bank Policy Research Working Paper WPS 1446, Washington, DC: World Bank.

Lees, S. H. (1973). Hydraulic development as a process of response. *Human Ecology*, **2**(November), 159–175.

Le-Huu, T., and Nguyen-Duc, L. (2003). *Mekong Case Study*. Paris: UNESCO IHP Technical Documents in Hydrology, PCCP Series No. 10.

Leitman, S. (2005). Negotiations of a water allocation formula for the Apalachicola–Chattahoochee–Flint basin. In *Adaptive governance*, edited by J. T. Scholz and B. Stifftel. Washington, DC: Resources for the Future, pp. 74–88.

LeMarquand, D. (1976). Politics of international river basin cooperation and management. *Natural Resources Journal*, **16**, 883–901.

LeMarquand, D. G. (1977). *International rivers: The politics of cooperation*. Vancouver: University of British Columbia, Westwater Research Centre.

LeMarquand, D. G. (1990). International development of the Senegal River. *Water International*, **15**, 223–230.

LeMarquand, D. (1993). The international joint commission and changing Canada–United States boundary relations. *Natural Resources Journal*, **33**(1), 59–92.

Le Monde Diplomatique. (2000). Available at http://mondediplo.com/2000/02/10boucaud?var_recherche = salween. Updated February 2000.

Leopold, A. (1949). *A Sand County almanac*. Oxford, UK: Oxford University Press.

Lesotho and South Africa, Governments of. (1986). Treaty on the Lesotho Highlands Water Project between the Government of the Republic of South Africa and the Government of the Kingdom of Lesotho.

Lesotho Highlands Development Authority. (2004). Lesotho Highlands Water Project. *Accessed* at http://www.lhwp.org.ls/news/default.htm. Updated October 11.

Lesotho Highlands Development Authority. (2006). Overview of the Lesotho Highlands Water Project. Accessed at http://www.lhwp.org.ls/overview/default.htm December 19.

Lesotho Highlands Water Project. (2002). Phase I, Policy for instream flow requirements, Available at http://www.lhwp.org.ls.

Leuchtenburg, W. E. (1952). Roosevelt, Norris and the "Seven Little TVAs". *Journal of Politics*, **14**, 418–441.

Lewicki, R. J., and Litterer, J. A. (1985). *Negotiation*. Homewood, IL: Irwin.

Lewicki, R., Litterer, J., Minton, J., and Saunders, D. (1994). *Negotiation.* Boston: Irwin.

Li., R. (1999). Yellow River. *River Basin Management: Proceedings of the UNESCO International Workshop,* The Hague, October 27–29, 1999, edited by E. Mostert. IHP-V Technical Document in Hydrology No. 31.

Libiszewski, S. (1995). Water disputes in the Jordan basin region and their role in the resolution of the Arab–Israeli conflict. *Occasional Paper,* **13**. Zurich: Center for Security Studies and Conflict Research.

Lincoln, W. F. (1986). *The course in collaborative negotiations.* Denver, CO: Denver National Center for Dispute Resolution.

Linnerooth-Bayer, J. (1986). Negotiated river basin management: Implementing the Danube Declaration. In *Management of International River Basin Conflicts,* edited by E. Vlachos, A. Webb, and I. Murphy. Proceedings of a Workshop held at the Institute for Applied Systems Analysis, Laxenberg, Austria, September 22–25, 1986.

Linnerooth-Bayer, J. (1993). Current Danube River Events and Issues. *Transboundary Resource Report,* Winter.

Linnerooth-Bayer, J. (1994). The Danube River basin: International cooperation for sustainable development. *Transboundary Resource Report,* **8** (6) Spring.

Llamas, M. R., and Custodio, E. (editors). (2002). *Intensive use of groundwater. Challenges and opportunities.* The Netherlands: Balkema Publishers.

London, J. B., and Miley, H. W., Jr. (1990). The interbasin transfer of water: An issue of efficiency and equity. *Water International,* **15,** 231–235.

Lonergan, S., Gustavson, K., and Carter, B. (2000). The index of human insecurity. *AVISO Bulletin,* **6,** Ottowa, ON: GECHS, pp. 1–11.

Lopez, A. (2004). *Environmental conflicts and regional cooperation in the Lempa River Basin: The role of the Trifinio Plan as a regional institution.* EDSP Working Paper No. 2, Environment, Development, and Sustainable Peace Initiative.

Lowi, M. (1993). *Water and power: The politics of a scarce resource in the Jordan River basin.* Cambridge, UK: Cambridge University Press.

Lowi, M., and Rothman, J. (1993). Arabs and Israelis: The Jordan River. In *Culture and negotiation: The resolution of water disputes,* edited by G. O. Faure and J. Z. Rubin. London: Sage Publications, pp. 156–175.

MacChesney, B. (1959). Judicial decisions: Lake Lanoux case. *American Journal of International Law,* **53,** 156–171.

MacDonnell, L. J. (1988). Natural resources dispute resolution: An overview. *Natural Resources Journal,* **28**(1), 5–20.

Main, Chas. T., Inc. (1953). *The unified development of the water resources of the Jordan Valley region.* Knoxville: Tennessee Valley Authority.

Mageed, Y. (1994). The Nile Basin: Lessons from the past. In *International waters of the Middle East,* edited by A. Biswas. Oxford: Oxford University Press.

Mandel, R. (1992). Sources of international river basin disputes. *Conflict Quarterly,* **12**(4, Fall), 25–56.

Margat, J. (1989). The sharing of common water resources in the European Community (EC). *Water International,* **14,** 59–61.

Martin, R. (1956). *TVA the first twenty years.* Tuscaloosa, AL: University of Alabama Press.

Martin, R. (1960). *River basin administration and the Delaware.* Syracuse, NY: Syracuse University Press.

Marty, F. (2001). *Managing international rivers: Problems, politics, and institutions.* Bern: Peter Lang.

Matsumoto, K. (2002). Transboundary groundwater and international law: Past practices and current implications. Unpublished Master's thesis, Oregon State University, Corvallis, OR. Available at http://www.transboundarywaters.orst.edu/publications/Matsumoto_abstract.htm.

Matthews, O. P. (1984). *Water resources: Geography and law.* Washington, DC: Association of American Geographers.

McCaffrey, S. C. (1996a). An assessment of the work of the International Law Commission. *Natural Resources Journal,* **36**(2), 297–318.

McCaffrey, S. C. (1996b). The Harmon Doctrine one hundred years later: Buried, not praised. *Natural Resources Journal,* **36**(3), 549–590.

McCaffrey, S. (1997). Water scarcity: Institutional and legal responses. In *The scarcity of water: Emerging legal and political responses,* edited by E. H. P. Brans, E. S. de Haan, A. Nollkaemper, and J. Rinzema. The Hague: Kluwer Law International, pp. 43–58.

McCaffrey, S. (1999). International groundwater law: Evolution and context, *Groundwater: Legal and Policy Perspectives, Proceedings of a World Bank Seminar.* Washington, DC: World Bank.

McCaffrey, S. (2001a). The contribution of the UN convention on the law of the non-navigational uses of international watercourses. *International Journal Global Environmental Issues,* **1**(3/4), 250–263.

McCaffrey, S. (2001b). *The law of international watercourses: Non-navigational uses,* 2nd ed. Oxford: Oxford University Press.

McCaffrey, S. (2003). The need for flexibility in freshwater treaty regimes. *Natural Resources Forum,* **27**(22), 156–163.

McClurg, S., and Sudeman, R. S. (2003). Public and stakeholder education to improve groundwater management. In *Intensive use of Groundwater, Challenges and Opportunities,* edited by R. Llama and E. Custodio. Lisse, The Netherlands: A. A. Balkema.

McDonald, A. (1988). *International river basin negotiations: Building a data base of illustrative successes.* October WP HH 096. Luxembourg, Austria: International Institute for Applied Systems Analysis.

McKay, J. M. (2005). Water institutional reform in Australia. *Water Policy,* 7(2), 35–35.

McKinney, D. C. (1997) Sustainable water management in the Aral Sea basin. *Water Resources Update,* Universities Council on Water Resources, **102,** Winter, 14–24.

McKinney, D. C. (2004). Cooperative management of transboundary water resources in Central Asia. In *In the tracks of Tamerlane – Central Asia's path into the 21st Century,* edited by D. Burghart and T. Sabonis-Helf. Washington DC: National Defense University Press.

Medzini, A., and Wolf, A. (2004). Towards a Middle East at peace: Hidden issues in Arab–Israeli hydropolitics. *International Journal of Water Resources Development,* **20**(2), 193–204.

Mehta, J. (1986). The Indus water treaty. In *The management of international river basin conflicts,* edited by E. Vlachos, A. Webb, and I. L. Murphy. Washington, DC: Graduate Program in Science, Technology, and Public Policy, George Washington University.

Mejia, A., Lopez Zayas, L. A., Tafflesse S., and Amore, L. (2004). The Guaraní Aquifer system: A key element for an integrated water resources management strategy in La Plata basin, Powerpoint Presentation: *Diving in to implementation,* Slide 8. World Bank, World Water Week.

Mekong River Commission. (1999). *River Basin Management: Proceedings of the UNESCO International Workshop,* The Hague, October 27–29, 1999, edited by E. Mostert. IHP-V Technical Document in Hydrology No. 31.

Mekong River Commission. (2007). *Mekong River Commission.* Available at http://www.mrcmekong.org.

Mestre, E. (2001). *The design of River Basin Organization in Mexico.* River Basin Management Workshop, World Bank, Washington, DC, August 28, 2001.

Mestre, E. (2002). The Lerma Chapala. In *Integrated water management at river basin level: An institutional development perspective,* edited by G. Alearts and G. LeMoigne. Washington, DC: RFF for the World Bank.

Mestre, E. (2004). *Training course in building river basin organizations,* for the World Bank, Abuja Nigeria, June 2004.

Meyer, T. A. (1987). Innovative approaches to transportation of water by tanker. *Non-Conventional Water Resources Use in Developing Countries,* Natural Resources/Water Series NP 22, United Nations, pp. 119–135.

Middle East Desalination Research Center. (2007). Available at http://www.medrc.org. Updated January 19, 2007.

Middle East Newsfile. (1998). OIC offers to mediate between Turkey, Syria. October 18.

Mideast Mirror. (1997). Turks attacks northern Iraq after renewing mandate of U.S.-led force. **11**(2), January 2.

Mideast Mirror. (1998). Turkey-Syria crisis: Saudis to join mediation efforts. **12**(195), October 9.

Mideast Mirror. (2000). Turkey's "water weapon." **14**(61), March 29.

Milich, L., and Varady, R. G. (1999). Openness, sustainability, and public participation: New designs for transboundary river basin institutions. *Journal of Environment and Development,* **8**(3), 258–306.

Miller, C. (editor). (2001). *Fluid arguments: Five centuries of western water conflict.* Tucson: University of Arizona Press.

Millington, P. (2002). The Murray Darling. In *Integrated water management at river basin level: An institutional development perspective,* edited by G. Alearts and G. LeMoigne. Washington, DC: RFF for the World Bank.

Mirza, M. (2003). The Ganges water-sharing treaty: Risk analysis of the negotiated discharge. *International Journal of Water,* **2**(1), 57–74.

Mitrany, D. (1975). *The functional theory of politics*. New York: St. Martin's Press.

Mnookin, R. H., Peppet, S. R., and Tulumello, A. S. (2000). *Beyond winning: Negotiating to create value in deals and disputes*. Cambridge, MA: Belknap Press.

Moench, M. (2004). Groundwater: The challenge of monitoring and management. In *The world's water 2004–2005*, edited by P. H. Gleick. Washington, DC: Island Press, pp. 79–100.

Molle, F. (2004). Defining water rights: By prescription or negotiation? *Water Policy*, **6**, 207–227.

Montville, J., and Delli Priscoli, J. (1994). *Report on preventing Slovak–Hungarian ethnic violence*. Washington, DC: Center for Strategic and International Studies.

Moore, C. (1985). *Executive training course in conflict management*, edited by J. Delli Priscoli and C. Moore. Alexandria, VA: Institute for Water Resources, USACE.

Moore, C. W. (1986). *The mediation process: Practical strategies for resolving conflict*. San Francisco: Jossey-Bass.

Moore, C. (1991). *Corps of Engineers uses mediation to settle hydropower dispute*. ADR Case Study No. 6, September. Alexandria and Ft. Belvoir, VA: IWR USACE.

Moore, C. W. (2003). *The mediation process: Practical strategies for resolving conflict*, 3rd ed. San Francisco: Jossey-Bass.

Moore, C., and Delli Priscoli, J. (editors). (1989). *Alternative dispute resolution (ADR) procedures*. Boulder, CO: CDR Associates (for the Executive Seminar for the U.S. Army Corps of Engineers).

Morgan, T. (2002). Turkey to pump water to Cyprus. *BBC World News*. January 11.

Morris, C. (2000). Turkey to dip into water market. *BBC WorldNews*. March 28.

Mostert. E, van Beek E., Bouman N. W. M, Hey, E., Savenije, H. H. G., and Thissen, W. A. H. (editors). (1999). *River Basin Management: Proceedings of the UNESCO International Workshop*, The Hague, October 27–29, 1999, IHP-V Technical Document in Hydrology No. 31. Paris: UNESCO.

Mostert, E. (2003). The challenge of public participation. *World Policy*, **5**, 179–197.

Muckleston, K. (2003). *International management in the Columbia River system*. Paris: UNESCO IHP Technical Documents in Hydrology, PCCP Series No. 12.

Mumme, S. (2000). Minute 242 and beyond: Challenges and opportunities for managing transboundary groundwater on the Mexico-United States border. *Natural Resources Journal*, 40 (2), Spring, pp. 341–378.

Mumme, S. (2004). *Advancing binational cooperation in the transboundary aquifer management on the U.S.–Mexico Border*. Paper presented at Groundwater in the West Conference, University of Colorado at Boulder.

Munich Re Group. (2000). *Topics, 2000: Natural catastrophes, the current position*. Münchener Rückversicherung-Gesellschaft, Munich, Germany. Available at http://www.munichre.com/publications/302-02354_en.pdf?rdm=80335.

Murray–Darling Basin Commission. (2007). Available at http://www.mdbc.gov.au. Updated January 20, 2007.

Mutayoba, W., Saidi, F., and Hirji, R. (2001). *The development of river basin management in Tanzania: Early experience and lessons*. World Bank Symposium, August 28, 2001.

Myers, N. (1993). *Ultimate security: The environmental basis of political stability*. New York: Norton.

Nachtnebel, H. P. (1999). The Danube. *River Basin Management: Proceedings of the UNESCO International Workshop*. The Hague, October 27–29, 1999, edited by E. Mostert. IHP-V Technical Document in Hydrology No. 31.

Nachtnebel, H. P. (2002). The Danube Commission and its environmental program. In *Integrated water management at river basin level: An institutional development perspective*, edited by G. Alearts and G. LeMoigne. Washington, DC: RFF for the World Bank.

Naff, T., and Dellapenna, J. (2002). Can there be confluence? A comparative consideration of Western and Islamic fresh water law. *Water policy*, **4**, 465–489.

Naff, T., and Matson, R. (1984). *Water in the Middle East: Conflict or cooperation*. Boulder and London: Westview Press.

Nagle, W. J., and Ghose, S. (1990). *Community participation in World Bank supported projects*. Strategic Planning and Review Department Discussion Paper 8. Washington, DC: World Bank, p. 4.

Nagy, L. (1987). *Perspective of cooperation on international river basins*. International Symposium on Water for the Future, Rome, 6–11 April, 1987, pp. 343–353.

Nakayama, M. (1997). Successes and failures of international organizations in dealing with international waters. *International Journal of Water Resources Development*, **13**(3), 367–382.

Nakayama, M. (editor). (2003). *International waters in Southern Africa*. Tokyo: United Nations University Press.

Nandalal, K. D. W., and Simonovic, S. P. (2003). *State-of-the-art report on systems analysis methods for resolution of conflicts in water resources management*. Paris: UNESCO IHP Technical Documents in Hydrology, PCCP Series No. 4.

Natchkov, I. (2002). The Danube Commission and its environmental program. In *Integrated water management at river basin level: An institutional development perspective*, edited by G. Alearts and G. LeMoigne. Washington, DC: RFF for the World Bank.

National Academy of Sciences. (1968). *Water and choice in the Colorado basin: An example of alternatives in water management*. Washington, DC: U.S. Government Printing Office.

National Environmental Policy Act of 1969 (NEPA); 42 U.S.C. 4321–4347. Available at http://ceq.eh.doe.gov/nepa/regs/nepa/nepaeqia.htm.

Natural Resources Law Center. (1997). *The watershed source book*. Boulder: University of Colorado.

Newfoundland, Statutes of. (1999). An act to provide for the conservation, protection, wise use and management of the water resources of the province.

News24.com. (2005). Thousands flee water fights. January 24, 2005. Available at http://www.news24.com/News24/Africa/News/0,6119,2-11-1447_1651487,00.html.

Ng, I. (2001). Tough talks, then progress on KL pact. *Straits Times*, September 5.

Ng, T. (2006). HK, Guangdong sign water supply deal. *China Daily*. April 4.

Nichols, J. (1974). *The Milagro beanfield war: A novel*. New York: Holt, Rinehart, and Winston.

Nicol, A. (2003). *The Nile: Moving beyond cooperation*. Paris: UNESCO IHP Technical Documents in Hydrology, PCCP Series No. 16.

Niem, N. T. (2000). Strategy for flood preparedness and mitigation in Vietnam. *International European-Asian Workshop, Ecosystems and Floods 2000, Hanoi Vietnam*, FAO Rome, June 27–29.

Nile Basin Organization. (1994). Nile Basin initiative. In *International waters of the Middle East*. Oxford: Oxford University Press. Available at www.nilebasin.org.

Nishat, A., and Pasha, M. (2001). A review of the Ganges Treaty of 1996. *Globalization and Water Resources Management: The Changing Values of Water*. AWRA/IWLRI – University of Dundee International Speciality Conference. August 6–8.

Nitze, W. (1991). *Greenhouse warming: Negotiating a global regime*. Washington, DC: World Resources Institute.

NUS Consulting Group. (2005). *2004–2005 international water report and cost survey*. New York: NUS Consulting Group.

NUS Consulting Group. (2006). *2005–2006 international Water Report and Cost Survey*. New York: NUS Consulting Group.

OAC. (1999). River basin management in Nigeria, In *River Basin Management: Proceedings of the UNESCO International Workshop*, The Hague, October 27–29, 1999, edited by E. Mostert. IHP-V Technical Document in Hydrology No. 31.

Officer, L. H., and Williamson, S. H. (2006). Computing "real value" over time with a conversion between U.K. Pounds and U.S. Dollars, 1830–2005, *MeasuringWorth.com*, August.

O'Hara, S. (editor). (2003). *Drop by drop: Water management in the Southern Caucuses and Central Asia*. Open Society Institute: Local Government and Public Service Reform Initiative, Budapest.

Oleson, J. P. (1984). *Greek and Roman mechanical water-lifting devices: The history of a technology*. Toronto: University of Toronto Press.

Oliver, J. I. (1992). *French river basin management*. Draft Water Policy Paper, World Bank, Washington, DC.

Omernik, J. M. (2003). The misuse of hydrologic unit maps for extrapolation, reporting, and ecosystem management. *Journal of the American Water Resources Association*, **39**, 563–573.

Omernik, J. M., and Bailey, R. G. (1997). Distinguishing between watersheds and ecoregions. *Journal of the American Water Resources Association*, **33**, 1–15.

Onn, L. P. (2005). *Water management issues in Singapore*. Institute of Southeast Asian Studies Paper presented at Water in Mainland Southeast Asia Conference. November 29–December 2. Siem Reap, Cambodia.

Onta, P. R., Gupta, A. D., and Loof, R. (1996). Potential water resources development in the Salween River basin. In *Asian international waters: From Ganges–Brahmaputra to Mekong*, edited by A. K. Biswas and T. Hashimoto. Oxford: Oxford University Press.

Organization for the Development of the Senegal River (1972). Senegal River Basin, Guinea, Mali, Mauritania, Senegal, pilot case studies, A focus on real-world examples, *Senegal, Mali and Mauritania*; pp. 448–461.

Organization of American States (OAS). (2000). *Strategic action program for the Bermejo Binational River basin*, Executive Summary, Washington DC, March 2000.

Organization of American States (OAS). (2001). *InterAmercian dialogue on water management*, Foz do Iguacu, Parana, Brasil, September 2–6, 2001.

Organization of American States (OAS). (2004). *Moving forward the water agenda: Issues to consider in Latin America*, Unit for Sustainable Development and Environment, Policy Series May (2), p. 1.

Ostrom, E. (1990). *Governing the commons: The evolution of institutions for collective action*. New York: Cambridge University Press.

Ostrom, E. (1992). The rudiments of a theory of the origins, survival, and performance of common-property institutions. In *Making the commons work: Theory, practice and policy*, edited by D. W. Bromley. San Francisco: ICS Press.

Ostrom, V. (1971). *Institutional arrangement for water resources development*. Arlington, VA: National Water Commission.

Ouyahia, M. A. (2006). *Public–private partnerships for funding municipal drinking water infrastructure: What are the challenges?* Government of Canada Discussion Paper. PRI Project, Sustainable Development.

Ozawa, C., and Susskind, L. (1985). Mediating science-intensive policy disputes. *Journal of Policy Analysis and Management*, **5**(1), 23–39.

Page, B. (2003). Has widening participation in decision-making influenced water policy in the UK? *Water Policy*, **5**, 313–329.

Painter, A. (1995). Resolving environmental conflicts through mediation. In *Water quantity/quality management and conflict resolution: Institutions, processes and economic analyses*, edited by A. Dinar and E. T. Loehman. Westport, CT: Praeger.

Paisley, R. (2003). Adversaries into partners: International water law and the equitable sharing of downstream benefits. *Melbourne Journal of International Law*, **3**, 280–300.

Palmer, R., Werick, W. J., MacEwan, A., and Woods, A. W. (1999). Modeling water resources opportunities, challenges, and trade-offs: The use of shared vision modeling for negotiation and conflict resolution. *Preparing for the 21st Century, Proceedings of the 26th Annual Water Resource Planning and Management Conference*, ASCE, Reston, VA.

Paoletto, G., and Uitto, J. I. (1996). The Salween River: Is international development possible? *Asia Pacific Viewpoint*, **37**(3, December), 269–282.

Paul, S. (1987). *Community participation in development projects: The World Bank experience*. World Bank Discussion Paper, Washington, DC.

Paul, S. (1991). *The Bank's work on institutional development in sectors, emerging tasks, and challenges*. World Bank Country Economics Department, Public Sector Management and Private Sector Development Division, Washington, DC.

Paul, T. V. (1994). *Asymmetric conflicts: War initiation by weaker powers*. Cambridge, UK: Cambridge University Press.

Peabody, N. S., III. (editor) (1991). *Water policy innovations in California: Water resources management in a closing water supply system*. Discussion paper No. 2, Winrock International Institute for Agriculture Development, Water Resource and Irrigation Policy Program, Center for Economic Policy Studies, December 1991.

Pearce, F. (2003). Conflict Looms over India's Colossal River Plan. *New Scientist*. Available at http://www.newscientist.com. Updated January 19, 2007.

Permanent Court of Arbitration. (1991). *New directions*. The Hague.

Peters, J. (1994). *Building bridges: The Arab–Israeli multilateral talks*. London: Royal Institute of International Affairs.

Plan Director Global Binacional de Protección (1995). Prevención de Inundaciones y Aprovechamiento de los Recursos del Lago Titicaca, Río Desaguadero, Lago Poopo y Lago Salar de Coipasa (Sistema TDPS). Comisión de las Comunidades Europeas, Repúblicas de Perú y Bolivia.

Postel, S., and A. Wolf. (2001). Dehydrating conflict. *Foreign Policy* (September/October), 60–67.

Postel, S. (1992). *Last oasis: Facing water scarcity*. New York: Norton.

Postel, S. (1999). *Pillar of sand: Can the irrigation miracle last?* New York: Norton.

Pritchett, C. H. (1943). *The Tennessee Valley Authority*. Chapel Hill, NC: University of North Carolina Press.

Public Awareness and Water Conservation. (2007). *Water Care*. Available at www.watercare.org. Updated January 19, 2007.

Puri, S. (2003). Transboundary aquifer resources: International water law and hydrogeological uncertainty. *Water International*, **28**(2), 276–279.

Puri, S., and El Naser, H. (2002). Intensive use of groundwater in transboundary aquifers. In *Intensive use of groundwater*, edited by Llamas, M. R. and Custudio, E. Balkema: The Netherlands, pp. 415–438.

Puri, S., Gaines, L., Wolf, A., and Jarvis, T. (2002). Lessons from intensively used transboundary river basin agreements for transboundary aquifers. *Proceedings of the Symposium on Intensive Use of Groundwater: Challenges and Opportunities*, Valencia, Spain, December 10–14, 2002.

Puri, S., Appelgren, B., Arnold, G., Aureli, A., Burchi, S., Burke, J., Margat, J., Pallas, P., and von Igel, W. (2001). *Internationally shared (transboundary) aquifer resources management, their significance and sustainable management: A framework document*. IHP-VI, International Hydrological Programme, Non-Serial Publications in Hydrology SC-2001/WS/40 Paris, France: UNESCO.

Racellis, M. (1992). People's participation: The use of UNICEFs experience. *World Bank Workshop on Participatory Development*, Washington, DC, February 26–27, 1992.

Radosevich, G., and Olson, D. (2002). Institutional change for managing the Tarim River basin China. In *Integrated water management at river basin level: An institutional development perspective*, edited by G. Alearts and G. LeMoigne. Washington, DC: RFF for the World Bank.

Raiffa, H. (1982). *The art and science of negotiation*. Cambridge, MA: Belknap Press.

Ramu, K., and Herman, T. (2002). River basin management corporation: An Indonesian approach. In *Integrated water management at river basin level: An institutional development perspective*, edited by G. Alearts and G. LeMoigne. Washington, DC: RFF for the World Bank.

Rangeley, W., and Kirmani, S. (1992). *International inland waters: Concepts for a more proactive role for the World Bank. Draft Background paper*, Washington, DC: World Bank.

Raskin, P. D., Hansen, E., and Margolis, R. M. (1996). Water and sustainability global patterns and long-range problems. *Natural Resources Forum*, **20**(1), 1–15.

Ravnborg, H. Munk (2004). From water "wars" to water "riots"? Lessons about trans-boundary water-related conflict and cooperation. In *Water and conflict conflict prevention and mitigation in water resources management*, edited by H. Munk Ravnborg. Copenhagen: Danish Institute for International Studies (DIIS) Report 2004, p. 2.

REC Caucasus (2001). *News from REC Caucus: Water*. Regional Environment Centre for the Caucasus, August (3).

Reisner, M. (1986). *Cadillac Desert: The American West and its disappearing water*. New York: Viking.

Remans, W. (1995). Water and war. *Humantäres Völkerrecht*, **8**(1), 1–14.

Rende, M. (2004). *Water transfer from Turkey to water stressed countries in the Middle East*. Paper presented at the Water for Life in the Middle East 2nd Israeli–Palestinian International Conference, October 10–14. Turkey.

Reuss, M. (editor). (2002) Learning from the Dutch: Technology, management, and water resources development. *Technology and Culture*, **43**(3, July), 465–472.

Revollo, M. (2001). Case report: Management issues in the Lake Titicaca and Lake Poopo system: Importance of developing a water budget. *Lakes and Reservoirs: Research and Management*, **6**(3), 225.

Revollo, M., Cruz, M., and Rivero, A. (2003). Lake Titicaca. *Lake Basin Management Initiative*. Available at http://www.worldlakes.org/lakedetails.asp?lakeid=8592. Updated January 19, 2007.

Rinaudo, J. D. (2002). Corruption and allocation of water: The case of public irrigation in Pakistan. *Water Policy*, **4**, 405–422.

Ringler, C. (2001). Optimal allocation and use of water resources in the Mekong River basin: Multicountry and intersectoral analyses. *Development Economics and Policy*, **20**. Peter Lang. GmbH. Europaischer Verlag der Wissenschapften, Frankfurt aim Main.

Rogers, P. (1969). A game theory approach to the problems of international river basins. *Water Resources Research*, **5**(4), 49–760.

Rogers, P. (1991). *International river basins: Pervasive unidirectional externalities*. Presented at a conference on The Economics of Transnational Commons, Universita di Siena, Italy, April 25–27.

Rogers, P. (1992a). *Comprehensive water resources management: A concept paper*. Washington, DC: World Bank.

Rogers, P. (1992b). *A note on economic benefits of cooperation on international rivers development, draft memo*, World Bank Water Policy Study. World Bank: Washington, DC.

Rogers, P. (1993). The value of cooperation in resolving international river basin disputes. *Natural Resources Forum*, (May), 117–131.

Rogers, P. (2002). *Water governance in Latin America and the Caribbean*. Washington DC: Inter American Development Bank, February.

Ronteltap, M., Rieckermann, J., and Daebel, H. (2004). *Management efforts at Lake Titicaca. The science and politics of international freshwater management*. Zurich: Swiss Federal Institute of Technology.

Rosegrant, M. W., and Cai, X. (2002). Global water demand and supply projections, Part 2. Results and prospects to 2025. *Water International*, **27**(2), 170–182.

Rosenne, S. (1995). *The World Court. What it is and how it works. Legal aspects of international organization, Bd. 16*. Dordrecht, the Netherlands: Martinus Nijhoff Publishers.

Roth, E. (2001). *Water pricing in the EU: A review*. Publication 2001/002, Brussels: European Environmental Bureau.

Rothfelder, J. (2003). Water rights, conflict, and culture. *Water Resources Impact*, **5**(2), 19–21.

Rothman, J. (1989). Supplementing tradition: A theoretical and practical typology for international conflict management. *Negotiation Journal*, **5**(3).

Rothman, J. (1995). Pre-Negotiation in water disputes: Where culture is core. *Cultural Survival Quarterly*, **19**(3), 19–22.

Rothman, J. (1997). *Resolving Identity-Based Conflicts in Nations, Organizations, and Communities*. San Francisco: Jossey-Bass.

Rudquist, A. (1992). *SIDA's experience with popular participation*. World Bank workshop on participatory development, Washington, DC, February 26–27, 1992.

Sabatier, P. A., Focht, W., Lubell, M., and Trachtenberg, Z. (editors). (2005). *Swimming upstream: Collaborative approaches to watershed management*. Cambridge, MA: MIT Press.

Sadoff, C. W., and Grey, D. (2002). Beyond the river: The benefits of cooperation on international rivers. *Water Policy*, **4**(5), 389–404.

Sadoff, C. W., and Grey, D. (2005). Cooperation on international rivers: A continuum for securing and sharing benefits. *Water International*, **30**(4, December), 420–427.

Sadria, M. (1998). Iranian strategies in central Asia. In *Central eurasian water crisis: Caspian, Aral and Dead Seas*, edited by Kobori and Glantz Tokyo, New York, Paris: United Nations University Press.

Saleth, R. M., and Dinar, A. (1998). *Institutional response to water challenge: A cross country perspective*. Report to the Water Sector Performance Studies, World Bank, March 25, 1998.

Saleth, R. M., and Dinar, A. (2005). Water institutional reforms: Theory and practice. *Water Policy*, **7**, 1–19.

Salman, L. (1989). *Benefitting assessments: An approach described*. World Bank Workshop on Participatory Development, Washington, DC, February 26–27, 1992.

Salman, S. M. A., and Boisson de Chazournes, L., (editors). (1998). *International watercourses: Enhancing cooperation and managing conflict*. Washington, DC: World Bank, Technical Paper No. 414.

Salman, S. M. A., and Uprety, K. (2002). Conflict and cooperation on South Asia's international rivers: A legal perspective. Washington, DC: World Bank Publications.

Samson, P., and Charrier, B. (1997). *International freshwater conflict: Issues and prevention strategies*. Geneva: Green Cross International.

Sanjinés, J. (1996). *Informe de gestión*. Proinsa, La Paz, Bolivia.

Savenije, H. H. G. (2001). *Why water is not an ordinary economic good*. IHE Delft Value of Water Research Report Series No. 9.

Savenije, H. H. G., and van der Zaag, P. (editors). (2000). Management of shared river basins; with special reference to the SADC and EU, *Water Policy*, **2**(1/2), 9–45.

Scanlon, J., Cassar, A., and Nemes, N. (2004). *Water as a human right?* IUCN Environmental Policy and Law Paper No. 51.

Schad, T. (1964). Legislative history of Federal River basin Planning Organization. In *Organization of methodology of river basin planning*, edited by C. E. Kinsvater. Atlanta: Georgia Institute of Technology, Water Resources Center.

Schama, F. S. (1995). *Landscape and memory*. London: Harper Collins, p. 288.

Schmida, L. (1983). *Keys to control: Israel's pursuit of Arab water resources*. Washington, DC: American Educational Trust.

Secretariat of the Interim Committee for Coordination of Investigations of the Lower Mekong Basin. (1989). *The Mekong Committee: A historical account (1957–89)*. Bangkok.

Segal, D. (2004). *Singapore's Water Trade with Malaysia and Alternatives*. Master's thesis, John F. Kennedy School of Government, Harvard University.

Selznick, P. (1953). *TVA and the grass roots*. Berkeley and Los Angeles: University of California Press.

Senegal River Basin Authority. (2006). Organisation de Mise en Valeur de la Vallee du Fleuve Senegal.Available at http://www.omvs-hc.org.

Serageldin, I. (1995). *New York Times*. The wars of the next century will be about water. August 10.

Sewell, W. R. Derrick (1966). *Comprehensive river basin planning: The Lower Mekong experience*. Madison, WI: University of Wisconsin Water Resources Center, June.

Shahin, M. (1986). Discussion of the paper "Ethiopian interests in the division of the Nile River waters." *Water International*, **11**(1, March), 16–20.

Shamir, U. (1999). Water and peace: How can international water agreements be influenced by the public? *Participatory Processes in Water Management (PPWM) – Proceedings of the Satellite Conference to the World Conference on Science: International Conference on Participatory Processes in Water Management*, June 28–30, 1999. Budapest, Hungary, edited by Gayer, J. UNESCO-IHP/IA2P/IWRA. Paris: UNESCO- IHP-V.

Shamir, Y. (2003). *Alternative dispute resolution approaches and their application*. Paris: UNESCO IHP Technical Documents in Hydrology, PCCP Series No. 7.

Shela, O. N. (1999). The management of shared river basins: The Case of the Zambesi. *River Basin Management: Proceedings of the UNESCO International Workshop*, The Hague, October 27–29, 1999, edited by E. Mostert. IHP-V Technical Document in Hydrology No. 31.

Sherk, G. W. (2003). East meets West: A tale of two water doctrines. *Water Resources Impact*, **5**(2), 5–8.

Shevchenko, M., Rodionov, V., and Kindler, J. (2002). The Volga River Management Organization. In *Integrated water management at river basin level: An institutional development perspective*, edited by G. Alearts and G. LeMoigne. Washington, DC: RFF for the World Bank.

Shiklomanov, I. A. (1993). World fresh water resources. In *Water in crisis. A guide to the world's fresh water resources*, edited by P. H. Gleick. New York: Oxford University Press, pp. 13–24.

Shmueli, D. F., and Shamir, U. (2001). Application of international law of water quality to recent Middle East water agreements. *Water Policy*, **3**, 405–423.

Shubik, M. (1984). *A game-theoretic approach to political economy: Volume 2 of game theory in the social sciences*. Cambridge, MA: MIT Press.

Shubik, M. (2002). Game theory and operations research: Some musings 50 years later. *Operations Research*, **50**, 192–196.

Shue, H. (1992). The unavoidability of justice. In *The international politics of the environment: Actors, interests, and institutions*, edited by A. Hurrell and B. Kingsbury. New York: Oxford University Press, pp. 373–397.

Shuval, H. (1992). Approaches to resolving the water conflicts between Israel and her neighbors – A regional water-for-peace plan. *Water International*, **17**(3, September), 133–143.

Simonovic, S. P. (1996). Decision support systems for sustainable management of water resources: General principles. *Water International*, **21**(4, December), 223–232.

Singapore Ministry of Foreign Affairs. (2006). Chronology of developments related to water issue. Accessed at http://www.mfa.gov.sg/internet/press/water/event.htm, December 20. Edited 2007.

Sluimer, G., and Xie, M. (2002). The Mekong Delta. In *Integrated water management at river basin level: An institutional development perspective*, edited by G. Alearts and G. LeMoigne. Washington, DC: RFF for the World Bank.

Smith, C. (2006). 2006 review. *Journal of the American Water Resources Association*, issue edited by Sabatier et al., (February), 257–258.

Smith, N. (1975). *Man and water: A history of hydro-technology*. London: Charles Scribner and Sons.

Smith, T. J. (2003). Native American water rights. *Water Resources Impact*, **5**(2), 16–18.

Song, J., and Whittington, D. (2004). Why have some countries on international rivers been successful negotiating treaties? A global perspective. *Water Resources Research*, **40**(W05S06).

SOPAC. (2006). Water Resources. Accessed at http://www.sopac.org.fj/tiki/tiki-index.php?page = CLP+Water+Resources+Assessment+and+Sanitation. December 21. Edited June 3, 2007.

Spector, B. I. (2000). Motivating water diplomacy: Finding the situational incentives to negotiate. *International Negotiation*, **5**(2), 223–236.

Sprout, H., and Sprout, M. (1957). Environmental factors in the study of international politics. *Journal of Conflict Resolution*, **1**, 309–328.

Stahl, K., and Wolf, A. T. (2003). Does hydro-climatic variability influence water-related political conflict and cooperation in international river basins? *Proceedings CD of the International Conference on Hydrology of the Mediterranean and Semi-Arid Regions*, Montpellier, France, April 1–4, 2003.

Starr, J. R. (1991). Water wars. *Foreign Policy*, **82**(Spring), 17–36.

Stauffer, T. (1982). The price of peace: The spoils of war. *American-Arab Affairs*, **1**, 43–54.

Stein, J. G. (1988). International negotiation: A multidisciplinary perspective. *Negotiation Journal*, July, 221–231.

Steiner, R. C., Hagen, E. R., and Ducnuigeen J. (2000). *Water supply demand and resources analysis in the Potomac River basin*. Rockville, MD: Interstate Commission on the Potomac River Basin.

Stern, P. C., and Druckman, D. (2000). Evaluating interventions in history: The case of international conflict resolution. *International Studies Review*, **2**(1), 33–63.

Stevens, G. (1965). *Jordan River partition*. Stanford: Hoover Institution.

Stork, J. (1983). Water and Israel's occupation strategy. *MERIP Reports*, **116**(13/6), 19–24.

Stucki, P. (2005). *Water wars or water peace? Rethinking the nexus between water scarcity and armed conflict*. Geneva, Switzerland: Programme for Strategic and International Security Studies (PSIS) Occasional Paper No. 3/2005.

Sullivan, C. A., Meigh, J. R., Giacomello, A. M., Fediw, T., Lawrence, P., Samad, M., Mlote, S., Hutton, C., Allan, J. A., Schulze, R. E., Dlamini, D. J. M., Cosgrove, W., Delli Priscoli, J., Gleick, P., Smout, I., Cobbing, J., Calow, R., Hunt, C., Hussain, A., Acreman, M. C., King, J., Malomo, S., Tate, E. L., O'Regan, D., Milner, S., and Steyl, I. (2003). The water poverty index: Development and application at the community scale. *Natural Resources Forum*, **27**(3, August), 189–199.

Sun Belt Water, Inc. (2006). Water – The Solution – Sun Belt Water, Inc. Acessed at http://www.sunbeltwater.com/. December 20.

Susskind, L., and Cruikshank, J. (1987). *Breaking the impasse: Consensual approaches to resolving public disputes*. New York: Basic Books.

Swain, A. (2004). *Managing water conflict: Asia, Africa and the Middle East*. London and New York: Routledge.

Swallow, B. M., Garrity, D. P., and van Noordwijk, M. (2001). The effects of scales, flows, and filters on property rights and collective action in watershed management. *Water Policy*, **3**, 457–474.

Tanzi, A. (1997). Codifying the minimum standards of the law of international watercourses: Remarks on part one and a half. *Natural Resources Forum*, **21**, 109–117.

Tanzeema, S., and Faisal, I. M. (2001). Sharing the Ganges: Acritical analysis of the water sharing treaties. *Water Policy*, **3**(1), 13–28.

TeBrake, W. H. (2002). Taming the waterwolf: Hydraulic engineering and water management in the Netherlands during the Middle Ages. *Technology and Culture*, **43**(3, July), 475–499.

Technical Review Middle East. (2001). Syria and Iraq hold talks. July 31, 2001.

Teclaff, L. (1967). *The river basin in history and law*. Buffalo, NY: W. S. Hein and The Hague: Marinus Nijihoff.

Teclaff, L. A. (1996). Evolution of the river basin concept in national and international water law. *Natural Resources Journal*, **36**(Spring), 359–352.

Teclaff, L. (2001). Fiat or custom: The checkered development of international water law. *Natural Resources Journal*, **31**(1), 45–73.

This American Life. (2003). *The Middle of Nowhere*. WBEZ Chicago. Radio recording accessed at http://www.thislife.org/. December 5.

Thomas, K. (1976). Conflict and conflict management. In *Handbook of industrial and organizational psychology*, edited by M. C. Dunnett. Chicago, IL: Rand-McNally.

Thomas, L. (1992). *The fragile species*. New York: Maxwell Macmillan International.

Tigran, Y. (2002). *Transboundary water resources and political tensions in south Caucasus*. Gland, Switzerland: Global Biodiversity Forum.

Tortajada, C., and Contreras-Moreno, N. (2005). Institutions for water management in Mexico. In *Water institutions: Policies, performance, and prospects*, edited by C. Gopalakrishnan, C. Tortajada, and A. K. Biswas. Berlin: Springer-Verlag, pp. 99–127.

Toset, H. P. W., Gleditsch, N. P., and Hegre, H. (2001). Shared rivers and interstate conflict. *Political Geography*, **19**(6), 971–996. Data available at http://www.prio.no/cwp/datasets.asp.

Toynbee, A. (1946). *A study of history*. Oxford: Oxford University Press.

Toynbee, A. (1958). Review: *Oriental Despotism*, by Karl Wittfogel, *American Political Science Review*, **52**(March), 195–198.

Trans-Caledon Tunnel Authority (TCTA). (2006a) LHWP funding. Accessed at http://www.tcta.co.za/article.jsp?article_id = 43, December 12.

Trans-Caledon Tunnel Authority (TCTA). (2006b). LHWP history. Accessed at http://www.tcta.co.za/article.jsp?article_id = 41, December 12.

Transboundary Freshwater Dispute Database (TFDD). (2006). Oregon State University. Available at http://www.transboundarywaters.orst.edu/.

Trolldalen, J. M. (1992). International river systems. In *International environmental conflict resolution: The role of the United Nations*, Oslo. Washington, DC: World Foundation for Environment and Development, pp. 61–91, 174–175.

Trolldalen, J. M., and Wennesland, T. (1998). (editors).*Strengthening the Capacity of the United Nations in prevention and resolution of international environmental conflicts*. Oslo, Norway: CESAR.

Trottier, J. (2000). Water and the challenge of Palestinian institution building. *Journal of Palestine Studies*, **29**(2, Winter), 35–50.

Trottier, J. (2003a). *The need for a multiscalar analyses in the management of shared water resources*. Paris: UNESCO IHP Technical Documents in Hydrology, PCCP Series No. 6.4.

Trottier, J. (2003b). *Water wars: The rise of a hegemonic concept*. Paris: UNESCO IHP Technical Documents in Hydrology, PCCP Series No. 6.8.

Tsur, Y., and Easter, W. (1994). The design of institutional arrangements for water allocation. In *Water quantity/quality management and conflict resolution*, edited by A. Dinar and E. Loehman. Westport, CT: Praeger Publishers.

Tuan, N. D., and Thanh, L. D. (1999). River basin management in Vietnam. *River Basin Management: Proceedings of the UNESCO International Workshop*, The Hague, October 27–29, 1999, edited by E. Mostert. IHP-V Technical Document in Hydrology No. 31.

Turkish Daily News. (1999). Manavgat "Peace Water" ready for export. September 25.

Turkish Daily News. (2001). Turkey's water sale policy. August 3.

Turkish Daily News. (2005). Turkey launches water pipeline project to Turkish Cyprus. October. Accessed at http://www.tusiad.us/specific_page.cfm?CONTENT_ID=570.

Turkish Daily News. (2006). Privatization plays a role in scrapping of water project. April 8.

Turner, B. L., II, Kasperson, R. E., Matson, P. A., McCarthy, J. J., Corell, R. W., Christensen, L., Eckley, N., Kasperson, J. X., Luers, A., Martello, M. L., Polsky, C., Pulsipher, A., and Schiller, A. (2003). A framework for vulnerability analysis in sustainability science. *Proceedings of the National Academy of Science (USA)*, **100**(14), 8074–8079.

Turton, A. R. (1999). W*ater and state sovereignty: The hydropolitical challenge for states in arid regions*. London: MEWREW Occasional Paper No. 5, Water Issues Study Group, School of Oriental and African Studies (SOAS), University of London.

Turton, A. R. (2001). *Hydropolitics and security complex theory: An African perspective*. 4th Pan-European International Relations Conference, University of Kent, Canterbury UK, September 8–10, 2001.

Turton, A. R. (2005). A critical assessment of the river basins at risk in the Southern African hydropolitical complex. Paper presented at the Workshop on the Management of International Rivers and Lakes, hosted by the Third World Centre for Water Management and the Helsinki University of Technology, Helsinki, Finland, August 17–19, 2005. (Forthcoming chapter in a book as yet untitled.) CSIR Report No. ENV-P-CONF 2005–001. Available online at http://www.awiru.co.za/pdf/7Critical%20Assessment%20of%20Basins%20at%20Risk%20in%20the%20Southern%20African%20Hydropolitical%20context.pdf

Turton, A. R., and Earle, A. (2005). Post-apartheid institutional development in selected southern African international river basins. In *Water institutions: Policies, performance, and prospects*, edited by C. Gopalakrishnan, C. Tortajada, and A. K. Biswas. New York: Springer.

Turton, A. R., Ashton, P., and Cloete, T. E. (editors). (2003). *Transboundary rivers, sovereignty and development: Hydropolitical drivers in the Okavango River basin*. Pretoria and Geneva: AWIRU & Green Cross International.

Ubilava, M. (2004). *Water management in south Caucasus*. Presented at Integrated Water Management of Transboundary Catchments: A Contribution from Transcat, March 24–26, Venice, Italy. Available at http://www.feemweb.it/transcat_conf/conf_papers/Ubilava.pdf.

Uitto, J. I., and Duda, A. M. (2001). Management of transboundary water resources: Lessons from international cooperation for conflict prevention. *Geographical Journal*, **168**(4), 365–378.

United Nations. (1975). *Management of international water resources: Institutional and legal aspects*. Report of the panel of experts on the legal and institutional aspects of international water resources development, New York. Natural Resources/Water Series No. 1.

United Nations. (1978). Register of international rivers. *Water Supply Management*, **2**(1), 1–58.

United Nations. (1994). *ILC draft articles on the non-navigational uses of international watercourses, 1994*. UN Doc. A/CN.4/L492 (1994). For history and commentary, see United Nations. *Yearbook of the ILC* from 1974–1991.

United Nations. (2002a). *Report of the World Summit on Sustainable Development*. Document A/CONF.199/20, Sales No. E.03.II.A.1. New York: United Nations.

United Nations. (2002b). United Nations Treaty Collection. Available at http://untreaty.un.org/English/treaty.asp.

United Nations. (2003). *Draft environmental performance review of Georgia: First Review*. Economic and Social Council, Special Session, February 18–19.

United Nations. (2005). Convention on the Law of the Non-navigational Uses of International Watercourses. Adopted by the General Assembly of the United Nations on 21 May 1997. Not yet in force. See General Assembly Resolution 51/229, annex, *Official Records of the General Assembly, Fifty-first Session, Supplement No. 49* (A/51/49). Available at http://untreaty.un.org/ilc/texts/instruments/english/conventions/8_3_1997.pdf

United Nations. (2007). United Nations Development Programme, Millennium Development Goals (MDGs). Available at http://www.undp.org/mdg/.

United Nations Development Programme (UNDP). (1987). *Restoring the heart of Nicosia: Master plan to the Year 2000*. November 1987.

United Nations Economic Commission for Europe (UNECE). (1998). UNECE Convention on Access to Information, Public Participation in Decision-making and Access to Justice in Environmental Matters (Aarhus Convention), adopted June 25, 1998, at the Fourth Ministerial Conference in the "Environment for Europe" process, Aarhus, Denmark. Available at http://www.unece.org/env/pp/.

United Nations Committee on Economic Development (UNCED). (1993). *Agenda 21: Earth Summit – The United Nations Programme of Action from Rio*. New York: United Nations Publications.

United Nations Educational, Scientific and Cultural Organization (UNESCO). (1991). *Hydrology and water resources of small islands: A practical guide*. A contribution to the International Hydrological Programme, IHP-111, Project 4.6.

United Nations Educational, Scientific and Cultural Organization (UNESCO). (2003). Lake Titicaca Basin, Bolivia and Peru. *The UN World Water Development Report: Water for People, Water for Life*. Available at http://www.unesco.org/water/wwap/wwdr/table_contents.shtml. Updated January 19, 2007.

United Nations Educational, Scientific and Cultural Organization – UNESCO-PCCP. (2007). From potential conflict to co-operation potential project, UNESCO International Hydrological Programme (IHP) contribution to the World Water Assessment Programme (WWAP). See Web site, http://www.unesco.org/water/wwap/pccp/.

United Nations Environment Programme (UNEP). (1996). Rapid integrated river basin assessments. *Proceedings UNEP Workshop*, Tarleton State University, 20–22 February Tarleton State, Stephenville, TX.

United Nations Environment Programme (UNEP). (1998). *Sourcebook of alternative technologies for freshwater augmentation in Latin America and the Caribbean*. Division of Technology, Industry, and Economics. Technical Publication Series 8.

United Nations Environment Programme (UNEP) and Oregon State University. (2002). *Atlas of international freshwater agreements*. Nairobi: UNEP Press.

United Nations Environment Programme (UNEP), United Nations Development Programme (UNDP), and Organization for Security and Co-operation in Europe (OSCE). (2003). Environment and security: Transforming risks into cooperation, The case of Central Asia and South Eastern Europe. Switzerland (UNEP), Slovakia (UNDP), and Austria (OSCE). Available at http://www.iisd.org/pdf/2003/envsec_cooperation.pdf.

United Nations Environment Programme (UNEP) and the Woodrow Wilson Center. (2004). *Understanding environmental conflict and cooperation*. Nairobi: UNEP-DEWA.

United Nations Food and Agriculture Organization (FAO). (1978). *Systematic index of international water resources treaties, declarations, acts and cases, by basin: Volume I*. Legislative Study No. 15.

United Nations Food and Agriculture Organization (FAO). (1984). *Systematic index of international water resources treaties, declarations, acts and cases, by basin: Volume II*. Legislative Study No. 34.

United Nations Food and Agriculture Organization (FAO). (2003). *Groundwater management – The search for practical approaches*. A joint publication of the FAO Land and Water Development Division, the International Atomic Energy Agency, the United Nations Department of Economic and Social Affairs and the United Nations Educational, Scientific and Cultural Organization (UNESCO) with contributions from the Institute for Social and Environmental Transition Boulder, Colorado, the International Association of Hydrogeologists, Kenilworth, United Kingdom. Water Report 25. Rome: Food and Agriculture Organization of the United Nations.

United Nations International Strategy for Disaster Reduction (UN/ISDR). (2004). *Living with risk: A global review of disaster reduction initiative Inter-Agency Secretariat of the International Strategy for Disaster Reduction (UN/ISDR)*. Geneva, Switzerland. http://www.unisdr.org/eng/about_isdr/bd-lwr-2004-eng-p.htm.

United Nations. *State of the world population 2004: the Cairo consensus at Ten – population, reproductive health and the global effort to end poverty* (1994). New York: United Nations Population Fund.

United Press International. (2003). Turkish prime minister to visit Syria. 3 January.

United States Army Corps of Engineers (USACE). (1991). *Water in the sand: A survey of Middle East water issues*. Draft. Corps Water Management Systems (CWMS). Washington, DC: U.S. Army Corps of Engineers, Institute for Water Resources (IWR).

United States Army Corps of Engineers (USACE). (2000). *Columbia River Treaty briefing*. Institute for Water Resources, Alexandria, VA, April 24, 2000.

United States Army Corps of Engineers (USACE). (2004). *Flood control: Value to the nation*. Brochure for the U.S. Congress, Alexandria, VA: Institute for Water Resources.

United States Army Corps of Engineers (USACE). (2006). *Shared vision planning*. Alexandria, VA: Institute for Water Resources.

United States Army Corps of Engineers (USACE). (2007). Alexandria, VA: Institute for Water Resources. Accessed on December 21, 2007. Available at http://www.iwr.uasce.army.mil/.

United States National Water Commission. (1974). *Summary of the National Water Commission*. Arlington, VA: United States National Water Commission

United States Water Resources Council. (1967). *Alternative institutional arrangements for managing river basin operations*. Washington, DC: United States Water Resources Council.

Uphoff, N. (1992). *Learning from Gal Oya. Possibilities for participatory development and post-Newtonian social science*. Ithaca, NY: Cornell University Press.

Ury, W. L. (1987). Strengthening international mediation. *Negotiation Journal*, July, 225–229.

Ury, W. (1991). *Getting past no: Negotiating your way from confrontations to cooperation*. New York: Bantam Books.

Ury, W. L., Brett, J. M., and Goldberg, S. B. (1988). *Getting disputes resolved: Designing systems to cut the costs of conflict*. San Francisco: Jossey-Bass.

U.S. Water News Online. (2002). Israel and Turkey strike 20-year water deal. August.

U.S. Water News Online. (2006). Israel, Turkey put landmark water agreement into deep freeze. April.

Utton, A. (editor). (1992). *Borders and water: North American issues*. International Transboundary Center, Santa Fe, NM: University of New Mexico.

Utton, R. (1996). Which rule should prevail in international water disputes: That of reasonableness or that of no harm? *Natural Resources Journal*, **36**, 635–641.

Vadas, R. G. (1999). The São Francisco. *River Basin Management: Proceedings of the UNESCO International Workshop*, The Hague, October 27–29, 1999, edited by E. Mostert. IHP-V Technical Document in Hydrology No. 31.

van Dam, P. J. E. M. (2002). Ecological challenges, technological innovations: The modernization of sluice building in Holland, 1300–1600, *Technology and Culture*, **43**(3), 500–520.

Van der Zaag, P., Seyam, I. M., and Savenije, H. H. G. (2002). Towards measurable criteria for the equitable sharing of international water resources. *Water Policy*, **4**, 19–32.

Valente, M. (2002). South America: MERCOSUR vows to take over huge water reserve, Mercosur Article, *Inter Press Service* (IPS)/Global Information Network, July 22, pp. 1–2.

Varady, R. G., and Iles-Shih, M. (2005). Global water initiatives: What do the experts think? Report on a survey of leading figures in the "World of Water." In *Impacts of mega-conferences on global water development and management*, edited by A. K. Biswas. New York: Springer Verlag.

Vaz, A. C. (1999). Incomati. *River Basin Management: Proceedings of the UNESCO International Workshop*, The Hague. October 27–29, 1999, edited by E. Mostert. Technical Document in Hydrology No. 31, IHP-V.

Vaz, A. C. and Pereira, A. L. (1999). The Incomati and Limpopo, *Savenije*.

Vlachos, E. (1990). Water, peace and conflict management. *Water International*, **15**, 185–188.

Vlachos, E. (1991). Water awareness and implementing action in local communities. In *Water awareness in societal planning and decision-making*, edited by I. Johansson. Stockholm: Byggforskningsradet, pp. 229–242.

Vlachos, E. (1994). *Transboundary water conflicts and alternative dispute resolution*. Cairo: International Water Resources Association, VIIIth Congress.

Vlachos, E., Webb, A. C., and Murphy, I. (editors). (1986). *The management of international river basin conflicts*. Proceedings of a Workshop, Laxenburg, Austria, September 22–25, 1986. Washington, DC: George Washington University.

Vogel, C., and O'Brien, K. (2004). Vulnerability and global environmental change: Rhetoric and reality. *AVISO Bulletin No. 13*. Ottowa, ON: GECHS, pp. 1–8.

Walden, G. (2004). Conflicting water needs in the Klamath basin. *Water Resources Impact*, **6**(1), 4–6.

Warshall, P. (1985). The morality of molecular water. *Whole Earth Review*, **85**(Spring), 4–11.

Washington Institute for Near East Studies. (2003). *Turkish water to Israel?* PolicyWatch No. 782. August 14.

Water Exports NZ Limited. (2007). Accessed at http://www.waterexportsnz.com. January 13.

Water Bank. (2007). Accessed at http://www.waterbank.com/. January 13.

Waterbolk, II. T. (1962). The lower Rhine basin. In *Courses toward urban life: Archeological considerations of some cultural alternates*, edited by R. J. Braidwood and G. R. Willey. Chicago: Aldine Pub. Co.

Waterbury, J. (1979). *Hydropolitics of the Nile Valley*. New York: Syracuse University Press.

Waterbury, J. (1993). *Transboundary water and the challenge of international cooperation in the Middle East*. Presented at a symposium on water in the Arab World, Harvard University, October 1–3, 1993.

Waterbury, J. (2002). *The Nile basin: National determinants of collective action*. New Haven and London: Yale University Press.

Wendall and Schwan (1975). *Intergovernmental relations in water resources activities, for the National Water Commission of the United States*. Arlington, VA: U.S. Department of Commerce PB 210358.

Wescoat, J. L., Jr. (1992). Beyond the river basin: The changing geography of international water problems and international watercourse law. *Colorado Journal of International Environmental Law and Policy*, **3**, 301–330.

Wescoat, J. L., Jr. (1996). Main currents in early multilateral water treaties: A historical–geographic perspective, 1648–1948. *Colorado Journal of International Environmental Law and Policy*, **7**(1), 39–74.

Westing, A. H. (1986). *Global resources and international conflict: Environmental factors in strategic policy and action*. New York: Oxford University Press.

White, G. (1963). The Mekong River plan. *Scientific American*, **208**(4), 49–59.

White, G. (1969). *Strategies of American water management*. Ann Arbor, MI: University of Michigan Press.

White, G. F. (1974). *Natural hazards: Local, national, global*. New York: Oxford University Press.

Whittington, D. (2004). Visions of Nile basin development. *Water Policy*, **6**(1), 1–24.

Whitington, D., and McClelland, E. (1992). Opportunites for regional and international cooperation in the Nile Basin. *Water International*, **17**(4), 144–154

Whittington, D., and Sadoff, C. (2005). Water resources management in the Nile basin: The economic value of cooperation. *Water Policy*, **7**, 227–252.

Wilkenfeld, J., and Brecher, M. (1997). *A Study of crisis*. Ann Arbor, MI: University of Michigan Press.

Williamson, S. H. (2006). Exchange rate between the United States dollar and forty other countries, 1913–2005. EH.Net (supported by Economic History Association), Accessed at http://eh.net/hmit/exchangerates, December 4.

Wishart, D. (1989). An economic approach to understanding Jordan Valley water disputes. *Middle East Review*, **21**(4), 45–53.

Wishart, D. (1990). The breakdown of the Johnston negotiations over the Jordan waters. *Middle Eastern Studies*, **26**(4, October), 536–546.

Wittfogel, K. A. (1956). The hydraulic civilizations. In *Man's role in changing the face of the Earth*, edited by W. L. Thomas. Chicago: University of Chicago Press, pp. 152–164.

Wolf, A. T. (1993). Guidelines for a water-for-peace plan for the Jordan River watershed. *Natural Resources Journal*, **33**(3, July), 797–839.

Wolf, A. T. (1994). A hydropolitical history of the Nile, Jordan, and Euphrates River basins. In *International waters of the Middle East*, edited by A. Biswas. Oxford: Oxford University Press.

Wolf, A. T. (1995a). Rural nonpoint source pollution control in Wisconsin: The limits of a voluntary program? *Water Resources Bulletin*, **31**(6, December), 1009–1022.

Wolf, A. T. (1995b). *Hydropolitics along the Jordan River: Scarce water and its impact on the Arab–Israeli conflict*. Tokyo: United Nations University Press.

Wolf, A. T. (1995c). International water dispute resolution: The Middle East Multilateral Working Group on water resources. *Water International*, **20**(3), 141–150.

Wolf, A. T. (1997). International water conflict resolution: Lessons from comparative analysis. *International Journal of Water Resources Development*, **13**(3, September), 333–356.

Wolf, A. T. (1998). Conflict and cooperation along international waterways. *Water Policy*, **1**(2), 251–265.

Wolf, A. T. (1999a). Criteria for equitable allocations: The heart of international water conflict. *Natural Resources Forum*, **23**(1), 3–30.

Wolf, A. T. (1999b). The transboundary freshwater dispute database project. *Water International*, **24**(2, June), 160–163.

Wolf, A. T. (2000). Indigenous approaches to water conflict negotiations and implications for international waters. *International Negotiation: A Journal of Theory and Practice*, **5**(December), 357–233.

Wolf, A. T. (2001). *Transboundary waters: Sharing benefits, lessons learned.* Thematic Paper, Bonn Freshwater Conference, Bonn Germany November.

Wolf, A. T. (2002a). *Conflict and cooperation: Survey of the past and reflections for the future.* Geneva: Green Cross International.

Wolf, A. T. (editor). (2002b). *Conflict prevention and resolution in water systems.* Cheltenham, UK: Elgar.

Wolf, A. T. (editor). (2006). *Hydropolitical vulnerability and resilience along international waters.* (Five volumes: Africa, Latin America, North America, Asia, and Europe). Nairobi: UN Environment Programme.

Wolf, A. T. (editor). (2009, in press). *Sharing water, sharing benefits: working towards effective transboundary water resources management: A graduate/professional skills-building workbook.* Paris and Washington DC: UNESCO and The World Bank. Available at: http://www.unesco.org/water/wwap/pccp/.

Wolf, A. T., and Dinar, A. (1994). Middle East hydropolitics and equity measures for water-sharing agreements. *Journal of Social, Political, and Economic Studies*, **19**(4, Spring), 69–94.

Wolf, A. T., Natharius, J. A., Danielson, J. J., Ward, B. S., and Pender, J. K. (1999). International river basins of the world. *International Journal of Water Resource Development*, **15**(4), 387–427.

Wolf, A. T., Stahl, K., and Macomber, M. F. (2003a). Conflict and cooperation within international river basins: The importance of institutional capacity. *Water Resources Update*, **125**, 31–40.

Wolf, A. T., Yoffe, S. B., and Giordano, M. (2003b). International waters: Identifying basins at risk. *Water Policy*, **5**(1), 29–60.

Wolf, A. T., Kramer, A., Carius, A., and Dabelko, G. D. (2005). Managing water conflict and cooperation. In *Worldwatch Institute, State of the world 2005: Redefining global security.* Washington, DC: Worldwatch Institute.

World Bank. (1993). *Water resources management.* Washington, DC: World Bank.

World Bank. (1999). *Initiating and sustaining water sector reforms: A synthesis.* New Delhi, India: Allied Publisher.

World Bank. (2001). Senegal River Basin Water And Environmental Management Project. Project Brief, World Bank. Available at http://www.gefweb.org/Documents/Council_Documents/GEF_C18/Regional_Senegal_River_Basin.pdf. Updated 19 January 2007.

World Bank. (2002). China country water resources assistance strategy, 2002.

World Bank. (2004). Fueling cooperation: A regional approach to poverty reduction in the Senegal River basin. World Bank. Available at http://www.gefweb.org/Documents/Council_Documents/GEF_C18/Regional_Senegal_River_Basin.pdf. Updated January 19. 2007.

World Bank. (2006). *Water and Growth*, for WWF, Mexico City, March 22, 2006.

World Development Report. (1992). *Development and the environment.* Washington, DC: World Bank Group.

World Press Review. (1995). Next, wars over water? November 1995, pp. 8–13.

World Water Council. (2003). *Financing for all.* Marseilles: World Water Council.

World Water Assessment Program. (2003). 7 pilot case studies of water related stress in river basins, aquifers, cities and countries. *World Water Development Report.* March.

Worster, D. (1992). *Rivers of empire: Water, aridity, and the growth of the American West.* New York: Oxford University Press.

Wouters, P. (1996). An assessment of recent developments in international watercourse law through the prism of the substantive rules governing use allocation. *Natural Resources Journal*, **36**, 417–439.

Wouters, P. (2000). *Codification and progressive development of international water law.* Dordrecht: Kluwer.

Wouters, P. (2001). The legal response to international water scarcity and water conflicts. The UN watercourses convention and beyond. *Water Policy International*, UK. German Yearbook of International Law, 2000, pp. 292–336.

Wouters, P. (2003). Personal communication.

Wouters, P. (2003). *Sharing transboundary waters: User's guide and legal report.* Dundee, Scotland: University of Dundee, Knowledge and Research Project No. R8039.

WRM. (2000). *Bulletin*, April (33), Available at http://www.wrm.org.uy/bulletin/33/index.rtf.

Xia, J, Huang, G. H, Chen, Z., and Rong, X. (2001) An integrated planning framework for managing flood endangered regions in the Yangtze River Basin, *Water International*, **26**(2), 153–161.

Yardley, J. (2004). Dam building threatens China's Grand Canyon, *New York Times*, March 10.

Yian, N. G. (2001). A deal for the future. *Today*, September 5.

Yoffe, S. B. (2001). Basins at risk: Conflict and cooperation over international freshwater resources. Unpublished Ph.D. dissertation, Oregon State University, Corvallis, OR. Available at http://www.transboundarywaters.orst.edu.

Yoffe, S. B., Wolf, A. T., and Giordano, M. (2003). Conflict and cooperation over international freshwater resources: Indicators of basins at risk. *Journal of the American Water Resources Association*, (October), 1109–1126.

Yoffe, S., Fiske, G., Giordano, M., Giordano, M., Larson, K., Stahl, K., and Wolf, A. T. (2004). Geography of international water conflict and cooperation: Data sets and applications. *Water Resources Research*, **40**(W05S04), doi:10.1029/2003WR002530.

Young, O. R. (1989). *International cooperation: Building institutions for natural resources and the environment.* Ithaca: Cornell University Press.

Young, O. (1992). *International cooperation: Building regimes for natural resources and the Environment.* Ithaca, New York, Cornell University Press.

Young, G. J., Dooge, J. C., and Rodda, J. C. (1994). *Global Water Resource Issues.* Cambridge, UK: Cambridge University Press.

Zaman, M. (1982). The Ganges basin and the water dispute. *Water Supply and Management*, **6**(4), 321–328.

Zartman, I. W. (1991). The structure of negotiation. In *International Negotiation*, edited by V. Kremenyuk, pp. 65–77.

Zartman, I. W. (1992). International environmental negotiation: Challenges for analysis and practice. *Negotiation Journal*, **8**(2), pp. 113–124.

Zartman, I. W. (1993). A skeptic's view. In *Culture and negotiation: The resolution of water disputes*, edited by G. O. Faure and J. Z. Rubin. Thousand Oaks, CA: Sage Publications, pp. 17–21.

Zawahri, N. A. (2006). *Institutional design and cooperation between adversaries: Accounting for the effectiveness of international river commissions.* Cleveland, OH: Cleveland State University.

Zilleßen, H. (1991). Alternative dispute resolution – Ein neuer Verfahrensansatz zu Optimierung politischer Entscheidungen. Lokale Konfliktregelung durch kooperative Verhandlung und Vermittlung (Mediation), in T. Bühler (Stiftung Mitarbeit) (Hrsg.): *Demokratie vor Ort. Modelle und Wege der lokalen Bürgerbeteiligung*, Beiträge zur Demokratieentwicklung von unten, Bd. 2, S, pp. 126–146.

Author Index

Subject Index

upstream/downstream conflict, 11
urbanization, 5, 30
Uruguay, 44, 160, 187, 208
USSR. *See* Soviet Union
utilitarian model, 87
Uzbekistan, 172, 280, 294
    Aral Sea and, 16, 173
    hydroelectric projects in, 22

Vatican, 47
Venezuela, 159
Vietnam, 13, 63, 156, 267, 279, 283
    allocations and, 315
    ECAFE report, 218
    Geneva Accords, 217
    irrigation and, 17
    Mekong and, 156, 216. *See* Mekong River
    NGOs and, 156
    NWRC and, 156
    Paris Peace Agreement, 221
    RBOs and, 156
    stakeholder participations, 156
virtual water, 19, 23, 41, 50
Vistula River, 5, 286
visualization method, 50
voice, participation and, 85
Volga River, 5
vulnerability, and conflict, 22

wars, xxiii, 10
    conflict, cooperation and, 13
    dataset on, 26
    desalination and, 23
    geographic scale, 32
    intensity of conflict, 32
    large-n studies, 26
    preventive diplomacy, 28
    water wars, 5, 9, 102
wastewater, 15, 118, 258, 259
water
    civilization and, 9, 31
    costs of, xxi, 3
    ecological values, 4
    ethos of, 18
    per capita data, 70
    physical parameters, 19
    rainfall, xxi
    symbolism of, 29, 30, 91, 121
    unique characteristics, 250
    values of, 249
    watercourse, defined, xxiii, 55, 124
    watersheds, defined, xxiii
Water Environment Federation, 7
Water for Peace program, 34, 35, 119, 156
Water Framework Directive (WFD), 180
water quality, 6, 15, 65, 274. *See also* pollution
Water Quality 2000, 142
Water Resources Planning Act (1965), 142
Water Resources Protection Act (1999), 260
WaterBank, 264
Waternet Project, 228
WCU. *See* World Conservation Union
WEF. *See* World Economic Forum

wetlands, 44, 47, 110, 154, 155, 175
WFD. *See* Water Framework Directive
Wheeler Mission, 219, 222
wheeling, of resources, 109
White Mission, 222
wildlife, 21
Works Progress Administration (WPA), 8
World Bank, 5, 46, 75, 156
    ADR and, 52
    Africa Regional Office, 84
    Bangladesh and, 36
    Black initiative, 192, 195
    Brazil and, 262
    Curu River basin, 161
    expert boards, 45
    financial assistance from, 194
    Guaraní Aquifer, 54
    Indus River and, 156, 192, 194, 314
    intersectoral dialogue, 118
    intervention by, 47
    Maqarin Dam and, 201
    new sector strategies of, 7
    Nile Basin Initiative, 20, 54
    privatization and, 262
    Senegal and, 36
    technical expertise and, 46
    Third World and, 7
    water related portfolio, 6
World Commission on Water, 53
World Conservation Union (WCU), 176
World Court, 195
World Economic Forum (WEF), 8
World Resources Institute, 24
World Summit on Sustainable Development (WSSD), 53
World Trade Organization (WTO), 5, 116, 117
World Water Council, i, 53, 118
World Water Forums, 53, 54, 117
World Water Vision, 53
World Wide Fund for Nature (WWF), 176
WPA. *See* Works Progress Administration
WSSD. *See* World Summit on Sustainable Development
WTO. *See* World Trade Organization
WWF. *See* World Wide Fund for Nature

Yacyreta Treaty, 209
Yarmuk River, 75, 199, 201, 223, 269
    allocations and, 314
    Bunger Plan, 199
    hydropower and, 205
    Johnston negotiations and, 201
    Jordan and, 198
    negotiations over, 201, 204
    Sea of Galilee and, 201
    Syria and, 199
    water quality, 280, 291
Yellow River, 157, 158
Yemen, 68, 293
Yugoslavia, 269, 270, 298

Zambezi River, 5, 47, 79, 277, 282
Zambia, 153, 277, 278, 282
zero-sum systems, 5, 33, 36, 37, 43, 74, 89
Zimbabwe, 21, 277, 278